SECOND EDITION

Finite Mathematics:

A MODELING APPROACH

SECOND EDITION

Finite Mathematics:

A MODELING APPROACH

J. CONRAD CROWN

MARVIN L. BITTINGER

Indiana University—
Purdue University at Indianapolis

ADDISON-WESLEY PUBLISHING COMPANY

Reading, Massachusetts • Menlo Park, California

London • Amsterdam • Don Mills, Ontario • Sydney

SPONSORING EDITOR: Steve Quigley
PRODUCTION EDITOR: Mary Cafarella
DESIGNER: Marshall Henrichs
ILLUSTRATOR: Robert Gallison
COVER DESIGN: Richard Hannus
COVER PHOTOGRAPH: Marshall Henrichs

Library of Congress Cataloging in Publication Data

Crown, J   Conrad.
   Finite mathematics.

   Authors' names in reverse order in previous ed.
Includes index.
   1. Mathematics—1961- 2. Mathematical
models. I. Bittinger, Marvin L., joint author.
II. Title.
QA39.2.B57   1981   510   80-19472
ISBN 0-201-03145-0

Reprinted with corrections, December 1982

ISBN 0-201-03145-0
  FGHIJK-VB-898765

Preface

The subject of this book, Finite Mathematics, is an extension of basic algebra to areas that have applications in the economic, behavioral, social, and life sciences. While the basic contents can be covered in a variety of one-term courses, there is sufficient material for a two-term course by covering it all and/or by slower pacing.

1. Approach.

The approach to the material is characterized by two features. First, examples are given before the theory is presented, in order to give the student an intuitive feel for the material. The theory that follows is presented at a level which will contribute to the student's understanding.

Second, at the outset of each section, behavioral objectives are stated, which then are implemented by detailed examples in the text. These are immediately followed by margin exercises permitting the student to become actively involved in the development of the topic and reinforcing the material in the examples.

2. Modularized text.

The text has been organized into modules so that instructors will have the flexibility to develop course syllabi in accordance with their own preferences and institutional requirements. To facilitate this, the prerequisites for each chapter are given in the table at the top of the next page.

Part	Chapter	Short title	Chapters required as prerequisite											Appendixes	
			1	2	3	4	5	6	7	8	9	10	11	A	B
I	1	Basics													
	2	Math of Finance		–											
	3	Systems of Equations	X		–										
II	4	Linear Programming	X		X	–									
	5	Simplex	X		X	X	–								
	6	Networks						–							
III	7	Sets							–						
	8	Probability							X	–					
	9	Statistics							X	X	–				
IV	10	Markov Chains	X		X				X	X	X	–			
	11	Games	X		X				X	X	X		–		
Appendix	A	Logic												–	
	B	Programming	X												–

As an additional aid in choosing topics, optional sections are marked in the Contents and in more detail in the text by an asterisk. A sample syllabus is provided in the *Instructor's Manual.*

3. What is New in the Second Edition.

Chapter 2 on Mathematics of Finance has been added.

Chapter 3. The presentation of matrices has been filled out.

Chapter 4. The text examples of linear programs have been simplified and a table introduced which facilitates the formulation of linear programs from their presentation as word problems. A set of simple problems has been added.

Chapter 5. The section on the simplex algorithm has been rewritten to clarify the presentation, and text examples with less "messy" fractions have been used. More problems have been added (to the Exercise Sets) with solutions involving easier-to-handle fractions. A section on the two-phase method and artificial variables has been added. The section on duality (which can be covered independently from the two-phase method) has been revised to include the use of dual programs in checking solutions and a discussion of the economic interpretation of duality. The condensed tableau is discussed now only briefly, and special cases are treated using the extended rather than the condensed tableau. Sections on the Transportation and Assignment Problems have been added.

Chapter 7. At the end of the section on combinations, a set of exercises has been introduced some of which involve permutations and some of which involve combinations, to enable the student to practice the skill of deciding which is which.

Chapter 8 on Probability has been rewritten to organize and present the material more clearly. The *Birthday Problem* has been extended to a consideration of the probability for three or more (and also for four or more) people to have the same birthday. Consideration is also given to the effect of leap years and multiple births.

Chapter 10. A simple criterion for determining the regularity of many Markov chains has been introduced.

Appendix B on Programming in BASIC has been added.

Challenge Exercises. Those exercises which in varying degrees are more challenging than the others are set off at the end of the exercise sets by a thin line that begins with a triangle.

4. Chapter Contents.

The following is a short statement of the distinguishing features of each chapter.

Chapter 1 contains the basic concepts of algebra, in particular those aspects needed in the following chapters. It may be omitted for students with sufficient background. The pretest, provided for only this chapter, can be used to determine whether Chapter 1 should be studied.

Chapter 2 is a new chapter on the mathematics of finance. It is a coverage of sequences and series as applied to financial problems. This chapter is mainly intended for business and economics students. It can be omitted without loss of continuity.

Chapter 3 is a continuation of basic algebra. The echelon method is introduced at this point since we have found from class testing that it provides a good way of motivating work with matrices, rather than introducing matrices and then solving systems of equations without really using the basic matrix properties just developed. The concept of row operations is actually simpler with the echelon tableau than with "equivalent" (but not equal) matrices. Furthermore, the *echelon tableau* as presented can be simply extended to become the *simplex tableau.* The pivoting operation of the echelon method is essentially the same as the pivoting operation of the simplex algorithm. Only the choice of pivot and the termination procedure differs.

Chapter 4 contains the graphical approach to linear programs. Incidentally, while originally the "linear program" was the solution, we have used "linear program" for the whole formulation, rather than the more cumbersome phrase "linear programming problem." One distinctive feature of this chapter is a section on post-optimality analysis, using geometric techniques.

Chapter 5 contains the algebraic or simplex approach to linear programs. We use the extended tableau and describe briefly the condensed tableau. Students find the extended tableau easier to learn than the condensed although the latter is used on high-speed computing machines and the printed outputs frequently published. After describing the simplex method for solving maximum-type linear programs, we show how to solve minimum-type linear programs by two methods either of which can be taught independently of the other. First, we describe the two-phase method which uses artificial variables. Then we describe the concept of duality and its use in solving minimum-type linear programs. The use of duality to solve minimum-type linear programs not only involves less work than does the two-phase method, but also has other advantages. These are described in two new sections. One shows how to use dual programs to check a solution (the students really appreciate this) and the other provides some insight into the economic interpretation of duality. All the examples so far are uncomplicated by degeneracy, nonuniqueness, or any of the other peculiarities which can confuse someone just learning the subject. We treat these special cases later, separately.

New sections on *Transportation* and *Assignment Problems* have been included. These two sections rely on the preceding discussion of linear programming only for their conceptual information. Because of the special structure of these problems, the solutions themselves are obtained by methods that are relatively simple and independent of the simplex algorithm. The *Transportation Problem* is solved by a variation of the steppingstone method, and the *Assignment Problem* is solved by the Hungarian method.

Chapter 6 on networks introduces the student to some real-world problems that can be solved very simply. With a little introduction by the instructor, this chapter can be taught *before* Chapters 4 and 5, rather than after them.

Chapter 7 contains an introduction to set theory, since this is needed in the following chapters. It is a separate chapter, since an attempt to integrate it with Chapter 8 yielded a chapter too heavy with new ideas. The present development seemed to permit a more logical development that is easier for the student to grasp.

Chapter 8 develops the concept of probability in steps, starting with simple concepts and building up to the more complex, rather than starting with the most general and then considering special cases.

Chapter 9 continues probability into statistics. While hypergeometric probability concepts are included in most courses, the name "hypergeometric" is usually omitted. We have used the name, purely for convenience, in order to distinguish it from binomial and geometric probabilities. Negative binomial probabilities have been included as a natural extension of geometric probabilities. The normal probability distribution is included in a simple, natural way.

Chapter 10 on Markov chains introduces the concept of ergodic transition matrices, since the irregularity of many transition matrices can be easily determined by their nonergodic character.

Chapter 11 on game theory begins with the development of matrix games from game trees. The present brief encounter with game trees is intended to provide some insight and motivation for matrix games. Also, we deal vary briefly with nonzero-sum games, since we cannot do justice to that subject in the present text. For an expanded treatment of the use of trees in game theory and of nonzero-sum games, see Luce and Raiffa, *Games and Decisions* (Wiley, 1957). Our treatment of voting coalitions uses not only the standard Shapley index, but also the new Banzhaf index, which was introduced in legal rather than mathematical journals. It is a simple but interesting and useful aspect of game theory.

Appendix A contains the fundamentals of Logic. For those who want to include it in a course, it would be most advantageous if it were studied before Set Theory, Chapter 7. However, the chapter on set theory and those following have been developed in the mainstream, in order to make these chapters independent of this appendix.

Appendix B on Programming starts with flowcharting. This introduces the student to the nature of computation in an essentially algebraic and geometric setting. Then we proceed to introduce BASIC. We have chosen BASIC over FORTRAN or ALGOL or any of the other high-level languages since the essentials of BASIC can be learned in a relatively short time while the other languages require more extensive study to be of value.

5. Tests and exercises.

Each chapter ends with a chapter test and there is a comprehensive final examination. All the answers to these tests are in the book. Two alternate forms of the tests appear in the *Instructor's Manual.* Great care has been given to constructing exercises, most of which are based on the behavioral objectives. The first exercises in each set are quite easy, while later ones become progressively more difficult. The exercises are also arranged in matching pairs; that is, any odd-numbered exercise is very much like the one that immediately follows. In effect, this makes each exercise set *two* exercise sets. The odd-numbered exercises have answers in the book, while the even-numbered exercises do not. This allows for various ways of assigning exercises. If the instructor wants students to have all the answers, the odd-numbered exercises are assigned. If the instructor does not want students to have the answers, the even-numbered exercises are assigned. If the instructor wants some of each, that option also exists. All margin exercises have answers in the text. It is recommended that students do all of these, stopping to do them when the text so indicates.

ACKNOWLEDGMENTS.

The authors wish to express their appreciation to many people who helped with the development of the book: to their own students for providing suggestions and criticisms so willingly during the extensive class testing; and to Judy Beecher of Indiana University—Purdue University at Indianapolis for her careful reading and checking of the manuscript. Special thanks to Michael C. Gemignani, Chairman, Department of Mathematical Sciences at Indiana University—Purdue University at Indianapolis, for his support and cooperation with the class testing.

For their thorough reviewing, the authors also wish to thank Audrey B. DeMello (Southeastern Massachusetts University), John W. Hooker (Southern Illinois University at Carbondale), Henry C. Howard (University of Kentucky), Richard S. Montgomery (The University of Connecticut at Groton), Donald E. Myers (The University of Arizona), and Jeanne Smith (Saddleback Community College).

Indianapolis, Indiana　　　　　　　　　　　　　　　　　　J. C. C.
October 1980　　　　　　　　　　　　　　　　　　　　　　M. L. B.

Contents

*Indicates an optional section in the sense that subsequent material is not dependent on it.

PART I

CHAPTER ONE

Basic Concepts

CHAPTER 1 PRETEST*

1. Graph

$$4x - 5y = 20$$

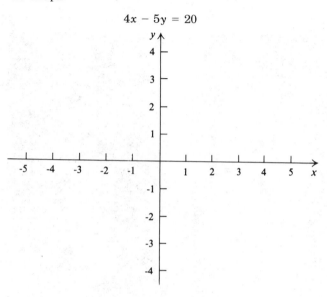

2. Solve

$$-\tfrac{7}{8}x + 5 = \tfrac{1}{4}x - 2$$

3. Solve

$$8 + x < 5x - 7$$

4. Graph this system and classify it as consistent or inconsistent, dependent or independent.

$$5x + 10y = 15$$
$$3x + 6y = 9$$

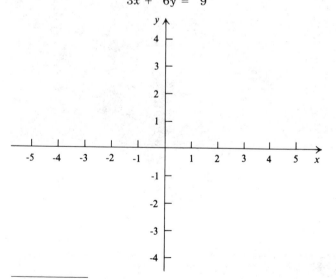

5. Decide whether $(2, -3)$ is a solution of the system:

$$2x + 3y = -5$$
$$x + y = 5$$

---

* Answers to the Pretest are in the back of the book. Should you get 75% of the Pretest correct, you could probably move on to Chapter 2.

6. Solve graphically

$$x + y = 5$$
$$x - y = 1$$

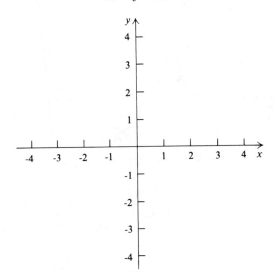

7. Solve

$$2x - y = -3$$
$$x + y = 9$$

8. Solve

$$3y - 3x = 11$$
$$9y - 2x = 5$$

9. Solve

$$5x - 15y = -10$$
$$3x - 9y = 7$$

OBJECTIVES

You should be able to

a) Solve equations like

$-5x + 7 = 8x + 4.$

b) Solve inequalities like

$-5x + 7 > 8x + 4.$

c) Solve applied problems.

1. Solve $-\frac{5}{6}x + 10 = \frac{1}{2}x + 2$

Equations

Basic to the solution of many equations are these two simple principles.

THE ADDITION PRINCIPLE. If an equation $a = b$ is true, then the equation $a + c = b + c$ is true for any number c.

THE MULTIPLICATION PRINCIPLE. If an equation $a = b$ is true, then the equation $ac = bc$ is true for any number c.

Example 1   Solve $-\frac{7}{8}x + 5 = \frac{1}{4}x - 2$.

Solution   We first multiply by 8 to clear the fractions.

$$8(-\tfrac{7}{8}x + 5) = 8(\tfrac{1}{4}x - 2) \qquad \text{(Multiplication Principle)}$$

$$8(-\tfrac{7}{8}x) + 8 \cdot 5 = 8(\tfrac{1}{4}x) - 8 \cdot 2 \qquad \left(\begin{array}{l}\text{Distributive Laws: } a(b + c) = ab + ac \\ \qquad\qquad\qquad a(b - c) = ab - ac\end{array}\right)$$

$$-7x + 40 = 2x - 16 \qquad \text{(Simplifying)}$$

$$40 = 9x - 16 \qquad \begin{array}{l}\text{(Addition Principle: We add } 7x \text{ to} \\ \text{get the variable term on one side)}\end{array}$$

$$56 = 9x \qquad \text{(Addition Principle: We add 16)}$$

$$\tfrac{1}{9} \cdot 56 = \tfrac{1}{9} \cdot 9x \qquad \text{(We multiply by } \tfrac{1}{9})$$

$$\tfrac{56}{9} = x$$

The solution is $\frac{56}{9}$, or $6\frac{2}{9}$. The reader can check this by substituting $\frac{56}{9}$ for x in the original equation.

DO EXERCISE 1.

To solve applied problems, we first translate to mathematical language. This is usually an equation, but quite often in this text it will be an inequality, or several inequalities. If we have translated to an equation, we solve the equation and check in the original problem to see whether the solution of the equation is a solution of the problem.

Example 2   An investment is made at 8%, compounded annually. It grows to $783 at the end of 1 year. How much was originally invested?

Solution   We first translate to an equation.

$$\underbrace{\text{(Original Investment)}}_{p} + 8\%\underbrace{\text{(Original Investment)}}_{p} = 783$$

$$+ 8\% = 783$$

Now we solve the equation.

$$p + 8\%p = 783$$

$$1 \cdot p + (0.08)p = 783$$

$$(1 + 0.08)p = 783 \qquad \text{(Factoring, or distributive law in reverse)}$$

$$1.08p = 783$$

$$p = \frac{783}{1.08} = 725$$

Check: $725 + 8\% \times 725 = 725 + 0.08 \times 725 = 725 + 58 = 783$

Thus the original investment was $725.

DO EXERCISE 2.

Inequalities

Principles for solving inequalities are similar to those for solving equations. We can add the same number to both sides of an inequality. Both sides of an inequality can also be multiplied by the same nonzero number, but if that number is negative, we must reverse the inequality sign. Let us see why this is necessary. Consider the true inequality

$$-3 < 5. \qquad (1)$$

Let us multiply both members by 2. We get another true inequality

$$-6 < 10.$$

Now let us multiply both members in (1) by -4.

$$12 < -20$$

This time the inequality is false. However, if we reverse the inequality symbol (use $>$ instead of $<$), we will get a true inequality

$$12 > -20.$$

The following is a reformulation of the inequality-solving principles.

If the inequality $a < b$ is true, then

  i) $a + c < b + c$ **is true, for any** c;
 ii) $a \cdot c < b \cdot c$, **for any** *positive* c;
iii) $a \cdot c > b \cdot c$, **for any** *negative* c.

Similar principles hold when $<$ is replaced by $\leqslant$, and when $>$ is replaced by $\geqslant$.

2. After a 5% gain in weight, an animal weighs 693 lb. What was its original weight?

3. Solve $3x > 14 - 2x$

4. Solve $15 - 7x \geq 10x - 4$

5. In Example 5, determine, as an inequality, the number of suits the firm must sell so that its revenue will be more than $70,000.

Example 3 Solve $\qquad 4x < 12 - 3x$.

Solution $\qquad\qquad 4x < 12 - 3x$

$$4x + 3x < 12 \qquad \text{(Adding } 3x)$$

$$7x < 12$$

$$\tfrac{1}{7} \cdot 7x < \tfrac{1}{7} \cdot 12 \qquad \text{(Multiplying by } \tfrac{1}{7})$$

$$x < \tfrac{12}{7}$$

Any number less than $\tfrac{12}{7}$ is a solution.

DO EXERCISE 3.

Example 4 Solve $18 - 8x \leq 5x - 4$.

Solution $\qquad\qquad 18 - 8x \leq 5x - 4$

$$-8x \leq 5x - 22 \qquad \text{(Adding } -18)$$

$$-13x \leq -22 \qquad \text{(Adding } -5x)$$

$$-\tfrac{1}{13}(-13x) \geq -\tfrac{1}{13}(-22) \qquad \begin{array}{l}\text{(Multiplying by } -\tfrac{1}{13} \text{ and} \\ \text{reversing the inequality sign)}\end{array}$$

$$x \geq \tfrac{22}{13}$$

Any number greater than or equal to $\tfrac{22}{13}$ is a solution.

DO EXERCISE 4.

Example 5 A clothing firm determines that its total revenue (money coming in) from the sale of x suits is

$$20x + 50 \qquad \text{(dollars)}.$$

Determine, as an inequality, the number of suits the firm must sell so its total revenue will be more than $40,000.

Solution We translate to an inequality and solve.

$$20x + 50 > 40,000$$

$$20x > 39,950 \qquad \text{(Adding } -50)$$

$$x > 1997.5 \qquad \text{(Multiplying by } \tfrac{1}{20})$$

Thus its total revenue will exceed $40,000 when it sells more than 1997.5 suits. Since it is impossible (or not probable) that it could sell half of a suit, it is reasonable that it must sell 1998 or more suits for its revenue to exceed $40,000.

DO EXERCISE 5.

EXERCISE SET 1.1

Solve

1. $-8x + 9 = 4x - 70$

2. $-7x + 10 = 5x - 11$

3. $5x - 2 + 3x = 2x + 6 - 4x$

4. $5x - 17 - 2x = 6x - 1 - x$

5. $x + 0.5x = 210$

6. $x + 0.8x = 216$

7. $x + 0.05x = 210$

8. $x + 0.08x = 216$

Applied Problems

9. After a 7% gain in weight, an animal weighs 363.8 lb. What was its original weight?

10. After a 6% gain in weight an animal weighs 508.8 lb. What was its original weight?

11. An investment is made at 7%, compounded annually. It grows to $856 at the end of 1 year. How much was originally invested?

12. An investment is made at 9%, compounded annually. It grows to $708.50 at the end of 1 year. How much was originally invested?

Solve

13. $x + 6 \geqslant 5x - 6$

14. $3 - x \geqslant 4x + 7$

15. $3x - 3 + 3x < 1 - 7x - 9$

16. $5x - 5 + x < 2 - 6x + 8$

17. $-5x \leqslant 6$

18. $-7x > 4$

Applied Problems

19. A firm determines that the total revenue from the sale of x units of a product is

$$5x + 100 \qquad \text{(dollars)}.$$

Determine, as an inequality, the number of units that must be sold so its total revenue will be more than $22,000.

20. A firm determines that the total revenue from the sale of x units of a product is

$$3x + 1000 \qquad \text{(dollars)}.$$

Determine, as an inequality, the number of units that must be sold so its total revenue will be more than $22,000.

21. To get a C, or better, in a course, a student's average must be greater than or equal to 70%. On the first three tests the student scores 65%, 83%, and 82%. Determine, as an inequality, what possible scores on the 4th test will yield a C, or better.

22. To get a B, or better, in a course a student's average must be greater than or equal to 80%. On the first three tests the student scores 78%, 90%, and 92%. Determine, as an inequality, what possible scores on the 4th test will yield a B, or better.

1.2 GRAPHS OF LINEAR EQUATIONS IN TWO VARIABLES

Points and Graphs

Each point in the plane corresponds to an ordered pair of numbers. Note that the pair $(2, 5)$ is different from the pair $(5, 2)$. This is why we call $(2, 5)$ an *ordered pair*. The number 2 is called the *first*

OBJECTIVE

You should be able to graph a linear equation in two variables.

6. Graph these ordered pairs. $(0, 3)$, $(3, 0)$, $(-1, 2)$, $(4, 2)$, and $(-3, -2)$.

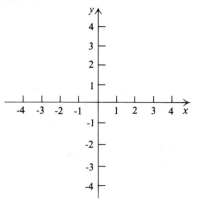

coordinate and the number 5 is called the *second coordinate*. Together these are called the *coordinates of a point*. The horizontal line is called the *x-axis*, or *first axis*, and the vertical line is called the *y-axis*, or *second axis*.

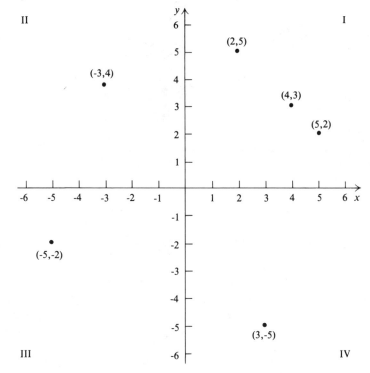

DO EXERCISE 6.

Note in the preceding drawing that, in region I (called the first *quadrant*), both coordinates of a point are positive. In region IV (the fourth quadrant), the first coordinate is positive, but the second is negative.

DO EXERCISE 7.

7. What can be said about the coordinates of a point in the second quadrant? third quadrant?

Graphs of Equations

A *solution* of an equation in two variables like

$$3x + y = 5$$

is an ordered pair like $(-1, 8)$ such that when x is replaced by -1 and y is replaced by 8 we get a true equation.

$$3x + y = 5$$

$3(-1) + 8$	5
$-3 + 8$	
5	

Here the "order" in a pair corresponds to the alphabetical order of the variables.

DO EXERCISE 8.

The *graph* of an equation is the geometric representation of all the solutions. It could be a line, curve (or curves), or some other configuration. To draw a graph we plot enough points to get an idea of the shape of the graph.

Example 1   Graph $y = -2x + 1$.

Solution

x	0	1	2	-1	-2
y	1	-1	-3	3	5

← We choose these numbers arbitrarily (since y is expressed in terms of x).

← We find these numbers by substituting in the equation.

For example, when $x = -2$, $y = -2(-2) + 1 = 4 + 1 = 5$. This yields the pair $(-2, 5)$. We plot all the pairs from the table. We see that we can draw a line to complete the graph.

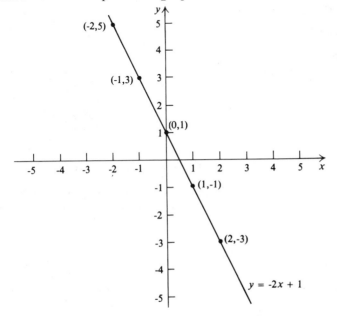

8. Decide whether each pair is a solution.

$$x - 2y = 6$$

a) $(-2, -4)$

b) $(3, 0)$

9. Graph $y = 2x + 1$

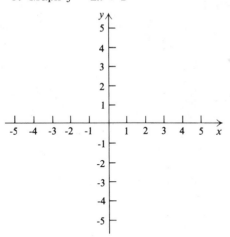

10. Find the y and x intercepts of the following equations.

a) $2x - 3y = 6$

b) $y - 2x = 0$

c) $5x + 3y = -15$

DO EXERCISE 9.

Linear Equations

Equations like the one in Example 1 that have straight lines for their graphs are called *linear equations;* for example:

$$3x + 2y = 6, \qquad 150x - 2y = 7, \qquad x = 40y + 1000.$$

Note that the variables all appear to the first power. There are no products of variables, and there is no division by a variable.

However, equations need not be linear, for example:

$$xy = -5, \qquad \frac{1}{x} - 14 = y, \qquad x^2 + y^2 = 16.$$

A general form of a linear equation is

$$ax + by = c.$$

In this text, we will restrict our attention to linear equations and linear inequalities.

From geometry we know that a straight line is determined by two points. Thus, in order to graph a linear equation we just need to know the coordinates of two points. The points where a line crosses the axes are called the *intercepts*. Since these points are the easiest to determine we use them in graphing.

Example 2   Graph $3x + 2y = 6$.

Solution

a) Any point on the y-axis has first coordinate, or x-coordinate, 0. Therefore, to find the y-intercept, we substitute 0 for x, and solve for y:

$$3 \cdot 0 + 2y = 6$$
$$0 + 2y = 6$$
$$2y = 6$$
$$y = 3 \qquad \text{The y-intercept is } (0, 3).$$

b) Any point on the x-axis has second coordinate, or y-coordinate, 0. Therefore, to find the x-intercept, we substitute 0 for y, and solve for x:

$$3x + 2 \cdot 0 = 6$$
$$3x + 0 = 6$$
$$3x = 6$$
$$x = 2 \qquad \text{The x-intercept is } (2, 0).$$

c) Next, we plot the points $(0, 3)$ and $(2, 0)$ and draw a line through them. If one intercept turns out to be the origin $(0, 0)$, then the other intercept is also $(0, 0)$. In such cases, one would have to pick some other value for x and solve again for y.

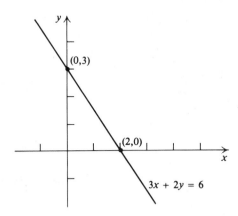

In any case, it might be wise to use a third point as a check against possible computation errors.

The y-intercept is found by setting $x = 0$ and solving for y.
The x-intercept is found by setting $y = 0$ and solving for x.

DO EXERCISES 10 THROUGH 12.

Variables with Subscripts

For the remainder of this text, we will use a variable x_1 to represent the x, or first coordinate. The "1" is called a *subscript*. Similarly, x_2 will represent the y, or second coordinate. Note the connection: x_1 is the "first" coordinate and x_2 is the "second" coordinate. An equation like

$$4x - 6y = -10$$

can now be written

$$4x_1 - 6x_2 = -10.$$

Graphing is done the same way as before, except that the axes are labeled x_1 and x_2.

Example 3 Graph $2x_1 + 3x_2 = 6$.

Solution

a) Any point on the x_2-axis (the vertical axis) has first coordinate, or

11. Graph $2x - 3y = 6$

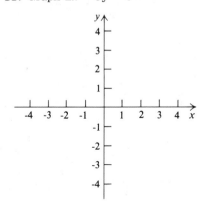

12. Graph $y - 2x = 0$

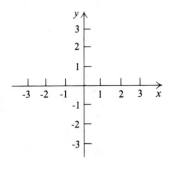

13. Graph $2x_1 - 3x_2 = 6$

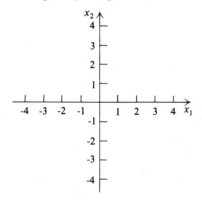

x_1 coordinate, 0. Therefore, to find the x_2-intercept, we substitute 0 for x_1, and solve for x_2:

$$2 \cdot 0 + 3x_2 = 6$$
$$0 + 3x_2 = 6$$
$$3x_2 = 6$$
$$x_2 = 2 \qquad \text{The } x_2\text{-intercept is } (0, 2).$$

b) Any point on the x_1-axis has second coordinate, or x_2 coordinate, 0. Therefore, to find the x_1-intercept, we substitute 0 for x_2, and solve for x_1:

$$2x_1 + 3 \cdot 0 = 6$$
$$2x_1 + 0 = 6$$
$$2x_1 = 6$$
$$x_1 = 3 \qquad \text{(The } x_1\text{-intercept is } (3, 0).$$

c) We plot $(0, 2)$ and $(3, 0)$ and draw a line through them.

14. Using the same set of axes, graph the following equations

a) $x_2 - 2x_1 = 0$

b) $x_2 - 2x_1 = -4$

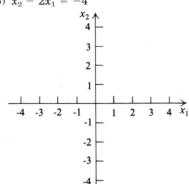

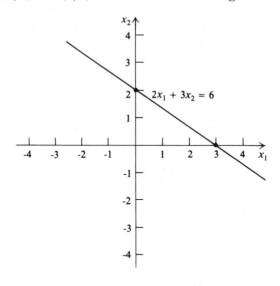

DO EXERCISES 13 AND 14.

Horizontal and Vertical Lines

Let us consider graphs of equations like $x_2 = b$ and $x_1 = a$.

Example 4 Graph $x_2 = 4$.

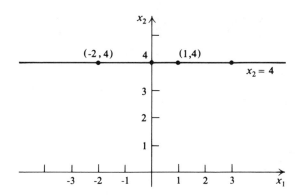

Solution The graph consists of all ordered pairs whose second coordinate is 4. To see how a pair such as $(-2, 4)$ could be a solution, we can consider the above equation in the form

$$0x_1 + x_2 = 4.$$

Then

$$0x_1 + x_2 = 4.$$

$0(-2) + 4$	4
$0 + 4$	
4	

Thus $(-2, 4)$ is a solution.

Example 5 Graph $x_1 = -3$.

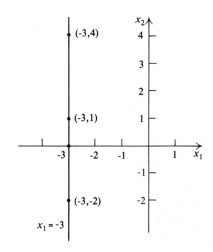

15. Graph $x_2 = -2$

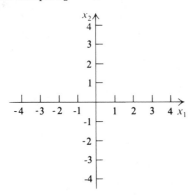

16. Graph $x_1 = 3$

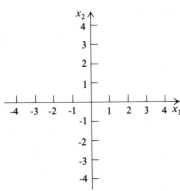

Solution The graph consists of all ordered pairs whose first coordinate is -3.

In general,

The graph of $x_2 = b$ is a horizontal line.
The graph of $x_1 = a$ is a vertical line.

DO EXERCISES 15 AND 16.

Families of Parallel Lines

Let us consider

$$2x_1 - x_2 = c$$

for various values of c.

Example 6 Using the same set of axes, graph

$$2x_1 - x_2 = 1,$$
$$2x_1 - x_2 = 0,$$
$$2x_1 - x_2 = -3.$$

Solution

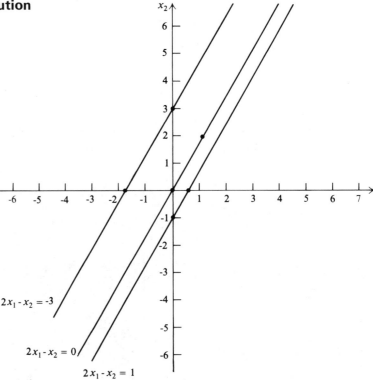

Note that the lines are parallel. In general, if we consider graphs of

$$ax_1 + bx_2 = c,$$

where a and b are fixed and c varies, the graphs form a *family* of parallel lines.

DO EXERCISE 17.

17. Using the same set of axes graph

$$x_1 + x_2 = 2,$$
$$x_1 + x_2 = 0,$$
$$x_1 + x_2 = -1.$$

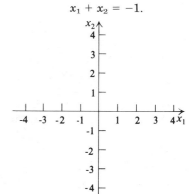

EXERCISE SET 1.2

Graph

1. $4x_1 + 5x_2 = 20$	**2.** $4x_1 - 5x_2 = 20$	**3.** $2x_1 - x_2 = 4$	**4.** $2x_1 + x_2 = 4$
5. $3x_2 + 4x_1 = 12$	**6.** $4x_2 - 3x_1 = 12$	**7.** $x_2 - x_1 = 0$	**8.** $x_2 + x_1 = 0$
9. $2x_1 + 3x_2 = 5$	**10.** $2x_1 - 3x_2 = 5$	**11.** $x_2 = 5$	**12.** $x_2 = 4$
13. $x_1 = -2$	**14.** $x_1 = -1$	**15.** $x_2 = 0$	**16.** $x_1 = 0$
17. $x_2 = -3.5$	**18.** $x_1 = 4.5$		

19. Using the same set of axes, graph each equation.

$$x_2 - x_1 = 0,$$
$$x_2 - x_1 = 2,$$
$$x_2 - x_1 = -3.$$

20. Using the same set of axes, graph each equation.

$$x_2 + 2x_1 = 0.$$
$$x_2 + 2x_1 = 3,$$
$$x_2 + 2x_1 = -2.$$

1.3 SYSTEMS OF EQUATIONS IN TWO VARIABLES—GRAPHICAL SOLUTION

Systems of Equations in Two Variables

A pair of linear equations

$$a_1x_1 + b_1x_2 = c_1$$
$$a_2x_1 + b_2x_2 = c_2$$

is called a *system* of two linear equations in two variables. A *solution* of a system is an ordered pair which is a solution of *both* equations.

OBJECTIVES

You should be able to

a) Decide whether an ordered pair is a solution of a system of equations.
b) Graph a system of equations and classify it as consistent or inconsistent, dependent or independent.
c) Solve a system of equations graphically.

Decide whether $(-2, 1)$ is a solution of each system.

18.

$$x_1 + x_2 = -1$$
$$-3x_1 - x_2 = 5$$

19.

$$2x_1 - x_2 = -5$$
$$3x_1 + 2x_2 = 3$$

Example 1 Decide whether $(2, 3)$ is a solution of the system

$$x_1 - x_2 = -1$$
$$4x_1 + 2x_2 = 14.$$

Solution We substitute 2 for x_1 and 3 for x_2 in each equation.

$x_1 - x_2 = -1$		$4x_1 + 2x_2 = 14$	
$2 - 3$	-1	$4 \cdot 2 + 2 \cdot 3$	14
-1		$8 + 6$	
		14	

We see that $(2, 3)$ is a solution of both equations, so it is a solution of the system.

DO EXERCISES 18 AND 19.

The graph of each equation in a system is a line. Given two lines, the following can happen:

1. The lines have no point in common—they are parallel.

2. The lines have exactly one point in common.

3. The lines are the same—they have infinitely many points in common.

A system of two linear equations in two variables is:

Consistent **if it has one or more solutions.**
Inconsistent **if it has** *no* **solution.**
Linearly dependent **if it is possible to multiply one equation by a constant and obtain the other.**
Linearly independent **if it is not possible to multiply one equation by a constant and obtain the other equation.**

Let us look at the three possibilities for lines to intersect, and describe the system in terms of the preceding terminology.

Example 2 The graph of the system

$$x_2 = 2x_1 - 1$$
$$x_2 = 2x_1 + 1$$

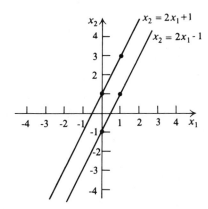

is shown above. Note that the lines are parallel. Thus the system has no solution—it is inconsistent. There is no way to obtain one equation from the other by multiplying by a constant, so the system is independent.

Example 3 The graph of the system

$$x_2 = 2x_1 - 1$$
$$3x_2 = 6x_1 - 3$$

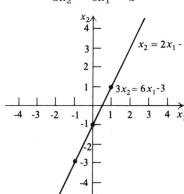

is shown above. Note that the lines are the same. Thus the system has infinitely many solutions—it is consistent. We obtain the second equation from the first by multiplying by 3. Thus the system is dependent.

Graph each system and classify it as consistent or inconsistent, dependent or independent.

20. $2x_1 - x_2 = 1$

$-6x_1 + 3x_2 = -3$

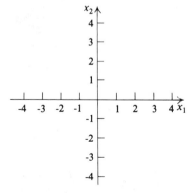

21. $x_1 - 4x_2 = -4$

$-x_1 + 4x_2 = 8$

22. $x_2 + x_1 = 3$

$x_2 - x_1 = 1$

Example 4 The graph of the system

$$x_1 - 2x_2 = 0$$
$$-2x_1 + x_2 = 2$$

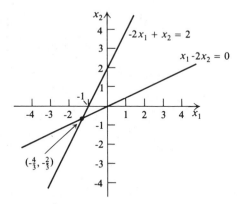

is shown above. Note that the lines intersect at exactly one point—it is consistent. There is no way to obtain one equation from the other by multiplying by a constant, so the system is independent.

DO EXERCISES 20 THROUGH 22.

Graphical Solution of Systems of Linear Equations

We can solve systems of equations graphically.

Example 5 Solve graphically.

$$x_2 - x_1 = 1$$
$$x_2 + x_1 = 5$$

Solution We graph the two equations. (See top of page 21.)

The point of intersection appears to be $(2, 3)$. We can check this as follows.

$x_2 - x_1 = 1$		$x_2 + x_1 = 5$	
$3 - 2$	1	$3 + 2$	5
1		5	

Thus, the solution is $(2, 3)$. Note that this procedure is subject to error, especially when fractional solutions are involved (see Example

4). Algebraic procedures will be developed in the next section and in Chapter 3 for obtaining exact answers.

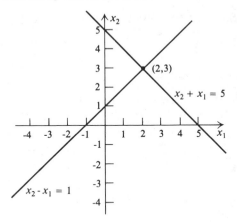

DO EXERCISE 23.

23. a) Solve the system in Margin exercise 22, graphically.

b) Solve graphically

$$x_2 + x_1 = 0$$

$$2x_1 - x_2 = -6$$

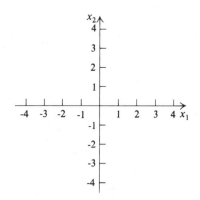

EXERCISE SET 1.3

Decide whether $(3, -2)$ is a solution of each system

1. $x_1 + x_2 = 1$

$x_1 - x_2 = 6$

2. $2x_1 + x_2 = 4$

$x_1 - 2x_2 = 7$

Graph each system and classify it as consistent or inconsistent, dependent or independent.

3. $x_1 + x_2 = 1$

$x_1 - x_2 = 6$

4. $x_2 - 2x_1 = 1$

$x_2 - 2x_1 = 3$

5. $2x_1 + 3x_2 = 1$

$-x_1 - 1.5x_2 = -\frac{1}{2}$

6. $2x_1 - 4x_2 = 8$

$-\frac{1}{2}x_1 + x_2 = -2$

7. $x_1 + 3x_2 = 4$

$x_1 + 3x_2 = 6$

8. $2x_1 + x_2 = 4$

$x_1 - 2x_2 = 7$

Solve graphically

9. $x_2 + 3x_1 = 5$

$2x_2 - x_1 = -4$

10. $2x_1 - x_2 = 4$

$5x_1 - x_2 = 13$

11. $2x_1 - 4x_2 = 7$

$x_1 - 2x_2 = 5$

12. $3x_2 - 6x_1 = 10$

$x_2 - 2x_1 = -1$

13. $x_2 - 4 = 0$

$x_1 + 5 = 0$

14. $x_1 + 3 = 0$

$x_2 - 1 = 0$

24. Solve, using the substitution method

$$-2x_1 + x_2 = 2$$
$$x_1 + x_2 = 6$$

25. Solve using the substitution method

$$4x_2 + 3x_1 = 4$$
$$x_2 + 2x_1 = 6$$

1.4 SYSTEMS OF EQUATIONS IN TWO VARIABLES—ALGEBRAIC SOLUTION

Here we consider algebraic procedures for finding exact solutions to systems of equations.

The Substitution Method

SUBSTITUTION METHOD. Solve one of the equations for one of the variables. Then substitute the resulting expression in the other equation and solve for the second variable.

Example 1 Solve

$$x_1 - 2x_2 = 0 \tag{1}$$
$$-2x_1 + x_2 = 2 \tag{2}$$

Solution Since Eq. (2) has x_2 with coefficient 1, it is easiest to solve that equation for x_2.

$$x_2 = 2 + 2x_1 \tag{3}$$

Now we substitute $2 + 2x_1$ for x_2 in Eq. (1).

$$x_1 - 2(2 + 2x_1) = 0$$

We now have an equation in one variable, x_1. We solve for x_1 using the addition and multiplication principles.

$$x_1 - 4 - 4x_1 = 0$$
$$-4 - 3x_1 = 0$$
$$-3x_1 = 4$$
$$x_1 = -\tfrac{4}{3}$$

We now substitute $-\tfrac{4}{3}$ for x_1 in Eq. (3) to find x_2. We could also substitute in either of the original equations, but it is faster to use Eq. (3) since x_2 has coefficient 1 on one side.

$$x_2 = 2 + 2(-\tfrac{4}{3}) = 2 - \tfrac{8}{3} = -\tfrac{2}{3}$$

The solution is $(-\tfrac{4}{3}, -\tfrac{2}{3})$. The reader should check this.

Always check! It is very easy to do with systems of linear equations.

DO EXERCISES 24 AND 25.

The Addition Method

The *addition method* for solving systems of equations makes use of the *addition* and *multiplication principles* for solving equations. The idea, just as with substitution, is to obtain an equation with one variable.

Example 2 Solve

$$3x_2 + 5x_1 = 17 \tag{1}$$

$$2x_2 - 5x_1 = 3 \tag{2}$$

Solution We add the equations as follows.

$$3x_2 + 5x_1 = 17$$

$$\underline{2x_2 - 5x_1 = 3}$$

$$5x_2 \qquad = 20 \qquad \text{(Adding the "sides" of the equa-}$$
$$5x_2 = 20 \qquad \text{tions, using the addition principle)}$$

$$x_2 = 4$$

We substitute in Eq. (1) to find x_1 (we could use Eq. (2) also).

$$3 \cdot 4 + 5x_1 = 17$$

$$12 + 5x_1 = 17$$

$$5x_1 = 5$$

$$x_1 = 1$$

The solution is $(1, 4)$. The reader should check this.

DO EXERCISE 26.

Note in Example 2 that the term $5x_1$ of Eq. (1) and the term $-5x_1$ of Eq. (2) add to 0. Thus when we added we eliminated x_1. In the following examples, we first multiply one or both of the equations in order to create a situation like Example 2.

Example 3 Solve

$$9x_1 - 2x_2 = -4 \tag{1}$$

$$3x_1 + 4x_2 = 1 \tag{2}$$

Solution We first multiply Eq. (1) by 2, then add.

$$18x_1 - 4x_2 = -8 \qquad \text{(Multiplying by 2, using the}$$
$$\underline{3x_1 + 4x_2 = 1} \qquad \text{multiplication principle)}$$

$$21x_1 \qquad = -7 \qquad \text{(Adding, using the addition}$$
$$21x_1 = -7 \qquad \text{principle)}$$

$$x_1 = \frac{-7}{21} = -\frac{1}{3}$$

26. Solve

$$-3x_2 + 2x_1 = 0$$

$$3x_2 - 4x_1 = -1$$

27. Solve

$$8x_1 - 3x_2 = -31$$
$$2x_1 + 6x_2 = 26$$

28. Solve

$$3x_1 + 2x_2 = -1$$
$$4x_1 + 3x_2 = 2$$

We substitute $-\frac{1}{3}$ for x_1 in Eq. (2) and solve for x_2.

$$3(-\tfrac{1}{3}) + 4x_2 = 1$$
$$-1 + 4x_2 = 1$$
$$4x_2 = 2$$
$$x_2 = \tfrac{2}{4} = \tfrac{1}{2}$$

The solution is $(-\frac{1}{3}, \frac{1}{2})$.

DO EXERCISE 27.

Example 4 Solve

$$3x_1 + 5x_2 = 7 \tag{1}$$
$$5x_1 + 3x_2 = -23 \tag{2}$$

Solution We multiply Eq. (1) by 5 and Eq. (2) by -3.

$$15x_1 + 25x_2 = 35 \qquad \text{(Multiplying by 5)}$$
$$\underline{-15x_1 - 9x_2 = 69} \qquad \text{(Multiplying by } -3)$$
$$16x_2 = 104$$
$$x_2 = \tfrac{104}{16} = \tfrac{13}{2}$$

We substitute $\frac{13}{2}$ for x_2 in Eq. (1) and solve for x_1.

$$3x_1 + 5(\tfrac{13}{2}) = 7$$
$$3x_1 + \tfrac{65}{2} = 7$$
$$3x_1 = 7 - \tfrac{65}{2}$$
$$3x_1 = \tfrac{14}{2} - \tfrac{65}{2}$$
$$3x_1 = -\tfrac{51}{2}$$
$$x_1 = (-\tfrac{51}{2})\tfrac{1}{3} = -\tfrac{17}{2}$$

The solution is $(\frac{13}{2}, -\frac{17}{2})$.

DO EXERCISE 28.

Example 5 Solve

$$4x_1 + 6x_2 = -8 \tag{1}$$
$$-2x_1 - 3x_2 = 4 \tag{2}$$

Solution We multiply Eq. (2) by 2 and add

$$4x_1 + 6x_2 = -8$$
$$\underline{-4x_1 - 6x_2 = 8} \qquad \text{(Multiplying by 2)}$$
$$0 = 0 \qquad \text{(Adding)}$$

We get the true equation $0 = 0$. This will happen for any ordered pair that is a solution of one of the equations. Thus we have an infinite number of solutions. If we had multiplied Eq. (2) by -2, we would have gotten Eq. (1), which is another way of verifying that we have an infinite number of solutions, because the system is linearly dependent.

Example 6 Solve

$$3x_2 + x_1 = 10$$
$$6x_2 + 2x_1 = 23$$

Solution We multiply Eq. (1) by -2 and add

$$-6x_2 - 2x_1 = -20 \qquad \text{(Multiplying by } -2)$$
$$\underline{6x_2 + 2x_1 = 23}$$
$$0 = 3 \qquad \text{(Adding)}$$

Since we get the false equation $0 = 3$, the system has no solution. We could check this by graphing the system—the lines would be *parallel*.

DO EXERCISES 29 AND 30.

The substitution and addition methods, when applied correctly, will yield the solution(s). Overall, the fastest method is usually the addition method, which is the basis for other procedures that we will develop later in the text. For this reason, it is better to practice using it more than the substitution method.

29. Solve

$$5x_2 + 3x_1 = 14$$
$$10x_2 + 6x_1 = 29$$

30. Solve

$$2x_1 - 6x_2 = -10$$
$$3x_1 - 9x_2 = -15$$

EXERCISE SET 1.4

Solve, using the substitution method

1. $x_2 + 4x_1 = 5$
$-3x_2 + 2x_1 = 13$

2. $4x_2 + x_1 = 8$
$5x_2 + 3x_1 = 3$

3. $5x_1 + x_2 = 8$
$3x_1 - 4x_2 = 14$

4. $2x_1 - 3x_2 = 8$
$4x_1 + x_2 = 2$

Solve, using the addition method

5. $3x_1 + 5x_2 = 28$
$5x_1 - 3x_2 = 24$

6. $4x_1 + 3x_2 = 17$
$6x_1 + 5x_2 = 27$

7. $5x_1 - 4x_2 = -3$
$7x_1 + 2x_2 = 6$

8. $-2x_1 + 4x_2 = 3$
$3x_1 - 7x_2 = 1$

9. $4x_1 + 2x_2 = 11$
$3x_1 - x_2 = 2$

10. $5x_1 - 3x_2 = -2$
$4x_1 + 2x_2 = 5$

11. $9x_1 - 2x_2 = 5$
$3x_1 - 3x_2 = 11$

12. $3x_1 + 4x_2 = 7$
$-5x_1 + 2x_2 = 10$

13. $3x_2 - 6x_1 = 15$
$4x_2 - 8x_1 = 20$

14. $8x_1 + 4x_2 = 20$
$6x_1 + 3x_2 = 14$

15. $5x_1 + 10x_2 = 20$
$2x_1 + 4x_2 = 9$

16. $2x_1 - 4x_2 = 8$
$5x_1 - 10x_2 = 20$

17. Eight times a certain number added to five times a second number is 184. The first number minus the second number is -3. Find the numbers.

18. One number is 4 times another number. Their sum is 175. Find the numbers.

19. *Business.* One day a business sold 20 pairs of gloves. The cloth gloves brought $4.95 per pair and the pigskin gloves sold for $7.50 per pair. The business took in $137.25. How many of each kind were sold?

20. *Business.* A store sold 30 sweatshirts. They sold white ones for $8.95 and red ones for $9.50. They took in $272.90. How many of each color did they sell?

21. *Biomedical.* Solution A is 2% alcohol. Solution B is 6% alcohol. A lab technician wants to mix the two to get 60 liters of a solution that is 3.2% alcohol. How many liters of each should the owner use?

22. *Agriculture.* A gardener has two kinds of solutions containing weedkiller and water. One is 5% weedkiller and the other is 15% weedkiller. The gardener needs 100 liters of a 12% solution and wants to make it by mixing. How much of each solution should be used?

Recall the formula for simple interest $I = Prt$, where I is interest, P is principal, r is rate, and t is time in years.

23. *Business.* Two investments are made totaling $4800. In the first year they yield $604 in simple interest. Part of the money is invested at 12% and the rest at 13%. Find the amount invested at each rate of interest.

24. *Business.* For a certain year $9500 is received in interest from two investments. A certain amount is invested at 13% and $10,000 more than this is invested at 14%. Find the amount invested at each rate.

25. A boat travels 46 km downstream in 2 hr. It travels 51 km upstream in 3 hr. Find the speed of the boat and the speed of the stream.

26. An airplane travels 3000 km with a tail wind in 3 hr. It travels 3000 km with a head wind in 4 hr. Find the speed of the airplane and the speed of the wind.

▶

Solve for (x, y)

27. $\sqrt{2}x + \pi y = 3$

$\pi x - \sqrt{2}y = 1$

28. $ax - by = a^2$

$bx + ay = ab$

The symbol ▦ indicates an exercise or problem facilitated by use of a calculator.

Solve

29. ▦ $4.026x - 1.448y = 18.32$

$0.724y = -9.16 + 2.013x$

30. ▦ $4.83x + 9.06y = -39.42$

$-1.35x + 6.67y = -33.99$

OBJECTIVES

You should be able to:

a) given a function of one variable and several inputs, find the outputs.

b) given a function of two variables and several input pairs, find the outputs.

1.5 FUNCTIONS AND MATHEMATICAL MODELS

Functions of One Variable

A *function* is a special kind of relation between two or more variables.

A *function of one variable* is a relation that assigns to each "input" number a unique "output" number. The set of all input numbers is called the *domain*. The set of all output numbers is called the *range*.

Example 1 Squaring numbers is a function. We can take any number x as an input. We square that number to find the output, x^2.

Inputs	Outputs
-5	25
2	4
t	t^2
x_1	x_1^2
$\sqrt{r}$	r

The domain of this function is the set of all real numbers, because any real number can be squared.

DO EXERCISE 31.

It is customary to use letters such as f and g to represent functions. Suppose f is a function and x is a number in its domain. For the input x, we can name the output as

$f(x)$,   read "f of x,"   or   "the value of f at x."

If f is the squaring function, then $f(-5)$ is the output for the input -5. Thus

$$f(-5) = (-5)^2 = 25.$$

Example 2   The squaring function is given by

$$f(x) = x^2.$$

Find $f(-4)$, $f(2)$, $f(1)$, $f(r)$, $f(\sqrt{r})$, and $f(\frac{1}{4})$.

Solution

$$f(-4) = (-4)^2 = 16, \qquad f(r) = r^2,$$
$$f(2) = 2^2 = 4, \qquad f(\sqrt{r}) = (\sqrt{r})^2 = r,$$
$$f(1) = 1^2 = 1, \qquad f(\tfrac{1}{4}) = (\tfrac{1}{4})^2 = \tfrac{1}{16}.$$

DO EXERCISE 32.

Example 3   A function f subtracts the square of an input from three times the input. A description of f is given by

$$f(x) = 3x - x^2.$$

Find $f(5)$, $f(-1)$, and $f(x_1)$.

Solution

$$f(5) = 3 \cdot 5 - 5^2 = 15 - 25 = -10,$$
$$f(-1) = 3(-1) - (-1)^2 = -3 - 1 = -4,$$
$$f(x_1) = 3x_1 - x_1^2.$$

31. The operation of "cubing" is a function. That is, the operation of going from x to x^3 is a function defined for all real numbers. Complete this table.

Inputs	Outputs
-2	
0	
t	
x_1	
$\sqrt[3]{k}$	

32. The cubing function is given by

$$f(x) = x^3.$$

Find $f(-2)$, $f(5)$, $f(\frac{1}{2})$, $f(t)$, and $f(\sqrt[3]{t})$.

33. A function f adds the square of an input to four times the input. A description of f is given by

$$f(x) = 4x + x^2.$$

Find $f(2)$, $f(-5)$, and $f(x_1)$.

34. For $P(x_1, x_2) = 4x_1 + 6x_2$,

a) Find $P(25, 10)$ and interpret its meaning.

b) Find $P(0, 18)$ and interpret its meaning.

35. For

$$f(x_1, x_2) = x_1^2 - x_2^2 - x_1,$$

find $f(4, -5)$ and $f(1, 2)$.

DO EXERCISE 33.

Functions of Two Variables

Suppose a one-product firm produces x items of its product at a profit of \$4 per item. Then its total profit P is given by

$$P(x) = 4x.$$

This is a function of *one* variable.

Suppose a two-product firm produces x_1 items of one product at a profit of \$4 per item and x_2 items of a second at a profit of \$6 per item. Then its total profit P is a function of the *two* variables x_1 and x_2 and is given by

$$P(x_1, x_2) = 4x_1 + 6x_2.$$

This is a function of *two* variables which assigns to the input pair (x_1, x_2) a unique output number $4x_1 + 6x_2$.

A function of *two* variables is a relation f that assigns to each input pair (x_1, x_2) a unique output number $f(x_1, x_2)$.

Example 4   For $P(x_1, x_2) = 4x_1 + 6x_2$, find $P(15, 20)$.

Solution $P(15, 20)$ is defined to be the value of the function found by substituting 15 for x_1 and 20 for x_2:

$$P(15, 20) = 4 \cdot 15 + 6 \cdot 20 = 60 + 120 = 180.$$

For the two-product firm, this means that by selling 15 items of the first product and 20 items of the second, it will make a profit of \$180.

DO EXERCISE 34.

Example 5   A function f is given by

$$f(x_1, x_2) = x_1^2 + x_2^2 - 2x_1x_2.$$

Find $f(-2, 3)$ and $f(8, 0)$.

Solution

$$f(-2, 3) = (-2)^2 + 3^2 - 2(-2)3 = 4 + 9 + 12 = 25,$$
$$f(8, 0) = 8^2 + 0^2 - 2 \cdot 8 \cdot 0 = 64 + 0 - 0 = 64.$$

DO EXERCISE 35.

Functions can arise without a specific formula. For example, it would probably be agreed that a student's score S on a test is a function of mental ability A and the amount of time studied t, even though it may be difficult or impossible to determine a formula for S. A statistical experiment in psychology may provide an approximating formula.

Mathematical Models

Suppose we use mathematical language to describe a problem. Then we say that we have a *mathematical model*. For example, the numbers

$$0, 1, 2, 3, \text{and so on}$$

provide a mathematical model for situations where counting is the main ingredient. Should consideration of parts of objects be involved, the above numbers are no longer appropriate, and we would use a larger set that contains fractions. In finite mathematics, we will consider many types of mathematical models. In Chapter 4, we will use inequalities, equations and functions in determining, for example, how to maximize profit, or minimize cost. In Chapter 11, we will use a model to analyze the strengths of various voting coalitions. If Congress ever does away with the electoral college, such a model may have played a significant role.

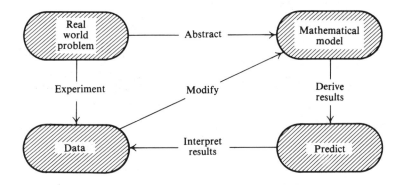

Mathematical models are abstracted from real-world situations. Procedures within the mathematical model then give results that we hope will allow us to predict what will happen in that real-world situation. To the extent that these predictions are inaccurate or the results of experimentation do not conform to the model, the model needs to be modified. This is shown in the diagram.

The diagram seems to indicate that mathematical modeling is an ongoing, possibly ever-changing, process. This is often the case. For example, finding a mathematical model that will enable accurate prediction of population growth is not a simple problem. Apparently any model that we might devise will have to be altered, as further information is acquired.

36. Find the cost of driving a car at 55 mph; at 70 mph. How much more does it cost to drive a car at 70 mph than at 55 mph?

The following graph is a mathematical model describing voter participation as a function of age. Even though no numbers are given on the axes, we gain an insight into voting patterns just by noting the way the graph rises and falls.

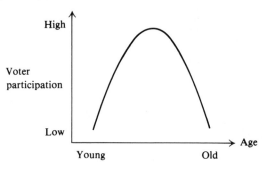

The following are some mathematical models provided by functions.

Example 6 The operating cost C, in cents per mile, of driving a car at speed x, in miles per hour, is given by

$$C(x) = 0.02x^2 - 1.3x + 30.$$

DO EXERCISE 36.

Surely this model is in need of continual revision due to inflation.

Example 7 *The Homans Model.* In psychology, the intensity I of interaction (or communication) between people in a certain group has the following as a model.

$$I(F, A) = c_1 F + c_2 A,$$

where F is the amount of friendliness among members of the group, A is the amount of activity carried on by members of the group, and c_1 and c_2 are fixed constants that change from group to group.

Example 8 An object falling in a vacuum will fall s feet in t seconds, where

$$s(t) = \tfrac{1}{2}gt^2$$

and g is a constant (about 32.2).

Suppose one wanted to consider an object falling through air. Friction and wind speed would affect the above formula, but physicists often still use it as an approximation.

EXERCISE SET 1.5

1. A function f is given by

$$f(x) = 5x - 1.$$

This function takes an input x, multiplies it by 5, and subtracts 1.

a) Complete this table.

Inputs	Outputs
6	
4.1	
4.01	
4.001	
4.0001	

b) Find $f(7)$, $f(-1)$, $f(-4)$, and $f(k)$.

3. A function g is given by

$$g(x) = x^2 + 4.$$

This function takes an input x, squares it, and adds 4.
Find $g(-3)$, $g(0)$, $g(-1)$, $g(9)$, and $g(u)$.

5. A function f is given by

$$f(x) = (x + 4)^2.$$

This function takes an input x, adds 4 to it, and then squares the result.
Find $f(3)$, $f(-6)$, $f(0)$, and $f(\frac{1}{2})$.

7. For $f(x) = x^2 + 2x - 3$, find $f(0)$, $f(-2)$, and $f(3)$.

9. For $f(x) = 2x^3 - x^2$, find $f(4)$.

11. A clothing firm determines that its total revenue (money coming in) from the sale of x suits is given by the function

$$R(x) = 20x + 50,$$

where $R(x)$ is the revenue, in dollars, from the sale of x suits. Find $R(1)$, $R(10)$, and $R(1000)$.

2. A function f is given by

$$f(x) = 3x + 2.$$

This function takes an input x, multiplies it by 3, and adds 2.

a) Complete this table.

Inputs	Outputs
8	
7.1	
7.01	
7.001	
7.0001	

b) Find $f(4)$, $f(-3)$, $f(-1)$, and $f(t)$.

4. A function g is given by

$$g(x) = x^2 - 3.$$

This function takes an input x, squares it, and subtracts 3.
Find $g(-1)$, $g(0)$, $g(2)$, $g(-5)$, and $g(e)$,

6. A function f is given by

$$f(x) = (x - 3)^2.$$

This function takes an input x, subtracts 3 from it, and then squares the result.
Find $f(4)$, $f(-2)$, $f(0)$, and $f(\frac{1}{4})$.

8. For $f(x) = x^2 - 3x + 2$, find $f(0)$, $f(-2)$, and $f(3)$.

10. For $f(x) = 3x^2 - x^3$, find $f(2)$.

12. A firm determines that the total revenue from the sale of x units of a product is

$$R(x) = 2x + 1000,$$

where $R(x)$ is the revenue, in dollars, from the sale of x units. Find $R(1)$, $R(2)$, and $R(50)$.

13. The amount of money in a savings account at 8% compounded annually depends on the initial amount x and is given by

$$A(x) = x + 8\%x,$$

where $A(x)$ = amount in the account at the end of one year. Find $A(200)$ and $A(1000)$.

15. For $f(x_1, x_2) = 3x_1 - 4x_2$, find $f(-2, 5)$, $f(4, 0)$, and $f(10, -6)$.

17. For $f(x_1, x_2) = x_2^2 + 3x_1x_2$, find $f(-2, 0)$, $f(3, 2)$, and $f(-5, 10)$.

19. For $f(x_1, x_2) = x_1^2 - x_2^2$, find $f(-2, -3)$, $f(5, 0)$, and $f(0, 5)$.

21. For $f(x_1, x_2) = (3x_1 + 4x_2)^2$, find $f(-1, 0)$, $f(2, 2)$, and $f(-4, 5)$.

14. The population of a city is growing at the rate of 2% per year. The population P at the end of the year is given by

$$P(x) = x + 2\%x,$$

where x is the population at the beginning of the year. Find $P(100,000)$ and $P(2,000,000)$.

16. For $f(x_1, x_2) = 5x_1 - 2x_2$, find $f(2, -5)$, $f(0, 6)$, and $f(-4, -20)$.

18. For $f(x_1, x_2) = x_1^2 - 2x_1x_2$, find $f(0, -2)$, $f(2, 3)$, and $f(10, -5)$.

20. For $f(x_1, x_2) = x_1^2 + x_2^2$, find $f(-3, -2)$, $f(7, 0)$, and $f(0, 7)$.

22. For $f(x_1, x_2) = (4x_1 - x_2)^2$, find $f(-2, 0)$, $f(5, 5)$, and $f(6, -4)$.

OBJECTIVES

You should be able to

a) Given total revenue and total cost functions, find the total profit function, the break-even point, the profit values, and the loss values.

b) Given a demand and a supply function, find the equilibrium point.

*1.6 (Optional) BUSINESS AND ECONOMIC APPLICATIONS

Let us consider some business and economic applications modeled by linear functions. A *linear function* f is given by $f(x) = mx + b$.

Profit, Loss, and Break-Even Analysis

Example 1 An electronics firm is planning to market a new hand calculator. For the first year the *fixed costs* for setting up the new production line are $80,000. These are costs such as rent, maintenance, machinery, and so on, which must be absorbed before a calculator is ever produced. To produce x calculators, it costs $20 per calculator in addition to the fixed costs. That is, the *variable costs* are $20x$ dollars. These are costs that are directly related to producing the calculators, such as materials, wages, fuel, and so on. Then the *total cost* function $C(x)$ of producing x calculators in a year is given by a function C:

$$C(x) = \text{(Variable costs)} + \text{(Fixed costs)}$$
$$= \quad 20x \quad + \quad 80,000$$

The firm determines that its total revenue (money coming in) from the sale of x calculators is $36 per calculator. That is, *total revenue* is given by a function R:

$$R(x) = 36x$$

Total profit is given by a function P:

$$P(x) = \text{(Total revenue)} - \text{(Total cost)} = R(x) - C(x).$$

a) Determine $P(x)$.

b) The firm will *break even* at those values of x for which $P(x) = 0$, or $R(x) - C(x) = 0$, or $R(x) = C(x)$. Find the break-even values of x.

c) The firm will be operating at a *profit* where $P(x) > 0$, or $R(x) - C(x) > 0$, or $R(x) > C(x)$; that is, at values of x for which total revenue exceeds total cost. Find these profit values.

d) The firm will operate at a *loss* where $P(x) < 0$, or $R(x) - C(x) < 0$, or $R(x) < C(x)$; that is, at values of x for which total cost exceeds total revenue. Find these loss values.

Solution

a) $P(x) = R(x) - C(x) = 36x - (20x + 80{,}000) = 16x - 80{,}000.$

b) We solve $R(x) = C(x)$.

$$36x = 20x + 80{,}000$$
$$16x = 80{,}000$$
$$x = 5000$$

That is, the firm will break even by producing and selling 5000 calculators. Here there is only one break-even value. If curves are involved, there could be more than one break-even value.

c) We solve $R(x) > C(x)$.

$$36x > 20x + 80{,}000$$
$$16x > 80{,}000$$
$$x > 5000$$

That is, the firm will operate at a profit when it produces and sells more than 5000 calculators.

d) We solve $R(x) < C(x)$.

$$36x < 20x + 80{,}000$$
$$16x < 80{,}000$$
$$x < 5000$$

That is, the firm will operate at a loss when it produces and sells less than 5000 calculators.

It is instructive to look at the graphs of revenue, cost, and profit using the same axes and marking the break-even point and the regions of profit and loss.

37. Rework Example 1, where

$$C(x) = 20x + 100,000$$

and

$$R(x) = 45x$$

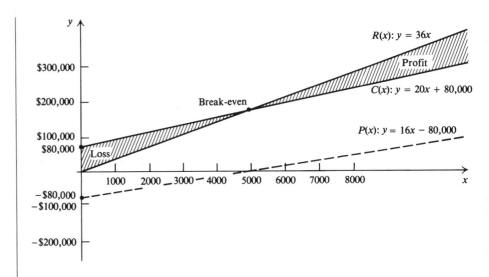

DO EXERCISE 37.

Supply and Demand

Demand Functions   Consider the following table and graph.

Demand schedule	
Price p (per bag)	Quantity D (number of 5-lb bags) in millions
$5	5
4	10
3	15
2	20
1	25

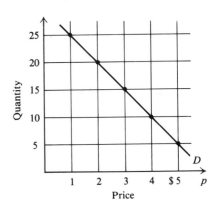

The table and graph show the relationship between the price p per bag of sugar and the quantity D of 5-lb bags that the consumer will buy at that price. Note that, as price per bag increases, the quantity demanded by the consumer decreases; and as price per bag decreases, the quantity demanded by the consumer increases. Thus it is natural to think of D as a function of p. Thus, for a *demand* function D, $D(p)$ is the number of units demanded by the consumer when the price per unit is p.

Supply Functions   Consider the following table and graph.

Supply schedule	
Price p (per bag)	Quantity D (number of 5-lb bags) in millions
$2	5
2.50	10
3	15
3.50	20
4	25

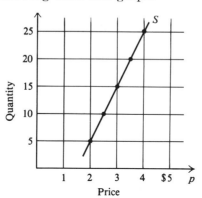

The table and graph show the relationship between the price p per bag of sugar and the quantity S of 5-lb bags which the seller is willing to supply at that price. Note that, as price per bag increases, the more the seller is willing to supply; and as the price per bag decreases, the less the seller is willing to supply. Again, it is natural to think of S as a function of p. Thus, for a *supply* function S, $S(p)$ is the number of units the seller will supply to the consumer at price p per bag.

Let us now look at these graphs together. Note that, as supply increases demand decreases, and as supply decreases demand increases. The point of intersection of the two graphs (p_E, q_E) is called the *equilibrium point*. The equilibrium price p_E (in this case $3 per bag) is where the quantity q_E (in this case 15 million bags) that the seller willingly supplies is the same as the amount the consumer willingly demands. The situation is analogous to a buyer and a seller haggling over the sale of an item. The equilibrium, or selling price, is what they finally agree on.

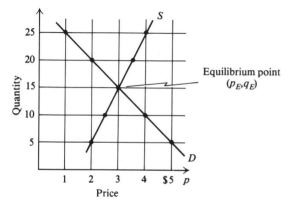

Equilibrium point (p_E, q_E)

38. Find the equilibrium point for the demand and supply functions

$$D(p) = 1000 - 60p,$$

$$S(p) = 200 + 4p.$$

Example 2 Find the equilibrium point for the demand and supply functions

$$D(p) = 300 - 49p, \qquad S(p) = 100 + p.$$

Solution

a) We first set $D(p) = S(p)$ and solve.

$$300 - 49p = 100 + p$$

$$-50p = -200$$

$$p = \frac{-200}{-50} = \$4.$$

Thus $p_E = \$4$ per unit.

b) To find q_E we substitute p_E into either $D(p)$ or $S(p)$. We use $S(p)$.

$$q_E = S(p_E) = S(4) = 100 + 4 = 104.$$

Thus the equilibrium quantity is 104, and the equilibrium point is ($\$4, 104$).

DO EXERCISE 38.

EXERCISE SET 1.6

For the following total cost and total revenue functions,

a) Determine the total profit function $P(x)$.
c) Find the profit values.

b) Find the break-even value.
d) Find the loss values.

1. $R(x) = 70x, \qquad C(x) = 25x + 360,000$

2. $R(x) = 65x, \qquad C(x) = 45x + 600,000$

3. $R(x) = 85x, \qquad C(x) = 30x + 49,500$

4. $R(x) = 60x, \qquad C(x) = 10x + 120,000$

5. A clothing firm is planning a new line of pant suits. For the first year, the fixed costs for setting up the production line are $10,000. Variable costs for producing each suit are $20. The sales department projects that 2000 suits can be sold during the first year. The contribution to revenue from each suit is to be $100.

a) Formulate a function $C(x)$ for the total cost of producing x suits.
b) Formulate a function $R(x)$ for the total revenue from the sale of x suits.
c) Formulate a function $P(x)$ for the total profit from the production and sale of x suits.
d) What profit or loss will the company realize if expected sales of 2000 suits occur?
e) Find the break-even value.
f) Find the profit values.
g) Find the loss values.

6. A clock manufacturer is planning a new type of wall clock. For the first year, the fixed costs for setting up the new production line are $22,500. Variable costs for producing each clock are estimated to be $40. The sales department projects that 3000 clocks can be sold during the first year. The revenue from each clock is to be $85.

a) Formulate a function $C(x)$ for the total cost of producing x clocks.
b) Formulate a function $R(x)$ for the total revenue from the sale of x clocks.
c) Formulate a function $P(x)$ for the total profit from the production and sale of x clocks.
d) What profit or loss will the company realize if expected sales of 3000 clocks occur?
e) Find the break-even value.
f) Find the profit values.
g) Find the loss values.

Find the equilibrium point for the following demand and supply functions.

7. $D(p) = 1000 - 10p$, $\ S(p) = 250 + 5p$

8. $D(p) = 2000 - 60p$, $\ S(p) = 460 + 94p$

9. $D(p) = 800 - 43p$, $\ S(p) = 210 + 16p$

10. $D(p) = 760 - 13p$, $\ S(p) = 430 + 2p$

11. $D(p) = 8800 - 30p$, $\ S(p) = 7000 + 15p$

12. $D(p) = 7500 - 25p$, $\ S(p) = 6000 + 5p$

CHAPTER 1 TEST

1. Graph

$$5x_1 + 4x_2 = 20$$

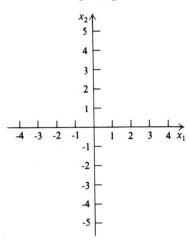

2. Solve

$$-\tfrac{2}{5}x + 13 = 12 - \tfrac{1}{2}x$$

3. Solve

$$6x - 5 \geq 20x + 3$$

4. An investment is made at 9%, compounded annually. It grows to $599.50 at the end of 1 year. How much was originally invested?

5. Graph this system and classify it as consistent or inconsistent, dependent or independent

$$5x_1 - 15x_2 = -10$$
$$3x_1 - 9x_2 = 6$$

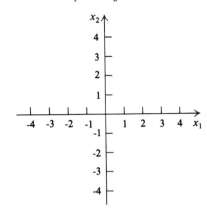

6. Decide whether $(-2, 3)$ is a solution of the system

$$x_1 + x_2 = 1$$
$$3x_1 + x_2 = -3$$

7. Solve graphically

$$x_1 - x_2 = 4$$

$$x_1 + x_2 = 2$$

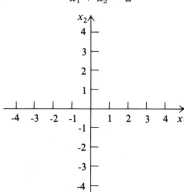

8. Solve, using the addition method

$$5x_1 + 4x_2 = 24$$

$$2x_1 - 3x_2 = 5$$

9. Solve, using the addition method

$$3x_1 - x_2 = 2$$

$$4x_1 + 2x_2 = 11$$

10. Solve, using the addition method

$$5x_1 + 10x_2 = 15$$

$$3x_1 + 6x_2 = 9$$

11. For $f(x) = x^2 + 3x$, find $f(1)$ and $f(-4)$

12. For $f(x_1, x_2) = 10x_1 - 3x_2$, find $f(-5, 20)$

*13. Find the equilibrium point for the demand and supply functions:

$$D(p) = 9800 - 35p, \qquad S(p) = 7000 + 5p.$$

CHAPTER TWO

Mathematics of Finance

OBJECTIVES

You should be able to

a) Find certain terms of a sequence given the nth term.

b) Identify the first term and the common difference of an arithmetic sequence.

c) Find a specified term of an arithmetic sequence.

d) Find the sum of the first n terms of an arithmetic sequence.

1. A sequence is given by

$$a_n = n^2 + 3.$$

Find the first 4 terms and the 15th term.

2.1 ARITHMETIC SEQUENCES*

Sequences

A *sequence* is an ordered set of numbers. Here is an example

$$3, 5, 7, 9, 11, \ldots$$

The three dots mean that there are more and more numbers in the sequence. A sequence that does not end is called *infinite*. A sequence that does end is called *finite*.

Each number is called a *term* of the sequence. The first term is 3, the second term is 5, the third term is 7, and so on. We can describe the terms as follows:

$$a_1 = 3,$$
$$a_2 = 5,$$
$$a_3 = 7,$$
$$a_4 = 9,$$

and so on, where the nth term is $a_n = 2n + 1$. That is, a sequence is a function whose domain is a set of consecutive natural numbers.

Instead of using $a(n)$ for the nth term, we are using a_n. We also call a_n the *general term*.

Example 1 A sequence is given by

$$a_n = 2^n$$

Find the first 4 terms and the 17th term.

Solution
$$a_1 = 2^1 = 2,$$
$$a_2 = 2^2 = 4,$$
$$a_3 = 2^3 = 8,$$
$$a_4 = 2^4 = 16,$$
$$\vdots$$
$$a_{17} = 2^{17} = 131{,}072.$$

DO EXERCISE 1.

---

* This chapter is intended principally for business and economics students. It can be omitted without loss of continuity although it provides a nice lead-in to the study of calculus. A hand calculator with an $\boxed{x^y}$ key will be quite helpful.

Arithmetic Sequences

Consider the sequence

$$3, 5, 7, 9, 11, \ldots$$

Note that the number 2 can be added to each term to obtain the next term. Sequences in which a certain number can be added to any term to get the next term are called *arithmetic sequences* (or *arithmetic progressions*). The number d which we add to one term to get the next is called the *common difference*. This is because we can subtract any term from the one that follows it and get d.

$$a_{k+1} - a_k = d \quad \text{for any } k \geqslant 1.$$

Examples The following are arithmetic sequences. Identify the first term and the common difference.

Solution

Sequence	First term	Common difference
2. $3, 5, 7, 9, 11, \ldots$	3	2
3. $34, 27, 20, 13, 6, -1, -8, \ldots$	34	-7
4. $\$5200, \$4687.50, \$4175, \ldots$	$\$5200$	$-\$512.50$

DO EXERCISES 2 THROUGH 5.

For an arithmetic sequence

The 1st term is $\quad a_1,$

the 2nd term is $\quad a_2 = a_1 + d,$

the 3rd term is $\quad a_3 = (a_1 + d) + d = a_1 + 2d,$

the 4th term is $\quad a_4 = [(a_1 + d) + d] + d = a_1 + 3d,$

and so on. Generalizing, we obtain the following.

The nth term of an arithmetic sequence is given by

$$a_n = a_1 + (n - 1)d, \quad \text{for any } n \geqslant 1.$$

Example 5 Find the 15th term of the sequence $4, 7, 10, 13, \ldots$

Solution First note that

$$a_1 = 4, \quad d = 3, \quad \text{and} \quad n = 15.$$

The following are arithmetic sequences. Identify the first term and the common difference.

2. $2, 5, 8, 11, 14, \ldots$

3. $19, 14, 9, 4, -1, -6, \ldots$

4. $\$6300, \$5953.25, \$5606.50, \ldots$

5. $2, 2\frac{1}{2}, 3, 3\frac{1}{2}, 4, 4\frac{1}{2}, \ldots$

6. Find the 18th term of the sequence

$$2, 6, 10, 14, \ldots$$

7. Find the 11th term of the sequence

$$\$6300, \$5953.25, \$5606.50, \ldots$$

Then using the formula

$$a_n = a_1 + (n - 1)d,$$

we have

$$a_{15} = 4 + (15 - 1)3 = 4 + 14 \cdot 3 = 4 + 42 = 46.$$

We could check this by writing out 15 terms of the sequence.

DO EXERCISES 6 AND 7.

Sum of the First *n* Terms of an Arithmetic Sequence

Suppose we add the first 4 terms of the sequence

$$3, 5, 7, 9, 11, \ldots$$

We get

$$3 + 5 + 7 + 9, \quad \text{or} \quad 24.$$

The sum of the first n terms of a sequence is denoted S_n. Thus, for the preceding sequence, $S_4 = 24$. We want to find a formula for S_n when the sequence is arithmetic. We can denote an arithmetic sequence as

$$a_1, (a_1 + d), (a_1 + 2d), \ldots, (a_n - 2d), (a_n - d), a_n.$$

Then S_n is given by

$$S_n = a_1 + (a_1 + d) + (a_1 + 2d) + \cdots + (a_n - 2d) + (a_n - d) + a_n. \tag{1}$$

If we reverse the order of addition we get

$$S_n = a_n + (a_n - d) + (a_n - 2d) + \cdots + (a_1 + 2d) + (a_1 + d) + a_1. \tag{2}$$

Suppose we add corresponding terms of each side of Eqs. (1) and (2). Then we get

$$2S_n = [a_1 + a_n] + [(a_1 + d) + (a_n - d)]$$
$$+ [(a_1 + 2d) + (a_n - 2d)] + \cdots + [(a_n - 2d) + (a_1 + 2d)]$$
$$+ [(a_n - d) + (a_1 + d)] + [a_n + a_1].$$

This simplifies to

$$2S_n = (a_1 + a_n) + (a_1 + a_n) + (a_1 + a_n) + \cdots + (a_1 + a_n).$$

Since there are n binomials $(a_1 + a_n)$ being added, it follows that $2S_n = n(a_1 + a_n)$, from which we get the following formula.

The sum of the first n terms of an arithmetic sequence is given by

$$S_n = \frac{n}{2}(a_1 + a_n).$$

Example 6   Find the sum of the first 100 natural numbers.

Solution   The sum is

$$1 + 2 + 3 + \cdots + 99 + 100$$

This is the sum of the first 100 terms of the arithmetic sequence for which

$$a_1 = 1, \qquad a_n = 100, \qquad \text{and} \qquad n = 100.$$

Then substituting in the formula

$$S_n = \frac{n}{2}(a_1 + a_n),$$

we get

$$S_{100} = \tfrac{100}{2}(1 + 100) = 50(101) = 5050.$$

DO EXERCISE 8.

The preceding formula is useful when we know a_1 and a_n, the first and last terms, but it often happens that a_n is not known. We thus need a formula in terms of a_1, n, and d.

Substituting $a_1 + (n - 1)d$ for a_n in the formula $S_n = \dfrac{n}{2}(a_1 + a_n)$, we get

$$S_n = \frac{n}{2}(a_1 + [a_1 + (n - 1)d]),$$

from which we get the following formula.

The sum of the first n terms of an arithmetic sequence is given by

$$S_n = \frac{n}{2}[2a_1 + (n - 1)d].$$

Example 7   Find the sum of the first 15 terms of the arithmetic sequence 4, 7, 10, 13, . . .

Solution   Note that

$$a_1 = 4, \qquad d = 3, \qquad \text{and} \qquad n = 15.$$

8. Find the sum of the first 200 natural numbers.   PG 43

Find the 15th term of the seq.
14, 9, 4, -1, -6
                                        difference
$a_{15} = 14 + (15 - 1) - 5$
$\quad = -56$

Find the sum of the First 40 terms
of the seq 50, 52.50, 55
$S_{40} = \frac{40}{2}[2(50) + (40-1) 2.5]$
$\quad 20[100 + (39) 2.5]$
$\quad 20[197.5] = 3950$

9. Find the sum of the first 16 terms of the sequence

$$1, 3, 5, 7, 9, \ldots$$

Then, substituting in the formula

$$S_n = \frac{n}{2}[2a_1 + (n - 1)d],$$

we get

$$S_{15} = \tfrac{15}{2}[2 \cdot 4 + (15 - 1)3] = \tfrac{15}{2}[8 + 14 \cdot 3] = \tfrac{15}{2}[8 + 42]$$
$$= \tfrac{15}{2}[50] = 375.$$

DO EXERCISE 9.

10. A family saves money in an arithmetic sequence. They save $500 the 1st year, $700 the 2nd, $900 the 3rd year, and so on, for 13 years. How much do they save in all?

Example 8 A family saves money in an arithmetic sequence. They save $600 the first year, $700 the second, and so on, for 20 years. How much do they save in all (disregarding interest)?

Solution The amount saved is the sum

$$\$600 + \$700 + \$800 + \cdots$$

Here the dots mean that the pattern continues, even though this is not an infinite sequence. In short, we need not bother to determine the last term. We can find the sum by noting that

$$a_1 = \$600, \qquad d = \$100, \qquad \text{and} \qquad n = 20.$$

Then, substituting in the formula

$$S_n = \frac{n}{2}[2a_1 + (n - 1)d],$$

we get

$$S_{20} = \tfrac{20}{2}[2 \cdot \$600 + (20 - 1)\$100] = 10[\$1200 + 19 \cdot \$100]$$
$$= 10[\$1200 + \$1900] = 10[\$3100] = \$31,000.$$

DO EXERCISE 10.

EXERCISE SET 2.1

In each of the following sequences, the nth term is given. Find the first 4 terms, and the 15th term.

1. $a_n = \dfrac{n}{n + 1}$
2. $a_n = n + \dfrac{1}{n}$
3. $a_n = \dfrac{n^2 - 1}{n^3 + 1}$
4. $a_n = (-\tfrac{1}{2})^n$

The following are arithmetic sequences. Identify the first term and the common difference.

5. $2, 7, 12, 17, \ldots$
6. $7, 3, -1, -5, \ldots$
7. $\$1.06, \$1.12, \$1.18, \$1.24, \ldots$

8. $\$214, \$211, \$208, \$205, \ldots$
9. $5, 4\tfrac{1}{3}, 3\tfrac{2}{3}, 3, 2\tfrac{1}{3}, \ldots$
10. $\tfrac{3}{2}, \tfrac{9}{4}, 3, \tfrac{15}{4}, \ldots$

11. Find the 12th term of the arithmetic sequence

$$3, 7, 11, \ldots$$

12. Find the 11th term of the arithmetic sequence

$$\$0.08, \$0.13, \$0.18, \ldots$$

13. Find the 13th term of the arithmetic sequence

$$\$1200, \$964.32, \$728.64, \ldots$$

14. Find the 10th term of the arithmetic sequence

$$\$200, \$198.32, \$196.64, \ldots$$

15. Find the sum of the first 300 natural numbers.

16. Find the sum of the first 400 natural numbers.

17. Find the sum of the first 20 terms of the sequence

$$6, 9, 12, 15, \ldots$$

18. Find the sum of the first 14 terms of the sequence

$$12, 8, 4, \ldots$$

19. Find a formula for the sum of the first n natural numbers:

$$1 + 2 + 3 + \cdots + n$$

20. Find a formula for the sum of the first n consecutive odd natural numbers starting with 1:

$$1 + 3 + 5 + \cdots + (2n - 1)$$

21. If a student saves 1¢ on October 1, 2¢ on October 2, 3¢ on October 3, etc., how much would be saved in October? (October has 31 days.)

22. If a student saves \$40 on September 1, \$60 on September 2, \$80 on September 3, how much would be saved in September? (September has 30 days.)

23. Find the sum of the first 8 terms of the arithmetic sequence

$$\$512.50, \$1025.00, \$1537.50, \ldots$$

24. Find the sum of the first 10 terms of the arithmetic sequence

$$\$78.90, \$157.80, \$236.70, \ldots$$

2.2 GEOMETRIC SEQUENCES

Geometric Sequences

Consider the sequence

$$2, 6, 18, 54, 162, \ldots$$

If we multiply each term by 3 we get the next term. Sequences in which each term can be multiplied by a certain number to get the next term are called *geometric*. We usually denote this number r. We refer to it as the *common ratio* because we can get r by dividing any term by the preceding term.

$$\frac{a_{k+1}}{a_k} = r, \quad \text{or} \quad a_{k+1} - a_k = d \quad \text{for any } k \geqslant 1.$$

Examples The following are geometric sequences. Identify the common ratio.

OBJECTIVES

You should be able to

a) Identify the common ratio of a geometric sequence.

b) Find the nth term of a geometric sequence.

c) Find the sum of the first n terms of a geometric sequence.

d) Find the sum of an infinite geometric series, if it exists.

The following are geometric sequences. Identify the common ratio.

11. $1, 5, 25, 125, \ldots$

12. $3, -9, 27, -81, \ldots$

13. $6000, $5100, $4335, $3684.75, \ldots$

14. $100, $109, $118.81, \ldots$

15. $1, \frac{1}{2}, \frac{1}{4}, \frac{1}{8}, \ldots$

16. Find the 9th term of the geometric sequence

$$2, 4, 8, 16, \ldots$$

Solution

Sequence	Common ratio
1. $3, 6, 12, 24, 48, 96, \ldots$	2
2. $3, -6, 12, -24, 48, -96, \ldots$	-2
3. $5200, $3900, $2925, $2193.75, \ldots$	0.75
4. $1000, $1080, $1166.40, \ldots$	1.08

DO EXERCISES 11–15.

If we let a_1 be the 1st term and r be the common ratio, then $a_1 r$ is the 2nd term, $a_1 r^2$ is the 3rd term, and so on. Generalizing, we obtain the following.

The nth term of a geometric sequence is given by

$$a_n = a_1 r^{n-1}, \quad \text{for any } n \geqslant 1.$$

Note that the exponent is 1 less than the number of the term.

Example 5   Find the 7th term of the geometric sequence 4, 20, 100, ...

Solution   First note that

$$a_1 = 4, \quad n = 7, \quad \text{and} \quad r = \frac{20}{4}, \text{ or } 5.$$

Then, using the formula

$$a_n = a_1 r^{n-1},$$

we have

$$a_7 = 4 \cdot 5^{7-1} = 4 \cdot 5^6 = 4 \cdot 15{,}625 = 62{,}500.$$

DO EXERCISE 16.

Example 6   Find the 10th term of the geometric sequence

$$64, 32, 16, 8, \ldots$$

Solution   First note that

$$a_1 = 64, \quad n = 10, \quad \text{and} \quad r = \frac{32}{64}, \text{ or } \frac{1}{2}.$$

Then using the formula

$$a_n = a_1 r^{n-1},$$

we have

$$a_{10} = 64 \cdot \left(\frac{1}{2}\right)^{10-1} = 64 \cdot \left(\frac{1}{2}\right)^9 = 2^6 \cdot \frac{1}{2^9} = \frac{1}{2^3} = \frac{1}{8}.$$

DO EXERCISE 17.

Sum of the First *n* Terms of a Geometric Sequence

We want to find a formula for the sum S_n of the first n terms of a geometric sequence

$$a_1, a_1 r, a_1 r^2, a_1 r^3, \ldots, a_1 r^{n-1}, \ldots$$

The sum S_n is given by

$$S_n = a_1 + a_1 r + a_1 r^2 + \cdots + a_1 r^{n-2} + a_1 r^{n-1}.$$

We want to develop a formula that allows us to find this sum without a great amount of adding. If we multiply both sides of the preceding equation by r, we have

$$rS_n = a_1 r + a_1 r^2 + a_1 r^3 + \cdots + a_1 r^{n-1} + a_1 r^n.$$

When we multiply S_n by -1, we get

$$-S_n = -a_1 - a_1 r - a_1 r^2 - \cdots - a_1 r^{n-2} - a_1 r^{n-1}.$$

Then, when we add rS_n and $-S_n$, we get

$$rS_n - S_n = a_1 r^n - a_1$$

or

$$(r - 1)S_n = a_1(r^n - 1),$$

from which we get the following formula.

The sum of the first *n* terms of a geometric sequence is given by

$$S_n = \frac{a_1(r^n - 1)}{r - 1}, \qquad \text{for any } r \neq 1.$$

Example 7   Find the sum of the first 7 terms of the geometric sequence $3, 15, 75, 375, \ldots$

Solution   First note that

$$a_1 = 3, \qquad n = 7, \qquad \text{and} \qquad r = \tfrac{15}{3}, \text{ or } 5.$$

Then, using the formula

$$S_n = \frac{a_1(r^n - 1)}{r - 1},$$

we have

$$S_7 = \frac{3(5^7 - 1)}{5 - 1} = \frac{3(78{,}125 - 1)}{4} = \frac{3(78{,}124)}{4} = 58{,}593.$$

17. Find the 6th term of the geometric sequence

$$3, 1, \tfrac{1}{3}, \tfrac{1}{9}, \ldots$$

18. Find the sum of the first 8 terms of the geometric sequence

$$2, 4, 8, 16, \ldots$$

19. Under the conditions of Example 8, how much would you make in October, which has 31 days?

DO EXERCISE 18.

Doubling Your Salary

Example 8   Suppose someone offered you a job during the month of September (30 days) under the following conditions. You will be paid $0.01 for the first day, $0.02 for the second, $0.04 for the third, and so on, doubling your previous day's salary each day. How much would you earn? (Would you take the job? Make a decision before reading further.)

Solution   The amount earned is the sum

$$\$0.01 + \$0.01(2) + \$0.01(2^2) + \$0.01(2^3) + \cdots + \$0.01(2^{29}),$$

where

$$a_1 = \$0.01, \quad n = 30, \quad \text{and} \quad r = 2.$$

Then, using the formula

$$S_n = \frac{a_1(r^n - 1)}{r - 1},$$

we have

$$S_{30} = \frac{\$0.01(2^{30} - 1)}{2 - 1}$$

$$\approx \$0.01(1,074,000,000 - 1) \quad \text{(Use a calculator to approximate } 2^{30})$$

$$\approx \$0.01(1,074,000,000)$$

$$\approx \$10,740,000.$$

Now would you take the job?

DO EXERCISE 19.

Note. One could find 2^{30} in various ways. It can be found directly on a calculator with an $\boxed{x^y}$ key or by expressing the power as, say, $2^{10} \cdot 2^{10} \cdot 2^{10}$. Then one could find 2^{10} and multiply that number by itself 3 times. Another way would be $2^5 \cdot 2^5 \cdot 2^5 \cdot 2^5 \cdot 2^5 \cdot 2^5$.

Infinite Geometric Series

Suppose we consider the sum of the terms of an infinite geometric sequence, such as 2, 4, 8, 16, 32, ... We get what is called an *infinite geometric series*

$$2 + 4 + 8 + 16 + 32 + \cdots$$

As n grows larger and larger, the sum of the first n terms, S_n, becomes larger and larger without bound. There are infinite series that get closer and closer to some specific number. Here is an example:

$$\frac{1}{2} + \frac{1}{4} + \frac{1}{8} + \frac{1}{16} + \cdots + \frac{1}{2^n} + \cdots$$

Let's consider S_n for some values of n.

$$S_1 = \tfrac{1}{2} \qquad\qquad\qquad = \tfrac{1}{2} \ = 0.5$$
$$S_2 = \tfrac{1}{2} + \tfrac{1}{4} \qquad\qquad = \tfrac{3}{4} \ = 0.75$$
$$S_3 = \tfrac{1}{2} + \tfrac{1}{4} + \tfrac{1}{8} \qquad\quad = \tfrac{7}{8} \ = 0.875$$
$$S_4 = \tfrac{1}{2} + \tfrac{1}{4} + \tfrac{1}{8} + \tfrac{1}{16} \quad = \tfrac{15}{16} = 0.9375$$
$$S_5 = \tfrac{1}{2} + \tfrac{1}{4} + \tfrac{1}{8} + \tfrac{1}{16} + \tfrac{1}{32} = \tfrac{31}{32} = 0.96875$$

Perhaps you have noticed that we can describe S_n as follows:

$$S_n = \frac{2^n - 1}{2^n}.$$

Note that the numerator is less than the denominator for all values of n, but as n gets larger and larger, the values of S_n get closer and closer to 1. We say that 1 is the *limit* of S_n and that 1 is the *sum* of the *infinite geometric series*. The sum of an infinite series, if it exists, is denoted S_∞. It can be shown (but we will not do it here) that the sum of the terms of a geometric series exists if and only if $|r| < 1$ (that is, the absolute value of the common ratio is less than 1).

We want to find a formula for the sum of an infinite geometric series. We first consider the sum of the first n terms:

$$S_n = \frac{a_1(r^n - 1)}{r - 1} = \frac{a_1 - a_1 r^n}{1 - r}.$$

For $|r| < 1$, it follows that values of r^n get closer and closer to 0 as n gets large. (Pick a number between -1 and 1 and check this by finding larger and larger powers on your calculator). As r^n gets closer and closer to 0, so does $a_1 r^n$, so S_n gets closer and closer to $a_1/(1 - r)$.

When $|r| < 1$, the sum of an infinite geometric series is given by

$$S_\infty = \frac{a_1}{1 - r}.$$

Example 9 Determine whether this infinite geometric series has a sum. If so, find it:

$$1 + 3 + 9 + 27 + \cdots$$

Determine whether each infinite geometric series has a sum. If so, find it.

20. $1 + 7 + 49 + 343 + \cdots$

21. $1 + (-1) + 1 + (-1) + \cdots$

22. $\frac{1}{2} + \frac{1}{4} + \frac{1}{8} + \frac{1}{16} + \frac{1}{32} + \cdots$

23. $625 + 250 + 100 + 40 + \cdots$

24. Rework Example 11 when the proportion is 95%.

Solution $|r| = |3| = 3$, and since $|r| \not< 1$ the series does *not* have a sum.

Example 10 Determine whether this infinite geometric series has a sum. If so, find it:

$$1 - \tfrac{1}{2} + \tfrac{1}{4} - \tfrac{1}{8} + \tfrac{1}{16} - \cdots$$

Solution

a) $|r| = |-\tfrac{1}{2}| = \tfrac{1}{2}$, and since $|r| < 1$ the series does have a sum.
b) The sum is given by

$$S_\infty = \frac{1}{1 - (-\frac{1}{2})} = \frac{2}{3}.$$

DO EXERCISES 20 THROUGH 23.

Example 11 *Economic multiplier.* The United States banking laws require most banks to maintain a reserve equivalent to a certain proportion of their outstanding deposits. This enables such banks, when they wish and when they can find borrowers, to loan out a certain proportion of the funds that have been deposited in them. Let us assume that this proportion is 0.90 (or 90%). Now suppose a corporation deposits $1000 in a bank which, subsequently, is able to loan the maximum legally possible amount, and this loan is redeposited elsewhere, and so on. What is the total effect of the $1000 on the economy?

Solution The total effect can be modeled as the sum of the infinite geometric series

$$\$1000 + \$1000(0.90) + \$1000(0.90)^2 + \$1000(0.90)^3 + \cdots$$

which is given by

$$S_\infty = \frac{\$1000}{1 - 0.90} = \$10,000.$$

The sum $10,000 is the result of what is referred to in economics as the *multiplier effect*.

DO EXERCISE 24.

EXERCISE SET 2.2

The following are geometric sequences. Identify the common ratio.

1. $7, 14, 28, 56, \ldots$

2. $5, -15, 45, -135, \ldots$

3. $12, -4, \frac{4}{3}, -\frac{4}{9}, \ldots$

4. $4, 2, 1, \frac{1}{2}, \frac{1}{4}, \ldots$

5. $\$5600, \$5320, \$5054, \$4801.30, \ldots$

6. $\$780, \$858, \$943.80, \$1038.18, \ldots$

7. Find the 8th term of the geometric sequence

$$1, 3, 9, \ldots$$

8. Find the 10th term of the geometric sequence

$$7, 35, 175, 875, \ldots$$

9. Find the 9th term of the geometric sequence

$$25, 5, 1, \tfrac{1}{5}, \tfrac{1}{25}, \ldots$$

10. Find the 10th term of the geometric sequence

$$64, 16, 4, 1, \tfrac{1}{4}, \tfrac{1}{16}, \ldots$$

11. Find the 12th term of the geometric sequence

$$\$1000, \$1080, \$1166.40, \ldots$$

Round to the nearest cent.

12. Find the 9th term of the geometric sequence

$$\$1000, \$1070, \$1144.90, \ldots$$

Round to the nearest cent.

13. Find the sum of the first 7 terms of the geometric sequence

$$8, 16, 32, \ldots$$

14. Find the sum of the first 8 terms of the geometric sequence

$$24, -48, 96, \ldots$$

15. Find the sum of the first 5 terms of the geometric sequence

$$\$1000, \$1000(1.08), \$1000(1.08)^2, \ldots$$

Round to the nearest cent.

16. Find the sum of the first 6 terms of the geometric sequence

$$\$200, \$200(1.06), \$200(1.06)^2, \ldots$$

Round to the nearest cent.

17. Suppose someone offered you a job during the month of February (28 days) under the following conditions. You will be paid $0.01 the 1st day, $0.02 the 2nd, $0.04 the 3rd, and so on, doubling your previous day's salary each day. How much would you earn?

18. In Exercise 17, how much would you earn during a February in a leap year (29 days)?

Determine whether each of the following infinite geometric series has a sum. If so, find it.

19. $4 + 20 + 100 + 500 + \cdots$

20. $-6 + 18 - 54 + 162 - \cdots$

21. $10 + 2 + \tfrac{2}{5} + \tfrac{2}{25} + \tfrac{2}{125} + \cdots$

22. $14 + 2 + \tfrac{2}{7} + \tfrac{2}{49} + \tfrac{2}{343} + \cdots$

23. $162 + 108 + 72 + 48 + \cdots$

24. $128 + 96 + 72 + 54 + \cdots$

25. $\$1000(1.08)^{-1} + \$1000(1.08)^{-2} + \$1000(1.08)^{-3} + \cdots$

26. $\$500(1.02)^{-1} + \$500(1.02)^{-2} + \$500(1.02)^{-3} + \cdots$

27. *Economics.* The government makes an $8,000,000,000 expenditure for a new type of aircraft. If 75% of this gets spent again, and 75% of that gets spent, and so on, what is the total effect on the economy?

28. Repeat Exercise 29 for $9,400,000,000 and 99%.

29. *Advertising effect.* A company is marketing a new product in a city of 5,000,000 people. They plan an advertising campaign which they think will induce 40% of the people to buy the product. They then estimate that if those people like the product, they will induce 40% (of the 40% of 5,000,000) more to buy the product, and those will induce 40% more to buy the product, and so on. In all, how many people will buy the product as a result of the advertising campaign? What percentage is this of the population?

30. Repeat Exercise 31 for 6,000,000 people and 45%.

OBJECTIVES

You should be able to

a) Use the straight-line method and prepare a depreciation schedule for a situation. Also, find a formula for the book values V_n, and find the common difference.

b) Use the double-declining balance method and prepare a depreciation schedule for a situation. Also, find a formula for the book values V_n.

c) Use the sum of the year's digits method, find the depreciation fractions for a situation, and prepare a depreciation schedule.

*2.3 (Optional) DEPRECIATION

A company buys an office machine for $5200 on January 1 of a given year. It is expected to last for 8 years, at which time its *trade-in*, or *salvage*, *value* will be $1100.

Over its lifetime it declines or *depreciates* $5200 − $1100, or $4100. The decline in value from $5200 to $1100 can occur in many ways, as shown in the table below.

Method (1) is called the *straight-line method*, Method (2) the *double-declining balance method*, and Method (3) the *sum of the year's digits method*. We shall consider each of these.

0 yrs.	1	2	3	4	5	6	7	8 yrs.	
$5200	$4687.50	$4175.00	$3662.50	$3150.00	$2637.50	$2125.00	$1612.50	$1100	(1)
$5200	$3900.00	$2925.00	$2193.75	$1645.31	$1233.98	$1100.00	$1100.00	$1100	(2)
$5200	$4288.89	$3491.67	$2808.34	$2238.90	$1783.34	$1441.67	$1213.89	$1100	(3)

25. For the following situation, find the total depreciation, annual depreciation, and rate of depreciation.

Item: Automobile.

Cost = $8700,
Expected life = 5 years,
Salvage value = $1600.

Straight-Line Depreciation

Suppose, for the machine above, the company figures the decline in value to be the *same* each year, that is $\frac{1}{8}$, or 12.5%, of $4100, which is $512.50. After 1 year the *book value*, or simply *value*, is

$$5200 - \$512.50, \text{ or } \$4687.50.$$

After 2 years it is

$$4687.50 - \$512.50, \text{ or } \$4175.00.$$

After 3 years it is

$$4175.00 - \$512.50, \text{ or } \$3662.50,$$

and so on.

For straight-line depreciation,

1. The total depreciation = Cost − Salvage value.

2. The annual depreciation $= \dfrac{\text{Cost} - \text{Salvage value}}{\text{Expected life}}.$

3. The rate of depreciation $= \dfrac{\text{Annual depreciation}}{\text{Total depreciation}}.$

DO EXERCISE 25.

A depreciation schedule gives a complete listing of the book values and total depreciation throughout the life of an item.

Example 1 Prepare a depreciation schedule for the following situation.

Item: Office machine.

$$Cost = \$5200,$$
$$Expected\ life = 8\ years,$$
$$Salvage\ value = \$1100.$$

Solution

Year	Rate of depreciation	Annual depreciation	Book value	Total depreciation
0			($5200)	
1	$\frac{1}{8}$ or (12.5%)	($512.50)	4687.50	$ 512.50
2	12.5%	512.50	4175.00	1025.00
3	12.5%	512.50	3662.50	1537.50
4	12.5%	512.50	3150.00	2050.00
5	12.5%	512.50	2637.50	2562.50
6	12.5%	512.50	2125.00	3075.00
7	12.5%	512.50	1612.50	3587.50
8	12.5%	512.50	(1100.00)	(4100.00)

The rate of depreciation is the same each year.

The annual depreciation is the same each year.

We find the book values by starting with the initial cost, $5200, and successively subtracting $512.50.

We find the total depreciations by starting with $512.50 after the first year and successively adding $512.50.

DO EXERCISE 26.

Why do we call this *straight-line depreciation*? If we make a graph of book values versus time, the values lie on a straight line.

26. Prepare a depreciation schedule for the situation in Margin Exercise 25.

Year	Rate of depreciation	Annual depreciation	Book value	Total depreciation
0				
1				
2				
3				
4				
5				

27. For the situation in Margin Exercise 25, find

a) A formula for the book values V_n.
b) The common difference.

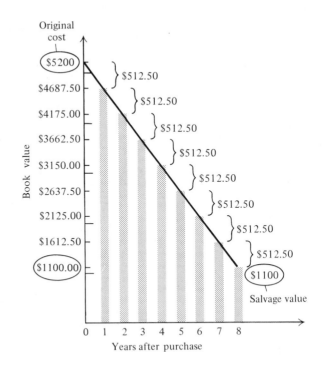

The book values V_n of an item n years after purchase form an arithmetic sequence for which

$$V_n = C - n\left(\frac{C - S}{N}\right),$$

where

$$C = \text{original cost of an item,}$$
$$N = \text{years of expected life,}$$
$$S = \text{salvage value.}$$

For the machine in Example 1,

$$V_n = \$5200 - n\left(\frac{\$5200 - \$1100}{8}\right) = \$5200 - (\$512.50)n,$$

and the common difference is $-\$512.50$.

DO EXERCISE 27.

Declining Balance Depreciation

A company buys a machine for $5200. It is expected to last for 8 years, at which time its salvage value will be $1100. The straight-line rate of depreciation would be $\frac{1}{8}$, or 12.5%. Depreciation can be deducted as a business expense when a business computes its taxes.

When a business is starting out it has many expenses and less income and therefore needs all the tax advantages it can get. For this and other reasons, the Internal Revenue Service allows certain assets to be depreciated at a rate which is larger than the straight-line rate, but *no more* than twice the straight-line rate. (Such a rate could be, for example, $1\frac{1}{4}$, $1\frac{1}{2}$, or 2 times the straight-line rate.) Suppose for the above, the rate is $2 \cdot \frac{1}{8}$, or 25%. This is called the *double-declining balance method*. Then the book value after 1 year is:

$5200 - (25\% \times \$5200)$ (We subtract 25% of the initial book value.)

$= \$5200 - (0.25 \times \$5200)$

$= \$5200 - \$1300$

$= \$3900.$

After 2 years it is

$3900 - (0.25 \times \$3900)$ (We subtract 25% of preceding book value.)

$= \$3900 - \$975$

$= \$2925.$

After 3 years it is

$2925 - (0.25 \times \$2925)$

$= \$2925 - \$731.25$

$= \$2193.75.$

After 4 years it is

$2193.75 - (0.25 \times \$2193.75)$

$= \$2193.75 - \$548.44$ (Rounded to the nearest cent)

$= \$1645.31,$

and so on.

Example 2 Prepare a depreciation schedule for the situation below. Use the double-declining balance method.

Item: Office machine.

$$\text{Cost} = \$5200,$$
$$\text{Expected life} = 8 \text{ years},$$
$$\text{Salvage value} = \$1100.$$

28. Prepare a depreciation schedule for the situation below. Use the double-declining balance method.

Item: Automobile.

Cost = $8700,
Expected life = 5 years,
Salvage value = $1600.

Year	Rate of depreciation	Annual depreciation	Book value	Total depreciation
0				
1				
2				
3				
4				
5				

Item: Auto
Cost: $8000
Ex Life: 4 yrs
Salvage Value $2000
Annual Dep $\frac{8,000 - 2000}{4}$, $\frac{6000}{4} = $1500

Str. Line

Year	Rate	Annual Dep	Book Value	Accum Dep
0			8000	
1	1/4	1500	6,500	1500
2	1/4		5,000	3000
3	1/4		3,500	4500
4	1/4		2,000	6000

Solution

Year	Rate of depreciation	Annual depreciation	Book value	Total depreciation
0			$5200	
1	$\frac{2}{8}$ or 25%	$1300.00	3900.00	$1300
2	25%	975.00	2925.00	2275
3	25%	731.25	2193.75	3006.25
4	25%	548.44	1645.31	3554.69
5	25%	411.33	1233.98	3966.02
6		133.98	1100.00	4100.00
7		0	1100.00	4100.00
8		0	1100.00	4100.00

The rate of depreciation is the same each year, twice the straight-line rate.

We find the annual depreciations when we multiply each successive book value by 0.25. For example, 0.25 × $5200 = $1300, and 0.25 × $3900 = $975.

We find the book values by starting with the initial cost, $5200, and successively subtracting 0.25 times the book value. For example, $5200 − (0.25 × $5200) = $3900. Then, $3900 − (0.25 × $3900) = $2925, and so on.

DO EXERCISE 28.

The book values V_n of an item n years after purchase form a geometric sequence

$$V_n = C\left(1 - \frac{m}{N}\right)^n, \qquad 0 < m \le 2,$$

where

$$C = \text{original cost of an item.}$$

This holds until V_n drops below the salvage value S.

---

* Note that

$$\$1233.98 - (0.25 \times \$1233.98) = \$1233.98 - \$308.50 = \$925.48,$$

but the book value cannot drop below the salvage value. Thus, after $1233.98 the next book value becomes $1100.00, and the annual depreciation for that year is $1233.98 − $1100.00, or $133.98.

DO EXERCISE 29.

Sum of the Year's Digits Depreciation

Another method of depreciation which allows larger amounts of depreciation in early years and smaller amounts in later years is the *sum of the year's digits method.* Each year a different rate of depreciation is used, which is a fraction.

Example 3 For the situation below,

a) Find the depreciation fractions;
b) Find the depreciation and book values after 1 year, 2 years.

Item: Office machine.

$$Cost = \$5200,$$
$$Expected \ life = 8 \ years,$$
$$Salvage \ value = \$1100.$$

Solution

a) To find the depreciation we first find the sum of the year's digits:

$$8 + 7 + 6 + 5 + 4 + 3 + 2 + 1 = \textcircled{36}.^*$$

The number 36 will be the denominator of each fraction. We then find the depreciation fractions (rates) by dividing each number in the sum by 36:

$$\frac{8}{36}, \ \frac{7}{36}, \ \frac{6}{36}, \ \frac{5}{36}, \ \frac{4}{36}, \ \frac{3}{36}, \ \frac{2}{36}, \ \frac{1}{36}.$$

b) The total depreciation is $5200 - $1100, or $4100. First year depreciation is

$$\frac{8}{36} \times \$4100 = \frac{8 \times \$4100}{36} = \frac{\$32,800}{36} = \$911.11 \qquad \text{(Rounded to the nearest cent.)}$$

---

*The sum

$$1 + 2 + 3 + \cdots + n$$

is the sum of the terms of an arithmetic sequence where $a_1 = 1$, and $a_n = n$. The sum is given by

$$S_n = \frac{n}{2}(a_1 + a_n) = \frac{n}{2}(1 + n)$$

$$= \frac{n(n+1)}{2}.$$

Thus,

$$8 + 7 + 6 + 5 + 4 + 3 + 2 + 1 = 1 + 2 + 3 + 4 + 5 + 6 + 7 + 8$$

$$= \frac{8(8+1)}{2} = \frac{8(9)}{2} = \frac{72}{2} = 36.$$

29. For the situation in Margin Exercise 28, find a formula for the book values V_n.

Sum of Year Digits

Year	Rate	Annual Dep	Book Value $6000	Accum Dep
0				
1	4/10	2400	5600	2400
2	3/10	1800	3800	4200
3	2/10	1200	2600	5400
4	1/10	600	2000	6000
10				

Total Dep is 8000 - 2000 = 6000

First Yr $\frac{4}{10} \times 6000 = \frac{4 \times 6000}{10} = 2400$

Sec. $\frac{3}{10} \times 6000 = \frac{3 \times 6000}{10} = 1800$

Third $\frac{2}{10} \times 6000 = \frac{2 \times 6000}{10} = 600$

30. For the situation below,

a) Find the depreciation fractions.

b) Find the depreciation and book values after 1 year, 2 years, and 3 years.

Item: Automobile.

$$\text{Cost} = \$8700,$$
$$\text{Expected life} = 5 \text{ years},$$
$$\text{Salvage value} = \$1600.$$

31. Prepare a depreciation schedule for the situation below. Use the sum of the year's digits method.

$$\text{Cost} = \$8700,$$
$$\text{Expected life} = 5 \text{ years},$$
$$\text{Salvage value} = \$1600.$$

Year	Rate of depreciation	Annual depreciation	Book value	Total depreciation
0				
1				
2				
3				
4				
5				

The book value after 1 year is

$$\$5200 - \$911.11, \text{ or } \$4288.89.$$

Second-year depreciation is

$$\frac{7}{36} \times \$4100 = \frac{7 \times \$4100}{36} = \frac{\$28,700}{36} = \$797.22.$$

The book value after 2 years is

$$\$4288.89 - \$797.22, \text{ or } \$3491.67.$$

DO EXERCISE 30.

Example 4 Prepare a depreciation schedule for the situation below. Use the sum of the year's digits method.

$$\text{Cost} = \$5200,$$
$$\text{Expected life} = 8 \text{ years},$$
$$\text{Salvage value} = \$1100.$$

Solution

Year	Rate of depreciation	Annual depreciation	Book value	Total depreciation
0			$5200	
1	$\frac{8}{36}$ or 22.2%	$911.11	4288.89	$ 911.11
2	$\frac{7}{36}$ or 19.4%	797.22	3491.67	1708.33
3	$\frac{6}{36}$ or 16.7%	683.33	2808.34	2391.66
4	$\frac{5}{36}$ or 13.9%	569.44	2238.90	2961.10
5	$\frac{4}{36}$ or 11.1%	455.56	1783.34	3416.66
6	$\frac{3}{36}$ or 8.3%	341.67	1441.67	3758.33
7	$\frac{2}{36}$ or 5.6%	227.78	1213.89	3986.11
8	$\frac{1}{36}$ or 2.8%	113.89	1100.00	4100.00

The rate of depreciation gets lower each year.

We find the annual depreciations first. To do this we multiply the total depreciation by each fraction. For example, $\frac{8}{36} \times \$4100 = \$911.11$, $\frac{7}{36} \times \$4100 = \$797.22$, and so on.

We find the book values by subtracting each annual depreciation in succession. For example, $\$5200 - \$911.11 = \$4288.89$, $\$4288.89 - \$797.22 = \$3491.67$, and so on.

DO EXERCISE 31.

EXERCISE SET 2.3

Use the straight-line method.

a) Prepare a depreciation schedule.
b) Find a formula for the book values V_n.
c) Find the common difference.

1. *Item:* Automobile.
Cost = $8000,
Expected life = 4 years,
Salvage value = $2000.

2. *Item:* Automobile.
Cost = $12,000,
Expected life = 3 years,
Salvage value = $4800.

3. *Item:* Postage machine.
Cost = $450,
Expected life = 8 years,
Salvage value = $0.

4. *Item:* Typewriter.
Cost = $2500,
Expected life = 6 years,
Salvage value = $0.

In Exercises 5 through 8, use the double-declining balance method.

a) Prepare a depreciation schedule.
b) Find a formula for the book values V_n.

5. (See Exercise 1.)   **6.** (See Exercise 2.)   **7.** (See Exercise 3.)   **8.** (See Exercise 4.)

In Exercises 9 through 12, use the sum of the year's digits method.

a) Find the depreciation fractions.
b) Prepare a depreciation schedule.

9. (See Exercise 1.)   **10.** (See Exercise 2.)   **11.** (See Exercise 3.)   **12.** (See Exercise 4.)

2.4 SIMPLE AND COMPOUND INTEREST

Simple Interest

You put $100 in a savings account for 1 year. This is called *principal*.

The *interest rate* is 8%. This means you get back 8% of the principal,

$$8\% \text{ of } \$100,$$

or

$$8\% \times \$100,$$

or

$$\$8.00,$$

in addition to the principal. The $8.00 is called *interest*.

OBJECTIVES

You should be able to

a) Given a principal P, an interest rate i, and a time t, find the amount to which P grows at simple interest.
b) Given a principal P, an interest rate i, and a time t, find the amount to which P grows at interest compounded n times a year.

32. Suppose $1000 is invested in a savings account at 6% simple interest. What is the amount in the account at the end of 3 months?

Hint. 3 months = $\frac{3}{12}$ yr. = $\frac{1}{4}$ yr.

The *amount* you get back is

$$(\text{Principal}) + (\text{Interest}), \quad \text{or } \$100 + \$8, \quad \text{or } \$108.$$

To find interest for a fraction t of a year (or for any time t), we compute the interest for 1 year and multiply by t. Thus, $100 principal invested at an interest rate of 8% for $\frac{1}{4}$ of a year, yields interest of

$$(8\% \times \$100) \times \tfrac{1}{4}, \quad \text{or } \$2.00.$$

We have the following formulas.

✳ **SIMPLE INTEREST.** **Principal P invested at simple interest rate i for time t, in years, yields interest I given by**

$$I = P \cdot i \cdot t.$$

AMOUNT. **The amount A to which principal P will grow at simple interest rate i, for t years, is given by**

$$A = P + Pit = P(1 + it).$$

Note that $I = Pit$ usually is written as $I = Prt$, but we are reserving the letter r for later use.

DO EXERCISE 32.

33. Under the conditions of Example 1, how much is due on $1700 left unpaid for 1 month?

Example 1 A loan charges 18% simple interest. How much is due on $1000 left unpaid for 1 month?

Solution $P = \$1000$, $i = 18\%$, or 0.18, and $t = \frac{1}{12}$ yr. Then the amount due is

$$✶ \; A = P(1 + it) = \$1000(1 + 0.18 \times \tfrac{1}{12}) = \$1000(1 + 0.015)$$
$$= \$1000(1.015) = \$1015.$$

DO EXERCISE 33.

Compound Interest

Suppose you invested $1000 at an interest rate of 8%, compounded annually. The amount A_1 in the account at the end of 1 year is given by

$$A_1 = \$1000(1 + 0.08) = \$1000(1.08) = \$1080.$$

Going into the second year you have a new principal of $1080, so by the end of 2 years you would have the amount A_2 given by

$$A_2 = \$1080(1 + 0.08) = \$1080(1.08) = \$1166.40.$$

Going into the third year you have a new principal of $1166.40, so by the end of 3 years you would have the amount A_3 given by

$$A_3 = \$1166.40(1 + 0.08) = \$1166.40(1.08) \approx \$1259.71.$$

Note the following:

$$A_1 = \$1000(1.08)^1,$$
$$A_2 = \$1000(1.08)^2,$$
$$A_3 = \$1000(1.08)^3.$$

The amounts A_n form a geometric sequence with common ratio 1.08. In general, suppose you invest a principal of P dollars at interest rate i, compounded annually. The amount A_1 in the account at the end of 1 year is given by

$$A_1 = P(1 + i) = Pr,$$

where, for convenience, $r = 1 + i$.

Going into the second year you would have a new principal of Pr dollars, so by the end of 2 years you would have the amount A_2 given by

$$A_2 = A_1 \cdot r = (Pr)r = Pr^2.$$

Going into the third year you have a new principal of Pr^2, so by the end of 3 years you would have the amount A_3 given by

$$A_3 = A_2 \cdot r = (Pr^2)r = Pr^3.$$

The amounts A_n form a geometric sequence with common ratio r, which is $1 + i$.

INTEREST COMPOUNDED ANNUALLY. If principal P is invested at interest rate i, compounded annually, in t years it will grow to the amount A given by

$$A = P(1 + i)^t.$$

Example 2 Suppose $1000 is invested at 5% compounded annually. How much is in the account at the end of 3 years?

Solution We substitute $1000 for P, 0.05 for i, and 3 for t in the equation $A = P(1 + i)^t$, and get

$$A = \$1000(1 + 0.05)^3 = \$1000(1.05)^3 = \$1000(1.157625)$$
$$= \$1157.625 \approx \$1157.63.$$

DO EXERCISE 34.

---

34. Suppose $2000 is invested at 6% compounded annually. How much is in the account at the end of 4 years?

2000 Invested @ 7%, 2 yrs

Simple in   i   Time

$A = 2000(1 + .07 [2])$

$2000(1 + .14)$

$2000(1.14) = 2280$

Compound Annualy

$A = 2000(1 + .07)^2$

$2000(1.07)^2$

$2000(1.1449) = 2289.80$

Compounded semi Any.

$A = 2000(1 + \frac{.07}{2})^{(2)(2)=4}$

$2000(1.035)^4 = 2295.04$

Compounded quar

$A = 2000(1 + \frac{.07}{4})^{(4)(2)=8}$

$2000(1.0175)^8$

$2000(1.1488) = 2297.76$

Annual Compount Interest

If interest is compounded quarterly, we can find a formula like the one above as follows:

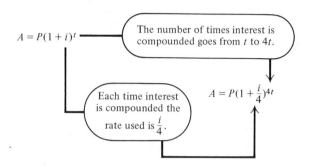

$A = P(1 + i)^t$

The number of times interest is compounded goes from t to $4t$.

Each time interest is compounded the rate used is $\frac{i}{4}$.

$A = P(1 + \frac{i}{4})^{4t}$

In general,

INTEREST COMPOUNDED n TIMES PER YEAR. **If principal P is invested at interest rate i, compounded n times per year, in t years it will grow to an amount A given by**

$$A = P\left(1 + \frac{i}{n}\right)^{nt}.$$

The number nt is the total number of payment periods.

Example 3 Suppose $1000 is invested at 8%. How much is in the account at the end of 3 years if interest is

a) simple?   b) compounded annually?
c) compounded semiannually?   d) compounded quarterly?
e) compounded daily?   f) hourly?

Solution
a) $A = P(1 + it) = \$1000(1 + 0.08 \times 3) = \$1000(1 + 0.24)$
$= \$1000(1.24) = \$1240.00.$

b) $A = P(1 + i)^t = \$1000(1 + 0.08)^3 = \$1000(1.08)^3$
$= \$1000(1.259712) \approx \$1259.71.$

c) $A = P\left(1 + \frac{i}{n}\right)^{nt} = \$1000\left(1 + \frac{0.08}{2}\right)^{2 \times 3} = \$1000(1 + 0.04)^6$
$= \$1000(1.04)^6 = \$1000(1.265319)$
$\approx \$1265.32.$

d) $A = P\left(1 + \dfrac{i}{n}\right)^{nt} = \$1000\left(1 + \dfrac{0.08}{4}\right)^{4\times3}$

$= \$1000(1 + 0.02)^{12}$

$= \$1000(1.02)^{12} = \$1000(1.268242)$

$\approx \$1268.24$

e) $A = P\left(1 + \dfrac{i}{n}\right)^{nt} = \$1000\left(1 + \dfrac{0.08}{365}\right)^{365\times3}$

$= \$1000(1 + 0.000219)^{1095}$

$= \$1000(1.000219)^{1095} = \$1000(1.270967)$

$\approx \$1270.97.$

f) $A = P\left(1 + \dfrac{i}{n}\right)^{nt} = \$1000\left(1 + \dfrac{0.08}{8760}\right)^{8760\times3}$

$= \$1000(1 + 0.00000913)^{26,280}$

$= \$1000(1.00000913)^{26,280} = \$1000(1.271168)$

$\approx \$1271.17.$

CALCULATOR NOTE. One can find these powers on a calculator with an x^y key, by a compound interest table such as the one in Table 1 at the back of the book, or by the method described earlier where the larger power is broken down to a product of smaller powers. The number of places on the calculator may affect the accuracy of the answer. Thus, you may occasionally find that your answers do not agree with those in the answer section which have been found on a calculator with a ten-digit readout. In general, if you are using a calculator, do all your computations and round only at the end.

Compare the amounts found in Example 3:

$1240, $1259.71, $1265.32, $1268.24, $1270.97, $1271.17.

Note that as the number of periods of compounding increases within a fixed time, the greater the *amount* becomes, but the *increase* gets less and less. If we keep using more compounding periods, the amount gets closer and closer to an amount found by *continuous compounding*, $1271.25. We will study continuous compounding in a later chapter.

DO EXERCISE 35.

Present Value

A representative of a financial institution is often asked to solve a problem like the following.

35. Suppose $1000 is invested at 6%. How much is in the account at the end of 2 years if interest is

a) simple
b) compounded annually?
c) compounded semiannually?
d) compounded quarterly?
e) compounded daily?

Example 4 Following the birth of a child, a parent wants to make an initial investment P which will grow to $10,000 by the child's twentieth birthday. Interest is compounded semiannually at 8%. What should the initial investment be?

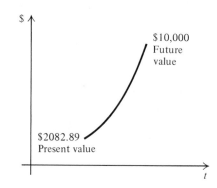

Solution Using the formula

$$A = P\left(1 + \frac{i}{n}\right)^{nt},$$

we find P such that

$$\$10,000 = P\left(1 + \frac{0.08}{2}\right)^{2 \times 20}$$

or

$$\$10,000 = P(1.04)^{40}.$$

Then

$$\$10,000 \approx P(4.801021),$$

and

$$P \approx \frac{\$10,000}{4.801021} \approx \$2082.89.$$

Thus a principal of $2082.89 would have to be invested at 8%, compounded semiannually, to grow to $10,000 in 20 years. The amount $2082.89 is called the *present value* of $10,000 for 20 years at 8% compounded semiannually. We can also say that the *future value* of $2082.89 is $10,000.

By solving $A = P(1 + i)^t$ and $A = P(1 + (i/n))^{nt}$ for P, we get general formulas for present value.

PRESENT VALUE. The present value P of an amount A at interest rate i, compounded annually, for t years is given by

$$P = A(1 + i)^{-t}.$$

For interest compounded n times per year, the present value P is given by

$$P = A\left(1 + \frac{i}{n}\right)^{-nt}.$$

If an appropriate calculator is not available, Table 2 can be helpful in computing present values.

DO EXERCISE 36.

36. Find the present value of $4000 for 5 years at 6%, compounded quarterly.

EXERCISE SET 2.4

Suppose $2000 is invested at the given simple interest rate and for the time indicated. What is the amount in the account?

1. 9%, 4 months   **2.** 10%, 8 months   **3.** 14%, 2 years   **4.** 11%, 3 years

5. Suppose $2000 is invested at 7%. How much is in the account at the end of 2 years if interest is

a) simple?
b) compounded annually?
c) compounded semiannually?
d) compounded quarterly?
e) compounded daily?

6. Suppose $1500 is invested at 10%. How much is in the account at the end of 3 years if interest is

a) simple?
b) compounded annually?
c) compounded semiannually?
d) compounded every 2 months?
e) compounded daily?

Find the present value of:

7. $1000 at 8% compounded annually for 3 years.

8. $1000 at 9% compounded annually for 4 years.

9. $1000 at 8% compounded quarterly for 3 years.

10. $1000 at 9% compounded semiannually for 4 years.

11. $10,000 at 6% compounded semiannually for 18 years.

12. $15,000 at 7% compounded semiannually for 18 years.

13. *Personal debt.* On the average every person in this country has a debt of $1000. How much will this debt be in 2 years at 7%, compounded annually?

14. *Personal debt.* In Exercise 13, how much will be due in 2 years at 12%, compounded annually?

15. *Inflation.* Inflation is based on what a person could buy in 1967 for $1. In 1980 what was bought for $1 in 1967 will cost $1(1 + 0.07)^{13}$, assuming a rate of inflation of 7%. How much is that cost?

16. *Inflation.* In 1984 what was bought for $1 in 1967 will cost $1(1 + 0.07)^{17}$, assuming a rate of inflation of 7%. How much is that cost?

17. *Finding the interest rate.* $2560 is invested at interest rate i, compounded annually. In 2 years it grows to $2890. What is the interest rate?

18. *Finding the interest rate.* $1000 is invested at interest rate i, compounded annually. In 2 years it grows to $1210. What is the interest rate?

OBJECTIVES

You should be able to

a) Find the effective annual yield of an amount invested at interest rate i compounded n times per year.

b) Find the annual percentage rate on a loan at a given add-on interest rate.

2.5 ANNUAL PERCENTAGE RATE

Effective Yield

Suppose $1000 is invested at 8%, compounded quarterly for 1 year. We know that this will grow to an amount

$$\$1000\left(1 + \frac{0.08}{4}\right)^4, \text{ or } \$1082.43,$$

which is an increase of 8.243%. This is the same as if $1000 were invested at 8.243% compounded once a year (simple interest). The 8.243% is called the *effective annual yield* or *annual percentage rate*, and the 8% is called the *nominal rate*. In general, if P is invested at interest rate i, compounded n times per year, then the effective annual yield is that number E satisfying

$$P(1 + E) = P\left(1 + \frac{i}{n}\right)^n.$$

Then

$$1 + E = \left(1 + \frac{i}{n}\right)^n$$

and

$$E = \left(1 + \frac{i}{n}\right)^n - 1.$$

37. Find the effective annual yield, when the nominal interest rate is 9%, compounded semiannually.

Example 1 Find the effective annual yield, when the nominal interest rate is 7%, compounded semiannually.

Solution

$$E = \left(1 + \frac{i}{n}\right)^n - 1$$

$$= \left(1 + \frac{0.07}{2}\right)^2 - 1 = (1.035)^2 - 1$$

$$= 1.071225 - 1 = 0.071225 \approx 7.123\%.$$

DO EXERCISE 37.

Add-on Interest

Consider a car loan.

Situation: Car loan of $1000 at 7% for 1 year

Question: Couldn't the borrower put the $1000 in a savings account at 7.5% and make money?

Car loans are examples of what lending institutions call *add-on-interest*. The nominal, or stated, interest rate is 7%. This is *not* the true rate, the *annual percentage rate*, APR. Lenders use the simple interest formula, $I = Prt$, and figure that the loan will earn interest of $1000 \times 0.07 \times 1$, or \$70. They "add on" the \$70, so you have to pay back \$1070. For simplicity, suppose you pay back the loan in 4 payments. Each payment is $1070 \div 4$, or \$267.50. Your loan decreases as follows:

$$\$1070, \quad \$802.50, \quad \$535, \quad \$267.50.$$

What's the catch? The lending institution *does not* allow you the full use of the \$1070 for the year. The average principal you have is

$$\frac{\$1070 + \$802.50 + \$535 + \$267.50}{4} = \$668.75.$$

How do we find APR? It is defined to be that interest rate such that

$$\begin{pmatrix} \text{Interest} \\ \text{for first} \\ \text{3 months} \end{pmatrix} + \begin{pmatrix} \text{Interest} \\ \text{for second} \\ \text{3 months} \end{pmatrix} + \begin{pmatrix} \text{Interest} \\ \text{for third} \\ \text{3 months} \end{pmatrix} + \begin{pmatrix} \text{Interest} \\ \text{for fourth} \\ \text{3 months} \end{pmatrix}$$

$$= \text{Total interest,}$$

or

$$(\$1070 \times \text{APR} \times \tfrac{1}{4}) + (\$802.50 \times \text{APR} \times \tfrac{1}{4})$$
$$+ (\$535 \times \text{APR} \times \tfrac{1}{4}) + (\$267.50 \times \text{APR} \times \tfrac{1}{4}) = \$70.$$

Factoring out APR, we get

$$[(\$1070 \times \tfrac{1}{4}) + (\$802.50 \times \tfrac{1}{4}) + (\$535 \times \tfrac{1}{4}) + (\$267.50 \times \tfrac{1}{4})] \cdot \text{APR}$$
$$= \$70$$

or

$$\left[\frac{\$1070 + \$802.50 + \$535 + \$267.50}{4} \right] \cdot \text{APR} = \$70.$$

Then

$$\$668.75 \cdot \text{APR} = \$70$$
$$\text{APR} = \frac{\$70}{\$668.75} \approx 10.5\%$$

In general,

APR = (Total interest) ÷ (Average principal).

For 12 payments in the above situation, the APR would have been 12.1%. In either case the true interest rate, or APR, is almost double

38. Find the APR. Assume a car loan at the given add-on interest rate for 1 year and 12 payments.

Loan = $4000,
Add-on rate = 9%.

the stated rate. You would not save money by putting the money in the bank. The Truth In Lending Law *requires* lenders to inform you of the APR.

DO EXERCISE 38.

EXERCISE SET 2.5

Find the effective yield.

1. 8%, compounded semiannually

2. 10%, compounded semiannually

3. 9%, compounded quarterly

4. 12%, compounded quarterly

5. 8%, compounded 6 times per year

6. 10%, compounded every 2 months

7. 8%, compounded daily

8. 10%, compounded daily

9. 8%, compounded hourly

10. 10%, compounded hourly

Find the *APR*. Assume that these are car loans at the given add-on interest rate for 1 year and 12 payments.

11. Loan = $1000,
Add-on rate = 8%.

12. Loan = $1000,
Add-on rate = 6%.

13. Loan = $2000,
Add-on rate = 10%.

14. Loan = $5000,
Add-on rate = 9%.

OBJECTIVES

You should be able to

a) Find the amount of an annuity where P dollars is being invested n times per year at interest rate i, compounded n times per year for N years.

b) Find what payment P will have to be made n times a year for N years at interest rate i, compounded n times per year, so that V dollars will have accumulated in N years.

2.6 ANNUITIES AND SINKING FUNDS

Annuities

An *annuity* is a series of equal payments made at equal time intervals. Rent payments are an example of an annuity. Fixed deposits in a savings account can also be an annuity. For example, suppose someone makes a sequence of deposits of $1000 each in a savings account on which interest is compounded annually at 8%. The total amount in the account, including interest, is called the *amount of the annuity*, or the *future value of the annuity*. Let us find the amount of the given annuity for a period of 5 years. The following time diagram can help. Note that we do not make a deposit until the end of the first year.

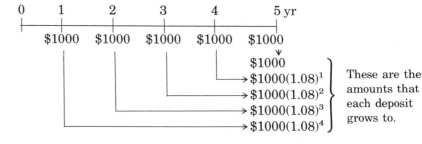

The amount of the annuity is the sum

$1000 + \$1000(1.08)^1 + \$1000(1.08)^2 + \$1000(1.08)^3 + \$1000(1.08)^4$.

This is the sum of the terms of a geometric sequence where

$$a_1 = \$1000, \qquad n = 5, \qquad \text{and} \qquad r = 1.08.$$

We can find this sum using the formula

$$S_n = \frac{a_1(r^n - 1)}{r - 1}.$$

We have

$$S_5 = \frac{\$1000(1.08^5 - 1)}{1.08 - 1} \approx \frac{\$1000(1.469328 - 1)}{0.08} \approx \$5866.60.$$

When equal deposits are made at equal time intervals which are the same as the periods of compounding and the first deposit is not made until the end of the first year, we have what is called an *ordinary annuity*. We shall consider only ordinary annuities.

In general, suppose we have an ordinary annuity in which P dollars are deposited each year for N years, at interest rate i, compounded annually. The first payment P, invested at the end of the first year, will be invested for $N - 1$ years and will grow to

$$P(1 + i)^{N-1}.$$

The second payment will be invested for $N - 2$ years and grow to

$$P(1 + i)^{N-2},$$

and so on; the next-to-last payment will be invested for 1 year and grow to

$$P(1 + i)^1,$$

and the last payment will not have time to grow, and will be

$$P.$$

Then the amount of the annuity will be the sum V given by

$$V = P + P(1 + i)^1 + P(1 + i)^2 + \cdots + P(1 + i)^{N-2} + P(1 + i)^{N-1}.$$

This is the sum of the terms of a geometric sequence where

$$a_1 = P, \qquad n = N, \qquad \text{and} \qquad r = 1 + i.$$

We can find this sum using the formula

$$S_n = \frac{a_1(r^n - 1)}{r - 1}.$$

39. Find the amount of an annuity where $200 per year is being invested at 6% compounded annually, for 16 years.

We have

$$V = \frac{P[(1 + i)^N - 1]}{(1 + i) - 1},$$

from which we get the following formula:

THE AMOUNT OF AN ANNUITY. The amount of an annuity, V, where P dollars are invested at the end of each of N years at interest rate i, compounded annually, is given by

$$V = \frac{P[(1 + i)^N - 1]}{i}$$

If interest is compounded n times per year and deposits are being made every compounding period, the formula for V is found by replacing i by $\frac{i}{n}$ and N by nN.

Example 1 Find the amount of an annuity where $1000 per year is being invested at 7%, compounded annually, for 15 years.

Solution

$$V = \frac{\$1000[(1 + 0.07)^{15} - 1]}{0.07} = \frac{\$1000[(1.07)^{15} - 1]}{0.07}$$

$$= \frac{\$1000[2.759032 - 1]}{0.07} \approx \$25{,}129.03.$$

DO EXERCISE 39.

40. Find the amount of an annuity where $200 every 3 months is being invested at 6%, compounded quarterly, for 16 years.

Example 2 Find the amount of an annuity where $1000 every 3 months is being invested at 7%, compounded quarterly, for 15 years.

Solution

$$V = \frac{\$1000\left[\left(1 + \dfrac{0.07}{4}\right)^{60} - 1\right]}{\dfrac{0.07}{4}} = \frac{\$1000[(1.0175)^{60} - 1]}{0.0175}$$

$$= \frac{\$1000[2.831816 - 1]}{0.0175} \approx \$104{,}675.20.$$

Note that much more money is being deposited in this annuity than in Example 1.

DO EXERCISE 40.

Sinking Funds

The following is an adaptation of a problem considered before.

Example 3   Following the birth of a child, a parent wants to make a deposit on each of the child's subsequent birthdays so that $10,000 will have accumulated by the child's 20th birthday. Interest will be compounded annually at 8%. What should each deposit be?

Solution   We can use the formula

$$V = \frac{P[(1 + i)^N - 1]}{i}.$$

We know that $V = \$10,000$, $i = 0.08$, and $N = 20$. Then we substitute,

$$\$10,000 = \frac{P[(1 + 0.08)^{20} - 1]}{0.08},$$

and solve for P:

$$\$10,000(0.08) = P[(1.08)^{20} - 1],$$

$$\$800 = P[4.660957 - 1] = P[3.660957],$$

$$P = \frac{\$800}{3.660957} \approx \$218.52.$$

Each birthday after the child's birth, the parent will need to deposit $218.52.

DO EXERCISE 41.

The situation in Example 3 illustrates what is called a *sinking fund.* Any financial arrangement in which periodic payments are made for the purpose of growing to a specific future amount is called a *sinking fund.* The word "sinking" is somewhat of a misnomer in that one is making deposits to get to a future amount. The word probably comes from "sinking" the future amount back to now to consider what deposits need to be made.

41. Following the birth of a child, a parent wants to make a deposit on each of the child's subsequent birthdays so that $10,000 will have accumulated by the child's 10th birthday. Interest will be compounded annually at 7%. What should each deposit be?

EXERCISE SET 2.6

Find the amount of an annuity where;

1. $1000 is being invested each year at 7%, compounded annually for 4 years.

2. $2000 is being invested each year at 9%, compounded annually for 5 years.

3. $1000 is being invested each year at 7%, compounded annually for 10 years.

4. $3000 is being invested each year at 9%, compounded annually for 10 years.

5. $2000 is being invested every 3 months at 8%, compounded quarterly for 5 years.

6. $300 is being invested every 3 months at 6%, compounded quarterly for 8 years.

7. $10 is being invested each month at 6%, compounded monthly for 8 years.

8. $20 is being invested each month at 7%, compounded monthly for 10 years.

9. A person decides to save money for retirement. $1000 is invested each year at 7.5%, compounded annually. How much will be in the retirement fund at the end of 30 years?

10. A person decides to save money for retirement, investing $500 each year at 8.5%, compounded annually. How much will be in the retirement fund at the end of 40 years?

11. A family expects to buy a new car 5 years from now. They decide to put away $50 a month. At 8% interest, compounded monthly, how much will they have at the end of 5 years?

12. A company decides to put away money for future expansion of their business. They save $1000 a month. At 9% interest, compounded monthly, how much will they have at the end of 6 years?

13. Due to increased business a company expects to have to buy a $10,000 machine 8 years from now. They decide to make deposits each year at 6.5%, compounded annually. What should each deposit be so that the company will accumulate the $10,000?

14. A young couple wants to have a down payment of $8000 to buy a new home in 9 years. They decide to make deposits each year at 5%, compounded annually. What should each deposit be so that they will accumulate the $8000?

15. A family expects to pay $7000 for a car 5 years from now. They decide to make a deposit each month. Interest at 6% will be compounded monthly. What should each deposit be so that they will have accumulated the $7000?

16. A family expects to pay $800 for a freezer 3 years from now. They decide to make a deposit each month. Interest at 5.5% will be compounded monthly. What should each deposit be so that they will have accumulated the $800?

Solve each formula for P.

17. $V = \dfrac{P[(1 + i)^N - 1]}{i}$

$P = \dfrac{V \frac{i}{n}}{\left[\left(1 + \frac{i}{n}\right)^{nN} - 1\right]}$

18. $V = \dfrac{P\left[\left(1 + \dfrac{i}{n}\right)^{nN} - 1\right]}{\dfrac{i}{n}}$

OBJECTIVES

You should be able to

a) Find what lump sum would have to be deposited now at interest rate k, compounded n times a year, so that P dollars can be withdrawn n times a year for N years.

b) Find the equal payment P required to pay off a loan of S dollars at interest rate k, compounded n times per year, in N years.

2.7 PRESENT VALUE OF AN ANNUITY AND AMORTIZATION

Present Value of an Annuity

The *present value of an annuity* is the sum of the present values of each payment of the annuity. Celebrities such as movie stars or athletes sometimes have years when they make lots of money, and then their incomes decline. Suppose a person wants to make a deposit in a lump sum right now, so that for each of the following 5 years $1000 can be drawn from the account. Interest is to be compounded annually at 8%. In effect, we can think of the deposit consisting of five different parts, the first being that amount which should be deposited now so that there will be $1000 one year from now, plus the second being that amount which should be deposited now so that there will be another $1000 two years from now, plus the third being that amount which should be deposited now so that there will be another $1000 three years from now, and so on. Each of these is a present value of $1000 a certain number of years from now. This can be shown in the following time diagram.

These are the
present values
of each withdrawal.

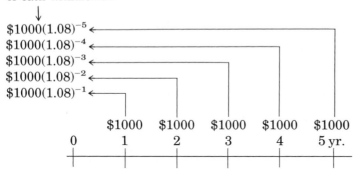

The present value of the annuity is the sum

$$\$1000(1.08)^{-1} + \$1000(1.08)^{-2} + \$1000(1.08)^{-3}$$
$$+ \$1000(1.08)^{-4} + \$1000(1.08)^{-5}.$$

This is the sum of a geometric sequence where

$$a_1 = \$1000(1.08)^{-1}, \qquad n = 5, \qquad \text{and} \qquad r = 1.08^{-1}.$$

We can find this sum using the formula

$$S_n = \frac{a_1(r^n - 1)}{r - 1}.$$

We have

$$S_5 = \frac{\$1000(1.08)^{-1}[(1.08^{-1})^5 - 1]}{1.08^{-1} - 1} = \frac{\$1000(1.08)^{-1}[(1.08)^{-5} - 1]}{1.08^{-1} - 1}.$$

Multiplying S_5 by 1 using $\dfrac{1.08}{1.08}$ will ease our calculations.

$$S_5 = \frac{1.08}{1.08} \times \frac{\$1000(1.08)^{-1}[(1.08)^{-5} - 1]}{1.08^{-1} - 1}$$

$$= \frac{\$1000[(1.08)^{-5} - 1]}{1 - 1.08} = \frac{\$1000[1 - (1.08)^{-5}]}{1.08 - 1}$$

$$= \frac{\$1000[1 - 0.680583]}{0.08} \approx \$3992.71.$$

In general, suppose a lump sum S is to be deposited now at interest rate i, compounded annually, so that P dollars can be withdrawn for each of the next N years; then S is the present value of the annuity and is given by the sum

$$P(1 + i)^{-1} + P(1 + i)^{-2} + \cdots + P(1 + i)^{-(N-1)} + P(1 + i)^{-N}.$$

42. What lump sum would have to be deposited now at 8%, compounded annually, so that withdrawals of $2000 can be made every year for 20 years?

We can find this sum using the formula

$$S_n = \frac{a_1(r^n - 1)}{r - 1},$$

where $a_1 = P(1 + i)^{-1}$, $n = N$, and $r = (1 + i)^{-1}$. We have

$$S = \frac{P(1 + i)^{-1}[(1 + i)^{-N} - 1]}{(1 + i)^{-1} - 1},$$

from which we get the following formula:

THE PRESENT VALUE OF AN ANNUITY

$$S = \frac{P[1 - (1 + i)^{-N}]}{i}.$$

If interest is compounded n times per year and withdrawals are to be made every compounding period, the formula for S is found by replacing i by $\dfrac{i}{n}$ and N by nN.

Example 1   What lump sum would have to be deposited now at 7%, compounded annually, so that withdrawals of $1000 can be made every year for 15 years?

Solution

$$S = \frac{\$1000[1 - (1.07)^{-15}]}{0.07} \approx \frac{\$1000[1 - 0.362446]}{0.07} \approx \$9107.91.$$

DO EXERCISE 42.

43. What lump sum would have to be deposited now at 8%, compounded quarterly, so that withdrawals of $500 can be made every 3 months for 10 years?

Example 2   What lump sum would have to be deposited now at 6%, compounded quarterly, so that withdrawals of $500 can be made every 3 months for 10 years?

Solution

$$S = \frac{\$500\left[1 - \left(1 + \dfrac{0.06}{4}\right)^{-40}\right]}{\dfrac{0.06}{4}} = \frac{\$500[1 - (1.015)^{-40}]}{0.015}$$

$$\approx \frac{\$500[1 - 0.551262]}{0.015} \approx \$14{,}957.93.$$

DO EXERCISE 43.

Amortization

In the formula for the present value of an annuity, what application is there for the situation where we know S, i, and N, and want to compute P?

Example 3 A person borrows $6000 at an interest rate of 14%, compounded monthly. The loan is to be paid off by 36 equal monthly payments over the next 3 years. How much is each payment?

Solution We can use the formula

$$S = \frac{P\left[1 - \left(1 + \dfrac{i}{n}\right)^{-nN}\right]}{\dfrac{i}{n}}.$$

We know that S = $6000, i = 0.14, n = 12, and N = 3. Then we substitute

$$\$6000 = \frac{P\left[1 - \left(1 + \dfrac{0.14}{12}\right)^{-36}\right]}{\dfrac{0.14}{12}},$$

and solve for P:

$$\$6000\left(\frac{0.14}{12}\right) = P[1 - 0.658646],$$

$$\$70.00 = P[0.341354],$$

$$P = \frac{\$70.00}{0.341354} \approx \$205.07.$$

DO EXERCISE 44.

One might ask why we cannot use the sinking-fund formula for this problem. In a sinking fund a payment is made and then allowed to grow at a certain interest rate. The payee gets the money back at the end. When one *amortizes* a loan, as in Example 3, the money is received at the outset. In Example 3, the borrower gets the $6000 at the outset.

When financial institutions loan money, they start computing interest right away. After 1 month a payment of $205.08 was made. In effect a part of the $6000, a present value, has grown to $205.08 during the first month and consists of principal and interest. Each monthly payment is such a payment of interest and principal; the amounts of each vary, but they always total $205.08. When we amortize a loan, we know the present value of an annuity and want to find the amount of each equal payment.

44. A family buys a house for $52,000, makes a down payment of $10,000 and borrows the remaining $42,000 at 10.75% interest, compounded monthly. The loan is to be paid off by 300 equal monthly payments over the next 25 years. How much is each payment?

EXERCISE SET 2.7

What lump sum would have to be deposited now at

1. 7%, compounded annually, so that withdrawals of $1000 can be made every year for 4 years?

2. 9%, compounded annually, so that withdrawals of $2000 can be made every year for 5 years?

3. 7%, compounded annually, so that withdrawals of $1000 can be made every year for 10 years?

4. 9%, compounded annually, so that withdrawals of $3000 can be made every year for 10 years?

5. 8%, compounded quarterly, so that withdrawals of $2000 can be made every 3 months for 5 years?

6. 6%, compounded semiannually, so that withdrawals of $300 can be made every 6 months for 8 years?

7. 6%, compounded monthly, so that withdrawals of $10 can be made every month for 8 years?

8. 7%, compounded monthly, so that withdrawals of $20 can be made every month for 10 years?

9. A family pays $4800 for a remodeling job. A down payment of $600 is made, and $4200 is borrowed at 12%, compounded monthly. The loan is to be paid off by 36 equal payments over the next 3 years. How much is each payment?

10. A family pays $10,000 for an addition to their home. A down payment of $1000 is made, and $9000 is borrowed at 10.5%, compounded monthly. The loan is to be paid off by 24 equal payments over the next 2 years. How much is each payment?

11. A family buys a house for $75,000. A down payment of $15,000 is made, and $60,000 is borrowed at 10.5%, compounded monthly. The loan is to be paid off by 360 equal payments over the next 30 years. How much is each payment?

12. A family buys a condominium for $40,000. A down payment of $10,000 is made, and $30,000 is borrowed at 11%, compounded monthly. The loan is to be paid off by 300 equal payments over the next 25 years. How much is each payment?

▶

Solve each formula for P.

13. $S = \dfrac{P[1 - (1 + i)^{-N}]}{i}.$

14. $S = \dfrac{P\left[1 - \left(1 + \dfrac{i}{n}\right)^{-nN}\right]}{\dfrac{i}{n}}.$

A *perpetuity* is an annuity in which payments will be made forever.

15. What lump sum would have to be deposited now at 8%, compounded annually so that withdrawals of $1000 can be made every year forever? *Hint.* Use the formula for the sum of an infinite geometric series. What other method can be used?

16. A family wishes to set up an endowed professorship at a college. It will pay a salary of $12,000 each year forever. What lump sum would have to be deposited now at 7.5%, compounded annually, to provide this salary?

CHAPTER 2 TEST

1. The following is an arithmetic sequence. Identify the first term and the common difference.

$$5, \quad 8, \quad 11, \quad \ldots$$

3. Find the sum of the first 20 terms of this arithmetic sequence.

$$\$1.00, \quad \$1.06, \quad \$1.12, \quad \ldots$$

5. Find the 10th term of the geometric sequence in Question 4. Round to the nearest cent.

7. Determine whether this infinite geometric series has a sum. If so, find it.

$$\$1000, \quad \$80, \quad \$6.40, \quad \ldots$$

Consider this situation for Questions 8 through 10.

Item. Automobile.

$$\text{Cost} = \$8500,$$
$$\text{Expected life} = 4 \text{ years},$$
$$\text{Salvage value} = \$2550.$$

8. Use the straight-line method,

a) Prepare a depreciation schedule.
b) Find a formula for the book values V_n.
c) Find the common difference.

10. Use the sum of the year's digits method,

a) Find the depreciation fractions.
b) Prepare a depreciation schedule.

12. Find the effective annual yield.

9%, compounded quarterly

14. Find the amount of an annuity where $2000 is being invested semiannually at 8.4%, compounded semiannually, for 7 years.

16. What lump sum would have to be deposited now at 5%, compounded annually, so that withdrawals of $10,000 can be made every year for 15 years?

2. Find the 20th term of the sequence in Question 1.

4. The following is a geometric sequence. Identify the common ratio.

$$\$100, \quad \$105, \quad \$110.25, \quad \ldots$$

6. Find the sum of the first 10 terms of the sequence in Question 4. Round to the nearest cent.

9. Use the double-declining balance method,

a) Prepare a depreciation schedule.
b) Find a formula for the book values V_n.

11. Suppose $1000 is invested at 6%. How much is in the account at the end of 3 years if interest is

a) simple?
b) compounded annually?
c) compounded semiannually?
d) compounded quarterly?

13. Find the APR. Assume a car loan at the given add-on interest rate for 1 year and 12 payments.

$$\text{Loan} = \$5000, \qquad \text{Add-on rate} = 11\%.$$

15. Due to increasing business, a company expects to have to pay $8000 for a machine 6 years from now. They decide to make deposits each year at 8%, compounded annually. What should each payment be so that the company will accumulate the $8000?

17. A family buys a condominium for $100,000. A down payment of $25,000 is made, and $75,000 is borrowed at 10%, compounded monthly. The loan is to be paid off by 240 equal payments over the next 20 years. How much is each payment?

Equations
and the
Echelon Method—Matrices

OBJECTIVE

You should be able to solve systems of equations using the echelon method.

1. Translate this tableau to the corresponding system of equations.

x_1	x_2	1
2	−6	$\frac{1}{4}$
−3	1	8

2. Translate this system to the corresponding echelon tableau.

$$2x_1 + 5x_2 = -24$$
$$5x_1 - 2x_2 = -2$$

3.1 THE ECHELON METHOD—UNIQUE SOLUTIONS

In your experience with solving a system such as

$$2x_1 + 6x_2 = 26$$
$$8x_1 - 3x_2 = -31,$$

you may have noticed that one actually works with the coefficients, or constants, and not the variables. It is helpful to just list the constants in what is called an *echelon* tableau:*

x_1	x_2	1
2	6	26
8	−3	−31

The *rows* of a tableau are horizontal. The *columns* are vertical.

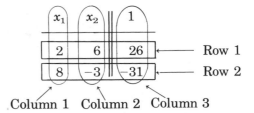

Note that the first column is labeled with the variable x_1, the second with the variable x_2, and the third with a 1. The double vertical lines correspond to "=." To see how to translate from a tableau to the corresponding system, we multiply the elements of each row by the column headings and add:

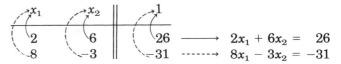

$$2x_1 + 6x_2 = 26$$
$$8x_1 - 3x_2 = -31$$

DO EXERCISES 1 AND 2.

The echelon method† for solving systems uses certain operations that correspond to operations on equations. We will carry out these operations in a way that may seem odd at first, but the reason for it will become apparent later. Before we formalize the method, let us consider an example. Compare each operation with the corresponding operation with the equations.

———————

* The word "echelon" is a French word meaning "a series of steps."
† Short for *reduced row echelon method.*

Example 1   Solve

$$2x_1 + 6x_2 = 26$$
$$8x_1 - 3x_2 = -31$$

Solution

Addition Method   *Echelon Method*

x_1	x_2	1
2	6	26
8	−3	−31

$$2x_1 + 6x_2 = 26$$
$$8x_1 - 3x_2 = -31$$

Our first goal is to get a 1 in the first row and first column.

1. Multiply the first equation by $\frac{1}{2}$:

1. Multiply the first row by $\frac{1}{2}$:

x_1	x_2	1
1*	3	13
8	−3	−31

$$x_1 + 3x_2 = 13$$
$$8x_1 - 3x_2 = -31$$

In the tableau we put a star (*) on the 1. It is called a *pivot* element. Our goal will always be to get 0's in the rest of a column where a pivot element occurs. To do this here, we multiply the first (pivot) row by −8 and add.

2. Multiply the first equation by −8 and add to the second. We leave the first equation as is: we are simply adding a multiple of it to the second equation.

2. Multiply the first (pivot) row by −8 and add to the second. We leave the first row as is in the tableau. We are simply adding a multiple of the pivot row to the second row.

x_1	x_2	1
1	3	13
0	−27	−135

$$x_1 + 3x_2 = 13$$
$$-27x_2 = -135$$

Our next goal is to get a 1 in the second row and second column; this will be a new pivot element.

3. Multiply the second equation by $-\frac{1}{27}$:

3. Multiply the second row by $-\frac{1}{27}$:

x_1	x_2	1
1	3	13
0	1*	5

$$x_1 + 3x_2 = 13$$
$$x_2 = 5$$

We put a star on the 1 to indicate that this is the new pivot element. Remember that we always want to get 0's in the rest of a column where a pivot element occurs. To do this here, we multiply the second (pivot) row by -3 and add.

4. Multiply the second equation by -3 and add to the first:

$$x_1 = -2$$
$$x_2 = 5$$

4. Multiply the second (pivot) row by -3 and add to the first:

x_1	x_2	1
1	0	-2
0	1	5

The solution is $(-2, 5)$. We can obtain this directly from the tableau by translating to the corresponding system of equations shown on the left.

THE ECHELON METHOD. In carrying out the echelon method, we obtain pivot elements of 1 diagonally from upper left to lower right. Then we get 0's in the rest of each column by adding multiples of the pivot row to the other rows. We use any of these operations in carrying out this procedure:

 i) **Interchange any two rows.**
 ii) **Interchange any two variable columns, provided we interchange the headings.**
 iii) **Multiply any row by a nonzero constant.**
 iv) **Add a multiple of the pivot row to another row.**

It is important to note that we can add *rows* only. We cannot add columns since this would mean adding unlike terms in the equations.

Example 2   Solve using the echelon method.

$$3x_1 - 4x_2 = -1$$
$$-3x_1 + 2x_2 = 0$$

Solution

1. We first obtain a tableau.

x_1	x_2	1
3	-4	-1
-3	2	0

2. We obtain the pivot element 1 in the first row and first column (upper left) by multiplying the first row by $\frac{1}{3}$.

x_1	x_2	1
1^*	$-\frac{4}{3}$	$-\frac{1}{3}$
-3	2	0

3. Next we *pivot*; that is, we obtain 0's in the rest of the first column. To do this we multiply the first (pivot) row by 3 and add to the second.

x_1	x_2	1
1	$-\frac{4}{3}$	$-\frac{1}{3}$
0	-2	-1

Based on your experience with the addition method, you may have been tempted at the outset to just add the first row to the second. While this is not incorrect, remember that we are proceeding in a special way to anticipate and ease later work.

4. We obtain the next pivot element by multiplying the second row by $-\frac{1}{2}$.

x_1	x_2	1
1	$-\frac{4}{3}$	$-\frac{1}{3}$
0	1^*	$\frac{1}{2}$

5. We pivot again, multiplying the second (pivot) row by $\frac{4}{3}$ and adding to the first.

x_1	x_2	1
1	0	$\frac{1}{3}$
0	1	$\frac{1}{2}$

The solution is $(\frac{1}{3}, \frac{1}{2})$, found by translating from the tableau to the system of equations $x_1 = \frac{1}{3}$ and $x_2 = \frac{1}{2}$.

DO EXERCISES 3 AND 4.

Now let us solve a system with three variables.

Example 3 Solve, using the echelon method.

$$2x_1 - x_2 + x_3 = -1$$
$$x_1 - 2x_2 + 3x_3 = 4$$
$$4x_1 + x_2 + 2x_3 = 4$$

Solution

1. We first write the tableau:

x_1	x_2	x_3	1
2	-1	1	-1
1	-2	3	4
4	1	2	4

3. Solve, using the addition and echelon methods together, as in Example 1.

$$2x_1 + 5x_2 = -24$$
$$5x_1 - 2x_2 = -2$$

4. Solve, using the echelon method.

$$9x_1 - 2x_2 = -4$$
$$3x_1 + 4x_2 = 1$$

2. We obtain a pivot element 1 in the first row and first column. We could do this by multiplying the first equation by $\frac{1}{2}$, but this would introduce fractions. An easier way to get the pivot element is to interchange the first and second rows.

x_1	x_2	x_3 || 1
1*	−2	3	4
2	−1	1	−1
4	1	2	4

3. Next we pivot to obtain 0's in the rest of the first column. We first multiply the first (pivot) row by −2 and add to the second row.

x_1	x_2	x_3 || 1
1*	−2	3	4
0	3	−5	−9
4	1	2	4

4. To complete the pivot, we multiply the first (pivot) row by −4 and add to the third row. (We usually work down the column in this way.)

x_1	x_2	x_3 || 1
1	−2	3	4
0	3	−5	−9
0	9	−10	−12

5. To obtain the next pivot, element we multiply the second row by $\frac{1}{3}$.

x_1	x_2	x_3 || 1
1	−2	3	4
0	1*	$-\frac{5}{3}$	−3
0	9	−10	−12

6. We pivot on the starred 1. We first multiply the second (pivot) row by 2 and add to the first.

x_1	x_2	x_3 || 1
1	0	$-\frac{1}{3}$	−2
0	1*	$-\frac{5}{3}$	−3
0	9	−10	−12

7. To complete the pivot, we multiply the second (pivot) row by -9 and add to the third.

x_1	x_2	x_3	1
1	0	$-\frac{1}{3}$	-2
0	1	$-\frac{5}{3}$	-3
0	0	5	15

8. To obtain the last pivot element, we multiply the third row by $\frac{1}{5}$.

x_1	x_2	x_3	1
1	0	$-\frac{1}{3}$	-2
0	1	$-\frac{5}{3}$	-3
0	0	1^*	3

9. We pivot on the starred 1. We first multiply the third (pivot) row by $\frac{1}{3}$ and add to the first.

x_1	x_2	x_3	1
1	0	0	-1
0	1	$-\frac{5}{3}$	-3
0	0	1^*	3

10. To complete the pivot, we multiply the third (pivot) row by $\frac{5}{3}$ and add to the second.

x_1	x_2	x_3	1
1	0	0	-1
0	1	0	2
0	0	1	3

The solution is found by translating from the tableau to the system of equations

$$x_1 = -1,$$
$$x_2 = 2,$$
$$x_3 = 3.$$

We can also say that the solution is the ordered triple $(-1, 2, 3)$.

In an actual computation we would record only the tableaux of Steps 1, 4, 7, and 10. The intermediate steps are recorded here so that the

5. Solve, using the echelon method.

$$3x_1 + 2x_2 + 2x_3 = 3$$
$$2x_1 - 4x_2 + x_3 = 0$$
$$x_1 + 2x_2 - x_3 = 5$$

student can follow the details. The actual computation would look like this:

x_1	x_2	x_3	1
2	-1	1	-1
1	-2	3	4
4	1	2	4
1^*	-2	3	4
0	3	-5	-9
0	9	-10	-12
1	0	$-\frac{1}{3}$	-2
0	1^*	$-\frac{5}{3}$	-3
0	0	5	15
1	0	0	-1
0	1	0	2
0	0	1^*	3

DO EXERCISE 5.

Application

Example 4 *Interest problem* Two investments are made that total $4800. For a certain year, these investments yield $412 in simple interest. Part of the $4800 is invested at 8% and the other part at 9%. Find the amount invested at each rate.

Solution Recall the formula for simple interest.

$$I = Pit.$$

Interest I is principal P times rate i times time t.

a) Let x_1 represent the amount invested at 8% and x_2 the amount invested at 9%. Then the interest from x_1 is $8\%x_1$, and the interest from x_2 is $9\%x_2$. Thus the $412 total interest is given by

$$8\%x_1 + 9\%x_2 = 412,$$

or

$$0.08x_1 + 0.09x_2 = 412.$$

b) Considering the total amount invested, we have

$$x_1 + x_2 = 4800.$$

c) We now have a system of equations:

$$0.08x_1 + 0.09x_2 = \ \ 412$$
$$x_1 + x_2 = 4800$$

We translate this system to an echelon tableau and solve.

x_1	x_2	1
0.08	0.09	412
1	1	4800

We first multiply the first row by 100 to clear the decimals.

x_1	x_2	1
8	9	41200
1	1	4800

To get the pivot element 1 in the first row, first column, we interchange the rows.

x_1	x_2	1
1*	1	4800
8	9	41200

We multiply the first row by -8 and add to the second.

x_1	x_2	1
1	1	4800
0	1*	2800

We already have a pivot element 1 in the second row, second column. We complete the pivot by multiplying the second row by -1 and adding to the first row.

x_1	x_2	1
1	0	2000
0	1	2800

6. Two investments are made that total \$3700. For a certain year, these investments yield \$297 in simple interest. Part of the \$3700 is invested at 7% and the other part at 9%. Find the amount invested at each rate.

Thus the solution is

$$x_1 = 2000,$$
$$x_2 = 2800,$$

so \$2000 is invested at 8% and \$2800 is invested at 9%.

DO EXERCISE 6.

EXERCISE SET 3.1

Solve, using the echelon method. Interchange rows and/or columns to avoid fractions when possible.

1. $\quad x_1 + 4x_2 = 5$
$\quad -3x_1 + 2x_2 = 13$

2. $\quad x_1 + 4x_2 = 8$
$\quad 3x_1 + 5x_2 = 3$

3. $\quad -x_1 + 3x_2 = 2$
$\quad 2x_1 - x_2 = 11$

4. $\quad 9x_1 - 2x_2 = 5$
$\quad 3x_1 - 3x_2 = 11$

5. $\quad 2x_1 - 5x_2 = 10$
$\quad 4x_1 + 3x_2 = 7$

6. $\quad 5x_1 - 3x_2 = -2$
$\quad 4x_1 + 2x_2 = 5$

7. $\quad x_1 + x_2 + x_3 = 9$
$\quad x_1 - x_2 - x_3 = -15$
$\quad x_1 + x_2 - x_3 = -5$

8. $\quad x_1 + x_2 + x_3 = 1$
$\quad x_1 + 2x_2 + 3x_3 = 4$
$\quad x_1 + 3x_2 + 7x_3 = 13$

9. $\quad x_1 - x_2 + 2x_3 = 0$
$\quad x_1 - 2x_2 + 3x_3 = -1$
$\quad 2x_1 - 2x_2 + x_3 = -3$

10. $\quad x_1 + 2x_2 - 3x_3 = 9$
$\quad 2x_1 - x_2 + 2x_3 = -8$
$\quad 3x_1 - x_2 - 4x_3 = 3$

11. $\quad 3x_1 + 2x_2 + 2x_3 = 3$
$\quad 2x_1 + 4x_2 - x_3 = 8$
$\quad 2x_1 - 4x_2 + x_3 = 0$

12. $\quad 4x_1 - x_2 - 3x_3 = 1$
$\quad 8x_1 + x_2 - x_3 = 5$
$\quad 2x_1 + x_2 + 2x_3 = 5$

13. $\quad 2x_1 - 3x_2 + x_3 - x_4 = -8$
$\quad x_1 + x_2 - x_3 - x_4 = -4$
$\quad x_1 - x_2 - x_3 - x_4 = -14$
$\quad x_1 + x_2 + x_3 + x_4 = 22$

14. $\quad 3x_1 - 2x_2 + 2x_3 + x_4 = -6$
$\quad x_1 - x_2 + 4x_3 + 3x_4 = -2$
$\quad x_1 + x_2 + x_3 + x_4 = -5$
$\quad 2x_1 + 2x_2 - 2x_3 - 2x_4 = -10$

15. *Business.* Two investments are made that total \$8800. For a certain year, these investments yield \$663 in simple interest. Part of the \$8800 is invested at 7% and part at 8%. Find the amount invested at each rate.

16. *Business.* Two investments are made that total \$15,000. For a certain year, these investments yield \$1432 in simple interest. Part of the \$15,000 is invested at 9% and part at 10%. Find the amount invested at each rate.

17. *Business.* For a certain year \$3900 is received in interest from two investments. A certain amount is invested at 5%, and \$10,000 more than this is invested at 6%. Find the amount invested at each rate. *Hint.* Express each equation in standard form $ax_1 + bx_2 = c$.

18. *Business.* For a certain year \$876 is received in interest from two investments. A certain amount is invested at 7%, and \$1200 more than this is invested at 8%. Find the amount invested at each rate.

19. *Business.* One day a campus bookstore sold 30 sweatshirts. White ones cost \$9.95 and yellow ones cost \$10.50. In dollars, \$310.60 worth of sweatshirts were sold. How many of each color were sold?

20. *Business.* One week a business sold 40 scarves. White ones cost \$4.95 and designed ones cost \$7.95. In dollars, \$282 worth of scarves were sold. How many of each kind were sold?

21. *Biology—Nutrition.* Soybean meal is 16% protein; corn meal is 9% protein. How many pounds of each should be mixed together to get 350 pounds of a mixture that is 12% protein?

22. *Chemistry.* A chemist has one solution of acid and water that is 25% acid and a second that is 50% acid. How many gallons of each should be mixed together to get 10 gallons of a solution that is 40% acid?

23. *Business.* A person receives \$212 per year in simple interest from three investments totalling \$2500. Part is invested at 7%, part at 8%, and part at 9%. There is \$1100 more invested at 9% than at 8%. Find the amount invested at each rate.

24. *Business.* A person receives \$306 per year in simple interest from three investments totalling \$3200. Part is invested at 8%, part at 9%, and part at 10%. There is \$1800 more invested at 10% than at 8%. Find the amount invested at each rate.

25. *Curve fitting.* Find numbers a, b, and c, such that the data points $(1, 4)$, $(-1, -2)$, and $(2, 13)$ are solutions of the quadratic function

$$y = ax^2 + bx + c.$$

26. *Curve fitting.* Find numbers a, b, and c, such that the data points $(1, 4)$, $(-1, 6)$, and $(-2, 16)$ are solutions of the quadratic function

$$y = ax^2 + bx + c.$$

27. *Business—Predicting earnings.* A business earns \$1000 in its first month, \$2000 in the second month, and \$8000 in the third month.

a) Find a quadratic function $y = ax^2 + bx + c$, which fits the data (see Exercise 25), where y = earnings, and x = month;

b) Use the model in (a) to predict the earnings for the fourth month.

28. *Biomedical—Death rate as a function of sleep.* (This problem is based on a study by Dr. Harold J. Morowitz.)

Average number of hours of sleep, x	Death rate per 100,000 males, y
5	1121
7	626
9	967

a) Use the given data points to find a quadratic function $y = ax^2 + bx + c$ that fits the data.

b) Use the model to find the death rate of males who sleep 4 hr; 6 hr; and 10 hr.

3.2 THE ECHELON METHOD—SPECIAL CASES

OBJECTIVE

You should be able to solve systems of equations using the echelon method for special cases where systems may have no solution, infinitely many solutions, or where the number of variables may not be the same as the number of equations.

In the preceding section, each system had exactly one solution and the final tableau had a form like the following:

x_1	x_2	x_3	1
1	0	0	p
0	1	0	q
0	0	1	r

This is called the *reduced row echelon form*, or *echelon form*, for short. Here we consider special cases where systems have no solution, or infinitely many solutions, or the number of variables is not the same as the number of equations. All these cases can be analyzed in the same general way.*

---

* Homogeneous equations, that is, equations whose constant terms (righthand sides) are zero require *no* special consideration.

7. Solve, using the echelon method.

$$4x_1 - 2x_2 = 2$$
$$2x_1 - x_2 = -8$$

Example 1 Solve

$$6x_1 - 3x_2 = 21$$
$$4x_1 - 2x_2 = 19$$

Solution We translate to the echelon tableau, multiply the first row by $\frac{1}{6}$, and obtain

x_1	x_2	1
1^*	$-\frac{1}{2}$	$\frac{7}{2}$
4	-2	19

We carry out the pivot by multiplying the pivot row by -4 and adding to the second row.

x_1	x_2	1
1	$-\frac{1}{2}$	$\frac{7}{2}$
0	0	5

The pivoting is considered to be complete here even though we do not have 1's down the main diagonal.

Earlier we found the solution by translating back to a system of equations. If we do that here, we obtain

$$x_1 - \frac{1}{2}x_2 = \frac{7}{2}$$
$$0 = 5$$

But the second equation is false. Thus the system is inconsistent. It has *no solution*.

DO EXERCISE 7.

Example 2 Solve

$$6x_1 - 3x_2 = 21$$
$$4x_1 - 2x_2 = 14$$

Solution We translate to the echelon tableau, multiply the first row by $\frac{1}{6}$, and obtain

x_1	x_2	1
1^*	$-\frac{1}{2}$	$\frac{7}{2}$
4	-2	14

We carry out the pivot by multiplying the pivot row by -4 and adding to the second row.

x_1	x_2	1
1	$-\frac{1}{2}$	$\frac{7}{2}$
0	0	0

Translating back to a system of equations we obtain

$$x_1 - \tfrac{1}{2}x_2 = \tfrac{7}{2}$$
$$0 = 0$$

We know from previous work that this system is consistent and the second equation is linearly dependent upon the first. The system has infinitely many solutions. Every point on the line $x_1 - \frac{1}{2}x_2 = \frac{7}{2}$ is a solution. We can describe this by solving the equation for x_1. Then the solutions can be described by

$$x_1 = \tfrac{7}{2} + \tfrac{1}{2}x_2, \qquad x_2 = \text{any number.}$$

By selecting arbitrary values of x_2 and computing x_1 we find the following as some solutions:

$x_2 = 0$	$x_2 = 1$	$x_2 = -3$
$x_1 = \frac{7}{2}$	$x_1 = \frac{7}{2} + \frac{1}{2} = 4$	$x_1 = \frac{7}{2} - \frac{3}{2} = 2$

DO EXERCISE 8.

Example 3 Suppose we are trying to solve a system with 3 equations and 4 variables, and we reach this stage in the tableau:

x_1	x_2	x_3	x_4	1
1*	0	-2	0	2
2	1	2	0	1
3	0	-6	5	26

The pivot element is in the first row and first column. We carry out the pivoting by multiplying the first row by -2 and adding to the second and multiplying the first row by -3 and adding to the third. We obtain.

x_1	x_2	x_3	x_4	1
1	0	-2	0	2
0	1	6	0	-3
0	0	0	5	20

Note that pivoting in the second column is also complete.

8. Solve using the echelon method.

$$4x_1 - 2x_2 = 6$$
$$-2x_1 + x_2 = -3$$

List the solutions for $x_2 = 0$, $x_2 = 1$, and $x_2 = -4$.

9. Consider the tableau

x_1	x_2	x_3	x_4	1
1	4	3	0	22
-2	-8	-1	0	-4
0	0	-2	2	-22

a) Find the reduced echelon tableau.

b) Describe the solutions.

c) List three specific solutions.
Let $x_2 = 0, 1, -2$.

Consider now the third row and third column, where we have a 0. The only way to get a 1 there is to interchange the third and fourth columns and multiply by a constant. Instead of doing this we simply move on to the third row and fourth column. We multiply by $\frac{1}{5}$ and obtain

x_1	x_2	x_3	x_4	1
1	0	-2	0	2
0	1	6	0	-3
0	0	0	1*	4

Since the rest of the fourth column already has 0's, the pivoting is complete. This is a more general example of the *echelon form*. The solutions are found by translating back to a system of equations. We obtain

$$x_1 - 2x_3 = 2$$
$$x_2 + 6x_3 = -3$$
$$x_4 = 4$$

The solutions can be described by

$$x_1 = 2 + 2x_3$$
$$x_2 = -3 - 6x_3$$
$$x_3 = \text{any number}$$
$$x_4 = 4$$

The following are some solutions obtained by picking arbitrary values of x_3.

$x_1 = 2$	$x_1 = 4$	$x_1 = 0$
$x_2 = -3$	$x_2 = -9$	$x_2 = 3$
$x_3 = 0$	$x_3 = 1$	$x_3 = -1$
$x_4 = 4$	$x_4 = 4$	$x_4 = 4$

The echelon form has a general "staircase" description like any of the following, where # means that any type of number can be in that location.

x_1	x_2	x_3	x_4	x_5	1
1	0	#	0	#	#
0	1	#	0	#	#
0	0	0	1	#	#

x_1	x_2	x_3	1
1	0	0	#
0	1	0	#
0	0	1	#

x_1	x_2	x_3	1
1	0	0	#
0	1	0	#
0	0	0	#

These do not show all possibilities, but they give the general idea.

DO EXERCISE 9.

The following discussion concerns those situations where there is a row with all 0's to the left of the double vertical line.

i) **Any time we have a row with all 0's to the left of the double vertical line and a nonzero number to the right, the system has no solution.**

$$\begin{array}{cccc|c} \cdot & \cdot & \cdot & \cdot & \cdot \\ 0 & 0 & 0 & 0 & k \\ \cdot & \cdot & \cdot & \cdot & \cdot \end{array} \quad \text{(where } k \neq 0\text{)}$$

ii) **For a row of *all* 0's, we cannot determine the nature of the solutions without further analysis.**

$$\begin{array}{cccc|c} \cdot & \cdot & \cdot & \cdot & \cdot \\ 0 & 0 & 0 & 0 & 0 \\ \cdot & \cdot & \cdot & \cdot & \cdot \end{array}$$

This row corresponds to an equation that is linearly dependent. So up to this point the system would seem to be consistent. We shift that row to the bottom of the tableau and make further analysis of the upper part of the tableau. We may have exactly one solution, infinitely many solutions, or no solution, should Case (i) later occur.

Example 4   Suppose we are trying to solve a system of 4 equations in 3 variables, and we reach this stage in the tableau.

x_1	x_2	x_3	1
1	0	5	8
0	1	$-\frac{1}{4}$	-2
0	0	0	$-\frac{1}{2}$
0	0	-3	6

Because the third row has all 0's to the left of the double vertical line and a nonzero number to the right, the system has *no solution*. No further analysis is necessary.

DO EXERCISE 10.

Example 5   Suppose we are trying to solve a system of 4 equations in 3 variables and we reach this stage in the tableau.

x_1	x_2	x_3	1
1	0	5	8
0	1	$-\frac{1}{4}$	-2
0	0	0	0
0	0	-3	6

10. Suppose we are solving a system of 4 equations in 3 variables and we reach this stage in the tableau.

x_1	x_2	x_3	1
1	0	6	-9
0	0	0	4
0	-2	8	-6
0	5	10	20

Complete the solution, using the echelon method.

11. Suppose we are solving a system of 4 equations in 3 variables and we reach this stage in the tableau.

x_1	x_2	x_3	1
1	0	6	−9
0	0	0	0
0	−2	8	−6
0	5	10	20

Complete the solution, using the echelon method.

Since we have a row with all 0's, we interchange it with the fourth row.

x_1	x_2	x_3	1
1	0	5	8
0	1	$-\frac{1}{4}$	−2
0	0	−3	6
0	0	0	0

The pivot element is in the third row and third column. We multiply the third row by $-\frac{1}{3}$:

x_1	x_2	x_3	1
1	0	5	8
0	1	$-\frac{1}{4}$	−2
0	0	1^*	−2
0	0	0	0

Now we pivot. We multiply the third row by $\frac{1}{4}$ and add to the second, and we multiply the third row by −5 and add to the first, obtaining

x_1	x_2	x_3	1
1	0	0	18
0	1	0	$-\frac{5}{2}$
0	0	1	−2
0	0	0	0

We now find the solution by translating back to a system of equations:

$$x_1 = 18,$$
$$x_2 = -\tfrac{5}{2},$$
$$x_3 = -2,$$
$$0 = 0.$$

The last equation plays no role here. The solution can be stated as an ordered triple $(18, -\frac{5}{2}, -2)$.

In Example 5 the original system of equations was consistent and *linearly dependent*. If we have a system of more than two equations that is consistent and linearly dependent, then one of the equations is a multiple of the others or a sum of multiples of the others.

DO EXERCISE 11.

EXERCISE SET 3.2

Solve using the echelon method. Interchange rows and/or columns to avoid fractions when possible.

1. $\begin{aligned} x_1 - 3x_2 &= 2 \\ -2x_1 + 6x_2 &= -4 \end{aligned}$

2. $\begin{aligned} 3x_1 + 6x_2 &= 9 \\ x_1 + 2x_2 &= 3 \end{aligned}$

3. $\begin{aligned} x_1 - 3x_2 &= 2 \\ -2x_1 + 6x_2 &= -3 \end{aligned}$

4. $\begin{aligned} 3x_1 + 6x_2 &= 8 \\ x_1 + 2x_2 &= 3 \end{aligned}$

5. $\begin{aligned} 4x_1 + 12x_2 + 16x_3 &= 4 \\ 3x_1 + 4x_2 + 7x_3 &= 3 \\ x_1 + 8x_2 + 9x_3 &= 1 \end{aligned}$

6. $\begin{aligned} 2x_1 - 3x_2 + 7x_3 &= 2 \\ x_1 - 4x_2 + x_3 &= 6 \\ 4x_1 - 16x_2 + 4x_3 &= 24 \end{aligned}$

7. $\begin{aligned} 4x_1 + 12x_2 + 16x_3 &= 0 \\ 3x_1 + 4x_2 + 5x_3 &= 0 \\ x_1 + 8x_2 + 11x_3 &= 0 \end{aligned}$

8. $\begin{aligned} 2x_1 + x_2 - 3x_3 &= 0 \\ x_1 - 4x_2 + x_3 &= 0 \\ 4x_1 - 16x_2 + 4x_3 &= 0 \end{aligned}$

9. $\begin{aligned} x_1 + x_2 + 13x_3 &= 0 \\ x_1 - x_2 - 6x_3 &= 0 \end{aligned}$

10. $\begin{aligned} x_1 + x_2 - x_3 &= -3 \\ x_1 + 2x_2 + 2x_3 &= -1 \end{aligned}$

11. $\begin{aligned} 3x_1 - 9x_3 &= 3 \\ 2x_1 + x_2 - x_3 &= 6 \\ x_1 + 2x_2 + 7x_3 + x_4 &= 7 \end{aligned}$

12. $\begin{aligned} x_1 + x_2 + 12x_3 + 2x_4 &= 20 \\ -2x_1 - x_2 - 20x_3 - 3x_4 &= -31 \\ 3x_1 + 4x_2 + 40x_3 + 7x_4 &= 69 \end{aligned}$

13. $\begin{aligned} 2x_1 - 2x_2 + 18x_3 &= -14 \\ x_1 - 2x_2 + 13x_3 &= -4 \\ -2x_2 + 8x_3 &= 4 \\ 2x_1 + x_2 + 36x_3 &= 7 \end{aligned}$

14. $\begin{aligned} -2x_1 - 3x_2 - 4x_3 &= -13 \\ x_1 + 2x_3 &= 3 \\ x_1 + x_2 + 2x_3 &= 8 \\ x_1 + 3x_3 &= 5 \end{aligned}$

Complete the solution using the echelon method.

15.

x_1	x_2	x_3	x_4	1
1	2	-4	8	7
0	-3	9	12	18
0	3	-9	-12	-18

16.

x_1	x_2	x_3	x_4	1
1	0	5	0	6
0	1	-3	0	4
0	0	0	-2	10

17.

x_1	x_2	x_3	1
1	-1	-2	-5
0	0	0	0
0	0	0	0
0	-2	4	-8

18.

x_1	x_2	x_3	1
1	-2	9	6
0	0	0	0
0	3	4	5

19.

x_1	x_2	x_3	x_4	x_5	1
1	0	8	-3	0	6
0	1	4	2	0	4
0	0	0	0	-2	10

20.

x_1	x_2	x_3	x_4	x_5	1
1	0	8	-3	0	6
0	1	4	2	0	4
0	0	0	0	3	-6

3.3 BASIC MATRIX PROPERTIES

A company makes two types of stereos. Type I requires 65 transistors, 50 capacitors, and 4 dials. Type II requires 85 transistors, 42 capacitors, and 6 dials. We can represent this information as follows.

	Transistors	Capacitors	Dials
Type I	65	50	4
Type II	85	42	6

OBJECTIVES

You should be able to

a) Find the dimensions of a matrix.
b) Find the sum and difference of two matrices of the same dimensions.
c) Find the scalar product of a matrix A and a constant c.
d) Find the *transpose* of a matrix.

Find the dimensions of each matrix.

12. $\begin{bmatrix} -2 & 0 & 3 \\ \frac{1}{2} & 16 & 3 \end{bmatrix}$

13. $[-4 \quad 8 \quad 7]$

14. $\begin{bmatrix} -9 \\ 4 \\ 5 \end{bmatrix}$

15. $[-5]$

16. $\begin{bmatrix} -3 & 6 \\ 4 & 0 \end{bmatrix}$

This table forms a rectangular array of numbers called a *matrix*.

A *matrix* is a rectangular array of numbers. The elements of a matrix are enclosed in brackets.

We can also think of a matrix being obtained from the coefficients and constants in a system of equations. Thus, from the system

$$-2x_1 + 8x_2 = 3,$$
$$\tfrac{1}{2}x_1 + 16x_2 = 5,$$

we get the matrix

$$\begin{bmatrix} -2 & 8 & 3 \\ \frac{1}{2} & 16 & 5 \end{bmatrix}$$

This matrix has 2 *rows* and 3 *columns*.

A matrix with *m* rows and *n* columns has *dimensions m* × *n, read "m* by *n."*

Example 1 Find the dimensions of each matrix.

a) $[-2 \quad 3 \quad 4 \quad \tfrac{1}{4}]$ b) $\begin{bmatrix} 6 \\ 7 \\ -3 \\ -\frac{1}{2} \end{bmatrix}$ c) $\begin{bmatrix} -2 & \frac{1}{4} & 8 \\ 0 & 1 & 5 \\ -8 & 6 & 4 \end{bmatrix}$ d) $[8]$

Solution

a) The dimensions are 1×4
b) The dimensions are 4×1
c) The dimensions are 3×3. Such a matrix is called a *square* matrix since it has the same number of rows as columns.
d) The dimensions are 1×1.

We will usually drop the brackets from a 1×1 matrix. That is,

$$[8] = 8.$$

DO EXERCISES 12 THROUGH 16.

We will use capital letters to represent matrices. The elements of a matrix will be denoted by lower-case letters with subscripts. For example, with

$$A = \begin{bmatrix} a_{11} & a_{12} & a_{13} \\ a_{21} & a_{22} & a_{23} \end{bmatrix},$$

the element in the ith row and jth column is given by a_{ij}. The above is a 2×3 matrix. We may also denote it

$$A = [a_{ij}]_{2\times3}, \quad \text{or} \quad [a_{ij}].$$

DO EXERCISE 17.

Two matrices are *equal* if and only if they have the same dimensions and corresponding elements are equal. Formally if $A = [a_{ij}]_{m\times n}$ and $B = [b_{ij}]_{m\times n}$, then $A = B$ if and only if $a_{ij} = b_{ij}$ for each i and j, where i ranges from 1 to m and j ranges from 1 to n.

Examples

a) $\begin{bmatrix} 2^3 & 0 \\ 1-5 & 9 \end{bmatrix} = \begin{bmatrix} 8 & 0 \\ -4 & 9 \end{bmatrix}$ b) $\begin{bmatrix} -2 & 4 & 7 \\ 0 & 1 & 5 \end{bmatrix} \neq \begin{bmatrix} -2 & 4 \\ 0 & 1 \end{bmatrix}$

c) $\begin{bmatrix} 8 & -9 \\ -6 & 7 \end{bmatrix} \neq \begin{bmatrix} 8 & -9 \\ 6 & 7 \end{bmatrix}$

Example 2 Solve for p and r.

$$\begin{bmatrix} -6 & 3r-5 \\ 0 & p \end{bmatrix} = \begin{bmatrix} -6 & 14 \\ 0 & -9 \end{bmatrix}$$

Solution Since the matrices are equal, $p = -9$, and $3r - 5 = 14$, or $r = \frac{19}{3}$.

DO EXERCISES 18 AND 19.

A $1 \times n$ matrix is often referred to as a *row vector*, and an $m \times 1$ matrix is often referred to as a *column vector*.

Example 3 Which are row vectors and which are column vectors?

$A = [5 \quad -3 \quad 0]$, $B = \begin{bmatrix} 7 \\ -4 \\ 3 \end{bmatrix}$, $C = [-8 \quad 9 \quad 10 \quad 0 \quad \frac{1}{4}]$, $D = \begin{bmatrix} -3 \\ 1 \end{bmatrix}$

Solution The row vectors are A and C, and the column vectors are B and D.

DO EXERCISE 20.

17. Consider

$$[a_{ij}] = \begin{bmatrix} -3 & 0 & 6 \\ 4 & -6 & 7 \\ -1 & -2 & \frac{1}{2} \end{bmatrix}$$

a) Find a_{12}.

b) Find a_{22}.

c) Find a_{21}.

d) Find a_{32}.

18. Decide whether each pair of matrices is equal.

a) $\begin{bmatrix} 3^2 & -1 \\ 2-3 & 7 \end{bmatrix}, \begin{bmatrix} 6 & -1 \\ -1 & 7 \end{bmatrix}$

b) $[-2 \quad 9 \quad 8 \quad 10]$,
   $[\ 3 \quad -2 \quad 9 \quad 8]$

19. Solve for a and b.

$$\begin{bmatrix} a & 3 & -4 \\ 0 & 6 & -8 \end{bmatrix} = \begin{bmatrix} -6 & 3 & -4 \\ 4b-2 & 6 & -8 \end{bmatrix}$$

20. Which are row vectors and which are column vectors?

$$A = \begin{bmatrix} 5 \\ 4 \\ 6 \\ -1 \end{bmatrix}, \quad B = [-2 \quad 4],$$

$$C = [-2 \quad 4 \quad 0],$$

$$D = \begin{bmatrix} 5 \\ 1 \\ 3 \\ 2 \end{bmatrix}.$$

21. Find the transpose of each matrix.

$$A = \begin{bmatrix} -8 & 1 & -2 \\ -4 & 0 & -1 \\ 6 & 7 & 8 \end{bmatrix},$$

$$B = [-7 \quad 9 \quad 10 \quad \tfrac{1}{4}],$$

$$C = \begin{bmatrix} -20 \\ 41 \end{bmatrix},$$

$$D = \begin{bmatrix} -4 & 5 \\ 1 & 0 \\ 0 & 1 \end{bmatrix}.$$

The Transpose of a Matrix

The *transpose* of a matrix A, denoted A^T, is found by interchanging the rows and columns of A. That is, if $A = [a_{ij}]$, then $A^T = [a_{ji}]$.

Example 4 Find A^T, B^T, C^T, and D^T.

$$A = \begin{bmatrix} 2 & 4 & 6 \\ 9 & 8 & -2 \\ 0 & -1 & 4 \end{bmatrix}, \qquad B = \begin{bmatrix} -3 & 0 & 4 \\ 7 & 1 & 6 \end{bmatrix},$$

$$C = \begin{bmatrix} -4 \\ 3 \\ 2 \end{bmatrix}, \qquad D = [-1 \quad 2 \quad 3 \quad 0]$$

Solution

$$A^T = \begin{bmatrix} 2 & 9 & 0 \\ 4 & 8 & -1 \\ 6 & -2 & 4 \end{bmatrix}, \qquad B^T = \begin{bmatrix} -3 & 7 \\ 0 & 1 \\ 4 & 6 \end{bmatrix},$$

$$C^T = [-4 \quad 3 \quad 2], \qquad D^T = \begin{bmatrix} -1 \\ 2 \\ 3 \\ 0 \end{bmatrix}$$

DO EXERCISE 21.

You have probably discovered in the margin exercises and Example 4 that if the dimensions of A are $m \times n$, then the dimensions of A^T are $n \times m$. Also, the transpose of a row vector is a column vector, and the transpose of a column vector is a row vector. The latter is a convenient method of saving space. That is, instead of writing

$$A = \begin{bmatrix} x_1 \\ x_2 \\ x_3 \end{bmatrix}$$

we may write

$$A^T = [x_1 \quad x_2 \quad x_3], \qquad \text{or} \qquad A = [x_1 \quad x_2 \quad x_3]^T.$$

Addition of Matrices

The *sum* of two matrices of the same dimensions is the matrix whose elements are the sums of corresponding elements of the given matrices. Formally, if $A = [a_{ij}]$ and $B = [b_{ij}]$, then $A + B = [a_{ij} + b_{ij}]$.

Note that matrix addition is defined only for matrices of the same dimensions.

Example 5 Find $A + B$ and $B + A$.

$$A = \begin{bmatrix} -4 & 0 \\ 3 & \frac{1}{4} \end{bmatrix}, \quad B = \begin{bmatrix} 7 & -5 \\ 2 & \frac{1}{2} \end{bmatrix}$$

Solution

$$A + B = \begin{bmatrix} -4 + 7 & 0 + (-5) \\ 3 + 2 & \frac{1}{4} + \frac{1}{2} \end{bmatrix} = \begin{bmatrix} 3 & -5 \\ 5 & \frac{3}{4} \end{bmatrix},$$

$$B + A = \begin{bmatrix} 7 + (-4) & -5 + 0 \\ 2 + 3 & \frac{1}{2} + \frac{1}{4} \end{bmatrix} = \begin{bmatrix} 3 & -5 \\ 5 & \frac{3}{4} \end{bmatrix}.$$

DO EXERCISES 22 THROUGH 24.

For any matrices, A, B, and C, of the same dimensions,
$A + B = B + A$ (Addition of matrices is commutative)
$A + (B + C) = (A + B) + C$ (Addition of matrices is associative)

We give a proof of the commutative law,

$A + B = [a_{ij}] + [b_{ij}]$

$\quad = [a_{ij} + b_{ij}]$ (Definition of matrix addition)

$\quad = [b_{ij} + a_{ij}]$ (Addition of real numbers is commutative)

$\quad = [b_{ij}] + [a_{ij}]$ (Reverse of matrix addition)

$\quad = B + A.$

A *zero matrix* is denoted by the capital letter O. From Margin Exercise 24, we can conjecture that, if A and O have the same dimensions, then

$A + O = O + A = A.$ (**O** is the *additive identity*)

The product of a constant and a matrix, a *scalar product*, is defined as follows.

The *scalar product* of a number c (sometimes called a *scalar*) and a matrix A is the matrix obtained by multiplying the elements of A by c. Formally, if $A = [a_{ij}]$, then $cA = [ca_{ij}]$.

Example 6 Find $4A$ and $(-1)A$.

$$A = \begin{bmatrix} -2 & 0 \\ 1 & 5 \end{bmatrix}$$

22. Consider

$$A = \begin{bmatrix} 5 & -2 \\ 7 & -4 \end{bmatrix} \quad \text{and} \quad B = \begin{bmatrix} -8 & -4 \\ 2 & 4 \end{bmatrix}$$

a) Find $A + B$.

b) Find $B + A$.

c) Determine whether $A + B = B + A$.

23. Add:

$$\begin{bmatrix} -2 & 6 \\ 4 & 5 \\ 0 & \frac{1}{4} \end{bmatrix} + \begin{bmatrix} 2 & 9 \\ -3 & 7 \\ 1 & \frac{1}{4} \end{bmatrix}$$

24. Consider

$$A = \begin{bmatrix} 5 & -2 \\ 7 & -4 \end{bmatrix} \quad \text{and} \quad O = \begin{bmatrix} 0 & 0 \\ 0 & 0 \end{bmatrix}.$$

a) Find $A + O$.

b) Find $O + A$.

c) Determine whether $A + O = O + A$.

25. Consider

$$A = \begin{bmatrix} 6 & 1 & 0 \\ 2 & 1 & -5 \\ -3 & 9 & \frac{1}{2} \end{bmatrix}$$

Find:

a) $6A$ b) $(-1)A$

c) $A + (-1)A$ d) $(-1)A + A$

e) $-\dfrac{1}{30}A$ f) tA

26. Consider matrices A and B of Margin Exercise 22.

Find

a) $A - B$

b) $B - A$

c) Determine whether $A - B = B - A$.

Solution

$$4A = \begin{bmatrix} 4(-2) & 4 \cdot 0 \\ 4 \cdot 1 & 4 \cdot 5 \end{bmatrix} = \begin{bmatrix} -8 & 0 \\ 4 & 20 \end{bmatrix},$$

$$(-1)A = \begin{bmatrix} -1(-2) & -1 \cdot 0 \\ -1 \cdot 1 & -1 \cdot 5 \end{bmatrix} = \begin{bmatrix} 2 & 0 \\ -1 & -5 \end{bmatrix}$$

DO EXERCISE 25.

For real numbers we know that $-a$ represents the number we add to a to get 0. We call $-a$ the *additive inverse* of a. For matrices, $-A$ is the matrix we add to A to get the zero matrix O. You may have conjectured the following from Margin Exercise 25.

$$A + (-1)A = (-1)A + A = O$$

(The additive inverse of A, $-A$, is $(-1)A$.)

We subtract as follows.

The difference $A - B = A + (-1)B$. We subtract B from A by adding the additive inverse of B.

Note that A and B must have the *same dimensions* in order to subtract one from the other.

Example 7 Find $A - B$.

$$A = \begin{bmatrix} -3 & 5 \\ 4 & 8 \end{bmatrix}, \qquad B = \begin{bmatrix} 11 & -2 \\ 6 & 7 \end{bmatrix}$$

$$A - B = A + (-1)B = \begin{bmatrix} -3 & 5 \\ 4 & 8 \end{bmatrix} + (-1)\begin{bmatrix} 11 & -2 \\ 6 & 7 \end{bmatrix}$$

$$= \begin{bmatrix} -3 & 5 \\ 4 & 8 \end{bmatrix} + \begin{bmatrix} -11 & 2 \\ -6 & -7 \end{bmatrix}$$

$$= \begin{bmatrix} -14 & 7 \\ -2 & 1 \end{bmatrix}$$

Note that $A - B$ can be found directly by subtracting corresponding elements of B from those of A.

DO EXERCISE 26.

EXERCISE SET 3.3

Consider

$$A = \begin{bmatrix} 1 & 3 \\ 4 & 2 \end{bmatrix}, \qquad B = \begin{bmatrix} -2 & 0 \\ -2 & -1 \end{bmatrix}, \qquad C = \begin{bmatrix} -1 & -2 & -3 \\ 3 & 2 & 1 \end{bmatrix}, \qquad \text{and} \qquad D = \begin{bmatrix} 0 & 8 & -4 \\ 1 & 0 & -1 \end{bmatrix}.$$

Find:

1. The dimensions of A **2.** The dimensions of B **3.** The dimensions of C **4.** The dimensions of D

5. $A + B$ **6.** $B + A$ **7.** $C + D$ **8.** $D + C$

9. $3A$ **10.** $3B$ **11.** $-5C$ **12.** $-6D$

13. $A - B$ **14.** $B - A$ **15.** $C - D$ **16.** $D - C$

17. $A + C$ **18.** $A + D$ **19.** kC **20.** kD

21. $A + O$ **22.** $O + D$ **23.** A^T **24.** B^T

25. C^T **26.** D^T **27.** $A^T + B^T$ **28.** $C^T + D^T$

29. For $[a_{ij}] = \begin{bmatrix} -4 & 5 \\ 0 & 9 \\ 1 & 3 \end{bmatrix}$, find a_{11}, a_{12}, a_{31}, a_{22}, a_{32}, and a_{21}.

30. For $[b_{ij}] = \begin{bmatrix} -2 & -3 & 0 \\ \frac{1}{4} & \frac{1}{2} & 1 \end{bmatrix}$, find b_{11}, b_{21}, b_{23}, b_{22}, b_{13}, and b_{12}.

31. Find X^T where $X = \begin{bmatrix} x_1 \\ x_2 \\ x_3 \\ x_4 \end{bmatrix}$.

32. Find Y^T where $Y = \begin{bmatrix} y_1 \\ y_2 \\ y_3 \\ y_4 \end{bmatrix}$.

3.4 MATRIX MULTIPLICATION

Summation Notation

Consider this sum:

$$a_1 + a_2 + a_3 + a_4$$

We can denote this sum using *summation notation* which utilizes the Greek capital letter $\sum$ (sigma),

$$\sum_{i=1}^{4} a_i \quad \text{or} \quad \sum_{i=1}^{4} a_i$$

This is read "the sum of the numbers a_i from $i = 1$ to $i = 4$." To recover the original sum, substitute the numbers 1 through 4 successively into a_i and write plus signs between the results.

Example 1 Write summation notation for $2 + 4 + 6 + 8 + 10$.

Solution

$$2 + 4 + 6 + 8 + 10 = \sum_{i=1}^{5} 2i$$

OBJECTIVES

You should be able to

a) Write summation notation for certain sums.

b) Express a sum without summation notation.

c) Find the product AB of two matrices A and B, where the number of columns in A is the same as the number of rows in B.

d) Given the inverse of a coefficient matrix, use that inverse to solve systems of 2 equations in 2 variables and 3 equations in 3 variables.

e) Given matrices A and B, find matrices like

$$[A \quad B], \quad [A \quad B]^T, \quad \text{and} \quad \begin{bmatrix} A \\ B \end{bmatrix}.$$

Write in summation notation.

27. $1 + 4 + 9 + 16 + 25 + 36$

28. $t + t^2 + t^3 + t^4$

29. $p_1 + p_2 + p_3 + \cdots + p_{38}$

Express without using summation notation.

30. $\displaystyle\sum_{i=1}^{3} 2^i$

31. $\displaystyle\sum_{i=1}^{20} p_i q_i$

32. $\displaystyle\sum_{i=1}^{5} it^i$

Example 2 Write summation notation for

$$a_1 + a_2 + a_3 + a_4 + \cdots + a_{19}$$

The three dots indicate that we are not writing all the terms in between.

Solution

$$a_1 + a_2 + a_3 + a_4 + \cdots + a_{19} = \sum_{i=1}^{19} a_i$$

DO EXERCISES 27 THROUGH 29.

Example 3 Express $\sum_{i=1}^{4} 3^i$ without using summation notation.

Solution

$$\sum_{i=1}^{4} 3^i = 3^1 + 3^2 + 3^3 + 3^4$$

Example 4 Express $\sum_{i=1}^{30} a_i b_i$ without using summation notation.

Solution

$$\sum_{i=1}^{30} a_i b_i = a_1 b_1 + a_2 b_2 + \cdots + a_{30} b_{30}$$

DO EXERCISES 30 THROUGH 32.

Matrix Multiplication

The product of two matrices A and B will *not* be defined as the matrix whose elements are products of corresponding elements of A and B. Some motivation for the definition of matrix multiplication is based on converting an equation such as

$$2x_1 - 4x_2 + 7x_3 = 8$$

to a *product of a row vector and a column vector:*

$$[2 \quad -4 \quad 7] \begin{bmatrix} x_1 \\ x_2 \\ x_3 \end{bmatrix} = [8], \quad \text{or} \quad 8$$

The *product* of a row vector A, a matrix of dimensions $1 \times n$, and a column vector B, a matrix of dimensions $n \times 1$, is the 1×1 matrix (or scalar) whose element is the sum of products of corresponding elements A and B. Formally, if

$$A = [a_1 \quad a_2 \quad a_3 \cdots a_n] \quad \text{and} \quad B = \begin{bmatrix} b_1 \\ b_2 \\ b_3 \\ \cdot \\ \cdot \\ \cdot \\ b_n \end{bmatrix},$$

then

$$AB = a_1b_1 + a_2b_2 + a_3b_3 + \cdots + a_nb_n$$

or, using summation notation,

$$AB = \sum_{i=1}^{n} a_ib_i.$$

Example 5 Find each product.

a) $[2 \quad -4 \quad 7] \begin{bmatrix} -1 \\ 0 \\ 5 \end{bmatrix} = [2(-1) - 4 \cdot 0 + 7 \cdot 5] = [33], \quad \text{or} \quad 33$

b) $[\frac{1}{4} \quad -8] \begin{bmatrix} 12 \\ 1 \end{bmatrix} = [\frac{1}{4} \cdot 12 - 8 \cdot 1] = [-5], \quad \text{or} \quad -5$

c) $[-2 \quad 1 \quad 4 \quad -5] \begin{bmatrix} x_1 \\ x_2 \\ x_3 \\ x_4 \end{bmatrix} = [-2x_1 + x_2 + 4x_3 - 5x_4], \quad \text{or}$
$-2x_1 + x_2 + 4x_3 - 5x_4$

To multiply the row vector A and the column vector B, the number of elements in A must be the same as the number of elements in B. The following illustration should help you remember this.

$$\begin{array}{cc} A & B \\ 1 \times n & m \times 1 \end{array}$$
$$\longrightarrow n = m \longleftarrow$$

DO EXERCISES 33 THROUGH 35

Before we define multiplication of more general matrices, let us reconsider an example given earlier. A company makes two types of stereos. Type I requires 65 transistors, 50 capacitors, and 4 dials. Type II requires 85 transistors, 42 capacitors, and 6 dials. We can

Find each product.

33. $[4 \quad 5] \begin{bmatrix} a \\ b \end{bmatrix}$

34. $[4 \quad 5] \begin{bmatrix} -2 \\ 3 \end{bmatrix}$

35. $[3 \quad -2 \quad 4] \begin{bmatrix} -6 \\ 7 \\ -1 \end{bmatrix}$

36. Put in matrix form:

$$11x_1 + 2x_2 = -1$$
$$7x_1 - 13x_2 = 8$$

37. Write as separate equations:

$$\begin{bmatrix} 3 & -7 \\ -2 & 1 \end{bmatrix} \begin{bmatrix} y_1 \\ y_2 \end{bmatrix} = \begin{bmatrix} 4 \\ 5 \end{bmatrix}$$

represent this information using a matrix.

	Transistors	Capacitors	Dials
Type I	65	50	4
Type II	85	42	6

Suppose the company wanted to make 20 stereos of Type I and 30 stereos of Type II. It would determine the number of transistors needed as follows:

$$20 \cdot 65 + 30 \cdot 85 = 3850.$$

It would determine the number of capacitors needed as follows:

$$20 \cdot 50 + 30 \cdot 42 = 2260.$$

It would determine the dials needed as follows:

$$20 \cdot 4 + 30 \cdot 6 = 260.$$

The entire procedure could be done using matrices as follows:

$$[20 \quad 30] \cdot \begin{bmatrix} 65 & 50 & 4 \\ 85 & 42 & 6 \end{bmatrix}$$

$$= [20 \cdot 65 + 30 \cdot 85 \quad 20 \cdot 50 + 30 \cdot 42 \quad 20 \cdot 4 + 30 \cdot 6]$$

$$= [3850 \quad 2260 \quad 260]$$

Some further motivation for multiplication of matrices is based on converting a system of equations such as

$$2x_1 - 3x_2 = 7$$
$$4x_1 + 5x_2 = 9,$$

to matrix form:

$$\begin{bmatrix} 2 & -3 \\ 4 & 5 \end{bmatrix} \begin{bmatrix} x_1 \\ x_2 \end{bmatrix} = \begin{bmatrix} 7 \\ 9 \end{bmatrix}.$$

Multiplying these matrices, we obtain the original equations. Note the similarity with the echelon tableau, Section 3.1.

DO EXERCISES 36 AND 37.

The *product* of two matrices A and B is the matrix $C = [c_{ij}]$, where c_{ij} is obtained by multiplying the ith row of A (as a vector) by the jth column of B (as a vector).

Example 6 Multiply

$$[2 \quad 3] \begin{bmatrix} x & a \\ y & b \end{bmatrix} = [2x + 3y \quad 2a + 3b].$$

Mentally this product is found in two steps:

$$[① \quad ②]$$

where

① is [█] $\begin{bmatrix} █ & \end{bmatrix}$

and

② is [█] $\begin{bmatrix} & █ \end{bmatrix}$

Example 7 Multiply

$$[2 \quad 3]\begin{bmatrix} 1 & -2 \\ -8 & 4 \end{bmatrix} = [2 \cdot 1 + 3(-8) \quad 2(-2) + 3 \cdot 4] = [-22 \quad 8]$$

DO EXERCISE 38.

Example 8 Multiply

$$\begin{bmatrix} 2 & 3 \\ -6 & 7 \end{bmatrix}\begin{bmatrix} x & a \\ y & b \end{bmatrix} = \begin{bmatrix} 2x + 3y & 2a + 3b \\ -6x + 7y & -6a + 7b \end{bmatrix}$$

Mentally this product is found in four steps:

$$\begin{bmatrix} ① & ② \\ ③ & ④ \end{bmatrix}$$

① [█] $\begin{bmatrix} █ & \end{bmatrix}$ ② [█] $\begin{bmatrix} & █ \end{bmatrix}$

③ [█] $\begin{bmatrix} █ & \end{bmatrix}$ ④ [█] $\begin{bmatrix} & █ \end{bmatrix}$

Example 9 Multiply

$$\begin{bmatrix} 2 & 3 \\ -6 & 7 \end{bmatrix}\begin{bmatrix} 1 & -2 \\ -8 & 4 \end{bmatrix} = \begin{bmatrix} 2 \cdot 1 + 3(-8) & 2(-2) + 3 \cdot 4 \\ -6 \cdot 1 + 7(-8) & -6(-2) + 7 \cdot 4 \end{bmatrix} = \begin{bmatrix} -22 & 8 \\ -62 & 40 \end{bmatrix}$$

DO EXERCISE 39.

So that we can carry out the multiplication of the rows of A and the columns of B, the number of columns in A must be the same as the number of rows in B. That is, if A has dimensions $m \times n$ and B has dimensions $p \times q$, then in order to multiply we must have $n = p$. In such a case we say that the matrices are *conformable*. The product

38. Multiply.

a) $[5 \quad 6]\begin{bmatrix} c & g \\ d & h \end{bmatrix}$

b) $[5 \quad 6]\begin{bmatrix} -3 & 0 \\ 2 & -4 \end{bmatrix}$

39. Multiply.

a) $\begin{bmatrix} 5 & 6 \\ 3 & -1 \end{bmatrix}\begin{bmatrix} c & g \\ d & h \end{bmatrix}$

b) $\begin{bmatrix} 5 & 6 \\ 3 & -1 \end{bmatrix}\begin{bmatrix} -3 & 0 \\ 2 & -4 \end{bmatrix}$

matrix has dimensions $m \times q$. The following may help you remember this:

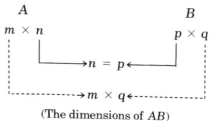

(The dimensions of AB)

Example 10 Find AB and BA, if possible.

$$A = \begin{bmatrix} -2 & 1 \\ 4 & 0 \\ -3 & -5 \end{bmatrix}, \qquad B = \begin{bmatrix} 2 & -1 & 0 & -7 \\ 4 & -3 & -1 & 0 \end{bmatrix}$$

Solution

$$AB = \begin{bmatrix} -2 \cdot 2 + 1 \cdot 4 & -2(-1) + 1(-3) & -2 \cdot 0 + 1(-1) & -2(-7) + 1 \cdot 0 \\ 4 \cdot 2 + 0 \cdot 4 & 4(-1) + 0(-3) & 4 \cdot 0 + 0(-1) & 4(-7) + 0 \cdot 0 \\ -3 \cdot 2 - 5 \cdot 4 & -3(-1) - 5(-3) & -3 \cdot 0 - 5(-1) & -3(-7) - 5 \cdot 0 \end{bmatrix}$$

$$= \begin{bmatrix} 0 & -1 & -1 & 14 \\ 8 & -4 & 0 & -28 \\ -26 & 18 & 5 & 21 \end{bmatrix}$$

BA cannot be found because B and A are not conformable; that is, the number of columns in B is not the same as the number of rows in A.

The products AB and BA of a row matrix A and a column matrix B are of special interest.

Example 11 Find AB and BA, if possible.

$$A = [-1 \quad 2 \quad -5], \qquad B = \begin{bmatrix} 4 \\ 0 \\ -2 \end{bmatrix}$$

Solution

A is a 1×3 matrix and B is a 3×1 matrix. Thus the product AB is a 1×1 matrix given by

$$AB = [-1 \quad 2 \quad -5] \begin{bmatrix} 4 \\ 0 \\ -2 \end{bmatrix} = [-1 \cdot 4 + 2 \cdot 0 - 5(-2)] = [6], \qquad \text{or} \qquad 6$$

B is a 3×1 matrix and A is a 1×3 matrix. Thus the product BA is a 3×3 matrix given by

$$BA = \begin{bmatrix} 4 \\ 0 \\ -2 \end{bmatrix} [-1 \quad 2 \quad -5] = \begin{bmatrix} 4(-1) & 4 \cdot 2 & 4(-5) \\ 0(-1) & 0 \cdot 2 & 0(-5) \\ -2(-1) & -2 \cdot 2 & -2(-5) \end{bmatrix}$$

$$= \begin{bmatrix} -4 & 8 & -20 \\ 0 & 0 & 0 \\ 2 & -4 & 10 \end{bmatrix}$$

Note that each element of BA is the product of two numbers.

DO EXERCISES 40 AND 41.

We have seen that matrix multiplication is not commutative. In some cases, there are matrices A and B such that $AB = BA$. In such cases we say that the matrices *commute*.

The notion of *conformability* is often used in many contexts with matrices. For example, matrices A and B are *equality conformable* if A and B have the same dimensions. A and B are *addition conformable* if A and B have the same dimensions. A and B are *multiplication conformable* if the number of columns in A is the same as the number of rows in B.

For any conformable matrices A, B, and C.

$(AB)C = A(BC) = ABC$ **(Multiplication of matrices is associative.)**

$A(B + C) = AB + AC$ **(Multiplication of matrices is distributive.)**

DO EXERCISE 42.

Identity Matrices

The letter I is used to represent square matrices such as

$$I = \begin{bmatrix} 1 & 0 \\ 0 & 1 \end{bmatrix}, \quad \text{and} \quad I = \begin{bmatrix} 1 & 0 & 0 \\ 0 & 1 & 0 \\ 0 & 0 & 1 \end{bmatrix}.$$

These square matrices have 1's extending from the upper left down to the lower right along what is called the *main diagonal*. The rest of the elements are 0.

40. Find AB and BA, if possible.

a) $A = \begin{bmatrix} -3 & 0 & 5 \\ -2 & 8 & 0 \end{bmatrix}$,

     $B = \begin{bmatrix} 1 & -2 & -3 \\ 1 & 0 & 1 \\ -4 & 2 & 6 \end{bmatrix}$

b) $A = [-2 \quad -3 \quad 1 \quad -1]$,

     $B = \begin{bmatrix} -6 \\ -8 \\ 0 \\ 5 \end{bmatrix}$

41. Consider

$$A = \begin{bmatrix} 3 & -8 \\ 1 & -5 \end{bmatrix}, \quad B = \begin{bmatrix} 0 & -1 \\ 4 & 0 \end{bmatrix}$$

a) Find AB.

b) Find BA.

c) Decide whether AB and BA are equal.

d) Is matrix multiplication commutative? Explain.

42. Given

$$A = \begin{bmatrix} 3 & -2 \\ 1 & 5 \end{bmatrix}, \quad B = \begin{bmatrix} 0 & -1 \\ 4 & 0 \end{bmatrix},$$

and

$$C = \begin{bmatrix} 5 & 2 \\ 1 & 1 \end{bmatrix}$$

a) Verify the associative law by matrix multiplication.

b) Verify the distributive law by matrix addition and multiplication.

43. Consider

$$A = \begin{bmatrix} x & a \\ y & b \end{bmatrix}, \quad X = \begin{bmatrix} x_1 \\ x_2 \end{bmatrix}$$

and

$$I = \begin{bmatrix} 1 & 0 \\ 0 & 1 \end{bmatrix}.$$

a) Find AI.

b) Find IA.

c) Compare AI and IA.

d) Find IX.

e) Find XI.

f) Compare IX and X.

DO EXERCISE 43.

You have just proved the following for the square matrix I of dimensions 2×2.

For any square matrix A of dimensions $n \times n$,

$$AI = IA = A \qquad (I \text{ is a } \textit{multiplicative identity}),$$

where I is the square matrix, described above, of dimensions $n \times n$.

We also have the following.

For any matrix A with exactly n rows,

$$IA = A,$$

where I is the square matrix, described above, of dimensions $n \times n$.

Matrix Inverses

The equation in real numbers, or *scalar* equation,

$$ax = b, \qquad \text{where } a \text{ is real},$$

can be solved for x by multiplying both sides of the equation by a^{-1}, the inverse of a:

$$a^{-1}ax = a^{-1}b.$$

Since $a^{-1}a = aa^{-1} = 1$, the multiplicative identity element such that $1x = x$, we obtain

$$x = a^{-1}b.$$

For scalar numbers

$$a^{-1} = \frac{1}{a}, \qquad \text{for } a \neq 0,$$

so that

$$x = \frac{b}{a}.$$

Let us see how we can use matrices to represent and solve systems of equations. Consider the system

$$3x_1 + 2x_2 = 1,$$
$$5x_1 + 3x_2 = -2.$$

We first express this system as a matrix equation:

$$\begin{bmatrix} 3 & 2 \\ 5 & 3 \end{bmatrix} \begin{bmatrix} x_1 \\ x_2 \end{bmatrix} = \begin{bmatrix} 1 \\ -2 \end{bmatrix}$$

To see that this is correct, multiply the matrices on the left and use the fact that matrices are equal if corresponding elements are equal. We let

$$A = \text{the } coefficient\ matrix = \begin{bmatrix} 3 & 2 \\ 5 & 3 \end{bmatrix}, \quad X = \begin{bmatrix} x_1 \\ x_2 \end{bmatrix}, \text{ and } B = \begin{bmatrix} 1 \\ -2 \end{bmatrix}.$$

Then

$$AX = B.$$

The solution of this *matrix* equation can*not* be written as the quotient

$$X = \frac{B}{A}.$$

Multiplying this equation by A on the *left*, yields

$$AX = B,$$

while multiplying the same equation by A on the *right*, yields

$$XA = B.$$

This implies that the two products AX and XA commute, that is

$$AX = XA.$$

This is not always true. As an example, consider

$$AX = B,$$

where

$$A = \begin{bmatrix} 3 & 2 \\ 5 & 3 \end{bmatrix}, \quad X = \begin{bmatrix} x_1 \\ x_2 \end{bmatrix}, \quad \text{and} \quad B = \begin{bmatrix} 1 \\ -2 \end{bmatrix}.$$

Then

$$AX = \begin{bmatrix} 3 & 2 \\ 5 & 3 \end{bmatrix}\begin{bmatrix} x_1 \\ x_2 \end{bmatrix} = \begin{bmatrix} 3x_1 + 2x_2 \\ 5x_1 + 3x_2 \end{bmatrix} = \begin{bmatrix} 1 \\ -2 \end{bmatrix}.$$

On the other hand

$$XA = \begin{bmatrix} x_1 \\ x_2 \end{bmatrix}\begin{bmatrix} 3 & 2 \\ 5 & 3 \end{bmatrix}$$

is not possible, since the matrices X and A in the product XA are not conformable.

Thus, **division by a matrix is not possible.**

However, we can replace division by multiplication by the *matrix inverse*. For example, consider solving

$$AX = B$$

44. Consider

$$A = \begin{bmatrix} 3 & 5 \\ 1 & 2 \end{bmatrix}, \qquad A^{-1} = \begin{bmatrix} 2 & -5 \\ -1 & 3 \end{bmatrix}$$

and

$$I = \begin{bmatrix} 1 & 0 \\ 0 & 1 \end{bmatrix}.$$

Find $A^{-1}A$ and AA^{-1} and compare to I.

45. Consider the system

$$3x_1 + 5x_2 = -1$$
$$x_1 + 2x_2 = 4$$

a) Express this system as a matrix equation.
b) What is the coefficient matrix A?
c) Given the inverse of the coefficient matrix A:

$$A^{-1} = \begin{bmatrix} 2 & -5 \\ -1 & 3 \end{bmatrix},$$

use it and the matrix equation to solve the system.

for X as we solved $ax = b$ for x. To do this we define the *multiplicative inverse*, or simply the *inverse*, of a square matrix A to be the square matrix A^{-1} with the property that (as for the scalar a)

$$A^{-1}A = AA^{-1} = I,$$

where I is the multiplicative identity element previously defined.

DO EXERCISE 44.

Now we multiply both sides of $AX = B$ by A^{-1} on the left and obtain

$$A^{-1}AX = A^{-1}B.$$

Since $A^{-1}A = I$ and $IX = X$, we obtain

$$X = A^{-1}B.$$

If we knew the inverse matrix A^{-1}, then we could find the solution of the system of equations by computing the product $A^{-1}B$.

In the present example, we give A^{-1} without explaining how we found it.

$$A^{-1} = \begin{bmatrix} -3 & 2 \\ 5 & -3 \end{bmatrix}$$

(A procedure for computing matrix inverses will be developed in Section 3.5.)

Compare the following:

$$AX = B \qquad \begin{bmatrix} 3 & 2 \\ 5 & 3 \end{bmatrix}\begin{bmatrix} x_1 \\ x_2 \end{bmatrix} = \begin{bmatrix} 1 \\ -2 \end{bmatrix}$$

$$X = A^{-1}B \qquad \begin{bmatrix} x_1 \\ x_2 \end{bmatrix} = \begin{bmatrix} -3 & 2 \\ 5 & -3 \end{bmatrix}\begin{bmatrix} 1 \\ -2 \end{bmatrix} = \begin{bmatrix} -7 \\ 11 \end{bmatrix}$$

From equality of matrices we can read off the solution. That is,

$$x_1 = -7 \quad \text{and} \quad x_2 = 11,$$

or, simply, the solution is $(-7, 11)$.

DO EXERCISE 45.

Let us relate this procedure to the echelon method. We consider the echelon tableau for the previous system:

x_1	x_2	1
3	2	1
5	3	-2

The entries to the left of the double vertical line make up the coefficient matrix A. The entries to the right make up the matrix B. When the echelon method is completed we have the tableau

$$
\begin{array}{cc||c}
x_1 & x_2 & 1 \\
\hline
1 & 0 & -7 \\
0 & 1 & 11
\end{array}
$$

The entries to the left of the double vertical line make up the identity matrix I. Those to the right make up the solution matrix. The steps of the echelon method have the same effect as multiplying by A^{-1} without actually knowing what it is.

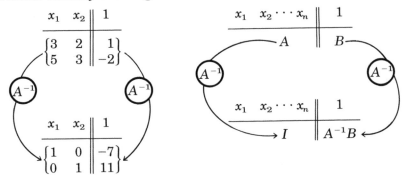

It is important to know the meaning and notation of matrix inverses. It is less important to know how to compute the inverse, although the preceding is the basis for such a method.

Augmented Matrices

For later purposes we need some additional notation. Let

$$X = [x_1 \quad x_2] \qquad \text{and} \qquad Y = [y_1 \quad y_2].$$

Then we can form the *augmented* matrix

$$[X \quad Y] = [x_1 \quad x_2 \vdots y_1 \quad y_2].$$

An augmented matrix is simply a particular way of putting two matrices together to form a new matrix. Sometimes, as above, a dashed line is used to indicate that $[X \quad Y]$ can be *partitioned* into X and Y.

Since the matrices X and Y have the same dimensions, we can also form the augmented matrix

$$\begin{bmatrix} X \\ Y \end{bmatrix} = \begin{bmatrix} x_1 & x_2 \\ \hline y_1 & y_2 \end{bmatrix}.$$

46. Consider

$$X = \begin{bmatrix} x_1 \\ x_2 \\ x_3 \end{bmatrix} \quad \text{and} \quad Y = \begin{bmatrix} y_1 \\ y_2 \\ y_3 \end{bmatrix}.$$

a) Find $[X \quad Y]$.

b) Find $\begin{bmatrix} X \\ Y \end{bmatrix}$.

47. Consider

$$A = \begin{bmatrix} -2 & 3 \\ 4 & 5 \end{bmatrix} \quad \text{and} \quad I = \begin{bmatrix} 1 & 0 \\ 0 & 1 \end{bmatrix}$$

a) Find $[A \quad I]$

b) Find $\begin{bmatrix} A \\ I \end{bmatrix}$.

48. Consider

$$A = \begin{bmatrix} 5 & 7 \\ -2 & 0 \end{bmatrix} \quad \text{and} \quad I = \begin{bmatrix} 1 & 0 \\ 0 & 1 \end{bmatrix}$$

a) Find $[AI]^T$.

b) Find $\begin{bmatrix} A \\ I \end{bmatrix}^T$.

When forming augmented matrices, the resulting array must be rectangular; that is, the matrices must conform. Thus, if $Z = [z_1 \quad z_2 \quad z_3]$, we can form the augmented matrix $[X \quad Z]$ but not $[\begin{smallmatrix} X \\ Z \end{smallmatrix}]$.

DO EXERCISE 46.

Suppose

$$A = \begin{bmatrix} 3 & 5 \\ 1 & -2 \end{bmatrix} \quad \text{and} \quad I = \begin{bmatrix} 1 & 0 \\ 0 & 1 \end{bmatrix}.$$

Then we can define an augmented matrix A' by

$$A' = [A \quad I] = \begin{bmatrix} 3 & 5 & 1 & 0 \\ 1 & -2 & 0 & 1 \end{bmatrix}$$

and another augmented matrix A'' by

$$A'' = \begin{bmatrix} A \\ I \end{bmatrix} = \begin{bmatrix} 3 & 5 \\ 1 & -2 \\ \hline 1 & 0 \\ 0 & 1 \end{bmatrix}.$$

DO EXERCISE 47.

Sometimes it is necessary to find the *transpose* of an augmented matrix. To do that, *first* form the augmented matrix and *then* form the transpose in the usual manner. Thus,

$$[X \quad Y]^T = \begin{bmatrix} x_1 \\ x_2 \\ \hline y_1 \\ y_2 \end{bmatrix} \quad \text{and} \quad \begin{bmatrix} X \\ Y \end{bmatrix}^T = \begin{bmatrix} x_1 & y_1 \\ x_2 & y_2 \end{bmatrix}.$$

DO EXERCISE 48.

EXERCISE SET 3.4

Find AB and BA, if possible.

1. $A = [-2 \quad 1], \quad B = \begin{bmatrix} 3 \\ -4 \end{bmatrix}$

2. $A = [-1 \quad -2], \quad B = \begin{bmatrix} 6 \\ -7 \end{bmatrix}$

3. $A = [2 \quad 0 \quad -4], \quad B = \begin{bmatrix} 9 \\ -5 \\ \frac{1}{4} \end{bmatrix}$

4. $A = [-3 \quad 6 \quad 8], \quad B = \begin{bmatrix} 5 \\ \frac{1}{2} \\ 1 \end{bmatrix}$

5. $A = \begin{bmatrix} 1 & 2 & 0 \\ -1 & 0 & 4 \\ 2 & 5 & 6 \end{bmatrix}$, $\quad B = \begin{bmatrix} 3 & -4 & 1 \\ 2 & -1 & 0 \\ -3 & 2 & 1 \end{bmatrix}$

6. $A = \begin{bmatrix} -1 & 0 & 0 \\ 0 & -1 & 0 \\ 0 & 0 & -1 \end{bmatrix}$, $\quad B = \begin{bmatrix} 2 & -5 & 1 \\ -4 & 4 & 3 \\ 5 & 6 & 9 \end{bmatrix}$

7. $A = [-4 \quad 1 \quad 3]$, $\quad B = \begin{bmatrix} -4 & 2 \\ 1 & 0 \\ 6 & -9 \end{bmatrix}$

8. $A = \begin{bmatrix} -2 & 3 \\ 1 & 0 \\ -5 & 4 \end{bmatrix}$, $\quad B = \begin{bmatrix} 2 \\ 3 \end{bmatrix}$

9. Find AB. $A = \begin{bmatrix} 1 & 0 & 2 \\ 5 & 0 & 1 \\ -1 & 0 & 4 \end{bmatrix}$, $\quad B = \begin{bmatrix} 0 & 0 & 0 \\ 1 & 3 & 7 \\ 0 & 0 & 0 \end{bmatrix}$.

10. Find two matrices A and B each of dimensions 2×2, such that $A \neq 0$ and $B \neq 0$ but $AB = 0$. *Hint.* See Exercise 9.

11. Write the system of equations of Exercise 17 in matrix form.

12. Write the system of equations of Exercise 18 in matrix form.

13. Write the system of equations of Exercise 21 in matrix form.

14. Write the system of equations of Exercise 22 in matrix form.

Write as separate equations:

15. $\begin{bmatrix} 1 & 2 \\ 4 & -3 \end{bmatrix} \begin{bmatrix} x_1 \\ x_2 \end{bmatrix} = \begin{bmatrix} -1 \\ 2 \end{bmatrix}$

16. $\begin{bmatrix} 2 & 4 \\ 3 & 5 \end{bmatrix} \begin{bmatrix} x_1 \\ x_2 \end{bmatrix} = \begin{bmatrix} -2 \\ -4 \end{bmatrix}$

In Exercises 17 through 22, a system of equations is given, together with the inverse of the coefficient matrix. Use the matrix inverse to solve the system.

17. $\begin{aligned} 11x_1 + 3x_2 &= -4, \\ 7x_1 + 2x_2 &= 5 \end{aligned}$ $\quad A^{-1} = \begin{bmatrix} 2 & -3 \\ -7 & 11 \end{bmatrix}$

18. $\begin{aligned} 8x_1 + 5x_2 &= -6, \\ 5x_1 + 3x_2 &= 2 \end{aligned}$ $\quad A^{-1} = \begin{bmatrix} -3 & 5 \\ 5 & -8 \end{bmatrix}$

19. $\begin{aligned} 4x_1 - 3x_2 &= 2, \\ x_1 + 2x_2 &= -1 \end{aligned}$ $\quad A^{-1} = \frac{1}{11}\begin{bmatrix} 2 & 3 \\ -1 & 4 \end{bmatrix}$

20. $\begin{aligned} 3x_1 + 5x_2 &= -4, \\ 2x_1 + 4x_2 &= -2 \end{aligned}$ $\quad A^{-1} = \frac{1}{2}\begin{bmatrix} 4 & -5 \\ -2 & 3 \end{bmatrix}$

21. $\begin{aligned} 3x_1 + x_2 \quad\;\; &= 2, \\ x_1 - x_2 + 2x_3 &= -4 \\ x_1 + x_2 + x_3 &= 5 \end{aligned}$ $\quad A^{-1} = \frac{1}{8}\begin{bmatrix} 3 & 1 & -2 \\ -1 & -3 & 6 \\ -2 & 2 & 4 \end{bmatrix}$

22. $\begin{aligned} x_1 \quad\;\; + x_3 &= -4, \\ 2x_1 + x_2 \quad\;\; &= -3 \\ x_1 - x_2 + x_3 &= 1 \end{aligned}$ $\quad A^{-1} = -\frac{1}{2}\begin{bmatrix} 1 & -1 & -1 \\ -2 & 0 & 2 \\ -3 & 1 & 1 \end{bmatrix}$

Consider

$$A = \begin{bmatrix} 0 & -1 \\ 1 & 2 \end{bmatrix} \quad \text{and} \quad B = \begin{bmatrix} -1 & 1 \\ 3 & 0 \end{bmatrix}.$$

23. Show that $(A + B)(A + B) \neq A^2 + 2AB + B^2$, where $AA = A^2$ and $BB = B^2$.

24. Show that $(A - B)(A + B) \neq A^2 - B^2$.

25. For $X = \begin{bmatrix} a \\ b \\ c \end{bmatrix}$ and $Y = \begin{bmatrix} e \\ f \\ g \end{bmatrix}$, find $\begin{bmatrix} X \\ Y \end{bmatrix}^{\mathrm{T}}$.

26. For $A = [-2 \quad 3 \quad 7]$ and $O = [0 \quad 0 \quad 0]$, find $[A \quad O]$.

Given

$$A = \begin{bmatrix} 2 & -1 & 3 \\ 4 & 1 & 0 \end{bmatrix} \quad \text{and} \quad B = \begin{bmatrix} 0 & 1 & -2 \\ 1 & -3 & 7 \end{bmatrix}.$$

27. Find $[A \quad B]$ and $[A \quad B]^T$.

28. Find $\begin{bmatrix} A \\ B \end{bmatrix}$ and $\begin{bmatrix} A \\ B \end{bmatrix}^T$.

29. Find $\begin{bmatrix} A^T \\ B^T \end{bmatrix}$. Is $[A \quad B]^T = \begin{bmatrix} A^T \\ B^T \end{bmatrix}$?

30. Find $[A^T \quad B^T]$. Is $\begin{bmatrix} A \\ B \end{bmatrix}^T = [A^T \quad B^T]$?

OBJECTIVE

You should be able to compute the inverse of a square matrix.

*3.5 (Optional) COMPUTATION OF THE MATRIX INVERSE

We now describe a way to compute a matrix inverse. We do this with an example and in the abstract. Suppose we wanted to find the inverse of the matrix

$$A = \begin{bmatrix} 2 & -1 & 1 \\ 1 & -2 & 3 \\ 4 & 1 & 2 \end{bmatrix}.$$

We are going to use an echelon tableau but a bit differently. Proceeding as if A were the coefficient matrix of some system of equations, we write the matrix A in the left side as we normally do, but on the right side we write the identity matrix and use 1's as headings:

x_1	x_2	x_3	1	1	1
2	-1	1	1	0	0
1	-2	3	0	1	0
4	1	2	0	0	1

x_1	$x_2 \cdots x_n$	1	1 $\cdots$ 1
	A		I

Now we proceed with the echelon method, but with one exception. We *never* interchange columns. In truth, we never did interchange columns in any of the examples of Sections 3.1 and 3.2, but we could have. Thus, we use only *row operations*, performing them on the entire augmented (or lengthened) rows. From the explanation in Section 3.4, we know that carrying out the echelon method has the same effect as multiplying by A^{-1}. But this time, when we multiply I on the right side of the tableau, we get A^{-1}.

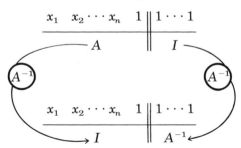

Suppose we wanted to find the inverse of the given matrix A. The procedure is to perform row operations to obtain I, or

$$\begin{bmatrix} 1 & 0 & 0 \\ 0 & 1 & 0 \\ 0 & 0 & 1 \end{bmatrix}$$

on the *left side* of the tableau. The resulting matrix appearing on the *right side* is the inverse matrix A^{-1}. We illustrate this as follows.

1. We first obtain a 1 in the first row, first column, by interchanging the first and second rows.

x_1	x_2	x_3	1	1	1
1*	−2	3	0	1	0
2	−1	1	1	0	0
4	1	2	0	0	1

2. Next we pivot to obtain 0's in the rest of the first column. We multiply the pivot row by −2 and add to the second row. We multiply the pivot row by −4 and add to the third row.

x_1	x_2	x_3	1	1	1
1	−2	3	0	1	0
0	3	−5	1	−2	0
0	9	−10	0	−4	1

3. Next we obtain a 1 in the second row, second column. We do this by multiplying the second row by $\frac{1}{3}$.

x_1	x_2	x_3	1	1	1
1	−2	3	0	1	0
0	1*	$-\frac{5}{3}$	$\frac{1}{3}$	$-\frac{2}{3}$	0
0	9	−10	0	−4	1

4. We pivot to obtain 0's in the rest of the second column. We multiply the pivot row by 2 and add to the first. We multiply the pivot row by −9 and add to the third:

x_1	x_2	x_3	1	1	1
1	0	$-\frac{1}{3}$	$\frac{2}{3}$	$-\frac{1}{3}$	0
0	1	$-\frac{5}{3}$	$\frac{1}{3}$	$-\frac{2}{3}$	0
0	0	5	−3	2	1

49. Use the echelon method to find the inverse.

$$A = \begin{bmatrix} 2 & -3 \\ 4 & 5 \end{bmatrix}$$

5. To get a 1 in the third row, third column, we multiply by $\frac{1}{5}$:

x_1	x_2	x_3	1	1	1
1	0	$-\frac{1}{3}$	$\frac{2}{3}$	$-\frac{1}{3}$	0
0	1	$-\frac{5}{3}$	$\frac{1}{3}$	$-\frac{2}{3}$	0
0	0	1^*	$-\frac{3}{5}$	$\frac{2}{5}$	$\frac{1}{5}$

6. We pivot to obtain 0's in the rest of the third column. We multiply the pivot row by $\frac{1}{3}$ and add to the first. We multiply the pivot row by $\frac{5}{3}$ and add to the second:

x_1	x_2	x_3	1	1	1
1	0	0	$\frac{7}{15}$	$-\frac{1}{5}$	$\frac{1}{15}$
0	1	0	$-\frac{2}{3}$	0	$\frac{1}{3}$
0	0	1	$-\frac{3}{5}$	$\frac{2}{5}$	$\frac{1}{5}$

Thus

$$A^{-1} = \begin{bmatrix} \frac{7}{15} & -\frac{1}{5} & \frac{1}{15} \\ -\frac{2}{3} & 0 & \frac{1}{3} \\ -\frac{3}{5} & \frac{2}{5} & \frac{1}{5} \end{bmatrix}$$

The reader can check this by doing the multiplication $A^{-1}A$.

If we do not obtain the identity matrix on the left, as would be the case when the system has no solution or infinitely many solutions, then A^{-1} does not exist.

This procedure will work for any square matrix that has an inverse.

DO EXERCISE 49.

EXERCISE SET 3.5

Find A^{-1}.

1. $A = \begin{bmatrix} 3 & 2 \\ 5 & 3 \end{bmatrix}$ **2.** $A = \begin{bmatrix} 3 & 5 \\ 1 & 2 \end{bmatrix}$ **3.** $A = \begin{bmatrix} 11 & 3 \\ 7 & 2 \end{bmatrix}$ **4.** $A = \begin{bmatrix} 8 & 5 \\ 5 & 3 \end{bmatrix}$

5. $A = \begin{bmatrix} 4 & -3 \\ 1 & 2 \end{bmatrix}$ **6.** $A = \begin{bmatrix} 3 & 5 \\ 2 & 4 \end{bmatrix}$

▶

7. $A = \begin{bmatrix} 3 & 1 & 0 \\ 1 & -1 & 2 \\ 1 & 1 & 1 \end{bmatrix}$ **8.** $A = \begin{bmatrix} 1 & 0 & 1 \\ 2 & 1 & 0 \\ 1 & -1 & 1 \end{bmatrix}$ **9.** $A = \begin{bmatrix} 1 & -1 & 2 \\ 0 & 1 & 3 \\ 2 & 1 & -2 \end{bmatrix}$ **10.** $A = \begin{bmatrix} 1 & -1 & 2 \\ 0 & 1 & 2 \\ 1 & -3 & -4 \end{bmatrix}$

CHAPTER 3 TEST

Put in matrix form and solve using the echelon method.

1. $7x_1 + 4x_2 = -21,$
$\quad 3x_1 + \ x_2 = \ -9$

2. $3x_1 - 2x_2 + 3x_3 = \ \ 24,$
$\quad x_1 + \ x_2 - \ x_3 = \ \ -7,$
$\quad 2x_1 + 3x_2 - 5x_3 = -32$

Write as separate equations and solve using the echelon method.

3. $\begin{bmatrix} 4 & -8 \\ 3 & -6 \end{bmatrix}\begin{bmatrix} x_1 \\ x_2 \end{bmatrix} = \begin{bmatrix} -20 \\ -15 \end{bmatrix}$

4. $\begin{bmatrix} 8 & -4 \\ 6 & -3 \end{bmatrix}\begin{bmatrix} x_1 \\ x_2 \end{bmatrix} = \begin{bmatrix} 20 \\ 16 \end{bmatrix}$

Complete the solution using the echelon method.

5.

x_1	x_2	x_3	x_4	1
1	0	6	0	5
0	1	-2	0	3
0	0	0	-4	-8

6.

x_1	x_2	x_3	1
1	3	-2	6
0	4	-1	8
0	4	1	2
0	8	0	-10

For Questions 7 through 10, consider

$$A = \begin{bmatrix} -3 & 2 \\ -5 & 1 \end{bmatrix} \quad \text{and} \quad B = \begin{bmatrix} 0 & -1 \\ 1 & 0 \end{bmatrix}.$$

Find:

7. $A + B$

8. AB

9. $A - B$

10. $-4A$

11. For

$$C = \begin{bmatrix} 2 \\ -3 \\ 4 \end{bmatrix}, \text{ find } C^T.$$

12. Find AB and BA, if possible.

$$A = \begin{bmatrix} 1 \\ -1 \\ 2 \end{bmatrix}, \quad B = \begin{bmatrix} 2 & -3 & 0 \\ 1 & 2 & 4 \end{bmatrix}.$$

In Questions 13 and 14, a system of equations is given, together with the inverse of the coefficient matrix. Use the matrix inverse to solve the system. Show your work.

13. $8x_1 + 3x_2 = \ \ 6,$
$\quad 4x_1 - 6x_2 = -2$
$\quad A^{-1} = \begin{bmatrix} \frac{1}{10} & \frac{1}{20} \\ \frac{1}{15} & -\frac{2}{15} \end{bmatrix}$

14. $x_1 + \ x_2 + \ x_3 = 10,$
$\quad x_1 + 2x_2 - \ x_3 = -5,$
$\quad 2x_1 - \ x_2 + 3x_3 = 20$
$\quad A^{-1} = \begin{bmatrix} -1 & \frac{4}{5} & \frac{3}{5} \\ 1 & -\frac{1}{5} & -\frac{2}{5} \\ 1 & -\frac{3}{5} & -\frac{1}{5} \end{bmatrix}$

15. For $X = \begin{bmatrix} p \\ q \end{bmatrix}$ and $Y = \begin{bmatrix} t \\ u \end{bmatrix}$, find $\begin{bmatrix} X \\ Y \end{bmatrix}^T$.

16. For a certain year, $850 is received in interest from two investments. A certain amount is invested at 8%, and $1300 more than this is invested at 10%. Find the amount invested at each rate.

*17. Use the echelon method to find A^{-1}.

$$A = \begin{bmatrix} 1 & 0 & -1 \\ -1 & 1 & -1 \\ -1 & 0 & 2 \end{bmatrix}$$

PART II

CHAPTER FOUR

Linear Programming—
An Introduction

During World War II the Army began to formulate certain *linear optimization* problems. Their solutions were called plans or *programs*. Subsequently, many other problems, particularly economic, were found to have a similar mathematical formulation. Such problems are called linear-programming problems or *linear programs* for short.

For example, consider the manager of a department store who sends his buyer to the "market." The buyer has a budget and can spend no more than a given amount of money. Furthermore, the goods must be brought back to the store in a company truck with a maximum cargo volume and a maximum cargo weight. That is, the total *volume* of the goods bought cannot exceed the cargo *volume* limit of the truck, nor can the total *weight* of the goods bought exceed the cargo *weight* limit of the truck. Each type of item bought can be marked up a certain percent. What items should the buyer buy to *maximize* the total value of the goods bought subject to the given constraints?

Such problems are characterized by the following features:

i) The specifications of the problem are related by *inequalities*, or *constraints*, rather than by equations. For example,

$$2x_1 + 3x_2 \leq 6 \quad \text{is an inequality,}$$

while

$$2x_1 + 3x_2 = 6 \quad \text{is an equation.}$$

ii) The inequalities or constraints of the problem are linear. For example, $2x_1 + 3x_2 \leq 6$ is a linear inequality; that is, all variables are to the first power and there are no divisions by a variable. Note that

$$2x_1^2 + 9x_2^2 + \frac{1}{x_1} \leq 36 \quad \text{is } not \text{ a linear inequality.}$$

iii) The *solution* of a linear program must satisfy these constraints. In addition some quantity, a *linear* function of the variables, must be maximized or minimized. For example, maximize f where $f = 3x_1 + 2x_2$, and, for short, we have written f in place of $f(x_1, x_2)$.

In the following sections we shall consider various aspects of the formulation and solution of linear programs.

4.1 GRAPHING A SYSTEM OF LINEAR CONSTRAINTS (INEQUALITIES) IN TWO VARIABLES

The first stage in the solution of a linear program is graphing the constraints (inequalities). Consider an inequality like any of the following:

$$a_1 x_1 + a_2 x_2 \geqslant b,$$

or

$$a_1 x_1 + a_2 x_2 \leqslant b,$$

or

$$a_1 x_1 + a_2 x_2 > b,$$

or

$$a_1 x_1 + a_2 x_2 < b,$$

where a_1, a_2, and b are constants.

Corresponding to any of these inequalities there is a *related equation*

$$a_1 x_1 + a_2 x_2 = b.$$

Its graph is a line that divides the plane into two half-planes. We graph this equation first.

For inequalities with

a) $\geqslant$ or $\leqslant$, use a solid line for the related equation;
b) $>$ or $<$, use a dashed line for the related equation.

Example 1   Graph $2x_1 + 3x_2 \leqslant 6$.

Solution   There are two steps. First, graph the related equation $2x_1 + 3x_2 = 6$.

Since the equality is included ($\leqslant$) in the present example, the line is solid in Fig. 4.1. See page 126.

Second, having divided the plane into two half-planes, we must decide which half-plane contains the solutions of the inequality. To do this, we need consider only one test point. If *any* point on one side of the graph of the related equation satisfies the inequality, then *all* points in the same half-plane satisfy the inequality. If any point on one side of the graph of the related equation does *not* satisfy the inequality, then all points in the *opposite* half-plane satisfy the inequality.

OBJECTIVES

You should be able, given a system of constraints, to:

a) Graph the solution set of the system.
b) Decide if the solution set is empty? nonempty?
c) Decide if the solution set is bounded—unbounded.
d) Decide which constraints if any, are redundant.
e) Decide which corners, if any, are degenerate.

1. Graph $x_1 + 5x_2 \leqslant 10$

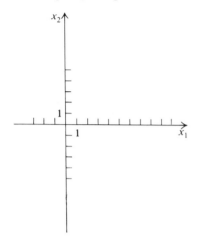

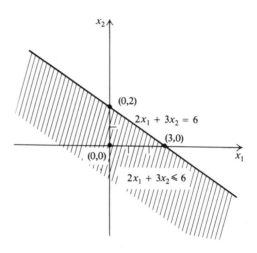

Figure 4.1

If the graph of the related equation does not include the origin $(0, 0)$, we use the origin as a test point. In the present example, we thus ask "Is $(0, 0)$ a solution of $2x_1 + 3x_2 \leqslant 6$?"

We replace x_1 by 0 and x_2 by 0:

$$
\begin{array}{c|c}
\multicolumn{2}{c}{2x_1 + 3x_2 \leqslant 6} \\
\hline
2 \cdot 0 + 3 \cdot 0 & 6 \\
0 + \quad 0 & \\
0 & \\
\end{array}
$$

Because $0 \leqslant 6$ is true, $(0, 0)$ is a solution.

Since $(0, 0)$ is in the lower half-plane, *all* points in the lower half-plane are solutions. Thus, the shaded half-plane of Fig. 4.1 represents the solutions of the inequality.

DO EXERCISE 1.

Example 2 Graph $2x_1 - x_2 > 0$.

The related equation $2x_1 - x_2 = 0$ is graphed, using a dashed line since the line is not included in the inequality. See Fig. 4.2. Here the line passes through the origin $(0, 0)$, so we use either $(1, 0)$ or $(0, 1)$ as a test point. For $(1, 0)$ we ask "Is $(1, 0)$ a solution of $2x_1 - x_2 > 0$?"

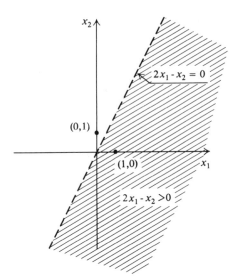

Figure 4.2

We replace x_1 by 1 and x_2 by 0.

$$2x_1 - x_2 > 0$$

$$\begin{array}{c|c} 2 \cdot 1 - 0 & 0 \\ 2 - 0 & \\ 2 & \end{array}$$

Because $2 > 0$ is true, $(1, 0)$ is a solution.

Since $(1, 0)$ is in the lower half-plane, all points in the lower half-plane are solutions of the inequality.

Alternatively, we could use $(0, 1)$ as a test point and ask "Is $(0, 1)$ a solution of $2x_1 - x_2 > 0$?"

We replace x_1 by 0 and x_2 by 1.

$$2x_1 - x_2 > 0$$

$$\begin{array}{c|c} 2 \cdot 0 - 1 & 0 \\ 0 - 1 & \\ -1 & \end{array}$$

Since $-1 > 0$ is false, $(0, 1)$ is not a solution.

Since $(0, 1)$ in the *upper* half-plane is *not* a solution, all points in the *lower* half-plane are solutions of the inequality.

2. Graph $x_1 - 5x_2 > 0$.

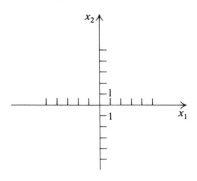

3. Graph $x_2 \geqslant 1$.

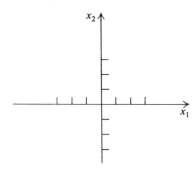

Either case tells us the solution set is the half-plane shaded in Fig. 4.2.

It should be noted that the inequality $2x_1 - x_2 > 0$ is equivalent to $x_2 - 2x_1 < 0$, if we recall that, when both sides of an inequality are multiplied by -1, the inequality sign is *reversed*.

DO EXERCISES 2 AND 3.

To graph a set, or *system*, of linear constraints, we first graph the solution set of each individual constraint using the same set of axes. The solution set of the *system* is that region, or set of ordered pairs, which satisfies *all* the constraints.

Example 3   Graph the solution set of the system of constraints

i) $x_1 - 2x_2 \leqslant 0$.
ii) $-2x_1 + \ x_2 \leqslant 2$.

These two constraints are each graphed in Fig. 4.3. A *pair of arrows* points in the direction of the half-plane representing solutions of each inequality. The region satisfying *both* constraints is indicated by the shading.

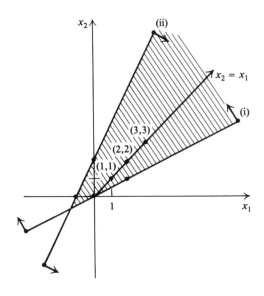

Figure 4.3

DO EXERCISE 4.

Example 4   Now let us add a third constraint to those of Example 3, so that we have

i) $x_1 - 2x_2 \leqslant 0,$
ii) $-2x_1 + x_2 \leqslant 2,$
iii) $x_1 + x_2 \leqslant 6.$

These are graphed in Fig. 4.4. The solution set of this system of *three* constraints is shaded. The solution set of Example 4 (Fig. 4.4) is *bounded.* This means simply that the solution set is confined to the boundary and interior of some polygon.

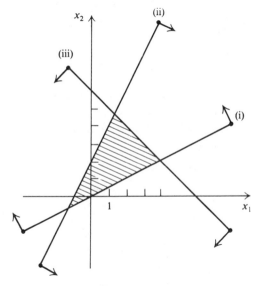

Figure 4.4

Not all solution sets are bounded. For example, the solution set of Example 3 (Fig. 4.3) is *unbounded.* This means that the solution set is *not* confined to the boundary and interior of some polygon. Note that in the direction of the arrow along the line $x_2 = x_1$, the boundary is *open*, so that in such a direction, given any solution one can find another solution farther out. In Example 3 any point (c, c) where $c \geqslant 0$ is a solution to the system of constraints (i) and (ii). The parameter c can become arbitrarily large and the point (c, c) will still be in the solution set.

DO EXERCISE 5.

4. Graph the solution set of the system of constraints

i) $x_1 + x_2 \geqslant 1$
ii) $-x_1 + x_2 \leqslant 2$

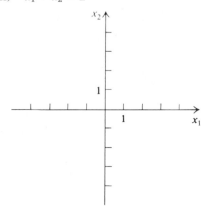

5. To the system of constraints of Margin Exercise 4, add

iii) $x_1 \leqslant 4$

and graph.

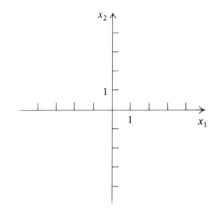

6. Add these constraints to those of Margin Exercise 5:

iv) $\qquad x_2 \geqslant 0$

v) $\qquad x_2 \leqslant 4$

and graph.

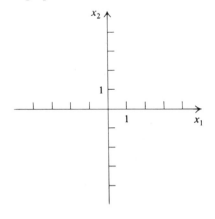

Example 5 Adding two more constraints, we have

i) $\qquad x_1 - 2x_2 \leqslant 0,$
ii) $-2x_1 + \quad x_2 \leqslant 2,$
iii) $\qquad x_1 + \quad x_2 \leqslant 6,$
iv) $\qquad x_1 \qquad \leqslant 2,$
v) $\qquad x_2 \leqslant 2.$

These are graphed in Fig. 4.5.

Figure 4.5

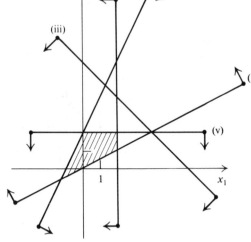

The solution set has now been sufficiently reduced so that constraint (iii) no longer affects the solution set and hence is considered *redundant*. Redundant constraints, when identified as such, may be disregarded for present purposes.

DO EXERCISE 6.

Example 6 Now let us replace constraint (iii) by (vi), so that we have:

i) $\qquad x_1 - 2x_2 \leqslant 0,$
ii) $-2x_1 + \quad x_2 \leqslant 2,$
iv) $\qquad x_1 \qquad \leqslant 2,$
v) $\qquad x_2 \leqslant 2,$
vi) $\qquad x_1 + \quad x_2 \leqslant 4.$

Thus, we obtain the graph in Fig. 4.6.

Figure 4.6

The *corners* (or vertices) of the solution set are each the intersection of two lines. Thus, the coordinates of the corners can be determined by the solution of the corresponding *two* related equations. For example, point a with coordinates $(-\frac{4}{3}, -\frac{2}{3})$ is the simultaneous solution of the equations related to the *two* constraints (i) and (ii). Similarly, point b with coordinates $(2, 1)$ corresponds to the intersection of the equations related to the *two* constraints (i) and (iv). Point c, on the other hand, with coordinates $(2, 2)$ is the simultaneous solution of the equations related to the *three* constraints (iv), (v), and (vi). Such points are called *degenerate* in linear programs. These degeneracies are important in later use but not here. For present purposes we consider constraint (vi) redundant. It may not be obvious from the graph which points are degenerate or which constraints are redundant. In this case, when we solved the pairs of related equations, the same solution would have occurred more than once, indicating a degeneracy.

DO EXERCISE 7.

Example 7   If constraint (vi) is replaced by (vii), so that we have:

i) $x_1 - 2x_2 \leqslant 0,$
ii) $-2x_1 + x_2 \leqslant 2,$
iv) $x_1 \leqslant 2,$
v) $x_2 \leqslant 2,$
vii) $x_1 + x_2 \geqslant 5,$

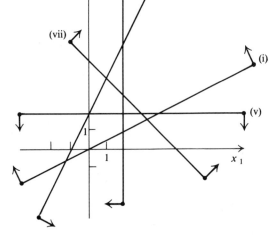

Figure 4.7

then we obtain the graph in Fig. 4.7. In this case there is *no* region common to all constraints (i), (ii), (iv), (v), and (vii). The solution set is then *empty*. This is not the same as saying that the solution set is $(0, 0)$, because if the point $(0, 0)$ were in the solution set, it would *not* be empty. Furthermore, the empty set is bounded.

DO EXERCISE 8.

7. Add the constraint

vi) $x_1 \leqslant 2$

to the preceding and graph.

a) Determine which points (if any) are degenerate.
b) Determine which constraints (if any) are redundant.

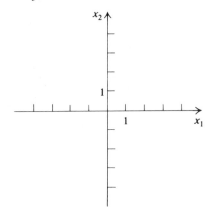

8. Replace constraint (vi) by

vii) $x_1 + 2 \leqslant 0$

and graph.

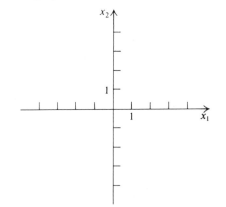

9. Which of the following areas are convex?

a)

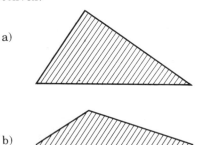

b)

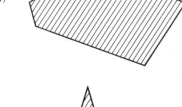

c)

d)

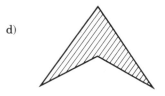

e)

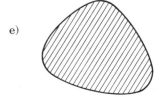

f)

Consider two points in the solution set shown shaded in Fig. 4.8. Any point on the line segment between these two points is also in the solution set.*

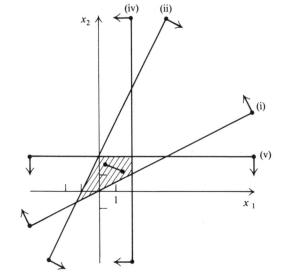

Figure 4.8

If the line segment between any pair of points in a set is also in the set, then such a set is called a *convex* set.

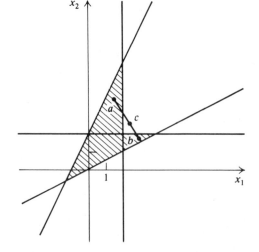

Figure 4.9

---

* Any point on the line segment *between* two points is called a *convex combination* of the two points; this concept is discussed further in Section 5.4.

An example of a set which is not convex is the shaded region of Fig. 4.9. In this case there are two points, such as a and b in the figure, such that the line segment *between* them includes points *outside* the given set, for example, point c. Thus, this set is *not* convex. It can be shown that

For any system of linear constraints, the corresponding solution set must be convex.

Knowing this is useful in determining the solution set of a graphed system of constraints. For example, the shaded region of Fig. 4.9 is not convex. Therefore, it could not possibly be a solution set for *any* system of linear constraints.

DO EXERCISE 9.

EXERCISE SET 4.1

Graph.

1. $5x_1 + x_2 > 10$ **2.** $3x_1 - 2x_2 \leqslant 12$ **3.** $3x_1 + 2x_2 \geqslant 6$

4. $x_1 - 4x_2 \leqslant 0$ **5.** $2x_1 + 5x_2 \leqslant 8$ **6.** $3x_1 + 7x_2 > 10$

Graph the following systems of constraints and shade the solution set.

7. $\begin{aligned} x_1 + x_2 &\leqslant 6, \\ x_2 &\leqslant 5, \\ x_1, x_2 &\geqslant 0. \end{aligned}$ **8.** $\begin{aligned} x_1 + 2x_2 &\leqslant 8, \\ x_1 &\leqslant 6, \\ x_1, x_2 &\geqslant 0. \end{aligned}$ **9.** $\begin{aligned} 3x_1 + 2x_2 &\leqslant 12, \\ x_1 + x_2 &\leqslant 5, \\ x_1, x_2 &\geqslant 0. \end{aligned}$ **10.** $\begin{aligned} 3x_1 + 5x_2 &\leqslant 15, \\ 3x_1 + 2x_2 &\leqslant 12, \\ x_1, x_2 &\geqslant 0. \end{aligned}$

11. $\begin{aligned} x_1 + 2x_2 &\leqslant 14, \\ 4x_1 + 3x_2 &\leqslant 26, \\ 2x_1 + x_2 &\leqslant 12, \\ x_1, x_2 &\geqslant 0. \end{aligned}$ **12.** $\begin{aligned} x_1 + 3x_2 &\leqslant 18, \\ 3x_1 + 2x_2 &\leqslant 19, \\ 2x_1 + x_2 &\leqslant 12, \\ x_1, x_2 &\geqslant 0, \end{aligned}$ **13.** $\begin{aligned} 3y_1 + y_2 &\geqslant 9, \\ y_1 + y_2 &\geqslant 7, \\ y_1 + 2y_2 &\geqslant 8, \\ y_1, y_2 &\geqslant 0. \end{aligned}$ **14.** $\begin{aligned} 2y_1 + y_2 &\geqslant 8, \\ 4y_1 + 3y_2 &\geqslant 22, \\ 2y_1 + 5y_2 &\geqslant 18, \\ y_1, y_2 &\geqslant 0. \end{aligned}$

15. $\begin{aligned} 2y_1 + y_2 &\geqslant 9, \\ 4y_1 + 3y_2 &\geqslant 23, \\ y_1 + 3y_2 &\geqslant 8, \\ y_1, y_2 &\geqslant 0. \end{aligned}$ **16.** $\begin{aligned} 5y_1 + 3y_2 &\geqslant 30, \\ 2y_1 + 3y_2 &\geqslant 21, \\ 3y_1 + 6y_2 &\geqslant 36, \\ y_1, y_2 &\geqslant 0. \end{aligned}$ **17.** $\begin{aligned} 4y_1 + y_2 &\geqslant 9, \\ 3y_1 + 2y_2 &\geqslant 13, \\ 2y_1 + 5y_2 &\geqslant 16, \\ y_1, y_2 &\geqslant 0. \end{aligned}$ **18.** $\begin{aligned} 4y_1 + y_2 &\geqslant 7, \\ y_1 + y_2 &\geqslant 4, \\ 2y_1 + 5y_2 &\geqslant 14, \\ y_1, y_2 &\geqslant 0. \end{aligned}$

For the following exercises

a) Graph the solution set of the system of constraints.   b) Is the solution set empty? nonempty?
c) Is the solution set bounded? unbounded?   d) Which constraints, if any, are redundant?
e) Which corners, if any, are degenerate?

Note. It may be difficult to answer (d) and (e) unless the equations are drawn carefully. In the next section, an alternative (algebraic) method will be used to determine the solution set more accurately.

Save your results for use at the end of the next section.

19. $x_1 + 2x_2 \leqslant 6,$
$\qquad 0 \leqslant x_1 \leqslant 5,$
$\qquad x_2 \geqslant -2$

20. $x_1 - x_2 \geqslant -4,$
$\qquad x_1 - x_2 \leqslant 6,$
$\qquad -2 \leqslant x_2 \leqslant 2$

21. $\qquad x_1 \geqslant -3,$
$\qquad x_1 - 2x_2 \leqslant 4,$
$\qquad x_2 - 3x_1 \leqslant 9,$
$\qquad 3x_1 + x_2 \leqslant 10,$

22. $\qquad 3x_1 \geqslant x_2,$
$\qquad 3x_2 \geqslant x_1,$
$\qquad x_1 + x_2 \geqslant 5,$
$\qquad 2x_1 + 3x_2 \leqslant 24$

▶

23. $-3x_1 + 2x_2 \geqslant 6,$
$\qquad 2x_1 + x_2 \leqslant -2,$
$\qquad x_1 + x_2 \geqslant 4,$
$\qquad 2x_1 + 7x_2 \leqslant 21$

24. $x_1 + 4x_2 \geqslant -4,$
$\qquad 2x_1 + x_2 \leqslant 2,$
$\qquad x_2 \geqslant 0,$
$\qquad x_1 \leqslant 5$

25. $\qquad x_1 \geqslant 0,$
$\qquad x_2 \geqslant 0,$
$\qquad x_1 + x_2 \geqslant 2,$
$\qquad x_1 - x_2 \leqslant 2,$
$\qquad x_2 \leqslant 6$

26. $-3x_1 + 4x_2 \leqslant 12,$
$\qquad 3x_1 + 2x_2 \leqslant 24,$
$\qquad x_1 \geqslant 0,$
$\qquad x_2 \geqslant 0,$
$\qquad x_2 \geqslant 6$

27. $x_1 + x_2 \leqslant 0,$
$\qquad 2x_1 - 3x_2 \leqslant 15,$
$\qquad x_2 \leqslant 5,$
$\qquad x_1 \geqslant 0,$
$\qquad 2x_1 + x_2 \geqslant 3$

28. $3x_1 + 2x_2 \geqslant 6,$
$\qquad x_1 \geqslant 1,$
$\qquad 0 \leqslant x_2 \leqslant 6,$
$\qquad 2x_1 + 3x_2 \leqslant 24,$
$\qquad 3x_1 + x_2 \leqslant 15$

29. $\qquad x_1 \geqslant 0,$
$\qquad x_2 \geqslant 0,$
$\qquad 5x_2 - 3x_1 \leqslant 15,$
$\qquad x_1 \leqslant 4x_2$
$\qquad 2x_1 - 5x_2 \leqslant 10$

30. $\qquad x_1 \geqslant 0,$
$\qquad x_2 \geqslant 0,$
$\qquad -7x_1 + x_2 \leqslant 7,$
$\qquad 4x_2 - 5x_1 \geqslant 20,$
$\qquad x_1 + x_2 \leqslant 10,$
$\qquad x_2 \leqslant 3$

OBJECTIVES

You should be able to

a) Find the corners of a system of constraints.
b) Find the optimum values of an objective function and the points at which they are attained, subject to a system of constraints.

4.2 DETERMINING THE OPTIMUM VALUE

We now find optimum values of some linear function of the variables. By this we mean the largest or smallest values of the function. The function being optimized is called the *objective function*. Let us again consider the solution set corresponding to constraints (i), (ii), (iv), and (v) of Example 5 of Section 4.1.

$\qquad$ i) $\quad x_1 - 2x_2 \leqslant 0,$
$\quad$ ii) $-2x_1 + x_2 \leqslant 2,$
$\quad$ iv) $\quad x_1 \qquad \leqslant 2,$
$\quad$ v) $\qquad x_2 \leqslant 2$

They are graphed in Fig. 4.8 and in Fig. 4.10.

The *corners* have been labeled a, b, c, and d.

We need to find the coordinates of the corners. We determine them algebraically. Point a (Fig. 4.10) is the intersection of the equations related to constraints (i) and (ii):

$\quad$ i') $\quad x_1 - 2x_2 = 0,$
$\quad$ ii') $-2x_1 + x_2 = 2.$

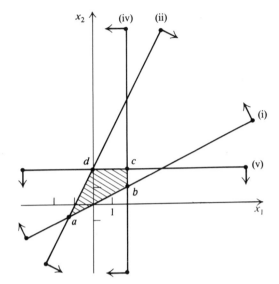

Figure 4.10

10. Use the echelon method and solve.

$$2x_1 - x_2 = 4,$$
$$3x_1 + 2x_2 = 12$$

We find the coordinates of point a, represented by (a_1, a_2), by solving the system of equations (i') and (ii'). We use the echelon method described in Section 3.1. Other methods of solution could be used, but we are practicing the echelon method for later use.

Example 1 Solve the system of equations (i') and (ii') and determine the coordinates of point a.

Solution The initial tableau is at the left.

x_1	x_2	1		x_1	x_2	1		x_1	x_2	1
1*	−2	0		1	−2	0		1	0	$-\frac{4}{3}$
−2	1	2		0	−3	2		0	1*	$-\frac{2}{3}$

Indicating the pivot by an * and pivoting (multiplying the first row by 2 and adding), we obtain the second tableau. Multiplying the second row by $-\frac{1}{3}$ to make the pivot element 1 and pivoting yields the final tableau. Thus, the coordinates of point a are $(a_1, a_2) = (-\frac{4}{3}, -\frac{2}{3})$.

DO EXERCISE 10.

By using the appropriate equations and solving algebraically (with the echelon method), we find the coordinates of the corners:

(i) and (ii) yield $(a_1, a_2) = (-\frac{4}{3}, -\frac{2}{3})$,
(i) and (iv) yield $(b_1, b_2) = (2, 1)$,
(iv) and (v) yield $(c_1, c_2) = (2, 2)$,
(ii) and (v) yield $(d_1, d_2) = (0, 2)$.

Example 2 Find the optimum (maximum and minimum) values and the points at which they are obtained, of the objective function* $f(x_1, x_2) = x_1 + x_2 = f$, for short, subject to the constraints:

i) $x_1 - 2x_2 \leqslant 0$,
ii) $-2x_1 + x_2 \leqslant 2$,
iv) $x_1 \leqslant 2$,
v) $x_2 \leqslant 2$.

Solution To do this, let f assume various values. Thus, $f = 6$ leads to the equation

$$6 = x_1 + x_2,$$

the graph of which is a straight line.

Similarly, $f = $ 4 leads to the equation $4 = x_1 + x_2$,
       $f = $ 2 leads to the equation $2 = x_1 + x_2$,
       $f = $ 0 leads to the equation $0 = x_1 + x_2$,
       $f = -2$ leads to the equation $-2 = x_1 + x_2$.

The graphs of these lines are shown in Fig. 4.11 together with the solution set (shaded area).

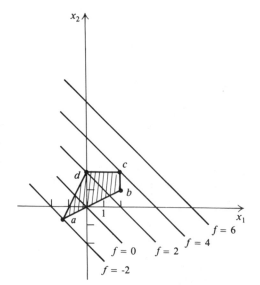

Figure 4.11

---

*A function of *two* variables is a relation f that assigns to each input pair (x_1, x_2) a unique output number $f(x_1, x_2)$. For example, for $f(x_1, x_2) = 4x_1 + 6x_2$, the function value $f(15, 20)$ is found by substituting 15 for x_1 and 20 for x_2:

$$f(15, 20) = 4(15) + 6(20) = 60 + 120 = 180.$$

From the figure we see that the line corresponding to $f = 6$ has *no* point in common with the solution set (shaded area). The lines corresponding to values of f between -2 and 4 ($-2 \leq f \leq 4$) *do* have points in common with the solution set. Of all such lines ($-2 \leq f \leq 4$), that corresponding to $f = 4$ has the maximum value of the objective function f. This maximum value is obtained at point c. Similarly, the minimum value of f is -2, obtained at point a.

It is important to note that the optima (maximum and minimum values) were obtained at the *boundary* of the solution set and, furthermore, at *corner points*. For linear programs, it can be shown that the optima will always be obtained at corner points.

To determine the optima we need evaluate the objective function *only at the corner points*. In the present example,

$$f(x_1, x_2) \quad = x_1 + x_2$$

$$
\begin{array}{llll}
f(-\frac{4}{3}, -\frac{2}{3}) = (-\frac{4}{3}) + (-\frac{2}{3}) = -2 & \text{(minimum)}, \\
f(2, 1) & = & 2 + 1 & = & 3, \\
f(2, 2) & = & 2 + 2 & = & 4 & \text{(maximum)}, \\
f(0, 2) & = & 0 + 2 & = & 2.
\end{array}
$$

Note. The notation $f(2, 1)$ represents the value of $f(x_1, x_2)$ when 2 is substituted for x_1 and 1 is substituted for x_2. Thus:

Maximum: $f(2, 2) = 4$,

Minimum: $f(-\frac{4}{3}, -\frac{2}{3}) = -2$.

Do Exercise 11.

Example 3   Find the optimum values of the objective function

$$f(x_1, x_2) = 2x_1 - x_2,$$

subject to the constraints:

i) $x_1 - 2x_2 \leq 0$,
ii) $-2x_1 + x_2 \leq 2$,
iv) $x_1 \quad\;\; \leq 2$,
v) $x_2 \leq 2$.

Solution   As before, let f assume various values, as indicated on Fig. 4.12. The maximum value of f occurs now at point b, while the minimum occurs all along the line segment from a to d. Evaluating f

11. Find the optima (maximum and minimum) values and the points at which they are obtained, for the objective function

$$f(x_1, x_2) = 2x_1 + 3x_2,$$

subject to the same constraints as in Example 2:

i) $x_1 - 2x_2 \leq 0$
ii) $-2x_1 + x_2 \leq 2$
iv) $x_1 \quad\;\; \leq 2$
v) $x_2 \leq 2$

Remember you need consider only corner points.

12. Find the optimum values and the points at which they are obtained, for the objectives function

$$f(x_1, x_2) = 2x_2 - x_1,$$

subject to the same constraints as in Margin Exercise 11.

at the corner points, we have:

$$f(x_1, x_2) = 2x_1 \quad - x_2$$

$f(-\frac{4}{3}, -\frac{2}{3}) = 2(-\frac{4}{3}) - (-\frac{2}{3}) = -2$	(minimum),
$f(2, 1) \quad = 2 \cdot 2 - 1 \quad = 3$	(maximum),
$f(2, 2) \quad = 2 \cdot 2 - 2 \quad = 2,$	
$f(0, 2) \quad = 2 \cdot 0 - 2 \quad = -2$	(minimum).

The minimum value -2 can be seen to be attained at *more than one point*, specifically at the points a and d. It turns out that -2 is also attained for any *convex combination* of the coordinates of these two points, that is, for any point on the line segment between them.

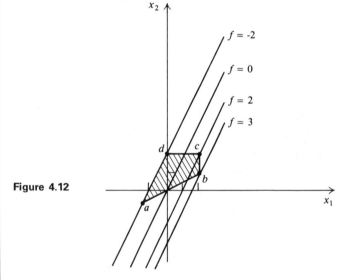

Figure 4.12

DO EXERCISE 12.

EXERCISE SET 4.2

Find the optimum value (maximum or minimum, as indicated) for each objective function and the point at which it is obtained for each set of constraints. Save your results for later use.

1. Maximize $f(x_1, x_2) = x_1 + 2x_2$, subject to the constraints (Exercise 7, Set 4.1):

$$x_1 + x_2 \leq 6,$$
$$x_2 \leq 5,$$
$$x_1, x_2 \geq 0.$$

2. Maximize $f(x_1, x_2) = x_1 + x_2$, subject to the constraints (Exercise 8, Set 4.1):

$$x_1 + 2x_2 \leq 8,$$
$$x_1 \leq 6,$$
$$x_1, x_2 \geq 0.$$

3. Maximize $f(x_1, x_2) = 5x_1 + 4x_2$, subject to the constraints (Exercise 9, Set 4.1):

$$3x_1 + 2x_2 \le 12,$$
$$x_1 + x_2 \le 5,$$
$$x_1, x_2 \ge 0.$$

5. Maximize $f(x_1, x_2) = 3x_1 + 4x_2$, subject to the constraints (Exercise 11, Set 4.1):

$$x_1 + 2x_2 \le 14,$$
$$4x_1 + 3x_2 \le 26,$$
$$2x_1 + x_2 \le 12,$$
$$x_1, x_2 \ge 0.$$

7. Minimize $f(y_1, y_2) = 3y_1 + 4y_2$, subject to the constraints (Exercise 13, Set 4.1):

$$3y_1 + y_2 \ge 9,$$
$$y_1 + y_2 \ge 7,$$
$$y_1 + 2y_2 \ge 8,$$
$$y_1, y_2 \ge 0.$$

9. Minimize $f(y_1, y_2) = 2y_1 + 5y_2$, subject to the constraints (Exercise 15, Set 4.1):

$$2y_1 + y_2 \ge 9,$$
$$4y_1 + 3y_2 \ge 23,$$
$$y_1 + 3y_2 \ge 8,$$
$$y_1, y_2 \ge 0.$$

11. Minimize $f(y_1, y_2) = 3y_1 + 5y_2$, subject to the constraints (Exercise 17, Set 4.1):

$$4y_1 + y_2 \ge 9,$$
$$3y_1 + 2y_2 \ge 13,$$
$$2y_1 + 5y_2 \ge 16,$$
$$y_1, y_2 \ge 0.$$

4. Maximize $f(x_1, x_2) = 2x_1 + x_2$, subject to the constraints (Exercise 10, Set 4.1):

$$3x_1 + 5x_2 \le 15,$$
$$3x_1 + 2x_2 \le 12,$$
$$x_1, x_2 \ge 0.$$

6. Maximize $f(x_1, x_2) = 5x_1 + 4x_2$, subject to the constraints (Exercise 12, Set 4.1):

$$x_1 + 3x_2 \le 18,$$
$$3x_1 + 2x_2 \le 19,$$
$$2x_1 + x_2 \le 12,$$
$$x_1, x_2 \ge 0.$$

8. Minimize $f(y_1, y_2) = 3y_1 + 4y_2$, subject to the constraints (Exercise 14, Set 4.1):

$$2y_1 + y_2 \ge 8,$$
$$4y_1 + 3y_2 \ge 22,$$
$$2y_1 + 5y_2 \ge 18,$$
$$y_1, y_2 \ge 0.$$

10. Minimize $f(y_1, y_2) = 9y_1 + 7y_2$, subject to the constraints (Exercise 16, Set 4.1):

$$5y_1 + 3y_2 \ge 30,$$
$$2y_1 + 3y_2 \ge 21,$$
$$3y_1 + 6y_2 \ge 36,$$
$$y_1, y_2 \ge 0.$$

12. Minimize $f(y_1, y_2) = 2y_1 + 3y_2$, subject to the constraints (Exercise 18, Set 4.1):

$$4y_1 + y_2 \ge 7,$$
$$y_1 + y_2 \ge 4,$$
$$2y_1 + 5y_2 \ge 14,$$
$$y_1, y_2 \ge 0.$$

Find the optimum (maximum and minimum) values for each objective function and the points at which they are obtained subject to the constraints given.

13. $f(x_1, x_2) = x_1 - x_2$, subject to the constraints (Exercise 19, Set 4.1):

$$x_1 + 2x_2 \le 6,$$
$$0 \le x_1 \le 5,$$
$$x_2 \ge -2.$$

14. $f(x_1, x_2) = 2x_1 + 3x_2$, subject to the constraints of Exercise 13.

15. $f(x_1, x_2) = x_1 - x_2$, subject to the constraints (Exercise 21, Set 4.1):

$$x_1 \geq -3,$$
$$x_1 - 2x_2 \leq 4,$$
$$x_2 - 3x_1 \leq 9,$$
$$3x_1 + x_2 \leq 10.$$

16. $f(x_1, x_2) = x_2 - 2x_1$, subject to the constraints of Exercise 15.

17. $f(x_1, x_2) = 3x_1 - x_2$, subject to the constraints (Exercise 22, Set 4.1):

$$3x_1 \geq x_2,$$
$$3x_2 \geq x_1,$$
$$x_1 + x_2 \geq 5,$$
$$2x_1 + 3x_2 \leq 24.$$

18. $f(x_1, x_2) = 3x_2 - x_1$, subject to the constraints of Exercise 17.

19. $f(x_1, x_2) = x_1 + x_2$, subject to the constraints (Exercise 23, Set 4.1):

$$-3x_1 + 2x_2 \geq 6,$$
$$2x_1 + x_2 \leq -2,$$
$$x_1 + x_2 \geq 4,$$
$$2x_1 + 7x_2 \leq 21.$$

20. $f(x_1, x_2) = x_1 + x_2$, subject to the constraints (Exercise 24, Set 4.1):

$$x_1 + 4x_2 \geq -4,$$
$$2x_1 + x_2 \leq 2,$$
$$x_2 \geq 0,$$
$$x_1 \leq 5.$$

21. $f(x_1, x_2) = 2x_1 + 3x_2$, subject to the constraints (Exercise 29, Set 4.1):

$$x_1 \geq 0,$$
$$x_2 \geq 0,$$
$$5x_2 - 3x_1 \leq 15,$$
$$x_1 \leq 4x_2,$$
$$2x_1 - 5x_2 \leq 10.$$

22. $f(x_1, x_2) = 3x_1 + 4x_2$, subject to the constraints (Exercise 28, Set 4.1):

$$3x_1 + 2x_2 \geq 6,$$
$$x_1 \geq 1,$$
$$0 \leq x_2 \leq 6,$$
$$2x_1 + 3x_2 \leq 24,$$
$$3x_1 + x_2 \leq 15.$$

23. $f(x_1, x_2) = 4x_1 + 3x_2$, subject to the constraints (Exercise 27, Set 4.1):

$$x_1 + x_2 \leq 0,$$
$$2x_1 - 3x_2 \leq 15,$$
$$x_2 \leq 5,$$
$$x_1 \geq 0,$$
$$2x_1 + x_2 \geq 3.$$

24. $f(x_1, x_2) = 2x_1 + 5x_2$, subject to the constraints (Exercise 30, Set 4.1):

$$x_1 \geq 0,$$
$$x_2 \geq 0,$$
$$-7x_1 + x_2 \leq 7,$$
$$4x_2 - 5x_1 \geq 20,$$
$$x_1 + x_2 \leq 10,$$
$$x_2 \leq 3.$$

4.3 FORMULATING MAXIMUM-TYPE LINEAR PROGRAMS AND SOLVING GRAPHICALLY

Example 1 Formulate (model) this problem.

A California vintner has available 660 lbs of Cabernet Sauvignon (CS) grapes, 1860 lbs of Pinot Noir (PN) grapes, and 2100 lbs of Barbera (B) grapes. The vintner makes a Pinot Noir (PN) wine which contains 20% CS, 60% PN, and 20% B grapes and sells for $3 a bottle and a Barbera (B) wine which contains 10% CS, 20% PN, and 70% B grapes and sells for $2 a bottle. Assuming that each bottle of wine requires 3 lbs of grapes, determine how many bottles of each type of wine the vintner should produce to maximize his income.

Solution This problem can be formulated using a table. The following steps show how to do this.

1. First, define the variables. Let

x_1 = the number of bottles of Pinot Noir wine to be produced,

x_2 = the number of bottles of Barbera wine to be produced, and

f = the income ($) obtained from the sale of all the wine.

Note that the variables x_1 and x_2 are defined in terms of *bottle* units while the grape supply is given in terms of *lb* units. Since one bottle of wine requires three lbs of grapes, the available supply of grapes in lb units must be divided by three to yield the available supply of grapes in bottle units, so that consistent units are used throughout the problem. Defining the variables carefully helps prevent formulation errors.

2. Set up a table with the following general headings, which actually vary depending on the specific problem. Across the top list the

	Composition			Number of units of supply available
	Product 1	Product 2	$\cdots$	
Number of units	$\bar{x}_1$	x_2	$\cdots$	
Ingredient 1 Ingredient 2 . . .				
Unit value				Objective

OBJECTIVES

You should be able to:

a) Formulate (model) a given problem as a maximum-type linear program.
b) Express the results of (a) in matrix form.
c) Solve maximum-type linear programs graphically.

products being manufactured and at the right write "Number of units of supply available." In the next row indicate the independent variables.

Look ahead to see how this is done for this problem.

3. In the first column list the ingredients used in making each product and at the bottom write "Unit value."

Look at the table to see how this is done for this problem.

4. Enter into the table columnwise the data describing the composition of each product and in the bottom row its unit value.

Note that the percents must be converted to fractions or decimals. The value, here price, could be expressed in either dollars or cents, but be consistent. See how this is done for this problem.

5. Enter into the column headed "supply available" the appropriate data for each ingredient. Indicate at the bottom of this column the "objective" of the problem.

Entering the data for this problem, we obtain the following table:

	Composition (bottles)		Number of bottles available
	PN Wine	B Wine	
Number of bottles	x_1	x_2	
CS Grapes	0.20	0.10	220
PN Grapes	0.60	0.20	620
B Grapes	0.20	0.70	700
$ Price per bottle	3	2	Maximize income

6. For each ingredient the corresponding constraint can be read rowwise from the table. Note the similarity to the echelon tableau.

Noting that the supply available cannot be exceeded, we obtain

$$\text{CS:}\quad 0.20x_1 + 0.10x_2 \leqslant 220,$$
$$\text{PN:}\quad 0.60x_1 + 0.20x_2 \leqslant 620,$$
$$\text{B:}\quad 0.20x_1 + 0.70x_2 \leqslant 700.$$

7. The next two constraints are not stated *explicitly* in the problem. Rather they are *implied* by the reality of the situation. Since one cannot produce a *negative* amount of wine, we constrain the amount of wine produced to be nonnegative; that is,

$$x_1 \geqslant 0 \quad \text{and} \quad x_2 \geqslant 0.$$

The constraint that a physical quantity be realistic and hence non-negative is called the *nonnegativity constraint*.

8. Read the objective function and problem objective from the bottom row.

Here the objective is to maximize income, so that we obtain

$$\text{Maximize } f, \quad \text{where } f = 3x_1 + 2x_2.$$

The formulation can be summarized as follows:

i) $0.20x_1 + 0.10x_2 \leqslant 220,$
ii) $0.60x_1 + 0.20x_2 \leqslant 620,$
iii) $0.20x_1 + 0.70x_2 \leqslant 700,$
iv, v) $x_1, x_2 \geqslant 0 ,$
vi) $\max f: f = 3 \; x_1 + \; 2 \; x_2.$

In summary, a *linear program* can be formulated, or modeled, as a system of linear inequalities, called *constraints* (both explicit and implicit), together with some linear function to be optimized—in this case maximized. That is, here we want to find the largest value of the objective function, subject to the given constraints, and the numerical values that yield it.

DO EXERCISE 13.

We can use matrices for clarification. The maximum-type linear program can be expressed in the form:

$$AX \leqslant B,$$

$$X \geqslant O,$$

$$\max f: f = CX,$$

where in Example 1,

$$A = \begin{bmatrix} 0.20 & 0.10 \\ 0.60 & 0.20 \\ 0.20 & 0.70 \end{bmatrix}, \quad X = \begin{bmatrix} x_1 \\ x_2 \end{bmatrix}, \quad B = \begin{bmatrix} 220 \\ 620 \\ 700 \end{bmatrix}, \quad C = [3 \;\; 2].$$

The matrix A is made up of the coefficients of the constraint system. The notation $X \geqslant O$ implies that *each* component of X is nonnegative. Similarly, the notation $AX \leqslant B$ implies that each component of AX is less than or equal to the corresponding component of B.

Thus, a maximum-type linear program is composed of three parts. First, there are the linear constraints, $AX \leqslant B$, particular to each problem, and second, the nonnegativity constraints, $X \geqslant O$, which are common to most problems. Third, there is a linear objective function $f = CX$, which is to be maximized.

13. Formulate (model) the problem below using a table. Save your results.

A furniture manufacturer produces chairs and sofas. The chairs require 20 feet of wood, 1 lb of foam rubber, and 2 square yards of material. The sofas require 100 feet of wood, 50 lbs of foam rubber, and 20 square yards of material. The manufacturer has in stock 1900 feet of wood, 500 lbs of foam rubber, and 240 square yards of material. If the chairs can be sold for $20 each and the sofas for $300 each, how many of each should be produced to maximize the income?

14. Go back to Exercise 13 and express the results in matrix form.

DO EXERCISE 14.

The linear program just formulated can be solved graphically using the techniques of the preceding sections.

A general procedure for solving linear programs in two variables is:

1) **Graph the constraints;**
2) **Determine the solution set (shade this region);**
3) **Label the corner points and determine their coordinates algebraically (with the echelon method);**
4) **Evaluate the objective function at all corner points; and**
5) **Determine by inspection the optimum value (maximum or minimum) and where it is attained.**

Example 2 (Maximum-type) Solve graphically the linear program of Example 1.

Solution The formulation of this linear program was found to be:

$$
\begin{aligned}
\text{i(CS):} && 0.20x_1 + 0.10x_2 &\le 220, \\
\text{ii(PN):} && 0.60x_1 + 0.20x_2 &\le 620, \\
\text{iii(B):} && 0.20x_1 + 0.70x_2 &\le 700, \\
\text{vi, iv, v: } \max f{:} f = && 3x_1 + \quad 2x_2; && x_1, x_2 &\ge 0.
\end{aligned}
$$

The constraints and solution set are shown in Fig. 4.13. Note that the nonnegativity constraints (iv, v) restrict consideration to the first quadrant.

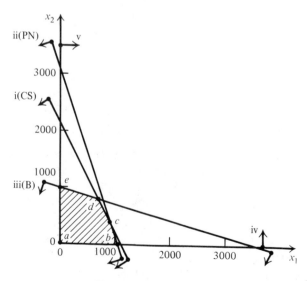

Figure 4.13

The coordinates of the corner points follow:

$$\text{iv and v yield } (a_1, a_2) = (0, 0),$$
$$\text{ii and v yield } (b_1, b_2) = \left(\frac{3100}{3}, 0\right),$$
$$\text{i and ii yield } (c_1, c_2) = (900, 400),$$
$$\text{i and iii yield } (d_1, d_2) = (700, 800),$$
$$\text{iii and iv yield } (e_1, e_2) = (0, 1000).$$

Evaluating the objective function at these points, we obtain

$$
\begin{aligned}
f(x_1, x_2) &= 3x_1 & &+ 2x_2, \\
f(0, 0) &= 3 \cdot 0 & &+ 2 \cdot 0 &&= 0, \\
f\left(\frac{3100}{3}, 0\right) &= 3 \cdot \left(\frac{3100}{3}\right) & &+ 2 \cdot 0 &&= 3100, \\
f(900, 400) &= 3 \cdot 900 & &+ 2 \cdot 400 &&= 3500, \\
f(700, 800) &= 3 \cdot 700 & &+ 2 \cdot 800 &&= 3700 && \text{(maximum)}, \\
f(0, 1000) &= 3 \cdot 0 & &+ 2 \cdot 1000 &&= 2000.
\end{aligned}
$$

Thus, the maximum *income* (not necessarily profit) that the vintner can obtain is \$3700 ($f = 3700$) by producing 700 bottles of PN wine ($x_1 = 700$) and 800 bottles of B wine ($x_2 = 800$). This maximum occurs at point d, so that all the CS and B grapes are used but there is an excess of PN grapes which can be put to some other use.

DO EXERCISE 15.

---

15. Solve graphically the linear program of Margin Exercise 13, whose formulation was

$$
\begin{aligned}
20x_1 + 100x_2 &\leqslant 1900, \\
x_1 + 50x_2 &\leqslant 500, \\
2x_1 + 20x_2 &\leqslant 240, \\
x_1, x_2 &\geqslant 0,
\end{aligned}
$$

$$\max f \colon f = 20x_1 + 300x_2.$$

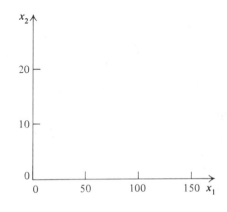

EXERCISE SET 4.3

In the following problems:

a) **Formulate the model;** b) **Express the results in matrix form;** c) **Solve graphically.**

1. *Business.* A clothier makes suits and dresses. Each suit requires 1 yd of polyester and 4 yds of wool while each dress requires 2 yds of polyester and 3 yds of wool. He has in stock 60 yds of polyester and 120 yds of wool. If a suit sells for \$120 and a dress sells for \$75, how many of each should he make to maximize his income?

2. *Business.* A manufacturer of hi-fi speakers makes two speaker assemblies. The inexpensive speaker assembly consists of one midrange speaker and one tweeter. It sells for \$25. The expensive speaker assembly consists of one woofer, one midrange speaker, and two tweeters. This one sells for \$150. The stock consists of 22 12″ woofers, 30 5″ midrange speakers, and 45 1½″ tweeters. How many of each type of speaker assembly should be made to maximize income?

3. *Business.* A nut dealer has 1800 lbs of peanuts, 1500 lbs of cashews, and 750 lbs of almonds, from which he wishes to make two mixtures. Mixture I sells for \$0.75 per lb and contains 60% peanuts, 30% cashews, and 10% almonds. Mixture II sells for \$2.00 per lb and contains 20% peanuts, 50% cashews, and 30% almonds. How many lbs of each should he mix to maximize his income?

4. *Business.* A tea merchant has 400 lbs cut black tea, 300 lbs of pekoe, and 240 lbs of orange pekoe tea, from which he blends two mixtures. Mixture A, selling for \$1.50 per lb contains 50% cut black tea, 30% pekoe, and 20% orange pekoe. Mixture B, selling for \$4.00 per lb, contains 50% pekoe and 50% orange pekoe. How many lbs of each mixture should he blend to maximize his income?

5. *Ecology.* A certain area of forest is populated by two species of animals (A1 and A2) and the forest supplies three kinds of food (F1, F2, and F3). Species A1 requires 1 unit of food F1, 2 units of food F2, and 2 units of food F3 while species A2 requires 1.2 units of food F1, 1.8 units of food F2, and 0.6 units of food F3. If the forest can normally supply a maximum of 600 units of food F1, 960 units of food F2, and 720 units of food F3, what is the maximum total numbers of these animals that the forest can support?

7. *Ecology.* In reference to Exercise 5, if there is a wet spring, so that the maximum available supply of food becomes 720 units of food F1, 960 units of food F2, and 600 units of food F3, what maximum number of animals can now be supported? What would happen to species A1?

6. *Ecology.* In reference to Exercise 5, if species A1 is valued at $150 and species A2 is valued at $120, how many animals of each species will maximize the value of the animal stock?

OBJECTIVE

You should be able to:

a) Formulate (model) a given problem as a minimum-type linear program.

b) Express results of (a) in matrix form.

c) Solve minimum-type linear programs graphically.

4.4 FORMULATING MINIMUM-TYPE LINEAR PROGRAMS AND SOLVING GRAPHICALLY

Example 1 Formulate (model) this problem.

Ecology, Nutrition, Business. A feed supplier mixes feed for a particular animal which requires at least 160 lbs of nutrient A, at least 24 lbs of nutrient B, and at least 28 lbs of nutrient C. The supplier has available soybean meal* which costs $15 per 100-lb sack and contains 40 lbs of nutrient A, 4 lbs of nutrient B, and 4 lbs of nutrient C and triticale† which costs $10 per 100-lb sack and contains 20 lbs of nutrient A, 4 lbs of nutrient B, and 6 lbs of nutrient C. How many sacks of each ingredient should be used to mix animal feed which satisfies the minimum requirements at minimum cost?

Solution This problem can be formulated as follows:

1. First, define the variables. Let

$$y_1 = \text{the number of 100-lb sacks of soybean meal,}$$

$$y_2 = \text{the number of 100-lb sacks of triticale, and}$$

$$f = \text{the \$ cost of the animal feed.}$$

2. As for maximum-type programs, set up a table with the following general headings. Across the top list the *sources* of materials and at

---

* Soybean meal is made from soybeans by extracting the oil, which is then used commercially for other purposes.
† *Tri±ti-cale:* a hybrid of wheat and rye.

the right write "Amount of component required." In the next row indicate the independent variables.

Look ahead to see how this is done for the present example.

	Composition			Amount of component required
	Source 1	Source 2	$\cdots$	
Number of units	y_1	y_2	$\cdots$	
Component 1 Component 2 . . .				
Unit cost				Objective

3. In the first column list the *components* of the sources of materials and at the bottom write "Unit cost."

Look at the table and see how this is done for this example.

4. Enter into the table columnwise the data describing the composition of each source material and its unit cost in the bottom row.

See how this is done for the present problem.

5. Enter into the column headed "Amount of component required" the appropriate data for each component. Indicate at the bottom of this column the "objective" of the problem.

Entering the data for this example, we obtain the following table:

	Composition (lbs per 100-lb sack)		Pounds required
	Soybean meal	Triticale	
Number of 100-lb sacks	y_1	y_2	
Nutrient A Nutrient B Nutrient C	40 4 4	20 4 6	160 24 28
$ Cost per 100-lb sack	15	10	Minimize cost

6. For each component the corresponding constraint can be read rowwise from the table by noting that the amount of the component must be at least satisfied:

$$A: 40y_1 + 20y_2 \geqslant 160,$$
$$B: 4y_1 + 4y_2 \geqslant 24,$$
$$C: 4y_1 + 6y_2 \geqslant 28.$$

7. As for the maximum-type programs, we have the implied non-negativity constraints:

$$y_1 \geqslant 0 \quad \text{and} \quad y_2 \geqslant 0.$$

In this example there is another implied constraint. Soybean meal and triticale are sold in 100-lb sacks. Hence, their number must be *integer*. Problems with this integer constraint (which is *not* linear) are called *integer programs*. (Their solutions can be found by considering the various integer solutions neighboring the linear program solution, but this involves much extra work.) We shall *ignore* this integer constraint and accept whatever numerical values are obtained from the solution to the *linear* program.

8. Read the objective function and problem objective from the bottom row.

Here the objective is to minimize the cost, so that we obtain

$$\text{Minimize } f, \quad \text{where } f = 15y_1 + 10y_2.$$

The formulation can be summarized as follows:

$$\begin{array}{lll} \text{i(A):} & 40y_1 + 20y_2 \geqslant 160, \\ \text{ii(B):} & 4y_1 + 4y_2 \geqslant 24, \\ \text{iii(C):} & 4y_1 + 6y_2 \geqslant 28, \\ \text{iv, v:} & y_1, y_2 \geqslant 0, \\ \text{vi:} & \min f\colon f = 15y_1 + 10y_2. \end{array}$$

In matrix notation, a minimum-type linear program can be written in the form

$$AY \geqslant B,$$
$$Y \geqslant O,$$
$$\min f\colon f = CY,$$

where, in the preceding example,

$$A = \begin{bmatrix} 40 & 20 \\ 4 & 4 \\ 4 & 6 \end{bmatrix}, \quad Y = \begin{bmatrix} y_1 \\ y_2 \end{bmatrix}, \quad B = \begin{bmatrix} 160 \\ 24 \\ 28 \end{bmatrix}, \quad C = [15 \quad 10].$$

Thus, a minimum-type linear program is also composed of three parts. First, there are the linear constraints, $AY \geqslant B$, particular to each problem, and second, the nonnegativity constraints, $Y \geqslant O$, which are common to most problems. Third, there is a linear objective function $f = CY$, which is to be minimized.

For convenience, we use the word "optimum" to refer to either a "maximum" or a "minimum," whichever is appropriate. Thus, in general, a linear program consists of a system of particular (linear) constraints, plus nonnegativity constraints and a linear objective function to be optimized.

DO EXERCISE 16.

Example 2 Solve graphically the linear program of Example 1.

Solution The formulation of this linear program is:

i(A): $\qquad 40y_1 + 20y_2 \geqslant 160,$
ii(B): $\qquad 4y_1 + 4y_2 \geqslant 24,$
iii(C): $\qquad 4y_1 + 6y_2 \geqslant 28,$
vi, iv, v: $\min f : f = 15y_1 + 10y_2; \quad y_1, y_2 \geqslant 0.$

The constraints and solution set are shown in Fig. 4.14. Again, the nonnegativity constraints (iv, v) restrict consideration to the first quadrant.

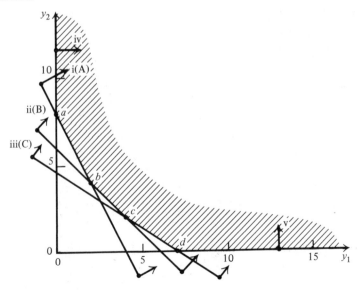

Figure 4.14

16. (a) Formulate (model) this problem using a table and (b) express the result in matrix form. Save your result.

An ore refining company has orders for 200 tons of iron, 500 tons of aluminum, and 100 tons of copper. They have available two kinds of ore. Type A contains 10% iron and 2% copper, and costs $10 per ton. Type B contains 20% aluminum and 1% copper, and costs $15 per ton. How many tons of each should be bought to minimize the cost?

17. Solve graphically the linear program of Margin Exercise 16, whose formulation was:

$$0.1y_1 + 0y_2 \geq 200,$$
$$0y_1 + 0.2y_2 \geq 500,$$
$$0.02y_1 + 0.01y_2 \geq 100,$$
$$y_1, y_2 \geq 0,$$

min f: $f = 10y_1 + 15y_2$.

The four corners a, b, c, and d of the solution set have coordinates as follows:

(iii) and (iv) yield $(a_1, a_2) = (0, 8)$,
(ii) and (iii) yield $(b_1, b_2) = (2, 4)$,
(i) and (ii) yield $(c_1, c_2) = (4, 2)$,
(i) and (v) yield $(d_1, d_2) = (7, 0)$.

Evaluating the objective function $f = 15y_1 + 10y_2$ at these points, we obtain:

$f(y_1, y_2) = 15y_1 + 10y_2$	
$f(0, 8) = 15(0) + 10(8) = 80$,	
$f(2, 4) = 15(2) + 10(4) = 70$	(minimum),
$f(4, 2) = 15(4) + 10(2) = 80$,	
$f(7, 0) = 15(7) + 10(0) = 105$.	

Thus, point b yields the minimum value of the objective function. The supplier should buy 2 sacks of soybean meal for each 4 sacks of triticale, to minimize his cost.

DO EXERCISE 17.

Example 3 (*Optional*) *Post-Optimality Analysis.* Now let us ask how high the cost of triticale can be increased before the supplier of Example 1 should change his feed mixture to keep his cost minimum?

Solution In Example 1 the objective function was

$$f_1 = 15y_1 + 10y_2.$$

If we now let

$$c_2 = \text{cost of triticale per 100-lb sack},$$

then our new objective function is

$$f_2 = 15y_1 + c_2y_2,$$

where c_2 remains to be determined.

Consider again the graph of the various constraints of Example 1, as shown in Fig. 4.15.

Here we have added the objective function f_1 which touches the feasible solution set (shaded area) at point b. As the cost c_2 of the triticale increases, the line representing the objective function rotates counter-clockwise about the outside of the feasible solution set until it touches point c. Then point c becomes the new optimum solution.

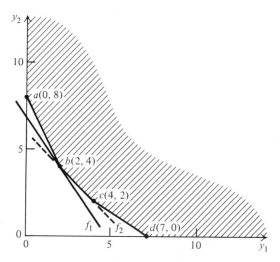

Figure 4.15

18. As in Example 3, how high would the cost of soybean meal per 100-lb sack have to be increased before the supplier should change his feed mixture to keep his cost minimum? And which point would then become optimum?

Putting the objective function

$$f_2 = 15y_1 + c_2y_2$$

through points $b(2, 4)$ and $c(4, 2)$, we obtain

$$c_2 = 15,$$

so that

$$f_2 = 15y_1 + 15y_2.$$

Thus, if the cost of triticale is increased beyond $15 per 100-lb sack, the optimum solution shifts from point b to point c. For any lesser increase in cost, point b remains optimum.

DO EXERCISE 18.

EXERCISE SET 4.4

a) Formulate (model) the following problems.   b) Express the results in matrix form.   c) Solve graphically.

1. *Nutrition, Business.* An animal feed to be mixed from soybean meal* and oats must contain at least 120 lbs of protein, 24 lbs of fat, and 10 lbs of mineral ash. Each 100-lb sack of soybean meal costs $15 and contains 50 lbs of protein, 8 lbs of fat, and 5 lbs of mineral ash. Each 100-lb sack of oats costs $5 and contains 15 lbs of protein, 5 lbs of fat, and 1 lb of mineral ash. How many sacks of each should be used to satisfy the minimum requirements at minimum cost?

2. *Nutrition, Business.* Suppose the oats in the preceding problem were replaced by alfalfa which costs $8 per 100 lbs and contains 20 lbs of protein, 6 lbs of fat, and 8 lbs of mineral ash. How much of each is now required to minimize the cost?

---

* See footnote on page 146.

3. *Nutrition, Business.* How is the formulation of Exercise 2 changed if the mineral requirement is doubled?

5. *Business, Transportation.* An airline with two types of airplanes, P1 and P2, has contracted with a tour group to provide accommodations for a minimum of each of 2000 first-class. 1500 tourist, and 2400 economy-class passengers. Airplane P1 costs $12 thousand per mile to operate and can accommodate 40 first-class, 40 tourist, and 120 economy-class passengers, while airplane P2 costs $10 thousand per mile to operate and can accommodate 80 first-class, 30 tourist, and 40 economy-class passengers. How many of each type of airplane should be used to minimize the operating cost?

7. *Business, Transportation.* If, instead of replacing P1 by P3, P2 is replaced by P3, how many of P1 and P3 should be used, to minimize the operating cost?

9. *Business, Transportation.* If the contract requirements are changed to a minimum of 1600 first-class, 2100 tourist, and 2400 economy-class passengers, how many of airplanes P1 and P2 are now required to minimize the operating cost?

11. *Business, Transportation.* As in Exercise 9 but with airplanes P1 and P3 ...?

4. *Nutrition, Business.* How is the formulation of Exercise 2 changed if the cost of alfalfa is increased to $10?

6. *Business, Transportation.* A new airplane P3 becomes available, having an operating cost of $15 thousand per mile and accommodating 40 first-class, 80 tourist, and 80 economy-class passengers. If airplane P1 of Exercise 5 were replaced by airplane P3, how many of P2 and P3 would be needed to minimize the operating cost?

8. *Business, Transportation.* If all three planes P1, P2, and P3 are used, how many of each should be used to minimize the cost? Formulate and put in matrix form only.

10. *Business, Transportation.* As in Exercise 9, but with airplanes P2 and P3 ...?

12. *Business, Transportation.* As in Exercise 9, but using all three-airplanes ...? Formulate and put in matrix form only.

▶——————————————————————

13. *Business.* In Exercise 1 of Set 4.3: If dresses are to be profitable, what must the minimum price of dresses be?

15. *Business.* In Exercise 3 of Set 4.3: How many *more* lbs of cashews can the nut dealer buy and still use the same mixtures profitably without having to buy more of the other nuts?

17. *Business, Nutrition.* In Exercise 1 of Set 4.4: What must the price of soybean meal be *raised* to before it becomes more economical to use a mixture with less soybean meal than oats?

19. *Business, Transportation.* In Exercise 5 of Set 4.4: What would the operating cost of plane P2 have to be *lowered* to in order for it to be more economical to use more P2's?

14. *Business.* In Exercise 2 of Set 4.3: To what value could the price of inexpensive speakers be *raised* before the production ratio of each should be changed?

16. *Business.* In Exercise 3 of Set 4.3. How many *fewer* lbs of peanuts can the nut dealer buy and still use the same mixtures profitably without having to buy less of the other nuts?

18. *Business, Nutrition.* In Exercise 1 of Set 4.4: What must the price of soybean meal be *lowered* to before it becomes more economical to use all soybean meal?

20. *Business, Transportation.* In Exercise 5 of Set 4.4: What would the operating cost of plane P2 have to be *raised* to in order for it to be more economical to use more P1's?

CHAPTER 4 TEST

1. Formulate the following problem, giving all the constraints, and write in matrix form.

A health food store manager is preparing two mixtures of breakfast cereal out of a supply of 100 lbs rolled oats, 10 lbs chopped almonds, 5 lbs chopped dried applies, 25 lbs chopped sunflower seeds, and 15 lbs monukka raisins. The first mixture contains 80% oats, 1% almonds, no dried apple, 12% sunflower seeds, and 7% raisins, and sells for $0.95 per lb. The second mixture contains 60% oats, 3% almonds, 4% dried apple, 24% sunflower seeds, and 9% raisins, and sells for $1.35 per lb. How much of each mixture should be made to maximize the income?

Given the following linear program:

i) $\qquad y_1 + 8y_2 \geqslant 24,$
ii) $\qquad 7y_1 + y_2 \geqslant 14,$
iii) $\qquad 2y_1 + 3y_2 \geqslant 18,$
$\qquad \min f: f = y_1 + y_2; y_1, y_2 \geqslant 0,$

3. Find the optimum feasible solution and explain how you found it.

2. Graph the constraints and shade the feasible solution.

*4. To what value could the requirement (right-hand side) of constraint (ii) be reduced before this constraint becomes redundant?

---

* Optional.

CHAPTER FIVE

Linear Programming—
The Simplex Algorithm

OBJECTIVE

You should be able to set up the initial simplex tableau for a maximum-type linear program, solve, and read off the solution.

The graphical method for solving linear programs, considered in Sections 4.3 and 4.4, can be used for problems with *two* variables and, with some difficulty, *three*. However, for problems where the number of variables might run into hundreds or thousands, algebraic techniques must be used. These have the advantage that they can be adapted to high-speed computers. The *simplex algorithm** is the basic technique that we will consider in this chapter.

5.1 MAXIMUM-TYPE LINEAR PROGRAMS—THE SIMPLEX ALGORITHM

Standard Form of a Linear Program

In this section we will consider the algebraic solution of maximum-type linear programs which are expressed in the following *standard form*.

a) $AX \leqslant B$ (This means that all constraints are in the form

$$a_1 x_1 + a_2 x_2 + \cdots + a_n x_n \leqslant b.)$$

b) $X \geqslant O$ (This means that all the variables $x_1, x_2, \ldots, x_n$ are nonnegative. The given variables are called *structural* variables.)

c) max x_0: $x_0 = CX$ (We are finding the maximum value of an objective function $x_0 = c_1 x_1 + c_2 x_2 + \cdots + c_n x_n$.)

(Here the objective function, previously denoted f, is denoted by the variable with the subscript "0", x_0)

We place one additional provision on the standard form, and this is the positivity constraint $B > O$, that is, *all* components of B are positive. This is assumed to avoid complications.

A set of values X which satisfies the constraints $AX \leqslant B$ together with the corresponding value of the objective function $x_0 = CX$ is called a *solution* and can be written X; x_0.

A solution which also satisfies the *nonnegativity constraint*, $X \geqslant O$, is called a *feasible solution*.

A feasible solution which is also a *corner point* of the region defined by $AX \leqslant B$ and $X \geqslant O$ is called a *basic feasible solution*.

A basic feasible solution which also *optimizes* the value of the objective function $x_0 = CX$ is called an *optimum basic feasible solution* or simply an *optimum feasible solution*.

---

* An *algorithm* is a special procedure. Here the simplex algorithm is used for solving linear programs.

Briefly, solution of a maximum-type linear program by the simplex algorithm involves the following steps:

1. Adding *slack variables* to convert the inequalities into equations,
2. Setting up the *initial simplex tableau* which is similar to the echelon tableau,
3. Finding an *initial feasible solution*, since the simplex algorithm proceeds from one feasible solution to a better one until the optimum is reached,
4. Introducing *basic and nonbasic variables* as the natural way to express *basic* feasible solutions,
5. *Choosing the proper pivot* element to advance the solution and maintain the nonnegativity of all variables,
6. *Pivoting*, which is done columnwise in exactly the same way as in the echelon method, and
7. Continuing and recognizing when the *algorithm terminates*.

We now consider these steps in detail.

Slack Variables

The first step of the simplex algorithm is to convert the inequalities of the problem into equations.

Example 1   Convert the formulation of Example 1 of Section 4.3 from a system of inequalities to a system of equations.

Solution   The problem in question had the formulation:

i) $$0.20x_1 + 0.10x_2 \leq 220,$$
ii) $$0.60x_1 + 0.20x_2 \leq 620,$$
iii) $$0.20x_1 + 0.70x_2 \leq 700,$$
iv) $$\max x_0: x_0 = 3.00x_1 + 2.00x_2,$$
v) $$x_1, x_2 \geq 0.$$

The first of these inequalities (i) can be made into an equation by adding a variable y_1 to the lefthand side. This produces the equation

$$0.20x_1 + 0.10x_2 + y_1 = 220.$$

The quantity y_1 is called a *slack variable* since it "takes up the slack" in the equation. It follows that y_1 is nonnegative, as are x_1 and x_2. To see this in an easy way, the inequality $4 + 1 \leq 7$ can be made into the equation $4 + 1 + 2 = 7$ where the slack is the quantity 2.

1. *Adding slack variables.* Convert the following formulation (Margin Exercise 13, Section 4.3) into a system of equations, and put in matrix form.

$$20x_1 + 100x_2 \leq 1900,$$

$$x_1 + 50x_2 \leq 500,$$

$$2x_1 + 20x_2 \leq 240,$$

$$x_1, x_2 \geq 0,$$

$$\max x_0: x_0 = 20x_1 + 300x_2.$$

For *each* constraint (i) through (iii), we add a *different* slack variable to the lefthand side, obtaining:

i') $\quad 0.20x_1 + 0.10x_2 + y_1 \qquad\qquad\qquad = 220,$

ii') $\quad 0.60x_1 + 0.20x_2 \qquad + y_2 \qquad\qquad = 620,$

iii') $\quad 0.20x_1 + 0.70x_2 \qquad\qquad + y_3 \qquad = 700,$

iv') $\quad \max x_0: -3x_1 - 2x_2 \qquad\qquad\qquad + x_0 = 0,$

v') $\qquad\qquad x_1, x_2; \quad y_1, \quad y_2, \quad y_3 \qquad\qquad \geq 0.$

Note that the objective function

iv) $$x_0 = 3x_1 + 2x_2$$

has been rewritten

iv') $$-3x_1 - 2x_2 + x_0 = 0.$$

We add a different slack variable each time, because what must be added to $0.20x_1 + 0.10x_2$ to get 220 may be different from what must be added to $0.60x_1 + 0.20x_2$ to get 620. Each *slack* variable must be nonnegative, as must the structural variables x_1 and x_2.

Note that the constraints (i) through (iii) can be recovered from the equations (i') through (iii') by setting each of the slack variables y_i, in turn, equal to 0 and replacing each = sign by $\leq$.

Suppose we wanted to graph the constraints. The equations related to the constraints are obtained by setting the slack variables in the equations (i') through (iii') equal to 0 and maintaining the equal signs. This is shown in Fig. 5.1.

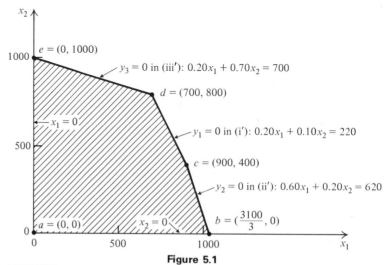

Figure 5.1

DO EXERCISE 1.

Initial Simplex Tableau

Now that we have converted the linear program from one using constraints to one using equations, we can prepare to solve it using the echelon tableau (see Sections 3.1 and 3.2). With a slight modification, the initial echelon tableau is the initial simplex tableau. See Example 2.

Example 2 Set up the initial *echelon* tableau and the initial *simplex* tableau for the problem of Example 1.

Solution The initial *echelon* tableau for the system of equations of Example 1 is

x_1	x_2	y_1	y_2	y_3	x_0	1
0.20	0.10	1	0	0	0	220
0.60	0.20	0	1	0	0	620
0.20	0.70	0	0	1	0	700
−3	−2	0	0	0	1	0

Since the x_0 column does not change in the course of the simplex algorithm, this column is usually omitted in the simplex tableau; however, the value of x_0 will always be in the lower righthand corner in the column headed "1".

Furthermore, the objective function in the *bottom row* plays a special role in the simplex algorithm and hence is set off from the constraints in the other rows with a horizontal line.* Thus, we obtain the *initial simplex* tableau:

x_1	x_2	y_1	y_2	y_3	1
0.20	0.10	1	0	0	220
0.60	0.20	0	1	0	620
0.20	0.70	0	0	1	700
−3	−2	0	0	0	0

---

* Alternately, the objective function is sometimes written in the top row. In either case it is distinguished from the constraints in the other rows by being set off with a horizontal line.

2. Set up the initial simplex tableau for the problem of Margin Exercise 1.

Matrix Formulation

Formally, in matrix notation, a problem with constraints of the form

$$AX \leqslant B, \qquad X \geqslant O,$$

$$\max x_0 : x_0 = CX,$$

or

$$-CX + x_0 = 0,$$

is converted, by addition of slack variables Y, into one of the form*

$$[A \quad I]\begin{bmatrix} X \\ Y \end{bmatrix} = B, \qquad \begin{bmatrix} X \\ Y \end{bmatrix} \geqslant O,$$

$$\max x_0 : x_0 = [C \quad O]\begin{bmatrix} X \\ Y \end{bmatrix}, \qquad \text{or} \qquad -[C \quad O]\begin{bmatrix} X \\ Y \end{bmatrix} + x_0 = 0.$$

In general this can be put into an initial simplex tableau of the form:

X^T	Y^T	1
A	I	B
$-C$	O	0

Note the negative sign in the bottom row.

Also the given equations can be recovered from the tableau in a straightforward manner.

DO EXERCISE 2.

Initial Feasible Solution

Recall that a feasible solution satisfies both $AX \leqslant B$ and $X \geqslant O$.

The simplex algorithm starts with one feasible solution and then generates a better one, if possible. The *initial feasible solution* is obtained in the following manner.

Example 3 Find an initial feasible solution from the initial simplex tableau of Example 2.

Solution In Section 4.3 we solved this linear program graphically. See Fig. 5.2.

---

* You may need to review augmented matrices, Section 3.4.

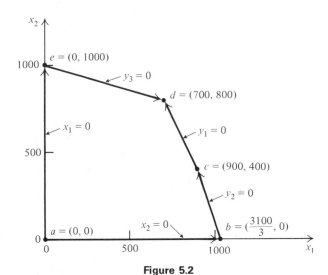

Figure 5.2

We showed previously (see Section 4.2) that the *optimum* feasible solution must occur at a *corner point*. Since we can use *any* feasible solution to start with, we choose that corner point which is simplest to find. This is the *origin* (point a), which becomes our *initial* feasible solution.

Formally, we note that our constraint equations can be written

$$[A \quad I]\begin{bmatrix} X \\ Y \end{bmatrix} = AX + IY = B.$$

Since $B > O$, for maximum-type problems we can always obtain an *initial* feasible solution by setting X (the structural variables) equal to zero and solving for Y (the slack variables):

$$X = O \quad \text{and} \quad Y = B.$$

Setting the structural variables (x_1, x_2) equal to zero in Eqs. (i') through (iv') of Example 1, we can determine the corresponding values of the slack variables (y_1, y_2, y_3) and objective function (x_0):

Initial feasible solution		
Structural variables	Slack variables	Objective function
$x_1 = 0$ $x_2 = 0$	$y_1 = 220$ $y_2 = 620$ $y_3 = 700$	$x_0 = 0$

3. Find an initial feasible solution from the initial simplex tableau of Margin Exercise 2.

This is a *feasible solution* since *all* variables (slack and structural) are nonnegative and satisfy the given constraints. We shall take this as our *initial* feasible solution.

DO EXERCISE 3.

The simplex algorithm starts with the initial feasible solution and proceeds to other feasible solutions. In particular, it proceeds from one corner point to an *adjacent* corner point. Thus, on Fig. 5.2 the solution will proceed from point a (the *initial* feasible solution) to point d (the *optimum* feasible solution) along one of two paths: $abcd$ or aed.

Basic Feasible Solutions

Consider the path $abcd$. In particular note that at

$$\text{point } a: \quad x_1 = x_2 = 0, \quad y_1 = 200, \quad y_2 = 620, \quad y_3 = 700;$$

$$\text{point } b: \quad x_2 = y_2 = 0, \quad x_1 = \frac{3100}{3}, \quad y_1 = \frac{40}{3}, \quad y_3 = \frac{1480}{3};$$

$$\text{point } c: \quad y_2 = y_1 = 0, \quad x_1 = 900, \quad x_2 = 400, \quad y_3 = 240;$$

$$\text{point } d: \quad y_1 = y_3 = 0, \quad x_1 = 700, \quad x_2 = 800, \quad y_2 = 40.$$

Also for the path aed, we have, at

$$\text{point } a: \quad x_1 = x_2 = 0, \quad y_1 = 220, \quad y_2 = 620, \quad y_3 = 700;$$

$$\text{point } e: \quad x_1 = y_3 = 0, \quad x_2 = 100, \quad y_1 = 120, \quad y_2 = 420;$$

$$\text{point } d: \quad y_3 = y_1 = 0, \quad x_1 = 700, \quad x_2 = 800, \quad y_2 = 40.$$

Note now that there are two structural variables, x_1 and x_2, and that at the origin (point a) they are both zero. As we progress from one corner to an adjacent corner, one of these variables becomes *nonzero* and a different variable becomes *zero*, so that in the present example there are always two variables that are zero in the solution at any stage.

This happens *automatically* provided we proceed along a path through adjacent corner points. Such *corner-point* solutions are important in the simplex algorithm and are called *basic* feasible solutions.

Basic and Nonbasic Variables

In general, we note that in the preceding initial feasible solution, some variables were *set* to *zero* and some variables were *computed*.

The variables which are *computed* are called *basic* variables and the variables *set to zero* are called *nonbasic*.

Basic variables can be identified from a simplex tableau as those heading a column of all zeroes except for one 1.

Nonbasic variables are the rest.

The *values* of the basic variables can be read from the *simplex* tableau simply by setting the nonbasic variables to zero and reading the tableau as in the *echelon* method. Recalling that the x_0-column is omitted from the simplex tableau, we find that the value of x_0 is the bottom element in the righthand column.

Initially, the slack variables y_1, y_2, y_3 are basic and the structural variables x_1, x_2 are nonbasic.

A basic feasible solution is written using basic and nonbasic variables. If we start with an *initial* feasible solution as in Example 3, it will be a *basic* feasible solution. Furthermore, each successive step of the simplex algorithm corresponds to a basic feasible solution.*

Choosing the Pivot

The essential difference between the echelon method and the simplex algorithm is in the way one chooses the pivot element. In the simplex algorithm we choose the pivot element to increase x_0 and to maintain feasibility.

Example 4   Find a pivot element for the initial simplex tableau of Example 2.

Solution   From Example 3, we see that the objective function can be written

$$x_0 = 0 + 3x_1 + 2x_2.$$

---

*The use of the terms *basic* and *nonbasic* variables simplifies the description of the simplex algorithm since each step is a *pivoting* operation (as in the echelon method) in which a *basic* variable is made *nonbasic* and a *nonbasic* variable is made *basic*. Thus, from one tableau to the next there is always a fixed number of basic variables and a fixed number of nonbasic variables. This interchanging the roles of the variables from basic to nonbasic and vice versa is done *automatically* as the pivoting is carried out and does *not* require any special consideration.

For each constraint there is a slack variable.

For each slack variable there is a basic variable.

Basic variables are *computed* and, with few exceptions, are greater than zero.

For each structural variable there is a nonbasic variable.

Nonbasic variables are *set to zero*.

Initially, $x_1 = x_2 = 0$ (nonbasic variables), so $x_0 = 0$. If either of these variables, x_1 or x_2, is increased, that is, made basic (nonzero), then x_0 will be increased. Therefore, either x_1 or x_2 could be chosen as the pivot column. When we choose either x_1 or x_2 as a pivot column and pivot, we increase the variable corresponding to that column.

Another way to determine the pivot column is to examine the bottom row of the simplex tableau. The equation

$$x_0 = 0 + 3x_1 + 2x_2$$

appears in the bottom row as

$$-3x_1 - 2x_2 + x_0 = 0.$$

Note the sign changes.

The bottom row entries to the left of the double vertical line are called *indicators*.

Initially, the indicators are $-3, -2, 0, 0, 0$.

We can pick our pivot column to be any column with a *negative indicator*. When we do this, we are picking some nonbasic variable to become basic. If there is no obvious reason to do otherwise, we shall choose that column with the *most negative* indicator as our pivot column.

Having chosen the pivot column (in this case the first column), we must now determine the pivot row. The intersection of the pivot column and pivot row determines the pivot element.

All nonbasic variables except the one involved in pivoting (the one heading the pivot column) remain nonbasic, so that x_2 is zero initially and remains zero.

Consider again the set of equations (i′) through (iv′) with $x_2 = 0$:

$$\left. \begin{aligned}
0.20x_1 + y_1 \qquad\qquad\qquad &= 220, \\
0.60x_1 \qquad + y_2 \qquad\qquad &= 620, \\
0.20x_1 \qquad\qquad + y_3 \qquad &= 700, \\
-3.00x_1 \qquad\qquad\qquad + x_0 &= 0.
\end{aligned} \right\} (E)$$

Setting *one* of the current basic variables to zero makes it nonbasic and permits us to determine the value of the other basic variables and the objective function.

The only question remaining is *which* basic variable to set to zero (that is, to make nonbasic).

If in Equations E we set *each* basic variable to zero in turn and solve for x_1, we obtain:

$$\text{for } y_1 = 0, \quad x_1 = \frac{220}{0.20} = 1100;$$

$$\text{for } y_2 = 0, \quad x_1 = \frac{620}{0.60} = 1033\tfrac{1}{3};$$

$$\text{for } y_3 = 0, \quad x_1 = \frac{700}{0.20} = 3500.$$

The simplex tableau is usually augmented with this *quotient* column headed "q."

x_1	x_2	y_1	y_2	y_3	1	q
0.20	0.10	1	0	0	220	$1100 \left(= \dfrac{220}{0.20} \right)$
0.60*	0.20	0	1	0	620	$1033\tfrac{1}{3} \left(= \dfrac{620}{0.60} \right)$ (Minimum)
0.20	0.70	0	0	1	700	$3500 \left(= \dfrac{700}{0.20} \right)$
-3	-2	0	0	0	0	

The pivot row is the row with *minimum* quotient. Here the pivot row is the second row.

The intersection of the pivot column (here the first column) and the pivot row (here the second row) yields the pivot element which is *starred* (*).

To summarize:

To determine the pivot row: Divide each element in the righthand column (above the bottom row) by the rowwise corresponding entry in the pivot column. The pivot row is the row with *minimum positive* quotient.

To determine the pivot element: Find the intersection of the pivot column and the pivot row. Star (*) this element.

DO EXERCISE 4.

Pivoting

The pivoting operation for one column is exactly the same in the simplex method as in the echelon method (Sections 3.1 and 3.2). The

4. Find a pivot for the initial simplex tableau of Margin Exercise 2.

results are written in a *new* tableau with the *same* headings:

i) Divide *all* elements in the pivot row by the pivot element. Thus, the entry replacing the pivot element will be 1 and the star (*) on this element is dropped.

ii) Add some multiple of the pivot row to each other row including the bottom row (usually a different multiple for each row) to create zeroes for *all* entries of the pivot column other than the pivot element, which is now a 1. Thus, a zero is also created in the *bottom* row in the pivot column.

Example 5   Starting with the initial simplex tableau of Example 2, pivot until the simplex algorithm terminates.

Solution   Pivoting, we obtain the second tableau:

x_1	x_2	y_1	y_2	y_3	1	q
0	$\frac{1}{30}$*	1	$-\frac{1}{3}$	0	$\frac{40}{3}$	400 (Minimum)
1	$\frac{1}{3}$	0	$\frac{5}{3}$	0	$\frac{3100}{3}$	3100
0	$\frac{19}{30}$	0	$-\frac{1}{3}$	1	$\frac{1480}{3}$	$\frac{14800}{19} = 778\frac{18}{19}$
0	-1	0	5	0	3100	

Here the quotient column to the right of the tableau should be ignored for the present.

In the initial tableau read horizontally from the pivot element to a column with a *1* entry in the pivot row. The basic variable heading this column becomes nonbasic (zero) through pivoting.

In this example y_2 was basic (= 620) and becomes nonbasic (= 0) as can be seen from the second tableau.

If we *interchange* the x_1 and y_2 columns, then we obtain:

y_2	x_2	y_1	x_1	y_3	1
$-\frac{1}{3}$	$\frac{1}{30}$*	1	0	0	$\frac{40}{3}$
$\frac{5}{3}$	$\frac{1}{3}$	0	1	0	$\frac{3100}{3}$
$-\frac{1}{3}$	$\frac{19}{30}$	0	0	1	$\frac{1480}{3}$
5	-1	0	0	0	3100

This tableau has the same *form* as our initial tableau provided we *reorder* the variables. However, it is *not* necessary to *physically* interchange the two columns. Solutions can be read from the tableaux in either form. Thus, it is less work simply to leave the columns in their original order and note that the *roles* of the variables have been interchanged; that is, x_1 was nonbasic and became basic, while y_2 was basic and became nonbasic. Thus, we shall speak of "interchanging the roles of x_1 and y_2." This should not be confused with physically interchanging the x_1 and y_2 columns which remain in their original location. The solution at this point is

Structural variables	Slack variables	Objective function
$x_1 = \frac{3100}{3}$, $x_2 = 0$	$y_1 = \frac{40}{3}$, $y_2 = 0$, $y_3 = \frac{1480}{3}$	$x_0 = 3100.$

or

Basic variables	Nonbasic variables	Objective function
$x_1 = \frac{3100}{3}$, $y_1 = \frac{40}{3}$, $y_3 = \frac{1480}{3}$	$x_2 = 0$, $y_2 = 0$	$x_0 = 3100.$

The solution is feasible (that is, it satisfies the given constraints including nonnegativity) and the objective function has been increased from a value of 0 to 3100.

Wrong Pivot

Let us examine what happens if the *wrong* pivot element is used. If, for example, we pivot using the first column and the *first* row, then we obtain

x_1	x_2	y_1	y_2	y_3	1
1	0.5	5	0	0	1100
0	−0.1	−3	1	0	−40
0	0.6	−1	0	1	480
0	−0.5	15	0	0	3300

The *negative* number in the righthand column is a reminder that the *wrong* pivot row has been used. In this case, we obtained $y_2 = -40$ which violates the nonnegativity constraint, which must be maintained.

Selecting the pivot row that corresponds to the *minimum* positive quotient guarantees that the solution does not violate the non-negativity constraint.

Termination

Returning our attention to the correct second tableau, we find that only the second column has a negative indicator, -1. Thus, the second column will be the next pivot column and we obtain a new set of quotients, as shown augmenting the second tableau. The minimum quotient is obtained for the first row. The pivot element is starred and the 1 in the pivot row corresponds to y_1. Thus the pivoting will interchange the roles of x_2 and y_1, that is, will make x_2 basic and y_1 nonbasic ($y_1 = 0$). Performing the pivoting, we obtain

x_1	x_2	y_1	y_2	y_3	1	q	
0	1	30	-10	0	400	—	
1	0	-10	5	0	900	180	
0	0	-19	6*	1	240	40	(Minimum)
0	0	30	-5	0	3500		

Again the quotient column here will be used to obtain the *next* tableau. The solution at this point is

$$x_1 = 900, \qquad y_1 = 0, \qquad x_0 = 3500.$$
$$x_2 = 400, \qquad y_2 = 0,$$
$$y_3 = 240,$$

The fourth column now has a negative indicator, -5, and thus becomes the next pivot column. The corresponding quotients are shown to the right of the tableau above. Note that the *negative* elements in the pivot column are disregarded in computing quotients. Such a pivot would make a variable *negative* and also would *decrease* the value of the objective function.

Pivoting to interchange the roles of y_2 and y_3, we obtain

x_1	x_2	y_1	y_2	y_3	1
0	1	$-\frac{5}{3}$	0	$\frac{5}{3}$	800
1	0	$\frac{35}{6}$	0	$-\frac{5}{6}$	700
0	0	$-\frac{19}{6}$	1	$\frac{1}{6}$	40
0	0	$\frac{85}{6}$	0	$\frac{5}{6}$	3700

The solution at this point is

$$x_1 = 700, \qquad y_1 = 0, \qquad x_0 = 3700.$$
$$x_2 = 800 \qquad y_2 = 40,$$
$$y_3 = 0$$

Since all indicators are nonnegative, there are no more possible pivot columns. Thus, the algorithm *terminates* and the current solution is *maximal*.

The path taken by the simplex algorithm can be followed in Fig. 5.2 as *abcd*. This particular path is a consequence of the choice of the initial pivot column. The alternate choice would have yielded the path *aed*.

Picking the most negative indicator to determine the pivot column is a good simple rule to minimize the number of steps required to achieve optimality but as can be seen in the present example, this is not always the best choice. Since there is no way to know beforehand the best possible pivot column, we use the "most negative" indicator as a good simple rule.

Summary

Let us summarize the steps in the simplex algorithm.

  I. Convert the inequalities of the problem into equations by adding slack variables.

 II. Set up the initial simplex tableau.

III. Carry out the pivoting operations as follows:

   1. To find the pivot column, pick the column with the most negative indicator. (Actually any column with a negative indicator will do.) If there is none, go to step IV.

   2. To find the pivot row, find the quotients of the entries in the righthand column by the rowwise corresponding entries in the pivot column. The pivot row is the row with the minimum positive quotient.

   3. Star the pivot element and carry out the pivoting operation, as in the echelon method, for this column.

   4. To find the feasible solution at any point in the procedure, translate the tableau to the corresponding system of equations, and let the nonbasic variables be 0. The basic variables can be identified by columns with a 1 for some element and all the other elements 0. The other columns are nonbasic.

IV. Look at the bottom row of the tableau. If all the indicators are nonnegative, there are no more possible pivot columns and the

5. Starting with the initial simplex tableau of Margin Exercise 2, pivot until the algorithm terminates. To check your answer, use more than one initial choice of pivot column and check your solution at each step against the graphical solution of Margin Exercise 15, Section 4.3.

algorithm terminates. The solution which can then be found using III(4) is maximal.

To further exemplify the procedure, let us return to the previous example, but this time start by choosing the other pivot column; that is, column two. Then pivoting would proceed as follows, starting with the initial tableau:

x_1	x_2	y_1	y_2	y_3	1	q
0.20	0.10	1	0	0	220	2200
0.60	0.20	0	1	0	620	3100
0.20	0.70*	0	0	1	700	1000 (Minimum)
−3	−2	0	0	0	0	

x_1	x_2	y_1	y_2	y_3	1	q
$\frac{6}{35}$*	0	1	0	$-\frac{1}{7}$	120	700 (Minimum)
$\frac{19}{35}$	0	0	1	$-\frac{2}{7}$	420	$773\frac{13}{19}$
$\frac{2}{7}$	1	0	0	$\frac{10}{7}$	1000	3500
$-\frac{17}{7}$	0	0	0	$\frac{20}{7}$	2000	

x_1	x_2	y_1	y_2	y_3	1
1	0	$\frac{35}{6}$	0	$-\frac{5}{6}$	700
0	0	$-\frac{19}{6}$	1	$\frac{1}{6}$	40
0	1	$-\frac{5}{3}$	0	$\frac{5}{3}$	800
0	0	$\frac{85}{6}$	0	$\frac{5}{6}$	3700

At this point the algorithm terminates. Following the solution from tableau to tableau, we see that on Fig. 5.2 the solution proceeded from point a to point e to point d, the same optimum as before. Note that the final *tableau* just obtained is *not* identical to the final tableau we had previously obtained. However, it is *row equivalent* (there are just row interchanges) and hence has the same solution.

Usually choosing the most negative indicator will tend to yield the solution in fewer steps but, as the present example illustrates, this need not be so. Alternate choice of pivots leads to *the* optimum solution along alternate paths.

DO EXERCISE 5.

EXERCISE SET 5.1

Exercises 1 through 6 are Exercises 1 through 6 from Set 4.2 where they were to be solved graphically. Here you are asked to solve these problems using the simplex algorithm and to check this solution with your graphical solution. Note that the simplex solution yields both structural and slack variables while the graphical solution yields only the structural variables.

1. Maximize $x_0(x_1, x_2) = x_1 + 2x_2$, subject to the constraints

$$x_1 + x_2 \leq 6,$$
$$x_2 \leq 5,$$
$$x_1, x_2 \geq 0.$$

2. Maximize $x_0(x_1, x_2) = x_1 + x_2$, subject to the constraints

$$x_1 + 2x_2 \leq 8,$$
$$x_1 \leq 6,$$
$$x_1, x_2 \geq 0.$$

3. Maximize $x_0(x_1, x_2) = 5x_1 + 4x_2$, subject to the constraints

$$3x_1 + 2x_2 \leq 12,$$
$$x_1 + x_2 \leq 5,$$
$$x_1, x_2 \geq 0.$$

4. Maximize $x_0(x_1, x_2) = 2x_1 + x_2$, subject to the constraints

$$3x_1 + 5x_2 \leq 15,$$
$$3x_1 + 2x_2 \leq 12,$$
$$x_1, x_2 \geq 0.$$

5. Maximize $x_0(x_1, x_2) = 3x_1 + 4x_2$, subject to the constraints

$$x_1 + 2x_2 \leq 14,$$
$$4x_1 + 3x_2 \leq 26,$$
$$2x_1 + x_2 \leq 12,$$
$$x_1, x_2 \geq 0.$$

6. Maximize $x_0(x_1, x_2) = 5x_1 + 4x_2$, subject to the constraints

$$x_1 + 3x_2 \leq 18,$$
$$3x_1 + 2x_2 \leq 19,$$
$$2x_1 + x_2 \leq 12,$$
$$x_1, x_2 \geq 0.$$

Use the simplex algorithm to solve the following problems. Check your answer by doing each problem twice, starting with different pivots, if possible.

7.
$$x_1 + 2x_2 \leq 60,$$
$$4x_1 + 3x_2 \leq 140,$$
$$\max x_0 \colon x_0 = 120x_1 + 75x_2; \quad x_1, x_2 \geq 0$$

8.
$$4x_1 + 5x_2 \leq 35,$$
$$3x_1 + x_2 \leq 18,$$
$$\max x_0 \colon x_0 = 3x_1 + 2x_2; \quad x_1, x_2 \geq 0$$

9.
$$3x_1 + 6x_2 \leq 90,$$
$$5x_1 + 3x_2 \leq 160,$$
$$x_1 + x_2 \leq 44,$$
$$\max x_0 \colon x_0 = 3x_1 + 2x_2; \quad x_1, x_2 \geq 0$$

10.
$$x_1 + 2x_2 \leq 16,$$
$$3x_1 + 2x_2 \leq 20,$$
$$4x_1 + x_2 \leq 20,$$
$$\max x_0 \colon x_0 = 2x_1 + x_2; \quad x_1, x_2 \geq 0$$

11.
$$5x_1 + 6x_2 \leq 60,$$
$$x_1 + x_2 \leq 11,$$
$$3x_1 + x_2 \leq 27,$$
$$\max x_0 \colon x_0 = 2x_1 + x_2; \quad x_1, x_2 \geq 0$$

12.
$$-x_1 + 2x_2 \leq 10,$$
$$3x_1 + 2x_2 \leq 18,$$
$$3x_1 + x_2 \leq 15,$$
$$\max x_0 \colon x_0 = x_1 + 2x_2; \quad x_1, x_2 \geq 0$$

13.
$$3x_1 + 4x_2 \leqslant 48,$$
$$x_1 + x_2 \leqslant 13,$$
$$2x_1 + x_2 \leqslant 22,$$
$$\max x_0 \colon x_0 = 7x_1 + 4x_2; \quad x_1, x_2 \geqslant 0$$

14.
$$x_1 + 3x_2 \leqslant 18,$$
$$2x_1 + x_2 \leqslant 11,$$
$$3x_1 + x_2 \leqslant 15,$$
$$\max x_0 \colon x_0 = 3x_1 + 4x_2; \quad x_1, x_2 \geqslant 0$$

15.
$$3x_1 - 2x_2 + x_3 \leqslant 8,$$
$$-4x_1 + 3x_2 + 2x_3 \leqslant 4,$$
$$3x_1 + x_2 - 6x_3 \leqslant 6,$$
$$\max x_0 \colon x_0 = \quad - 2x_2 + 5x_3; \quad x_1, x_2, x_3 \geqslant 0$$

16.
$$x_1 - 4x_2 + 3x_3 \leqslant 12,$$
$$x_1 - 2x_2 + 6x_3 \leqslant 4,$$
$$2x_1 + 7x_2 + x_3 \leqslant 17,$$
$$\max x_0 \colon x_0 = 5x_1 + x_2 + 2x_3; \quad x_1, x_2, x_3 \geqslant 0$$

17.
$$2x_1 + x_2 + 4x_3 \leqslant 24,$$
$$x_1 + x_2 + x_3 \leqslant 7,$$
$$2x_1 - x_2 + 3x_3 \leqslant 12,$$
$$\max x_0 \colon x_0 = x_1 + 2x_2 + 3x_3; \quad x_1, x_2, x_3 \geqslant 0.$$

18.
$$4x_1 + 3x_2 - 2x_3 \leqslant 5,$$
$$5x_1 - 2x_2 + 7x_3 \leqslant 11,$$
$$3x_1 - x_2 + 2x_3 \leqslant 3,$$
$$\max x_0 \colon x_0 = 9x_1 - 2x_2 + 11x_3; \quad x_1, x_2, x_3 \geqslant 0.$$

19. *Business.* A carpentry shop makes bookcases, desks, and tables. Each bookcase requires 5 man-hours of woodworking, 4 hours of finishing, 30 board feet of hardwood, and 15 board feet of inexpensive wood, and sells for $60. Each desk requires 10 man-hours of woodworking, 3 hours of finishing, 20 board feet of hardwood, and 20 board feet of inexpensive wood, and sells for $100. Each table requires 7 hours of woodworking, 2 hours of finishing, and 24 board feet of hardwood, and sells for $80. The available supply is 575 man-hours for woodworking, 220 man-hours for finishing, 1800 board feet of hardwood, and 1000 board feet of inexpensive wood. How many of each product should be made to maximize sales?

20. *Business.* A coffee merchant has a supply of 600 lbs of Mocha coffee, 2400 lbs of Columbian coffee, and 4800 lbs of Brazilian coffee. He sells 100% Mocha coffee as Turkish at $4.00 per lb. He sells a Mocha blend consisting of 25% Mocha and 75% Columbian coffee at $2.50 per lb. He sells a Columbian blend consisting of 10% Mocha, 60% Columbian, and 30% Brazilian coffee at $1.75 per lb. He sells a Brazilian blend consisting of 20% Columbian and 80% Brazilian coffee at $1.25 per lb. How many lbs of each blend should he prepare to maximize his sales?

OBJECTIVE

You should be able to solve a minimum-type linear program using artificial variables and the two-phase simplex algorithm.

5.2 THE TWO-PHASE METHOD AND MINIMUM-TYPE LINEAR PROGRAMS*

In Section 4.3 we showed that maximum-type linear programs could be formulated and put in the form

$$AX \leqslant B,$$
$$X \geqslant O,$$
$$\max x_0 \colon x_0 = CX.$$

_____

* This section can be omitted with no loss of continuity.

In Section 5.1 we showed how to solve such a linear program using the simplex algorithm provided B satisfied the positivity constraint $B > O$.

In Section 4.4 we showed that minimum-type linear programs could be formulated and put in the form

$$AY \geqslant B,$$
$$Y \geqslant O,$$
$$\min y_0 : y_0 = CY.$$

We can convert this minimum-type linear program into a maximum-type linear program by multiplying both sides of the inequality by -1. This reverses the sense of the inequality and we obtain:

$$-AY \leqslant -B,$$
$$Y \geqslant O,$$
$$\max -y_0 : -y_0 = -CY.$$

Here to find the minimum of y_0, we seek the maximum of $-y_0$.

Now if $-B > O$, then we could proceed as in Section 5.1. However, the positivity constraint $-B > O$ rarely holds in minimum-type problems and we must seek an alternate method.

The first step is to convert the inequalities of the problem into equations and the minimum-type linear program into a maximum-type linear program.

Example 1 Convert the formulation of Example 1 of Section 4.4 from a minimum-type linear program expressed as inequalities to a maximum-type linear program expressed as equations.

Solution The original formulation was

$$40y_1 + 20y_2 \geqslant 160,$$
$$4y_1 + 4y_2 \geqslant 24,$$
$$4y_1 + 6y_2 \geqslant 28,$$
$$\min y_0 : y_0 = 15y_1 + 10y_2; \quad y_1, y_2 \geqslant 0.$$

Note that the righthand side of each inequality is positive; that is, $B > O$.

For maximum-type linear programs for which the sense of the inequalities was $\leqslant$, we *added* slack variables to obtain equations. Here for minimum-type linear programs for which the sense of the ine-

6. Convert the minimum-type linear program of Margin Exercise 17 of Section 4.4 into a maximum-type program in equation form.

qualities is $\geqslant$, we *subtract* slack variables. Thus, we obtain

$$40y_1 + 20y_2 \quad - x_1 \qquad\qquad = 160,$$
$$4y_1 + 4y_2 \qquad\quad - x_2 \qquad = 24,$$
$$4y_1 + 6y_2 \qquad\qquad\quad - x_3 = 28,$$

where

$$y_1, y_2; \quad x_1, x_2, x_3 \geqslant 0.$$

We convert the problem to a maximum type by writing the objective function

$$\max -y_0 \colon -y_0 = -15y_1 - 10y_2.$$

Thus, starting with a minimum-type linear program of the form

$$AY \geqslant B,$$
$$\min y_0 \colon y_0 = CY; \quad Y \geqslant O,$$

we subtract slack variables and obtain it in the form

$$AY - X = B,$$
$$\max -y_0 \colon -y_0 = -CY; \quad Y \geqslant O, \quad X \geqslant O.$$

DO EXERCISE 6.

With a maximum-type linear program we could obtain an *initial* feasible solution by setting the structural variables (initially nonbasic) to zero and solving for the slack variables (initially basic). Doing this for the present problem, we obtain:

$$x_1 = -160, \qquad x_2 = -24, \qquad \text{and} \qquad x_3 = -28.$$

This violates the nonnegativity constraint ($X \geqslant O$) and hence is not a feasible solution (which must satisfy $AY \geqslant B$, $Y \geqslant O$, and $X \geqslant O$). To obtain an initial feasible solution, we introduce *artificial variables*.

Example 2 Introduce artificial variables to obtain an initial feasible solution to the problem of Example 1.

Solution In maximum-type linear programs we *add slack* variables to do two things: to convert inequalities into equations and to provide an initial feasible solution. In minimum-type linear programs we *subtract slack* variables to convert inequalities into equations and *add artificial* variables to provide an initial feasible solution. Thus, adding artificial variables z_1, z_2, and z_3 we obtain:

$$40y_1 + 20y_2 - x_1 \qquad\qquad + z_1 \qquad\qquad = 160,$$
$$4y_1 + 4y_2 \qquad - x_2 \qquad\qquad + z_2 \qquad\quad = 24,$$
$$4y_1 + 6y_2 \qquad\qquad - x_3 \qquad\qquad + z_3 = 28,$$

$$\max -y_0: -y_0 = -15y_1 - 10y_2; \quad y_1, y_2; \quad x_1, x_2, x_3; \quad z_1, z_2, z_3 \geqslant 0.$$

Initially, the structural and slack variables are nonbasic and, hence, zero and the artificial variables are basic. In the present example we have initially

$$y_1 = y_2 = 0,$$
$$x_1 = x_2 = x_3 = 0,$$
$$z_1 = 160, \quad z_2 = 24, \quad z_3 = 28.$$

Thus, adding artificial variables puts a minimum-type linear program into the form

$$AY - X + Z = B,$$
$$\max -y: -y_0 = -CY; \quad Y \geqslant O, \quad X \geqslant O, \quad Z \geqslant O.$$

DO EXERCISE 7a.

While the slack variables must be *nonnegative* $(x_1, x_2, x_3 \geqslant 0)$, the artificial variables must be *zero* $(z_1 = z_2 = z_3 = 0)$ in the *optimum* feasible solution. In order to force the artificial variables to become zero, we introduce an *artificial objective function*.

Example 3 Introduce an artificial objective function into the formulation of Example 2 to force the artificial variables to become zero and set up the initial simplex tableau.

Solution Consider the artificial objective function

$$\max z_0: z_0 = -(z_1 + z_2 + z_3).$$

It can be seen that z_0 has a *maximum* value $z_0 = 0$ when the artificial variables are all *minimum;* that is, zero $(z_1 = z_2 = z_3 = 0)$ since they cannot be negative.

Our formulation now takes the form

$$AY - X + Z = B,$$
$$\max -y_0: -y_0 = -CY; \quad Y \geqslant O, \quad X \geqslant O,$$
$$\max z_0: z_0 = -(z_1 + z_2 + z_3 + \cdots); \quad Z \geqslant O.$$

DO EXERCISE 7b.

7.

a) Add artificial variables to the formulation of Margin Exercise 6.

b) Set up the artificial objective function for this problem.

8. Set up the initial simplex tableau for the formulation of Margin Exercise 7.

We can now put this into a simplex tableau (remembering that the negatives of the coefficients of the objective function are put into the bottom rows):

	y_1	y_2	x_1	x_2	x_3	z_1	z_2	z_3	1
	40	20	−1	0	0	1	0	0	160
	4	4	0	−1	0	0	1	0	24
	4	6	0	0	−1	0	0	1	28
$-y_0$	15	10	0	0	0	0	0	0	0
z_0	0	0	0	0	0	1	1	1	0

Here we have put $-y_0$ and z_0 to the left of the two bottom rows to keep track of the two objective functions.

DO EXERCISE 8.

The simplex algorithm presented in Section 5.1 is the *second* phase of the *two-phase* method. We shall now consider the *first* phase which consists of two parts:

i) getting the tableau into standard form,
ii) eliminating the artificial variables.

Example 4 Put the simplex tableau of Example 3 into standard form and obtain an initial feasible solution.

Solution To get our tableau into standard form (that is, with basic and nonbasic variables), we must make the artificial variables basic. Basic variables head columns with one 1 and all the rest 0's. We can make all the artificial variables basic by subtracting in turn each row with a 1 in a column headed by an artificial variable from the bottom row. Thus, we obtain:

	y_1	y_2	x_1	x_2	x_3	z_1	z_2	z_3	1	q	
	40*	20	−1	0	0	1	0	0	160	4	(Minimum)
	4	4	0	−1	0	0	1	0	24	6	
	4	6	0	0	−1	0	0	1	28	7	
$-y_0$	15	10	0	0	0	0	0	0	0		
z_0	−48	−30	1	1	1	0	0	0	−212		

This can be seen to be in standard form with the initial feasible solution:

$$y_1 = y_2 = 0, \quad x_1 = x_2 = x_3 = 0, \quad z_1 = 160, \quad z_2 = 24, \quad z_3 = 28.$$

DO EXERCISE 9.

Example 5 Eliminate the artificial variables from the tableau of Example 4.

Solution To eliminate the artificial variables we first optimize the artificial objective function treating the $-y_0$ row like any other. To determine the pivot column we seek the most negative *artificial indicator* (indicator in artificial objective function) which will make an artificial variable nonbasic (zero).

The first column (y_1) has the most negative artificial indicator, -48. Computing quotients, we find that the first row has the minimum quotient and becomes the pivot row. The basic variable with a 1 in this row is z_1. Hence z_1 will become nonbasic and, since z_1 is an artificial variable, the first column–first row element is an acceptable pivot. Once an artificial variable becomes nonbasic, we drop it from the tableau so that it cannot again become basic.

Pivoting, we obtain:

	y_1	y_2	x_1	x_2	x_3	z_1	z_2	z_3	1
	1	$\frac{1}{2}$	$-\frac{1}{40}$	0	0	$\frac{1}{40}$	0	0	4
	0	2	$\frac{1}{10}$	-1	0	$\frac{1}{10}$	1	0	8
	0	4^*	$\frac{1}{10}$	0	-1	$\frac{1}{10}$	0	1	12
$-y_0$	0	$\frac{5}{2}$	$\frac{3}{8}$	0	0	$-\frac{3}{8}$	0	0	-60
z_0	0	-6	$-\frac{1}{5}$	1	1	$\frac{6}{5}$	0	0	-20

The second column (y_2) can now be taken as pivot. This will make z_3 nonbasic. Pivoting, we obtain

	y_1	y_2	x_1	x_2	x_3	z_2	z_3	1
	1	0	$-\frac{3}{80}$	0	$\frac{1}{8}$	0	$-\frac{1}{8}$	$\frac{5}{2}$
	0	0	$\frac{1}{20}$	-1	$\frac{1}{2}^*$	1	$-\frac{1}{2}$	2
	0	1	$\frac{1}{40}$	0	$-\frac{1}{4}$	0	$\frac{1}{4}$	3
$-y_0$	0	0	$\frac{5}{16}$	0	$\frac{5}{8}$	0	$-\frac{5}{8}$	$-\frac{135}{2}$
z_0	0	0	$-\frac{1}{20}$	1	$-\frac{1}{2}$	0	$\frac{3}{2}$	-2

9. Put the tableau of Margin Exercise 8 into standard form and obtain an initial feasible solution.

10.

a) Eliminate the artificial variables from the tableau of Margin Exercise 9.

The fifth column (x_3) is now taken as pivot making z_2 nonbasic. Pivoting, we obtain:

y_1	y_2	x_1	x_2	x_3	z_2	1
1	0	$-\frac{1}{20}$	$\frac{1}{4}$	0	$-\frac{1}{4}$	2
0	0	$\frac{1}{10}$	-2	1	2	4
0	1	$\frac{1}{20}$	$-\frac{1}{2}$	0	$\frac{1}{2}$	4

	y_1	y_2	x_1	x_2	x_3	z_2	1
$-y_0$	0	0	$\frac{1}{4}$	$\frac{5}{4}$	0	$\frac{5}{4}$	-70
z_0	0	0	0	0	0	1	0

Note that we have now pivoted out all artificial variables and have obtained a row of zeroes in the z_0 row. Thus, z_0 has been optimized and the z_0 row can be dropped. An initial feasible solution (without artificial variables) can now be obtained:

$$y_1 = 2, \qquad y_2 = 4, \qquad y_0 = 70, \qquad x_1 = x_2 = 0, \qquad x_3 = 4.$$

This concludes phase one of the simplex algorithm. If it is not possible to eliminate all the artificial variables, then there is *no* feasible solution to the given minimum problem.

DO EXERCISE 10a.

b) Apply phase two of the simplex algorithm to find the optimum feasible solution.

Example 6 Find the optimum solution to the given minimum problem of Example 1.

Solution Examining the indicators of the $-y_0$ row, we find that *in the present example* they are all nonnegative, so that we have also achieved optimality of y_0. If this were not so, we would simply continue with phase two of the simplex algorithm as we did in Section 5.1 to solve a maximum-type linear program (given some initial feasible solution).

Thus, the optimum feasible solution to our given minimum problem is

$$y_1 = 2, \qquad y_2 = 4, \qquad x_1 = x_2 = 0, \qquad x_3 = 4, \qquad y_0 = 70.$$

DO EXERCISE 10b.

Summary of the Two-Phase Method

I. Preconditioning

1. **Write the linear program such that the righthand side of each inequality is *positive*.**

2. Write the linear program in a form with an objective function to be *maximized*.

II. Phase One

1. For each inequality with sense $\leq$, add a slack variable.
2. For each inequality with sense $\geq$, subtract a slack variable and add an artificial variable.
3. For each equation, add an artificial variable.
4. Append to the given linear program an artificial objective function equal to the negative sum of all artificial variables. This artificial objective function is to be maximized.
5. Set up the initial simplex tableau with the two objective functions to be maximized.
6. Put the tableau in standard form by subtracting from the artificial objective function row the sum of the rows containing artificial variables.
7. Pivot until all artificial variables are nonbasic and can be dropped and there are all zeroes in the artificial objective function row which can then be dropped.

III. Phase Two

Continue as in Section 5.1.

EXERCISE SET 5.2

Exercises 1 through 6 are Exercises 7 through 12 from Set 4.2 where they were solved graphically. Here you are asked to solve these problems using artificial variables and the two-phase simplex method. Check each solution against the corresponding graphical solution.

1. Exercise 7, Set 4.2: Minimize $y_0(y_1, y_2) = 3y_1 + 4y_2$, subject to the constraints

$$3y_1 + y_2 \geq 9,$$
$$y_1 + y_2 \geq 7,$$
$$y_1 + 2y_2 \geq 8,$$
$$y_1, y_2 \geq 0.$$

2. Exercise 8, Set 4.2: Minimize $y_0(y_1, y_2) = 3y_1 + 4y_2$, subject to the constraints

$$2y_1 + y_2 \geq 8,$$
$$4y_1 + 3y_2 \geq 22,$$
$$2y_1 + 5y_2 \geq 18,$$
$$y_1, y_2 \geq 0.$$

3. Exercise 9, Set 4.2: Minimize $y_0(y_1, y_2) = 2y_1 + 5y_2$, subject to the constraints

$$2y_1 + y_2 \geq 9,$$
$$4y_1 + 3y_2 \geq 23,$$
$$y_1 + 3y_2 \geq 8,$$
$$y_1, y_2 \geq 0.$$

4. Exercise 10, Set 4.2: Minimize $y_0(y_1, y_2) = 9y_1 + 7y_2$, subject to the constraints

$$5y_1 + 3y_2 \geq 30,$$
$$2y_1 + 3y_2 \geq 21,$$
$$3y_1 + 6y_2 \geq 36,$$
$$y_1, y_2 \geq 0.$$

5. Exercise 11, Set 4.2: Minimize $y_0(y_1, y_2) = 3y_1 + 5y_2$, subject to the constraints

$$4y_1 + y_2 \geq 9,$$
$$3y_1 + 2y_2 \geq 13,$$
$$2y_1 + 5y_2 \geq 16,$$
$$y_1, y_2 \geq 0.$$

6. Exercise 12, Set 4.2: Minimize $y_0(y_1, y_2) = 2y_1 + 3y_2$, subject to the constraints

$$4y_1 + y_2 \geq 7,$$
$$y_1 + y_2 \geq 4,$$
$$2y_1 + 5y_2 \geq 14,$$
$$y_1, y_2 \geq 0.$$

In Exercise Set 4.4 various minimum-type linear programs were given to be formulated and solved graphically. Here you are asked to solve these problems using the two-phase simplex algorithm and to check each answer with your graphical solution.

7. Exercise 1, Set 4.4:

$$50y_1 + 15y_2 \geq 120,$$
$$8y_1 + 5y_2 \geq 24,$$
$$5y_1 + y_2 \geq 10,$$
$$\min y_0: y_0 = 15y_1 + 5y_2; \quad y_1, y_2 \geq 0.$$

8. Exercise 2, Set 4.4:

9. Exercise 3, Set 4.4:

$$50y_1 + 20y_3 \geq 120,$$
$$8y_1 + 6y_3 \geq 24,$$
$$5y_1 + 8y_3 \geq 20,$$
$$\min y_0: y_0 = 15y_1 + 8y_3; \quad y_1, y_3 \geq 0.$$

10. Exercise 4, Set 4.4:

11. Exercise 5, Set 4.4:

$$40y_1 + 80y_2 \geq 2000,$$
$$40y_1 + 30y_2 \geq 1500,$$
$$120y_1 + 40y_2 \geq 2400,$$
$$\min y_0: y_0 = 12y_1 + 10y_2; \quad y_1, y_2 \geq 0.$$

12. Exercise 6, Set 4.4:

13. Exercise 7, Set 4.4:

$$40y_1 + 40y_3 \geq 2000,$$
$$40y_1 + 80y_3 \geq 1500,$$
$$120y_1 + 80y_3 \geq 2400,$$
$$\min y_0: y_0 = 12y_1 + 15y_3; \quad y_1, y_3 \geq 0.$$

14. Exercise 8, Set 4.4:

15. Exercise 9, Set 4.4:

$$40y_1 + 80y_2 \geq 1600,$$
$$40y_1 + 30y_2 \geq 2100,$$
$$120y_1 + 40y_2 \geq 2400,$$
$$\min y_0: y_0 = 12y_1 + 10y_2; \quad y_1, y_2 \geq 0.$$

16. Exercise 10, Set 4.4:

17. Exercise 11, Set 4.4:

$$40y_1 + 40y_3 \geq 1600,$$
$$40y_1 + 80y_3 \geq 2100,$$
$$120y_1 + 80y_3 \geq 2400,$$
$$\min y_0: y_0 = 12y_1 + 15y_3; \quad y_1, y_3 \geq 0.$$

18. Exercise 12, Set 4.4. Check your answer using an alternate pivot.

The following problems involve mixed constraints. Those with only two structural variables can be checked by also solving graphically.

19.
$$x_1 + 2x_2 \leq 14,$$
$$4x_1 + 3x_2 \leq 26,$$
$$2x_1 + x_2 \leq 12,$$
$$3x_1 + 4x_2 \geq 12,$$
$$\max x_0: x_0 = 3x_1 + 4x_2; \quad x_1, x_2 \geq 0.$$

20.
$$x_1 + 3x_2 \leq 18,$$
$$3x_1 + 2x_2 \leq 19,$$
$$2x_1 + x_2 \leq 12,$$
$$x_1 + 2x_2 \geq 5,$$
$$\max x_0: x_0 = 5x_1 + 4x_2; \quad x_1, x_2 \geq 0.$$

21.
$$5x_1 + 6x_2 \leq 60,$$
$$x_1 + x_2 \leq 11,$$
$$3x_1 + x_2 \leq 27,$$
$$5x_1 + 2x_2 \geq 10,$$
$$3x_1 + 2x_2 \geq 6,$$
$$\max x_0: x_0 = 2x_1 + x_2; \quad x_1, x_2 \geq 0.$$

22.
$$-x_1 + 2x_2 \leq 10,$$
$$3x_1 + 2x_2 \leq 18,$$
$$3x_1 + x_2 \leq 15,$$
$$2x_1 + x_2 \geq 4,$$
$$2x_1 + 3x_2 \geq 6,$$
$$\max x_0: x_0 = x_1 + 2x_2; \quad x_1, x_2 \geq 0.$$

23.
$$3x_1 + 4x_2 \leq 48,$$
$$x_1 + x_2 \leq 13,$$
$$2x_1 + x_2 \leq 22,$$
$$5x_1 + 3x_2 \geq 30,$$
$$2x_1 + 7x_2 = 28,$$
$$\max x_0: x_0 = 7x_1 + 4x_2; \quad x_1, x_2 \geq 0.$$

24.
$$x_1 + 3x_2 \leq 18,$$
$$2x_1 + x_2 \leq 11,$$
$$3x_1 + x_2 \leq 15,$$
$$2x_1 + 5x_2 \geq 10,$$
$$5x_1 + 2x_2 = 12,$$
$$\max x_0: x_0 = 3x_1 + 4x_2; \quad x_1, x_2 \geq 0.$$

25.
$$3x_1 - 2x_2 + x_3 \leq 8,$$
$$-4x_1 + 3x_2 - x_3 \leq 4,$$
$$2x_1 - 3x_2 - 6x_3 \leq 6,$$
$$x_1 - x_2 + x_3 \geq 1,$$
$$x_1 + x_2 + x_3 = 5,$$
$$\max x_0: x_0 = -2x_2 + 5x_3; \quad x_1, x_2, x_3 \geq 0.$$

26.
$$x_1 - 4x_2 + 3x_3 \leq 12,$$
$$x_1 - 2x_2 + 5x_3 \leq 13,$$
$$2x_1 + 7x_2 + x_3 \leq 17,$$
$$x_1 + 2x_2 + x_3 \geq 2,$$
$$x_1 - x_2 + 3x_3 = 7,$$
$$\max x_0: x_0 = 5x_1 + x_2 + 2x_3; \quad x_1, x_2, x_3 \geq 0.$$

OBJECTIVES

a) Given a formulation of a linear program you should be able to formulate its dual.

b) You should be able to set up a minimum-type linear program in a simplex tableau, solve, and read off the solution.

c) You should be able to check your solution.

5.3 DUALITY AND MINIMUM-TYPE LINEAR PROGRAMS

Minimum-Type Linear Programs

In Section 5.1 we considered maximum-type linear programs which were expressed in the standard form

$$AX \leqslant B, \qquad X \geqslant O,$$
$$\max x_0 \colon x_0 = CX$$

with the positivity constraint $B > O$. Now suppose we wanted to solve a minimum-type linear program, which is expressed in the form

$$AY \geqslant B, \qquad Y \geqslant O,$$
$$\min y_0 \colon y_0 = CY.$$

One way to solve such programs is to use artificial variables as in Section 5.2. However, an alternate method involving the Duality Theorem can also be used. This method has several advantages. It involves less computation and has an economic interpretation (which is discussed later on).

Duality

THE DUALITY RELATIONSHIP. Given a *maximum*-type linear program[*]

$$AX \leqslant B, \qquad X \geqslant O,$$
$$\max x_0 \colon x_0 = CX,$$

its *dual*, a *minimum-type* linear program is

$$A^T Y \geqslant C^T, \qquad Y \geqslant O$$
$$\min y_0 \colon y_0 = B^T Y$$

where, as in Section 3.3, the superscript T refers to matrix transpose. Note that the dual is expressed using *new* variables Y, called dual variables.

Similarly, starting with a *minimum*-type linear program

$$AY \geqslant B, \qquad Y \geqslant O, \qquad \min y_0 \colon y_0 = CY,$$

we can *derive* its *dual*, a *maximum*-type linear program

$$A^T X \leqslant C^T, \qquad X \geqslant O, \qquad \max x_0 \colon x_0 = B^T X.$$

---

[*] In general, we do not need the restriction $B > O$. Furthermore, any equation of the form

$$a_1 x_1 + a_2 x_2 + \cdots = b$$

can be written as two inequalities

$$a_1 x_1 + a_2 x_2 + \cdots \leqslant b$$

and

$$a_1 x_1 + a_2 x_2 + \cdots \geqslant b \quad \text{or} \quad -a_1 x_1 - a_2 x_2 - \cdots = \leqslant -b.$$

This way of treating equations is done to avoid later complications.

The original physical linear program, whether maximum or minimum, is often called the *primal* and the corresponding *derived* program its *dual*. The dual frequently has significance in an applied problem.

THE DUALITY THEOREM. For a pair of primal and dual linear programs, the objective function x_0 attains its maximum if and only if the objective function y_0 attains its minimum and, furthermore,

$$\text{max } x_0 = \text{min } y_0.$$

Thus, to solve a minimum-type linear program, we first obtain its dual, a maximum-type linear program. Then, as we shall illustrate, we can obtain the solutions of the pair of dual programs at the same time.

When working with the simplex tableau, it is customary to call the *maximum*-type linear program the *primal* and the *minimum*-type linear program the *dual* regardless of the type of physical problem. Which meaning of primal or dual is intended can be determined from the context of its use.

Example 1   Given the maximum-type linear program of Example 1 of Section 4.3, with formulation

$$0.20x_1 + 0.10x_2 \leq 220,$$
$$0.60x_1 + 0.20x_2 \leq 620,$$
$$0.20x_1 + 0.70x_2 \leq 700,$$
$$\text{max } x_0: x_0 = \quad 3x_1 + \quad 2x_2; \quad x_1, x_2 \geq 0,$$

find its dual.

Solution   The minimum-type linear program dual to this is

$$0.20y_1 + 0.60y_2 + 0.20y_3 \geq 3,$$
$$0.10y_1 + 0.20y_2 + 0.70y_3 \geq 2,$$
$$\text{min } y_0: y_0 = 220y_1 + 620y_2 + 700y_3; \quad y_1, y_2, y_3 \geq 0.$$

DO EXERCISE 11.

Example 2   Given the minimum-type linear program (Example 1, Section 4.4) with the formulation

$$40y_1 + 20y_2 \geq 160,$$
$$4y_1 + \quad 4y_2 \geq \quad 24,$$
$$4y_1 + \quad 6y_2 \geq \quad 28,$$
$$\text{min } y_0: y_0 = 15y_1 + 10y_2; \quad y_1, y_2 \geq 0,$$

find its dual.

11. Formulate the linear program dual to the following program (Margin Exercise 15, Section 4.3):

$$20x_1 + 100x_2 \leq 1900,$$
$$x_1 + \quad 50x_2 \leq \quad 500,$$
$$2x_1 + \quad 20x_2 \leq \quad 240,$$
$$\text{max } x_0: x_0 = 20x_1 + 300x_2; \quad x_1, x_2 \geq 0.$$

12. Formulate the program dual to the following program (Margin Exercise 17 Section 4.4):

$$0.1y_1 + \quad 0y_2 \geqslant 200,$$

$$0y_1 + \quad 0.2y_2 \geqslant 500,$$

$$0.02y_1 + 0.01y_2 \geqslant 100,$$

$$\min y_0\colon y_0 = 10y_1 + \quad 15y_2; \quad y_1, y_2 \geqslant 0.$$

Solution   The maximum-type linear program dual to this is

$$40x_1 + \quad 4x_2 + \quad 4x_3 \leqslant 15,$$

$$20x_1 + \quad 4x_2 + \quad 6x_3 \leqslant 10,$$

$$\max x_0\colon x_0 = 160x_1 + 24x_2 + 28x_3; \quad x_1, x_2, x_3 \geqslant 0.$$

DO EXERCISE 12.

Example 3   Express the constraint set of Example 1 as a set of equations, set up the initial simplex tableau, and read off *both* the initial *primal* and *dual* solutions.

Solution   Adding slack variables, we obtain

$$0.20x_1 + 0.10x_2 + y_1 \qquad\qquad = 220,$$

$$0.60x_1 + 0.20x_2 \qquad + y_2 \qquad\quad = 620,$$

$$0.20x_1 + 0.70x_2 \qquad\qquad + y_3 \quad = 700,$$

$$\max x_0\colon \; -3x_1 \quad -2x_2 \qquad\qquad\qquad + x_0 = 0,$$

$$x_1, x_2; \quad y_1, \quad y_2, \quad y_3 \qquad \geqslant 0.$$

Note that we have used the same notation for the *slack* variables as we did for the *dual* variables.

The initial simplex tableau is, as before,

x_1	x_2	y_1	y_2	y_3	1
0.20	0.10	1	0	0	220
0.60	0.20	0	1	0	620
0.20	0.70	0	0	1	700
−3	−2	0	0	0	0

Recall that the x_0-column has been suppressed. As before, the initial *primal solution* is:

$$x_1 = 0, \qquad y_1 = 220, \qquad x_0 = 0.$$

$$x_2 = 0, \qquad y_2 = 620,$$

$$y_3 = 700,$$

Similarly, the dual (minimum) program can be expressed as a set of equations by *subtracting* slack variables:

$$0.20y_1 + 0.60y_2 + 0.20y_3 - x_1 \qquad\qquad = 3,$$

$$0.10y_1 + 0.20y_2 + 0.70y_3 \qquad - x_2 \qquad = 2,$$

$$\min y_0\colon 220y_1 + 620y_2 + 700y_3 \qquad\qquad - y_0 = 0,$$

$$y_1, y_2, y_3; \quad x_1, x_2 \geqslant 0,$$

where we have used the same notation for the dual slack variables as for the primal structural variables.

As with the primal, setting to zero the nonbasic dual variables $y_1 = y_2 = y_3 = 0$, we obtain for the basic dual variables

$$x_1 = -3, \quad \text{and} \quad y_0 = 0.$$
$$x_2 = -2,$$

Note that while the primal solution was nonnegative, the dual solution usually violates the nonnegativity constraint.

Alternately, the *dual solution* is obtained directly from the *bottom row* of the simplex tableau, so that initially

$$x_1 = -3,$$
$$x_2 = -2,$$
$$y_1 = y_2 = y_3 = 0,$$
$$y_0 = 0.$$

Note that $y_0 = x_0$ and that the x_0 column has been suppressed.

Note, further, that both primal and dual solutions use the same *notation* but are read off differently from the tableau.

DO EXERCISE 13.

Consider now setting up the initial simplex tableau for a minimum-type linear program.

Example 4   Given the minimum-type linear program of Example 2, find its dual. Then express this constraint set as a set of equations, set up the initial simplex tableau, and read off *both* the initial *primal* and *dual* solutions.

Solution   Starting now with the *primal*, that is, the maximum problem, we add slack variables and obtain:

$$40x_1 + 4x_2 + 4x_3 + y_1 \qquad\qquad = 15,$$
$$20x_1 + 4x_2 + 6x_3 \qquad + y_2 \qquad = 10,$$
$$\max x_0: -160x_1 - 24x_2 - 28x_3 \qquad\qquad + x_0 = 0,$$
$$x_1, x_2, x_3; \quad y_1, y_2 \geq 0.$$

Again, we have purposely used the same *notation* for the *slack* variables as for the *dual* variables (in this case, the original variables).

13. For the linear program of Margin Exercise 11, set up the initial simplex tableau and read off both the initial primal and dual solutions.

14. Express the constraint set of Margin Exercise 12 as a set of equations, set up the initial simplex tableau, and read off *both* the initial *primal* and *dual* solutions.

Putting this into a simplex tableau, we obtain:

x_1	x_2	x_3	y_1	y_2	1
40	4	4	1	0	15
20	4	6	0	1	10
-160	-24	-28	0	0	0

The primal solution can be read off as before:

$$x_1 = x_2 = x_3 = 0, \qquad y_1 = 15,$$
$$x_0 = 0, \qquad y_2 = 10.$$

Now, again the dual minimum program can be expressed as a set of equations by *subtracting* slack variables, obtaining;

$$40y_1 + 20y_2 - x_1 \qquad\qquad = 160,$$
$$4y_1 + 4y_2 \qquad - x_2 \qquad\qquad = 24,$$
$$4y_1 - 6y_2 \qquad\qquad - x_3 \qquad = 28,$$
$$\text{min } y_0\colon 15y_1 + 10y_2 \qquad\qquad - y_0 = \quad 0,$$
$$y_1, y_2; \quad x_1, x_2, x_3 \geqslant \quad 0,$$

where we have used the same *notation* for the dual slack variables as for the primal variables.

As with the primal, setting the nonbasic dual variables $y_1 = y_2 = 0$, we obtain, for the basic dual variables,

$$x_1 = -160, \qquad \text{and} \qquad y_0 = 0.$$
$$x_2 = -24,$$
$$x_3 = -28,$$

Alternately, we can read the dual solution directly from the bottom row of the simplex tableau, obtaining;

$$x_1 = -160, \qquad y_1 = y_2 = 0,$$
$$x_2 = -24, \qquad y_0 = 0.$$
$$x_3 = -28.$$

DO EXERCISE 14.

Thus, we see that the initial solution to the primal is

$$X = O, \qquad Y = B, \qquad x_0 = 0,$$

which is nonnegative, since we have constrained $B > O$. Similarly, the initial solution to the dual is

$$X = -C^T, \qquad Y = O, \qquad y_0 = 0,$$

which is generally not nonnegative.

The solution of the primal program has been considered in the preceding section. Consider now the solution to the dual.

Example 5 Solve the minimum-type problem of Example 2. The initial simplex tableau is given in Example 4.

Solution We proceed exactly the same way as for a maximum-type problem.

The most negative number in the bottom row is -160, so that x_1 is to become basic. Using x_1, we obtain the quotients to the right of the tableau.

x_1	x_2	x_3	y_1	y_2	1	q
40*	4	4	1	0	15	$\frac{3}{8}$ (Minimum)
20	4	6	0	1	10	$\frac{1}{2}$
-160	-24	-28	0	0	0	

The minimum quotient of $\frac{3}{8}$ implies that y_1 is to become nonbasic, or that the roles of x_1 and y_1 are to be interchanged. The pivot element is starred (*).

Pivoting to interchange the roles of x_1 and y_1, we obtain:

x_1	x_2	x_3	y_1	y_2	1	q
1	$\frac{1}{10}$	$\frac{1}{10}$	$\frac{1}{40}$	0	$\frac{3}{8}$	$\frac{15}{4}$
0	2	4*	$-\frac{1}{2}$	1	$\frac{5}{2}$	$\frac{5}{8}$ (Minimum)
0	-8	-12	4	0	60	

The *dual* solution at this point can be read from the bottom row and is

$$y_1 = 4, \qquad x_1 = 0, \qquad y_0 = 60.$$
$$y_2 = 0, \qquad x_2 = -8,$$
$$x_3 = -12,$$

Picking x_3 (most negative indicator) to become basic leads to y_2 (minimum-quotient) to become nonbasic.

15. Solve the minimum-type problem whose initial simplex tableau is given in Margin Exercise 14.

Pivoting to interchange the roles of x_3 and y_2, we obtain:

x_1	x_2	x_3	y_1	y_2	1	q
1	$\frac{1}{20}$	0	$\frac{3}{80}$	$-\frac{1}{40}$	$\frac{5}{16}$	$\frac{25}{4}$
0	$\frac{1}{2}^*$	1	$-\frac{1}{8}$	$\frac{1}{4}$	$\frac{5}{8}$	$\frac{5}{4}$ (Minimum)
0	-2	0	$\frac{5}{2}$	3	$\frac{135}{2}$	

The *dual* solution at this point is

$$y_1 = \tfrac{5}{2}, \qquad x_1 = 0, \qquad y_0 = \tfrac{135}{2}.$$
$$y_2 = 3, \qquad x_2 = -2,$$
$$x_3 = 0,$$

Pivoting now to interchange the roles of x_2 and x_3, we obtain:

x_1	x_2	x_3	y_1	y_2	1
1	0	$-\frac{1}{10}$	$\frac{1}{20}$	$-\frac{1}{20}$	$\frac{1}{4}$
0	1	2	$-\frac{1}{4}$	$\frac{1}{2}$	$\frac{5}{4}$
0	0	4	2	4	70

The *dual* solution at this point is

$$y_1 = 2, \qquad x_1 = 0, \qquad y_0 = 70.$$
$$y_2 = 4, \qquad x_2 = 0,$$
$$x_3 = 4,$$

This solution is nonnegative. It is therefore also minimal and the algorithm terminates.

Note that the nonnegativity of the dual solution corresponds to optimality of the primal.

DO EXERCISE 15.

Checking Solutions

To check either a maximum- or a minimum-type linear program, we substitute *both* primal and dual solutions into their respective programs written in *equation* form.*

---

* Actually, it is sufficient that if X is a primal feasible solution ($AX \leqslant B$, $X \geqslant O$) and Y is a dual feasible solution ($A^T Y \geqslant C^T$, $Y \geqslant O$), then the condition $CX = B^T Y$ insures that both X and Y are optimal. Effecting the check in *equation* form (rather than *constraint* form as in this footnote) checks slack variables as well as structural variables and hence is useful in *locating* any errors that may exist.

Example 6   Check the solution to the linear program of Example 4.

Solution   First, let us restate the programs in equation form and their solutions which can be obtained from the final tableau of Example 5.

Primal program:
$$40x_1 + 4x_2 + 4x_3 + y_1 = 15,$$
$$20x_1 + 4x_2 + 6x_3 + y_2 = 10,$$
$$\max x_0 \colon 160x_1 + 24x_2 + 28x_3 \qquad = x_0,$$
$$x_1, x_2, x_3; \quad y_1, y_2 \geqslant 0.$$

Primal solution:
$$x_1 = \tfrac{1}{4}, \quad x_2 = \tfrac{5}{4}, \quad x_3 = 0,$$
$$y_1 = y_2 = 0,$$
$$x_0 = 70.$$

Dual program:
$$40y_1 + 20y_2 - x_1 = 160,$$
$$4y_1 + 4y_2 - x_2 = 24,$$
$$4y_1 + 6y_2 - x_3 = 28,$$
$$\min y_0 \colon 15y_1 + 10y_2 \qquad = y_0,$$
$$y_1, y_2; \quad x_1, x_2, x_3 \geqslant 0.$$

Dual solution:
$$y_1 = 2, \quad y_2 = 4,$$
$$x_1 = x_2 = 0, \quad x_3 = 4,$$
$$y_0 = 70.$$

To check, we first note that all variables (structural and slack) are nonnegative. Next we evaluate the objective functions:

$$x_0 = 160(\tfrac{1}{4}) + 24(\tfrac{5}{4}) + 28(0) = 40 + 30 + 0 = 70, \qquad \text{OK.}$$
$$y_0 = 15(2) + 10(4) = 30 + 40 = 70, \qquad\qquad\qquad \text{OK.}$$

Next we substitute the primal solution into the primal equations:

$$40(\tfrac{1}{4}) + 4(\tfrac{5}{4}) + 4(0) + 0 = 10 + 5 + 0 + 0 = 15, \qquad \text{OK.}$$
$$20(\tfrac{1}{4}) + 4(\tfrac{5}{4}) + 6(0) + 0 = 5 + 5 + 0 + 0 = 10, \qquad \text{OK.}$$

Then we substitute the dual solution into the dual equations:

$$40(2) + 20(4) - 0 = 80 + 80 - 0 = 160, \qquad \text{OK.}$$
$$4(2) + 4(4) - 0 = 8 + 16 - 0 = 24, \qquad \text{OK.}$$
$$4(2) + 6(4) - 4 = 8 + 24 - 4 = 28. \qquad \text{OK.}$$

All equations check. Hence the solutions are correct.

DO EXERCISE 16.

16. Check your solution to Margin Exercise 12 using primal and dual solutions. (See Margin Exercise 15 for solutions.)

*Economic Interpretation of Duality (Optional)

Example 7   Formulate the linear program *dual* to that of Example 1 of Section 4.4.

Business, Nutrition. A feed supplier mixes feed for a particular animal which requires at least 160 lbs of nutrient A, at least 24 lbs of nutrient B, and at least 28 lbs of nutrient C. The supplier has available soybean meal which costs $15 per 100-lb sack and contains 40 lbs of nutrient A, 4 lbs of nutrient B, and 4 lbs of nutrient C and triticale which costs $10 per 100-lb sack and contains 20 lbs of nutrient A, 4 lbs of nutrient B, and 6 lbs of nutrient C. How many sacks of each ingredient should be used to mix animal feed which satisfies the minimum requirements at minimum cost?

Solution   The data from that problem was put into the following table:

	Composition (lbs per 100-lb sack)		Pounds required
	Soybean meal	Triticale	
Number of 100-lb sacks	y_1	y_2	
Nutrient A	40	20	160
Nutrient B	4	4	24
Nutrient C	4	6	28
$ Cost per 100-lb sack	15	10	Minimize cost

In formulating the dual we reinterpret the data. To start, we define dual variables:

$$x_1 = \text{\$ price per lb of nutrient A}$$
$$x_2 = \text{\$ price per lb of nutrient B} \bigg\} \text{ (if available in pure form),}$$
$$x_3 = \text{\$ price per lb of nutrient C}$$

where

$$x_1, x_2, x_3 \geqslant 0.$$

If we consider selling these nutrients separately (but still for animal feed), then we cannot obtain more for them than their cost combined as soybean meal or triticale; so that

$$40x_1 + 4x_2 + 4x_3 \leqslant 15 \quad \text{and} \quad 20x_1 + 4x_2 + 6x_3 \leqslant 10.$$

We now seek to determine the prices x_1, x_2, x_3 such that we maximize our income and still satisfy the animal's requirements (at no more than the cost of the food sources):

$$\max x_0 : x_0 = 160x_1 + 24x_2 + 28x_3.$$

It can be seen now that the program just formulated *is* the dual to the given program.

The optimal values of these dual variables x_1, x_2, x_3 are called *shadow prices* and are used in economic analyses.

Many other linear programs also have duals which have physical significance.

EXERCISE SET 5.3

Exercises 1 through 6 are Exercises 7 through 12 from Set 4.2 where they were solved graphically. Here you are asked to solve the problems using duality and the simplex algorithm and to check this solution with your graphical solution. Note that only the final solution checks and that intermediate solutions from the simplex tableaux have no corresponding points on the graph.

1. Exercise 7, Set 4.2 and Exercise 1, Set 5.2:

Minimize $y_0(y_1, y_2) = 3y_1 + 4y_2$, subject to the constraints

$$3y_1 + y_2 \geqslant 9,$$
$$y_1 + y_2 \geqslant 7,$$
$$y_1 + 2y_2 \geqslant 8,$$
$$y_1, y_2 \geqslant 0.$$

2. Exercise 8, Set 4.2 and Exercise 2, Set 5.2:

Minimize $y_0(y_1, y_2) = 3y_1 + 4y_2$, subject to the constraints

$$2y_1 + y_2 \geqslant 8,$$
$$4y_1 + 3y_2 \geqslant 22,$$
$$2y_1 + 5y_2 \geqslant 18,$$
$$y_1, y_2 \geqslant 0.$$

3. Exercise 9, Set 4.2 and Exercise 3, Set 5.2:

Minimize $y_0(y_1, y_2) = 2y_1 + 5y_2$, subject to the constraints

$$2y_1 + y_2 \geqslant 9,$$
$$4y_1 + 3y_2 \geqslant 23,$$
$$y_1 + 3y_2 \geqslant 8,$$
$$y_1, y_2 \geqslant 0.$$

4. Exercise 10, Set 4.2 and Exercise 4, Set 5.2:

Minimize $y_0(y_1, y_2) = 9y_1 + 7y_2$, subject to the constraints

$$5y_1 + 3y_2 \geqslant 30,$$
$$2y_1 + 3y_2 \geqslant 21,$$
$$3y_1 + 6y_2 \geqslant 36,$$
$$y_1, y_2 \geqslant 0.$$

5. Exercise 11, Set 4.2 and Exercise 5, Set 5.2:

Minimize $y_0(y_1, y_2) = 3y_1 + 5y_2$, subject to the constraints

$$4y_1 + y_2 \geqslant 9,$$
$$3y_1 + 2y_2 \geqslant 13,$$
$$2y_1 + 5y_2 \geqslant 16,$$
$$y_1, y_2 \geqslant 0.$$

6. Exercise 12, Set 4.2 and Exercise 6, Set 5.2:

Minimize $y_0(y_1, y_2) = 2y_1 + 3y_2$, subject to the constraints

$$4y_1 + y_2 \geqslant 7,$$
$$y_1 + y_2 \geqslant 4,$$
$$2y_1 + 5y_2 \geqslant 14,$$
$$y_1, y_2 \geqslant 0.$$

In Exercise Set 4.4, various minimum-type linear programs were given to be formulated and solved graphically. Here you are asked to solve these problems using duality and the simplex algorithm and to check your answer with your graphical solution.

7. Exercise 1, Set 4.4 and Exercise 7, Set 5.2:

$$50y_1 + 15y_2 \geq 120,$$
$$8y_1 + 5y_2 \geq 24,$$
$$5y_1 + y_2 \geq 10,$$
$$\min y_0: y_0 = 15y_1 + 5y_2; \quad y_1, y_2 \geq 0.$$

8. Exercise 2, Set 4.4 and Exercise 8, Set 5.2:

9. Exercise 3, Set 4.4 and Exercise 9, Set 5.2:

$$50y_1 + 20y_3 \geq 120,$$
$$8y_1 + 6y_3 \geq 24,$$
$$5y_1 + 8y_3 \geq 20,$$
$$\min y_0: y_0 = 15y_1 + 8y_3; \quad y_1, y_3 \geq 0.$$

10. Exercise 4, Set 4.4 and Exercise 10, Set 5.2:

11. Exercise 5, Set 4.4 and Exercise 11, Set 5.2:

$$40y_1 + 80y_2 \geq 2000,$$
$$40y_1 + 30y_2 \geq 1500,$$
$$120y_1 + 40y_2 \geq 2400,$$
$$\min y_0: y_0 = 12y_1 + 10y_2; \quad y_1, y_2 \geq 0.$$

12. Exercise 6, Set 4.4 and Exercise 12, Set 5.2:

13. Exercise 7, Set 4.4 and Exercise 13, Set 5.2:

$$40y_1 + 40y_3 \geq 2000,$$
$$40y_1 + 80y_3 \geq 1500,$$
$$120y_1 + 80y_3 \geq 2400,$$
$$\min y_0: y_0 = 12y_1 + 15y_3; \quad y_1, y_3 \geq 0.$$

14. Exercise 8, Set 4.4 and Exercise 14, Set 5.2:

15. Exercise 9, Set 4.4 and Exercise 15, Set 5.2:

$$40y_1 + 80y_2 \geq 1600,$$
$$40y_1 + 30y_2 \geq 2100,$$
$$120y_1 + 40y_2 \geq 2400,$$
$$\min y_0: y_0 = 12y_1 + 10y_2; \quad y_1, y_2 \geq 0.$$

16. Exercise 10, Set 4.4 and Exercise 16, Set 5.2:

17. Exercise 11, Set 4.4 and Exercise 17, Set 5.2:

$$40y_1 + 40y_3 \geq 1600,$$
$$40y_1 + 80y_3 \geq 2100,$$
$$120y_1 + 80y_3 \geq 2400,$$
$$\min y_0: y_0 = 12y_1 + 15y_3; \quad y_1, y_3 \geq 0.$$

18. Exercise 12, Set 4.4 and Exercise 18, Set 5.2. (You were not asked to prepare a graph in Exercise 12, Set 4.4.) Check your solution using an alternate pivot.

Use duality and the simplex algorithm to solve Exercises 19 through 26. Check your answer using primal and dual solutions.

19.
$$3y_1 + y_2 \geq 14,$$
$$4y_1 + 3y_2 \geq 34,$$
$$3y_1 + 4y_2 \geq 36,$$
$$\min y_0: y_0 = 7y_1 + 8y_2; \quad y_1, y_2 \geq 0.$$

20.
$$2y_1 + y_2 \geq 11,$$
$$y_1 + y_2 \geq 9,$$
$$y_1 + 2y_2 \geq 13,$$
$$\min y_0: y_0 = 3y_1 + 2y_2; \quad y_1, y_2 \geq 0.$$

21.
$$3y_1 + 2y_2 \geq 29,$$
$$4y_1 + 5y_2 \geq 55,$$
$$y_1 + 2y_2 \geq 18,$$
$$\min y_0: y_0 = 5y_1 + 4y_2; \quad y_1, y_2 \geq 0.$$

22.
$$5y_1 + y_2 \geq 15,$$
$$4y_1 + 2y_2 \geq 24,$$
$$5y_1 + 5y_2 \geq 35,$$
$$\min y_0: y_0 = 13y_1 + 5y_2; \quad y_1, y_2 \geq 0.$$

23.
$$2x_1 + 9x_2 + 4x_3 \geq 6,$$
$$3x_1 + 6x_2 - 2x_3 \geq 8,$$
$$x_1 + x_2 + x_3 \geq 3,$$
$$\min x_0: x_0 = 5x_1 + 2x_2 + x_3; \quad x_1, x_2, x_3 \geq 0.$$

24.
$$2x_1 + x_2 + 5x_3 \geq 5,$$
$$2x_1 - x_2 + 2x_3 \geq 3,$$
$$5x_1 + 3x_2 + x_3 \geq 7,$$
$$\min x_0: x_0 = 6x_1 + 2x_2 + 7x_3; \quad x_1, x_2, x_3 \geq 0.$$

25.
$$2y_1 + 3y_2 + y_3 \geq 2,$$
$$5y_1 + 2y_2 - 3y_3 \geq 4,$$
$$7y_1 + 6y_2 + 4y_3 \geq 5,$$
$$\min y_0: y_0 = 8y_1 + 9y_2 + 5y_3; \quad y_1, y_2, y_3 \geq 0.$$

26.
$$6y_1 + 3y_2 + 4y_3 \geq 2,$$
$$y_1 - 2y_2 - 5y_3 \geq 3,$$
$$2y_1 + 9y_2 + y_3 \geq 8,$$
$$\min y_0: y_0 = 12y_1 + 9y_2 + 8y_3; \quad y_1, y_2, y_3 \geq 0.$$

27. *Nutrition.* The calcium, iron, protein, and cost of various foods (per 100 g) is given in the accompanying table.

Food (100 g)	Calcium (mg)	Iron (mg)	Protein (g)	Cost (¢)
Eggs	54	2.7	12.8	22
Beef	7	6.6	18.6	36
Chicken	14	1.5	20.2	16
Bluefish	23	0.6	20.5	24
Whole-wheat bread	96	2.2	9.3	10
Cheddar cheese	570	0.7	20.5	32
Soy beans	260	10.0	34.9	10
Sunflower seeds	57	6.0	28.0	20

27. *(Cont.)* Using eggs, beef, cheese, and soy beans, what is the minimum-cost diet that will satisfy the minimum daily requirements?

Food (100 g)	Calcium (mg)	Iron (mg)	Protein (g)	Cost (¢)
Sesame seeds	72	7.7	23.4	18
Almonds	234	4.7	18.6	48
Cashews	38	3.8	17.2	48
Filberts	209	3.4	12.6	40
Millet	20	6.8	9.9	12
Minimum daily requirement	750	10	50*	

* This is an average value. Some nutritionists recommend more, some less.

28. Using data from the table of Exercise 27, select various combinations of 3 or 4 foods (other than soy beans) and determine the minimum-cost diet that will satisfy the minimum daily requirements. Answers may vary.

29. Using data from the table of Exercise 27, for eggs, beef, and cheese, what is the minimum-cost diet that will satisfy the minimum daily requirements?

OBJECTIVE

You should be able to solve a linear program and indicate all such special characteristics of both the primal and dual solutions as degeneracy, nonuniqueness, unboundedness, or emptiness. If the solution is not unique, find all basic solutions.

5.4 DEGENERACIES AND OTHER SPECIAL CASES

As was shown in the graphical method of solution of linear programs, solutions may be:

  i) unique,
 ii) nonunique,
iii) unbounded, or
iv) empty.

On a graph, such solutions may be easily recognized, but our present concern is with recognizing these situations from the simplex tableau.

Up to now, we have used the simplex algorithm to solve linear programs with unique solutions. We turn our attention now to those with nonunique solutions and to the related subject of degeneracies.

Degeneracies (see Section 4.1) occur when a basic variable has zero value. Thus, primal degeneracy occurs when a basic primal variable has zero value. Similarly, dual degeneracy occurs when a basic dual variable has zero value.

Degeneracies may occur in any part of the sequence of tableaux of the simplex algorithm. If primal degeneracy occurs before the final tableau has been obtained, special consideration is required in pivoting.

Degenerate Pivots

Example 1   Pivot in the following tableau:

x_1	x_2	x_3	y_1	y_2	y_3	1	q	
4	-1	2	1	0	0	2	—	
3	1^*	-1	0	1	0	0	0	Minimum
5	3	-2	0	0	1	1	$\frac{1}{3}$	
4	-2	1	0	0	0	0		

Solution   In this case, the only pivot column available is the second column. The minimum quotient (to the right of the tableau) has zero value corresponding to the second row. The pivot element has been starred. Such pivots (corresponding to zero quotients) are called *degenerate pivots.* We perform the pivot operation the same as we do with nondegenerate pivots, and obtain:

x_1	x_2	x_3	y_1	y_2	y_3	1	q
7	0	1	1	1	0	2	2
3	1	−1	0	1	0	0	—
−4	0	1*	0	−3	1	1	1 Min
10	0	−1	0	2	0	0	

Pivoting with a degeneracy permits the computation to proceed. The next pivot is nondegenerate:

x_1	x_2	x_3	y_1	y_2	y_3	1
11	0	0	1	4	−1	1
−1	1	0	0	−2	1	1
−4	0	1	0	−3	1	1
6	0	0	0	−1	1	1

This tableau contains no degeneracy and one can proceed as usual.

The degenerate pivot requires *no* modification of the rules other than noting that no modification of the rules is required.*

If there is a choice of pivots, one degenerate and one nondegenerate, choose the nondegenerate pivot, since the solution will be advanced and the degenerate pivot *may* be avoided.

DO EXERCISE 17.

Nonunique Solutions

Let us turn our attention to degeneracies in the *terminal* tableau.

Example 2   Obtain all possible solutions to the following linear

17. Pivot in the following tableau until the degeneracy has been removed or the final tableau reached:

x_1	x_2	x_3	y_1	y_2	y_3	1
3	−2	1	1	0	0	0
−7	1	3	0	1	0	2
2	3	6	0	0	1	7
6	3	−2	0	0	0	0

---

* Cycling (that is, obtaining a complete set of values previously obtained) occurs so rarely that techniques to avoid cycling have been omitted.

program:

$$4x_1 - x_2 + 2x_3 \leqslant 2,$$
$$3x_1 + x_2 + x_3 \leqslant 1,$$
$$5x_1 + 3x_2 + 2x_3 \leqslant 4,$$
$$\max x_0: x_0 = 4x_1 + 3x_2 + 3x_3; \quad x_1, x_2, x_3 \geqslant 0.$$

Solution Starting with the second column as pivot, we obtain the following tableaux:

x_1	x_2	x_3	y_1	y_2	y_3	1	q
4	−1	2	1	0	0	2	1
3	1*	1	0	1	0	1	1 Min.
5	3	2	0	0	1	4	$\frac{4}{3}$
−4	−3	−3	0	0	0	0	

x_1	x_2	x_3	y_1	y_2	y_3	1
7	0	3	1	1	0	3
3	1	1	0	1	0	1
−4	0	−1	0	−3	1	1
5	0	0	0	3	0	3

Thus, the primal solution is

$$y_1 = 3, \qquad x_1 = y_2 = x_3 = 0,$$
$$x_2 = 1,$$
$$y_3 = 1, \qquad x_0 = 3,$$

and the dual solution is

$$x_1 = 5, \qquad y_1 = x_2 = y_3 = 0,$$
$$y_2 = 3,$$
$$x_3 = 0, \qquad y_0 = 3.$$

Alternately, let us start pivoting with the *third* column:

x_1	x_2	x_3	y_1	y_2	y_3	1	q
4	−1	2	1	0	0	2	1 Min
3	1	1*	0	1	0	1	1 Min
5	3	2	0	0	1	4	2
−4	−3	−3	0	0	0	0	

x_1	x_2	x_3	y_1	y_2	y_3	1
-2	-3	0	1	-2	0	0
3	1	1	0	1	0	1
-1	1	0	0	-2	1	2
5	0	0	0	3	0	3

Thus, this primal solution is

$$y_1 = 0, \qquad x_1 = x_2 = y_2 = 0,$$
$$x_3 = 1,$$
$$y_3 = 2, \qquad x_0 = 3,$$

and the dual solution is

$$x_1 = 5, \qquad y_1 = x_3 = y_3 = 0,$$
$$x_2 = 0,$$
$$y_2 = 3, \qquad y_0 = 3.$$

It is apparent that we have obtained *two* different solutions to the linear program. Each such solution of a linear program in terms of *basic* and *nonbasic* variables is called a *basic feasible solution*.

If two solutions are represented by the vectors X_1 and X_2, then any combination

$$cX_1 + (1 - c)X_2, \qquad \text{where } 0 \leq c \leq 1,$$

is also a solution. Such a combination is called a *convex combination*.

In general, the convex combination of n points X_i can be written:

$$\sum_{i=1}^{n} c_i X_i, \qquad \text{where } 0 \leq c_i \leq 1 \quad \text{for all } i \text{ and } \sum_{i=1}^{n} c_i = 1.$$

Any convex combination of basic feasible solutions is also a solution.

How do we know whether one solution is unique? Dual degeneracy indicates primal nonuniqueness. Once we know that the solution is nonunique, we try alternative pivots until the other solutions have been obtained.

Similarly, primal degeneracy indicates dual nonuniqueness. This particular example is interesting because the terminal tableau just obtained exhibits both primal and dual degeneracies.

Let us, then, choose the other available pivot in the third column:

x_1	x_2	x_3	y_1	y_2	y_3	1	q	
4	−1	2*	1	0	0	2	1	Min.
3	1	1	0	1	0	1	1	Min.
5	3	2	0	0	1	4	2	
−4	−3	−3	0	0	0	0		

x_1	x_2	x_3	y_1	y_2	y_3	1	q	
2	$-\frac{1}{2}$	1	$\frac{1}{2}$	0	0	1	—	
1	$\frac{3}{2}$*	0	$-\frac{1}{2}$	1	0	0	0	Min.
1	4	0	−1	0	1	2	$\frac{1}{2}$	
2	$-\frac{9}{2}$	0	$\frac{3}{2}$	0	0	3		

x_1	x_2	x_3	y_1	y_2	y_3	1
$\frac{7}{3}$	0	1	$\frac{1}{3}$	$\frac{1}{3}$	0	1
$\frac{2}{3}$	1	0	$-\frac{1}{3}$	$\frac{2}{3}$	0	0
$-\frac{5}{3}$	0	0	$\frac{1}{3}$	$-\frac{8}{3}$	1	2
5	0	0	0	3	0	3

This gives us, for the primal solution:

$$x_3 = 1, \quad x_1 = y_2 = y_1 = 0,$$
$$x_2 = 0,$$
$$y_3 = 2, \quad x_0 = 3,$$

and for the dual solution

$$x_1 = 5, \quad x_3 = x_2 = y_3 = 0,$$
$$y_2 = 3,$$
$$y_1 = 0, \quad y_0 = 3.$$

This basic feasible solution is distinct from the previous one, allowing for *all* variables (basic and nonbasic). However, with regard to the original problem, they both yield the same solution to the original linear program,

$$x_1 = x_2 = 0, \quad x_3 = 1, \quad x_0 = 3,$$

differing only in the value of the slack variables.

They both also yield the same solution to the dual program,

$$y_1 = y_3 = 0, \qquad y_2 = 3, \qquad y_0 = 3,$$

where again the difference is in the slack variables.

Let us pivot now on the first column of the initial tableau:

x_1	x_2	x_3	y_1	y_2	y_3	1	q	
4	−1	2	1	0	0	2	$\frac{1}{2}$	
3*	1	1	0	1	0	1	$\frac{1}{3}$	Min.
5	3	2	0	0	1	4	$\frac{4}{5}$	
−4	−3	−3	0	0	0	0		

x_1	x_2	x_3	y_1	y_2	y_3	1
0	$-\frac{7}{3}$	$\frac{2}{3}$	1	$-\frac{4}{3}$	0	$\frac{2}{3}$
1	$\frac{1}{3}*$	$\frac{1}{3}*$	0	$\frac{1}{3}$	0	$\frac{1}{3}$
0	$\frac{4}{3}$	$\frac{1}{3}$	0	$-\frac{5}{3}$	1	$\frac{7}{3}$
0	$-\frac{5}{3}$	$-\frac{5}{3}$	0	$\frac{4}{3}$	0	$\frac{4}{3}$

We can now pivot with either the second or third column. Doing both in turn, we obtain:

x_1	x_2	x_3	y_1	y_2	y_3	1
7	0	3	1	1	0	3
3	1	1	0	1	0	1
−4	0	−1	0	−3	1	1
5	0	0	0	3	0	3

x_1	x_2	x_3	y_1	y_2	y_3	1
−2	−3	0	1	−2	0	0
3	1	1	0	1	0	1
−1	1	0	0	−2	1	2
5	0	0	0	3	0	3

Each of these basic feasible solutions has been previously obtained. Since all possible pivot choices have been exhausted, there are no more basic feasible solutions.

18. Obtain all possible basic solutions to the following linear program:

$$4x_1 + x_2 - 2x_3 \leqslant 2,$$
$$3x_1 \quad\quad + 4x_3 \leqslant 11,$$
$$6x_1 + 3x_2 + 2x_3 \leqslant 8,$$
$$\max x_0: x_0 = 10x_1 + 4x_2, \quad x_1, x_2, x_3 \geqslant 0.$$

A complete discussion of the number of basic feasible solutions is beyond the scope of this text; however, when nonuniqueness is indicated, the student should seek *two* basic solutions, unless there is evidence of more.

DO EXERCISE 18.

Unbounded Solutions

Now let us look at linear programs with unbounded solution sets. Consider the following tableau:

x_1	x_2	x_3	y_1	y_2	y_3	1
1	−4	2	1	0	0	2
3	0	4	0	1	0	9
−3	−6	2	0	0	1	5
1	−1	1	0	0	0	0

Here we must pivot on the second column, but *all* entries in this column (including the bottom one) are *nonpositive*. Thus, no pivot now can be found and the algorithm cannot be continued. The *dual* solution is clearly *infeasible*. Setting all primal nonbasic variables to zero except for that of the pivot column, we can write the basic solution as:

$$y_1 = 2 + 4x_2,$$
$$y_2 = 9,$$
$$y_3 = 5 + 6x_2,$$

and

$$x_0 = 0 + x_2.$$

Thus, as x_2 increases indefinitely, the basic feasible solution and the objective function become large without bound.

At *any* stage of the simplex algorithm, if there is a column with *all* nonpositive entries (and the primal solution is feasible), then the primal solution is unbounded and the dual solution is infeasible.

The corresponding situation where the dual solution is unbounded and the primal solution is infeasible cannot occur with the present constraint $B \geqslant O$, which corresponds to primal feasibility. Similarly, the possibility that both solution sets are empty, that is, that both are infeasible, also cannot occur with the constraint $B \geqslant O$.

Example 3   Show that the solution to the following linear program is unbounded:

$$2x_1 + x_2 - 4x_3 \leq 2,$$
$$3x_1 - x_2 + 2x_3 \leq 11,$$
$$3x_1 - 2x_2 + x_3 \leq 8,$$
$$\max x_0: x_0 = 6x_1 + 4x_2 + 2x_3; \quad x_1, x_2, x_3 \geq 0.$$

Solution   Setting up this program in a tableau and pivoting with the second column, we obtain:

x_1	x_2	x_3	y_1	y_2	y_3	1
2	1*	−4	1	0	0	2
3	−1	2	0	1	0	11
3	−2	1	0	0	1	8
−6	−4	−2	0	0	0	0

x_1	x_2	x_3	y_1	y_2	y_3	1
2	1	−4	1	0	0	2
5	0	−2	1	1	0	13
7	0	−7	2	0	1	12
2	0	−18	4	0	0	8

Since the third column has all nonpositive entries, the primal solution is unbounded (and the dual solution infeasible.)

Let us now ask what would happen if an alternate pivot were chosen, for example, the first column. In that case, we obtain:

x_1	x_2	x_3	y_1	y_2	y_3	1
2*	1	−4	1	0	0	2
3	−1	2	0	1	0	11
3	−2	1	0	0	1	8
−6	−4	−2	0	0	0	0

19. Show by pivoting that the following linear program has an unbounded solution:

$$4x_1 + x_2 - 2x_3 \leq 2,$$
$$3x_1 - x_2 \leq 11,$$
$$6x_1 - 2x_2 + 3x_3 \leq 8,$$

$$\max x_0: x_0 = 10x_1 + 4x_2 + 6x_3,$$

$$x_1, x_2, x_3 \geq 0.$$

x_1	x_2	x_3	y_1	y_2	y_3	1
1	$\frac{1}{2}$	-2	$\frac{1}{2}$	0	0	1
0	$-\frac{5}{2}$	8	$-\frac{3}{2}$	1	0	8
0	$-\frac{7}{2}$	7*	$-\frac{3}{2}$	0	1	5
0	-1	-14	3	0	0	6

Pivoting on the second column to interchange x_1 and x_2 yields the same result as before. Therefore, consider pivoting the third column. Thus, we obtain:

x_1	x_2	x_3	y_1	y_2	y_3	1
1	$-\frac{1}{2}$	0	$\frac{1}{14}$	0	$\frac{2}{7}$	$\frac{17}{7}$
0	$\frac{3}{2}$*	0	$\frac{3}{14}$	1	$-\frac{8}{7}$	$\frac{16}{7}$
0	$-\frac{1}{2}$	1	$-\frac{3}{14}$	0	$\frac{1}{7}$	$\frac{5}{7}$
0	-8	0	0	0	2	16

x_1	x_2	x_3	y_1	y_2	y_3	1
1	0	0	$\frac{1}{7}$	$\frac{1}{3}$	$-\frac{2}{21}$	$\frac{67}{21}$
0	1	0	$\frac{1}{7}$	$\frac{2}{3}$	$-\frac{16}{21}$	$\frac{32}{21}$
0	0	1	$-\frac{1}{7}$	$\frac{1}{3}$	$-\frac{1}{21}$	$\frac{56}{21}$
0	0	0	$\frac{8}{7}$	$\frac{16}{3}$	$-\frac{86}{21}$	$\frac{592}{21}$

No matter which way we go, we obtain a tableau with a column of nonpositive entries. If the algorithm can be continued, we ultimately obtain a column of nonpositive entries in a tableau that cannot be continued. Under these conditions, if the primal solution is feasible, it is also unbounded.

DO EXERCISE 19.

(Optional) The Condensed Tableau

The presence of the zeros in the unit matrix in the previous, "extended," tableau takes up space unnecessarily. On a high-speed computer, the saving in space obtained by omitting these zeros can be significant for real problems with large numbers of variables and constraints. Increased insight is also provided by the "condensed" tableau.

Example 4 Condense the (extended) tableau of Example 2 of Section 5.1.

Solution

x_1	x_2	y_1	y_2	y_3	1
0.20	0.10	1	0	0	220
0.60	0.20	0	1	0	620
0.20	0.70	0	0	1	700
−3	−2	0	0	0	0

Extended tableau

1	x_1	x_2	1
y_1	0.20	0.10	220
y_2	0.60	0.20	620
y_3	0.20	0.70	700
1	−3	−2	0

Condensed tableau

The *basic* variables of the *primal* program are written in a column at the left of the condensed tableau. The *dual* program can be read directly from the condensed tableau noting that the double *horizontal* line indicates the location of the equal sign (as does the double *vertical* line for the *primal* program). The student is invited to read off the primal and dual programs from the condensed tableau and check them with the results in Section 5.3.

To illustrate the process of condensing, consider the pair of dual programs

i)
$$AX \le B, \quad X \ge O,$$
$$\max x_0: x_0 = CX$$

ii)
$$A^T Y \ge C^T, \quad Y \ge O,$$
$$\min y_0: y_0 = B^T Y,$$

where, by the Duality Theorem, $\min y_0 = \max x_0$ provided these optima exist.

Adding slack variables (in the notation of dual variables), we obtain the matrix equations

$$[A \quad I]\begin{bmatrix} X \\ Y \end{bmatrix} = B; \quad X \ge O, \quad Y \ge O,$$

$$x_0 = [C \quad O]\begin{bmatrix} X \\ Y \end{bmatrix}.$$

These equations can be put into the following tableaux:

X^T	Y^T	1
A	I	B
$-C$	O	0

Extended tableau

1	X^T	1
Y	A	B
1	$-C$	0

Condensed tableau

Since the condensed tableau can be read horizontally or vertically, we can put a minimum-type linear program directly into condensed tableau form.

The actual computation of successive condensed tableaux is basically the same as for successive extended tableaux. The only difference is in the way the information is stored.

The student is invited to solve Example 2 of Section 5.1 using the condensed tableaux and check the work by condensing the successive extended tableaux given there. Note the similarities and differences in the storage of data.

EXERCISE SET 5.4

Solve the following, and indicate all such special characteristics of both the primal and dual solutions as degeneracy, nonuniqueness, unboundedness, or emptiness. If the solution is not unique, find all basic solutions.

1.
$$2x_1 + x_2 + 4x_3 \leq 24,$$
$$x_1 + x_2 + x_3 \leq 4,$$
$$2x_1 - x_2 + 3x_3 \leq 12,$$
$$\max x_0: x_0 = \tfrac{11}{4}x_1 + 2x_2 + 3x_3; \quad x_1, x_2, x_3 \geq 0.$$

2.
$$2x_1 + 5x_2 + 7x_3 \leq 6,$$
$$3x_1 + 2x_2 + 6x_3 \leq 9,$$
$$x_1 - 3x_2 + 4x_3 \leq 5,$$
$$\max x_0: x_0 = 2x_1 + 4x_2 + 5x_3; \quad x_1, x_2, x_3 \geq 0.$$

3.
$$x_1 - x_2 + 2x_3 \leq 3,$$
$$3x_1 - 4x_2 + 7x_3 \leq 10,$$
$$2x_1 - 3x_2 + 4x_3 \leq 9,$$
$$\max x_0: x_0 = 2x_1 - x_2 + 5x_3; \quad x_1, x_2, x_3 \geq 0.$$

4.
$$6x_1 + x_2 - 5x_3 \leq 12,$$
$$3x_1 - 4x_2 + 9x_3 \leq 9,$$
$$4x_1 - 5x_2 + x_3 \leq 8,$$
$$\max x_0: x_0 = 2x_1 + 3x_2 + 8x_3; \quad x_1, x_2, x_3 \geq 0.$$

5.
$$6x_1 + x_2 + 2x_3 \leq 12,$$
$$-2x_1 - x_2 + x_3 \leq 3,$$
$$6x_1 - x_2 + 2x_3 \leq 6,$$
$$\max x_0: x_0 = 2x_1 - 2x_2 + 4x_3; \quad x_1, x_2, x_3 \geq 0.$$

6.
$$3x_1 + 3x_2 - 2x_3 \leq 5,$$
$$5x_1 - 2x_2 + 3x_3 \leq 6,$$
$$3x_1 - x_2 + x_3 \leq 2,$$
$$\max x_0: x_0 = x_1 - 2x_2 + 2x_3; \quad x_1, x_2, x_3 \geq 0.$$

OBJECTIVE

You should be able to solve the transportation problem.

5.5 THE TRANSPORTATION PROBLEM

The transportation problem is a linear program concerned with minimizing the cost of distribution of a single product from a company's various factories (sources of supply) to its various distributors (points of demand).

Transportation problems can be solved with techniques previously described; however, the special structure of these problems permits special, more efficient, methods of solution.

Example 1 *Business.*   A company can produce 50 cases of corkscrews at one factory in Chicago and 80 cases at a second factory in Atlanta. The company sells these corkscrews through three distributors; one in New York with orders for 70 cases, a second in Denver with orders for 20 cases, and a third in San Francisco with orders for 40 cases.

The unit cost of transportation from factory to distributor is given in the following table:

Factory, i	Distributor, j			Supply capacity s_i
	New York 1	Denver 2	San Francisco 3	
Chicago 1	3	7	11	50
Atlanta 2	7	9	17	80
Demand required, r_j	70	20	40	Minimize cost

The problem is to determine how many cases should be shipped from each factory to each distributor to meet the demand with the supply available, at minimum cost. Set this problem up as a linear program and obtain an initial feasible solution.

Solution   To formulate this problem, let

$$x_{ij} = \text{the number of cases of corkscrews shipped}$$
$$\text{from factory } i \text{ to distributor } j.$$

Then, since a factory can ship *no more* than its capacity, we have

$$x_{11} + x_{12} + x_{13} \leqslant 50,$$
$$x_{21} + x_{22} + x_{23} \leqslant 80.$$

And, since each distributor must have *at least* enough stock to satisfy current orders, we have

$$x_{11} + x_{21} \geqslant 70,$$
$$x_{12} + x_{22} \geqslant 20,$$
$$x_{13} + x_{23} \geqslant 40.$$

Here the quantities x_{ij} are not only nonnegative

$$x_{ij} \geqslant 0, \quad \text{for all } i, j$$

but also *integer*, since only case lots can be shipped.

The cost of shipment, x_0, is given by

$$x_0 = 3x_{11} + 7x_{12} + 11x_{13} + 7x_{21} + 9x_{22} + 17x_{23},$$

and is to be minimized.

Note now that, in the present example, the total *supply available* from both factories is

$$50 + 80 = 130$$

and the total *demand required* by all three distributors is

$$70 + 20 + 40 = 130$$

so that the supply *just equals* the demand. In this case the inequalities in the preceding constraints can be replaced by equal signs, yielding

$$x_{11} + x_{12} + x_{13} = 50,$$
$$x_{21} + x_{22} + x_{23} = 80,$$

and

$$x_{11} + x_{21} = 70,$$
$$x_{12} + x_{22} = 20,$$
$$x_{13} + x_{23} = 40.$$

Let

$$s_i = \text{supply capacity of factory } i,$$
$$r_j = \text{demand requirement of distributor } j, \text{ and}$$
$$c_{ij} = \text{unit cost of shipping from factory } i \text{ to distributor } j.$$

Then, for the *balanced* transportation problem the supply just equals the demand:

$$\sum_i s_i = \sum_j r_j.$$

That is, the sum of the supplies from all factories equals the sum of the demands by all distributors and our formulation will always have the form

$$\sum_j x_{ij} = s_i, \quad \text{for all } i,$$

$$\sum_i x_{ij} = r_j, \quad \text{for all } j,$$

$$x_{ij} \geq 0 \quad \text{and integer}, \quad \text{for all } i \text{ and } j,$$

$$\min x_0 \colon x_0 = \sum_{ij} c_{ij} x_{ij}.$$

The *integer* constraint is new in our *formulation* of linear programs and such problems are called *integer* programs. They *usually* require special solution techniques; however, because of the special structure of transportation problems, we can solve them as linear programs and their solutions will *automatically* be integer. (This is not true of all integer programs.)

Equation constraints can be incorporated into the simplex algorithm by introducing slack and artificial* variables and solving using the two-phase method.* However, we can obtain the solution much more simply in the following manner. We use a modification of the table representing the given data with each cell now containing the following information:

$$D_j$$

c_{ij}	
F_i	s_i
	x_{ij}

$$r_j$$

To solve the transportation problem, we need an initial feasible solution which we shall obtain by the *minimum-cost assignment* method. This will provide us with a set of basic and nonbasic values of x_{ij}. The basic values we *write in* in the lower right corner of the cell and the nonbasic values (which are zero) we indicate with a *blank* (or a dash "—").

---

* See Section 5.2.

Minimum-Cost Assignment Method for an Initial Feasible Solution

At each stage of this method we seek the *minimum* value of c_{ij} (if there are more than one with this minimum value, choose either) and assign to this x_{ij} the maximum value which will neither overdraw the supply s_i of factory i (F_i) nor oversupply r_j the distributor j (D_j).

Thus, we obtain:

	D_1	D_2	D_3	
F_1	3 _50_	7	11	50
F_2	7 _20_	9 _20_	17 _40_	80
	70	20	40	

with the following assignment steps:

1. The minimum c_{ij} is $c_{11} = 3$. The maximum assignment that we can make for x_{11} without exceeding the supply $s_1 = 50$ of factory F_1 or exceeding the requirement $r_1 = 70$ of distributor D_1 is $x_{11} = 50$ and we write this number in the lower right corner of cell $F_1 D_1$.

2. The next minimum c_{ij} is $c_{12} = c_{21} = 7$. Since the supply from F_1 has been exhausted by D_1, we cannot make an assignment for x_{12}, so we leave its place in its cell blank. The maximum assignment we can make for x_{21} is 20 and we write this in its cell.

3. The next minimum c_{ij} is $c_{22} = 9$. The maximum assignment we can make for x_{22} is 20 and we write this in.

4. The next minimum c_{ij} is $c_{13} = 11$ but F_1 is exhausted, so no assignment can be made for x_{13} and we leave a blank.

5. The next minimum c_{ij} is $c_{23} = 17$. The maximum assignment we can make for x_{23} is 40 and we write this in.

At this point we have exhausted the supply of all the factories and met the requirements of all the distributors (but not necessarily at minimum cost) with *four* assignments.

For the balanced transportation problem (with $\sum_i s_i = \sum_j r_j$), one equation is linearly dependent on the others, so the number of linearly independent equations—which is equal to the number of basic variables (see Section 5.1)—is one less than the sum of the number of factories and distributors. In this case the number of basic variables is $2 + 3 - 1 = 4$. Since we have made four assignments, we have a *full* set of basic variables.

If we were able to make assignments which exhaust the supply and meet the requirements with *fewer* assignments, then we must make *zero* assignments until a full set of basic variables is obtained and write these zeroes into the cells.

The cost of this initial set of assignments is

$$x_0 = 3(50) + 7(20) + 9(20) + 17(40)$$
$$= 150 + 140 + 180 + 680$$
$$= 1150.$$

Thus, from the table our initial feasible solution is

Basic variables: $x_{11} = 50$, $x_{21} = 20$, $x_{22} = 20$, $x_{23} = 40$,

Nonbasic variables: $x_{12} = x_{13} = 0$,

Objective function: $x_0 = 1150$.

DO EXERCISE 20.

Steppingstone Algorithm

Example 2 Solve the transportation problem of Example 1.

Solution We solve the transportation problem using the *steppingstone* algorithm. As in the simplex algorithm, we make one nonbasic variable basic and one basic variable nonbasic.

To determine which nonbasic variable to make basic, we consider closed paths starting with and returning to a given nonbasic cell and passing through a series of basic cells each associated rowwise or columnwise with its predecessor.

Here there are two *nonbasic* cells: F_1D_2 and F_1D_3. We consider a circuit for each:

1. F_1D_2: F_1D_2—F_2D_2—F_2D_1—F_1D_1—F_1D_2,
2. F_1D_3: F_1D_3—F_2D_3—F_2D_1—F_1D_1—F_1D_3.

For each circuit we determine the change in cost obtainable by reassigning one unit of our commodity around this circuit. To do this, we add one unit to the starting nonbasic cell and subtract one unit from the next basic cell in our circuit, then add one unit to the next basic cell and subtract one unit from the next basic cell, and so on, until the circuit is complete.

20. For the following transportation problem with data given in the format of Example 1, find an initial feasible solution.

		j		
i	1	2	3	s_i
1	7	11	4	120
2	18	9	13	70
r_j	60	50	80	Min. cost

The change in cost obtained by doing this for each circuit is

1. $F_1 D_2$: $c_{12} - c_{22} + c_{21} - c_{11}$
$$= 7 - 9 + 7 - 3$$
$$= 2;$$

2. $F_1 D_3$: $c_{13} - c_{23} + c_{21} - c_{11}$
$$= 11 - 17 + 7 - 3$$
$$= -2.$$

Thus, for $F_1 D_2$ the unit cost is *increased* by 2, but for $F_1 D_3$ the unit cost is *decreased* by 2.

For each circuit representing a *decrease* in *unit* cost, we now seek the *maximum* number of units that can be reassigned around this circuit without disturbing the supply and requirement balances. This is the *minimum* value of x_{ij} of those cells of the circuit with *minus* signs associated with c_{ij}.

Thus, for $F_1 D_3$ we seek

$$\min \{x_{23}, x_{11}\} = \min \{40, 50\}$$
$$= x_{23} = 40.$$

The *overall* change in cost for this circuit is

$$x_{23}(c_{13} - c_{23} + c_{21} - c_{11})$$
$$= 40(-2)$$
$$= -80.$$

We select that circuit which yields the maximum *overall* decrease in cost. This determines the basic variable to become nonbasic, in this case x_{23}. We modify the previous solution accordingly to obtain the next feasible solution:

	D_1	D_2	D_3	
F_1	3	7	11	50
	10		40	
F_2	7	9	17	80
	60	20		
	70	20	40	

The new value of the objective function is

$$x_0 = 3(10) + 11(40) + 7(60) + 9(20)$$
$$= 30 + 440 + 420 + 180$$
$$= 1070; \text{ or}$$
$$x_0 = 1150 - 80$$
$$= 1070.$$

We now have two nonbasic cells: F_1D_2 and F_2D_3. The unit change in cost for the circuits for each is

1. F_1D_2: $c_{12} - c_{11} + c_{21} - c_{13}$
$$= 7 - 3 + 7 - 9$$
$$= 2;$$

2. F_2D_3: $c_{23} - c_{21} + c_{11} - c_{13}$
$$= 17 - 7 + 3 - 11$$
$$= 2.$$

No circuit is possible which yields a decrease in cost. Hence, the current solution is minimal:

Basic variables: $x_{11} = 10$, $x_{13} = 40$, $x_{21} = 60$, $x_{22} = 20$,

Nonbasic variables: $x_{12} = x_{23} = 0$,

Objective function: $x_0 = 1070$.

If the current solution were not minimal, we would continue as before.

DO EXERCISE 21.

21. Solve the transportation problem of Margin Exercise 20.

EXERCISE SET 5.5

Solve the following transportation problems for the minimum-cost assignment where the data is given in the format of Example 1.

1.

i	1	2	3	s_i
1	5	11	17	150
2	15	9	13	80
r_j	70	90	70	Min. cost

2.

i	1	2	3	s_i
1	3	11	9	190
2	5	8	7	130
r_j	110	90	120	Min. cost

3.

i	j 1	2	s_i
1	14	17	70
2	13	9	80
3	11	12	100
r_j	120	130	Min. cost

4.

i	j 1	2	s_i
1	21	17	80
2	13	15	90
3	19	11	70
r_j	110	130	Min. cost

5.

i	j 1	2	3	4	s_i
1	6	10	7	13	150
2	9	14	8	17	100
r_j	45	60	70	75	Min. cost

6.

i	j 1	2	3	4	s_i
1	11	17	13	21	150
2	19	14	9	15	200
r_j	70	80	90	110	Min. cost

7.

i	j 1	2	3	s_i
1	7	11	9	80
2	13	17	14	90
3	21	15	17	110
r_j	70	120	90	Min. cost

8.

i	j 1	2	3	s_i
1	11	7	13	90
2	9	17	15	70
3	17	14	21	100
r_j	80	110	70	Min. cost

Hint. If the basic cells are assigned as follows:

	1	2	3
1	x		x
2		x	x
3		x	

then for nonbasic cell 31, the appropriate route is

31—32—22—23—13—11—31.

9.

i	j 1	2	3	4	s_i
1	11	10	8	9	100
2	13	8	16	14	125
3	14	10	12	15	150
r_j	75	100	120	80	Min. cost

10.

i	j 1	2	3	4	s_i
1	21	17	13	14	70
2	15	19	23	10	90
3	17	12	25	28	140
r_j	40	90	75	95	Min. cost

5.6 THE ASSIGNMENT PROBLEM

The assignment problem is a special case of the balanced transportation problem as is shown in the following example.

OBJECTIVE

You should be able to solve the assignment problem.

Example 1 *Business Management.* The personal officer of a large company seeks to fill 5 job positions, each requiring a particular set of skills. There are 5 applicants for these jobs. In order to find the "best fit" of applicant to job, a test is given to each applicant. The qualifications a_{ij} for assigning applicant i to job j are given in the following table:

Applicant i	Job j 1	2	3	4	5
1	9	2	3	7	6
2	4	8	5	6	9
3	2	3	8	7	5
4	8	8	1	5	8
5	5	3	7	5	4

The problem is to find the "best fit" if the sum of the assigned qualifications is to be *maximized*. Show that this is a special case of the transportation problem.

Solution

Step 1: *Let* x_{ij} represent the assignment of applicant i to job j, where $x_{ij} = 1$ if such an assignment *is* made and $x_{ij} = 0$ otherwise; so that,

$$x_{ij} = 0 \quad \text{or} \quad 1 \quad \text{for all } i, j.$$

Since one person cannot do two jobs nor can one job have two people assigned to it, we have:

a) Each person is assigned to one and only one job, or

$$\sum_j x_{ij} = 1, \quad \text{for all } i; \quad \text{and}$$

b) Each job is filled by one and only one person, or

$$\sum_i x_{ij} = 1, \quad \text{for all } j.$$

Step 2: We could write the objective function as

$$\max a_0: a_0 = \sum_i \sum_j a_{ij} x_{ij}.$$

However, for analogy with the transportation problem and also for later use, we prefer to have this problem in *minimum* form. To do this we convert "goodness of fit," a_{ij}, to "badness of fit," b_{ij}, by subtracting each a_{ij} from $a_{i*j*} = \max\limits_{i,j} a_{ij}$ (in this case, 9) obtaining:

		\multicolumn{5}{c}{j}					Row minimum (b_{ij*})
		1	2	3	4	5	
	1	0	7	6	2	3	0
	2	5	1	4	3	0	0
i	3	7	6	1	2	4	1
	4	1	1	8	4	1	1
	5	4	6	2	4	5	2

Thus, rather than *maximizing* the "goodness of fit," we *minimize* the "badness of fit" and introduce a new objective function:

$$\min x_0: x_0 = \sum_i \sum_j b_{ij} x_{ij}.$$

This formulation can be seen to be in the form of a transportation problem by setting

$$s_i = 1, \quad \text{for all } i, \quad \text{and}$$
$$r_j = 1, \quad \text{for all } j.$$

Example 2   Solve the assignment problem of Example 1.

Solution   We could solve the assignment problem now as a transportation problem (see Section 5.5); however, the special structure of this problem permits a much simpler method of solution, known as the Hungarian Method based on a theorem by the Hungarian mathematician Dénes König. Continuing with the steps of Example 1, we have

Step 3: We use a star (*) to indicate an assignment. If we could find zeroes to star (make an assignment) such that there were one and only one star in each row and in each column, then such an assignment would minimize x_0. The first two rows have exactly one zero and, hence, one star in each but the remaining rows do not.

Let b_{ij*} be the minimum value of b_{ij} in row i. Since

$$\sum_j x_{ij} = 1,$$

if we subtract b_{ij*} from each value of b_{ij} in row i, this changes x_0 only by the *constant* b_{ij*}, and this has no effect on the optimum assignment.† Thus, we subtract from each value b_{ij} in a row the minimum value b_{ij*} in that row and do this for each row. We denote the result by b'_{ij}. This process is called *row reduction*.

Row-reducing the remaining three rows, we obtain:

		\multicolumn{5}{c}{j}				
		1	2	3	4	5
	1	0	7	6	2	3
	2	5	1	4	3	0
i	3	6	5	0	1	3
	4	0	0	7	3	0
	5	2	4	0	2	3
Column minimum (b'_{i*j})		0	0	0	1	0

Looking at the columns, we see that there are zeroes in each column except the fourth. Let b'_{i*j} be the minimum value of b'_{ij} in column j. Then, as for rows, we can *column-reduce* by subtracting b'_{i*j} from each value of b'_{ij} in column j and do this for each column without changing the optimum assignment.

---

† $x_0 = \sum_i (\sum_j b_{ij} x_{ij}) = \sum_i [\sum_j (b_{ij} - b_{ij*}) x_{ij} + b_{ij*} \sum_j x_{ij}] = \sum_i \sum_j (b_{ij} - b_{ij*}) x_{ij} + \sum_i b_{ij*}$.

Column-reducing the fourth row, we obtain the row- and column-reduced table with elements c_{ij} (ignore the *'s for the present):

		1	2	3	4	5
		j				
i	1	0*	7	6	1	3
	2	5	1	4	2	0*
	3	6	5	0	0*	3
	4	0	0*	7	2	0
	5	2	4	0*	1	3

Step 4: We now seek to star (*) zeroes to make assignments such that there is exactly one assignment (*) in each row and in each column.

i) Let us start by examining the first row. Since there is only one zero in this row, we star that element, c_{11}.

ii) Similarly, there is only one zero in the second row (c_{25}) and that zero in not in the same column as the previous zero. Thus, we star c_{25}.

iii) Similarly, there is only one zero in the fifth row (c_{53}) and that zero is not in the same column as previous starred zeroes (assignments). Thus, we star c_{53}.

iv) Looking now at columns, the first column has a starred zero. The second column has only one zero (c_{42}) and this is not in the same row as previous starred zeroes. Thus, we star c_{42}.

v) Similarly, the fourth column has only one zero (c_{34}) and this is not in the same row as previous starred zeroes. Thus, we star c_{34}.

Note. There may be some choice as to which assignment to make "next." At this point we have made five assignments with one and only one star in each row and in each column. Thus, we have found an optimum assignment.

$$x_{11} = x_{25} = x_{34} = x_{42} = x_{53} = 1,$$
$$x_{ii} = 0 \quad \text{for all other } i, j,$$

so that

$$a_0 = a_{11} + a_{25} + a_{34} + a_{42} + a_{53}$$
$$= 9 \ + 9 \ + 7 \ + 8 \ + 7$$
$$= 40.$$

We can just as well column-reduce first and then row-reduce. Since these problems are frequently degenerate, the solutions are not always unique. Hence column–row reduction may not yield the same assignment as row–column reduction but the value of a_0 would be the same.

DO EXERCISE 22.

It sometimes happens that an optimum assignment cannot be made from the reduced table. Consider the next example.

Example 3 Find the optimum assignment for the following *reduced* table:

		j				
		1	2	3	4	5
i	1	1	7	6	0	3
	2	5	0	4	2	0
	3	6	5	0	0	3
	4	0	0	7	2	0
	5	2	4	0	1	3

Solution Proceeding as in Step 4, we obtain the following assignments:

		j				
		1	2	3	4	5
i	1	1	7	6	0*	3
	2	5	0*	4	2	0
	3	6	5	0	0	3
	4	0*	0	7	2	0
	5	2	4	0*	1	3

i) First row: star c_{14},
ii) Fifth row: star c_{53},
iii) First column: star c_{41},
iv) Second column: star c_{22}.

No more zero assignments can be made without having more than one zero in some row or column and no zero in some other row or column. Hence, we proceed with Step 5.

Step 5. If an optimum assignment can *not* be made, then it *is* possible to cover all the zeroes in the table with a number of horizontal and vertical lines *less* than the number of assignments to be made. These lines may be drawn more than one way.

  i) Starting with the greatest number of zeroes, we draw a line through Row 4.

22. Solve the assignment problem of Example 1 using column reduction first and compare solutions.

ii) Next we draw a line through Row 2.

iii) Then we draw lines through Columns 3 and 4.

		j				
		1	2	3	4	5
	1	①	7	6	0	3
	2	5	0	4	2	0
i	3	6	5	0	0	3
	4	0	0	7	2	0
	5	2	4	0	1	3

All zeroes are now covered with 4 lines and this number is less than the number of required assignments, 5, so that further reduction is necessary.

Note. It is not always obvious how to draw these lines and some "playing around" may be necessary.

To obtain a better assignment, we must add (or subtract) some quantity to (from) certain rows and/or columns to change the location of zeroes. This corresponds to a pivoting operation in the simplex operation. We continue with *iv*.

iv) We now seek the *smallest uncovered* element in this reduced table and call this element c_{i*j*}. Here $c_{i*j*} = c_{11} = 1$ and is circled. (There may be more than one element equal to c_{i*j*} but this does not change the algorithm.)

Since we can add or subtract any constant from any row or column without changing the optimum assignment, we

v) subtract c_{i*j*} from each c_{ij} in each *uncovered row* and

vi) add c_{i*j*} to each c_{ij} in each *covered column*.

Alternately, it is simpler but equivalent to

v) subtract c_{i*j*} from each *uncovered element* and

vi) add c_{i*j*} to each *doubly covered element*.

Doing this, we obtain

		j				
		1	2	3	4	5
	1	0*	6	6	0	2
	2	5	0ᵃ	5	3	0ᵇ
i	3	5	4	0	0*	2
	4	0	0ᵇ	8	3	0ᵃ
	5	1	3	0*	1	2

Going back to Step 4:

i) Fifth row: star c_{53},
ii) Third row: star c_{34},
iii) First row: star c_{11}.
iv) Now we have a choice:

  a) star c_{22} and c_{45},   or
  b) star c_{25} and c_{42}.

Thus, we have obtained two optimal assignments:

  a) $x_{53} = x_{34} = x_{11} = x_{22} = x_{45} = 1$,   and
  b) $x_{53} = x_{34} = x_{11} = x_{25} = x_{42} = 1$,

where the value of the objective function is the same in either case but of no concern here.

If an optimal solution were *not* obtained at this point, then we would repeat Steps 5 and 4 until one is obtained.

DO EXERCISE 23.

23. Find the optimum assignment for the following reduced table:

		\multicolumn{5}{c}{j}			
		1	2	3	4	5
	1	3	7	0	6	3
	2	5	0	4	2	0
i	3	6	5	0	3	0
	4	0	0	7	2	0
	5	3	4	0	2	3

EXERCISE SET 5.6

Solve the following assignment problems where the "goodness of fit" is given in the tables.

1.

		\multicolumn{5}{c}{j}			
		1	2	3	4	5
	1	82	89	73	77	83
	2	79	85	72	78	84
i	3	84	82	81	69	78
	4	83	86	91	88	82
	5	86	85	87	81	76

2.

		\multicolumn{5}{c}{j}			
		1	2	3	4	5
	1	73	75	79	81	76
	2	69	77	72	83	78
i	3	77	82	87	85	81
	4	81	83	94	82	76
	5	71	69	73	82	72

3.

		\multicolumn{5}{c}{j}			
		1	2	3	4	5
	1	86	83	90	79	89
	2	79	80	81	84	83
i	3	85	84	78	86	82
	4	83	81	92	87	78
	5	84	88	87	93	89

4.

		\multicolumn{5}{c}{j}			
		1	2	3	4	5
	1	82	79	83	77	73
	2	83	85	80	73	74
i	3	69	82	73	74	77
	4	85	84	87	81	68
	5	74	77	81	70	82

5.

		\multicolumn{5}{c}{j}				
		1	2	3	4	5
	1	85	75	92	78	83
	2	67	89	77	76	87
i	3	73	86	91	84	79
	4	86	76	93	79	84
	5	82	91	85	85	86

6.

		\multicolumn{5}{c}{j}				
		1	2	3	4	5
	1	87	83	79	91	86
	2	62	73	70	69	80
i	3	82	79	68	83	81
	4	71	73	74	69	82
	5	86	81	81	90	87

7.

		\multicolumn{6}{c}{j}					
		1	2	3	4	5	6
	1	84	87	83	79	91	86
	2	62	73	70	69	80	75
i	3	77	82	79	68	83	81
	4	89	84	86	79	92	85
	5	71	73	74	69	82	80
	6	83	86	81	81	90	87

8.

		\multicolumn{6}{c}{j}					
		1	2	3	4	5	6
	1	73	75	79	81	76	75
	2	69	77	78	83	72	73
i	3	77	82	85	87	81	79
	4	83	81	82	94	76	88
	5	71	69	73	82	72	73
	6	74	75	77	76	70	71

CHAPTER 5 TEST

1. Using the simplex method, solve

$$x_1 + 2x_2 \leq 26,$$
$$x_1 + x_2 \leq 16,$$
$$5x_1 + 3x_2 \leq 70,$$
$$\max x_0: x_0 = 4x_1 + 3x_2; \quad x_1, x_2 \geq 0.$$

Show all work.

3. a) Write the linear program dual to that given in (1).
   b) Read off the dual solution from the final tableau of (1).
   c) Write the equations to be used in checking your solution.
   d) Check your solution.

2. Solve the following linear program using artificial variables and the two-phase method:

$$y_1 + 2y_2 \geq 26,$$
$$y_1 + y_2 \geq 16,$$
$$5y_1 + 3y_2 \geq 70,$$
$$\min y_0: y_0 = 4y_1 + 3y_2; \quad y_1, y_2 \geq 0.$$

4. Pivot until the next pivot is *not* degenerate and indicate the new pivot:

x_1	x_2	x_3	y_1	y_2	y_3	1
1	5	−1	1	0	0	2
−3	1	2	0	1	0	0
−4	3	2	0	0	1	5
2	−1	1	0	0	0	0

5. Read both primal and dual solutions from the following tableaux indicating any solutions that are degenerate, nonunique, unbounded, or infeasible. Read off *all* variables.

a)

x_1	x_2	x_3	x_4	x_5	x_6	1
1	5	−1	1	0	0	2
−3	1	2	0	1	0	0
−4	3	2	0	0	1	5
2	1	1	0	0	0	10

6. Solve the following transportation problem:

		D_j			
		1	2	3	s_i
F_i	1	27	17	18	45
	2	19	20	15	50
	3	18	16	23	65
	r_j	30	70	60	Min. cost

7. Solve the following assignment problem given the "goodness of fit":

b)

x_1	x_2	x_3	x_4	x_5	x_6	1
1	5	−1	1	0	0	2
2	1	−3	0	1	0	1
3	2	−4	0	0	1	5
2	1	−1	0	0	0	10

		j				
		1	2	3	4	5
	1	9	7	3	1	4
	2	8	8	7	3	4
i	3	7	9	6	4	5
	4	7	8	7	5	6
	5	8	6	7	3	5

CHAPTER SIX

Graphs and Networks

OBJECTIVE

You should be able to find a minimum and maximum spanning tree for a network, and the corresponding overall path lengths.

6.1 TREES AND MINIMUM SPANNING TREES

The Puzzle of the Königsberg Bridges

In the 1700's the people of the town of Königsberg, then in Prussia but now in Russia and called Kaliningrad, amused themselves by trying to solve the following problem. A river, in which there are two islands, flows through the town. There were seven bridges connecting the islands and the river banks, as shown in the following figure. The problem was to walk in such a way that each bridge is crossed exactly once.

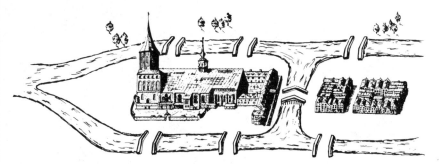

This problem can be considered using what is called a *network*.

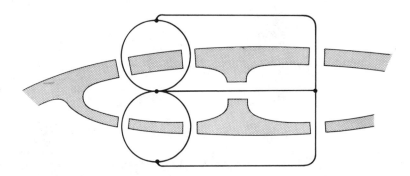

Problems such as this formed a basis of a branch of mathematics known as *graph theory*, or *network theory*. We will examine certain real applications of this theory. Incidentally, it turns out that it is impossible to take such a walk over the bridges, though we will not go into a proof here. You can convince yourself of this somewhat by trying to draw the network without lifting your pencil from the paper or retracing a segment or curve. Try it!

Basic Definitions

Consider the following figure, a *graph*:

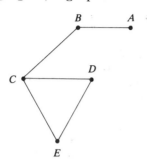

A *graph* is a set of points (here A, B, C, D, and E) called *nodes* and a set of lines (here AB, BC, CD, CE, and DE) called *arcs* joining pairs of nodes. The lines need not be straight, even though we usually draw them that way. It is assumed, for the present, that there is no more than one arc between any pair of nodes. A *network* is a *connected* graph; that is, a graph with each node connected to each other node along some path of arcs.

Example 1 Determine whether the following graphs are networks.

a)

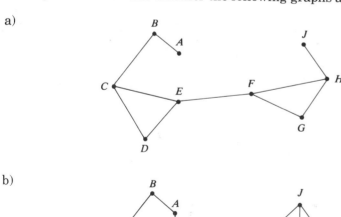

b)

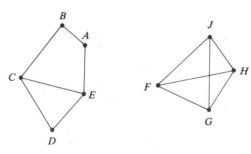

1. Determine whether the following graphs are networks.

a)

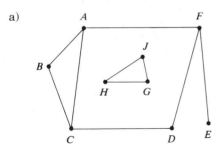

b)

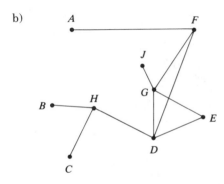

2. Determine all cycles of the graphs of Margin Exercise 1.

Solution

a) This graph *is* a network because each node is linked to each other node by some path of arcs. This is similar to a collection of cities where we ask "Can we get from each city to each other city by a sequence of roads?"

b) This graph is *not* a network. It consists of *two* separate networks not connected to each other. For example, there is no way to get from A to J along some path of arcs. This situation is similar to cities on two islands not connected by bridges.

DO EXERCISE 1.

A *cycle* is a path of arcs which starts at some node and ends at that same node without going over the same arc more than once. An example of a cycle is a car trip, which starts at some city and ends at that same city without traveling over the same road more than once. A cycle is described by listing the nodes or arcs in sequence in either direction.

Example 2 Determine all cycles of the graphs in Example 1.

Solution

a) There are 2 cycles: *CDEC* and *FGHF*. Note that *CDEC* and *CEDC* represent the same cycle but in opposite directions. We pick one direction and just list it that way.

b) There are 8 cycles: *ABCEA*, *CDEC*, *ABCDEA*, *FGHF*, *FGJF*, *FJHF*, *FGHJF*, *GJHG*.

DO EXERCISE 2.

A *tree* is a network without cycles, so that there is only one path between any pair of nodes. For example, the following network:

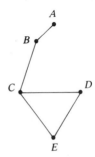

is *not* a tree since it contains the cycle *CDEC*. It also fails to be a tree because we can trace two paths, *ABCE* and *ABCDE*, between nodes A and E.

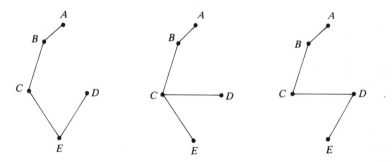

This network can be made into a tree by deleting any one of the arcs *CD*, *DE*, or *CE*, as shown above.

A tree obtained by deleting some arcs of a network is called a *spanning tree*. We have seen that more than one spanning tree *may* be obtained from a network.

Example 3   Form all possible spanning trees from the network of Example 1(a).

Solution   Each of the 2 cycles *CDEC* and *FGHF* can be broken 3 ways. Thus 9 spanning trees can be formed. Deleted arcs are shown dashed.

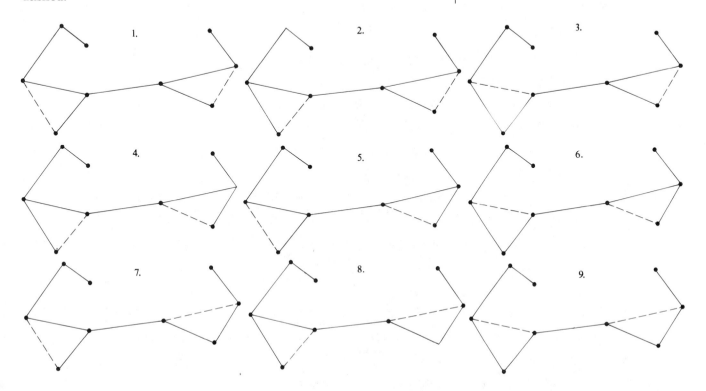

3. Form all possible spanning trees from the network of Margin Exercise 1(b).

DO EXERCISE 3.

If a length (or any other number) is assigned to each arc of a network, then the spanning tree with the minimum sum of arc lengths is called the *minimum spanning tree*. Similarly, that with the maximum sum of arc lengths is called the *maximum spanning tree*.

Example 4 *Business*. A telephone company wants to link some cities together with new telephone lines, the cost of which is proportional to the length of the lines; that is, the longer the lines the higher the cost. The distance between cities, with accessible paths (not necessarily straight distances), is indicated by arc lengths in the following network. Find the arcs the telephone lines must follow to minimize the arc length of this path and thus to minimize the construction cost to the telephone company.

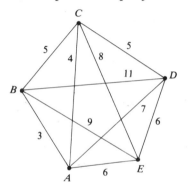

Solution The solution of this problem turns out to be the path length of the minimum spanning tree. It could be found by using a linear-programming model, but it is much easier to use network theory.

We could find the minimum spanning tree by listing all possible spanning trees and picking out the minimum, or minimums. However, this process can be shortened by using the following algorithm.

Kruskal's Algorithm. We are going to select arcs from the network one at a time to form a tree.

1. Look over the network and find the arc or arcs of shortest length. Select one of these.
2. Of the arcs not yet part of the tree, select one of shortest length which, when added to the tree, does not form a cycle.
3. Repeat Step 2 until there are $(n-1)$ arcs of the tree, where the original network has n nodes. Since some arcs have equal lengths, there may be more than one spanning tree.

Thus, in this example,

a) We start with arc AB of length 3.
b) Add arc AC of length 4.
c) We have two arcs of length 5: BC and CD. Since BC forms a cycle, $ABCA$, we omit it, and select CD.
d) We have two arcs of length 6: AE and DE. We can select either one of these, since neither forms a cycle, and either selection results in a different minimum spanning tree.
e) We have 5 nodes, so the minimum spanning tree will have 4 arcs. Since we now have 4 arcs, we stop.

Thus, we have either of:

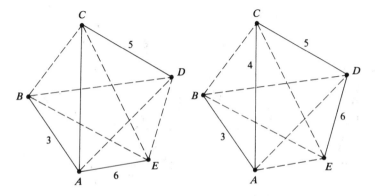

as minimum spanning trees, each with overall path length of $3 + 4 + 5 + 6$, or 18.

DO EXERCISE 4.

Example 5 A member of a microwave communications company has infiltrated the telephone company of Example 4, and wants to sabotage their plans by providing a *maximum* cost layout. Find this *maximum* spanning tree and its arc length.

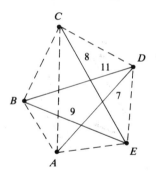

4. Find the minimum spanning tree and corresponding overall path length of:

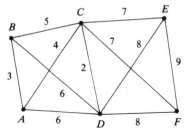

5. Find a maximum spanning tree for the network of Margin Exercise 4 and its overall path length.

Solution To do this, we use Kruskal's Algorithm, but this time we select the longest arc available at each step rather than the shortest. Thus we obtain the diagram at the bottom of page 229, with overall path length

$$11 + 9 + 8 + 7, \quad \text{or } 35.$$

DO EXERCISE 5.

EXERCISE SET 6.1

Find a minimum and maximum spanning tree for each of the following networks and the corresponding overall path length.

1.

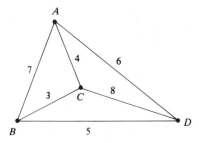

2.

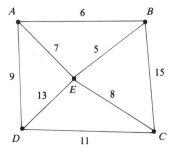

3.

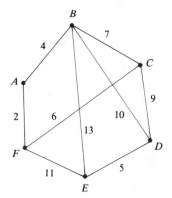

4.

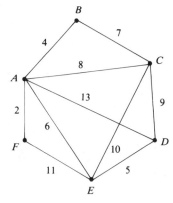

5. *Economics.* An underdeveloped country wishes to connect certain cities with roads, where the cost of the roads is large and the initial amount of traffic is small. As traffic builds up, more roads would probably be built, and finding minimum cost would be another problem.

The problem is to minimize the cost of linking the cities, where costs of roads are indicated by path lengths in the following network. Determine which roads are to be built and what this minimum cost might be. Numerals on arcs indicate road construction costs in millions of dollars.

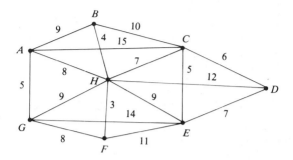

Also, find which roads are to be built by a corrupt engineer who wants to maximize costs and what the maximum cost might be.

6. As in Exercise 5, but with the following network.

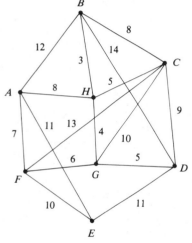

6.2 MINIMUM PATHS AND DIGRAPHS

Minimum Paths

In the minimum-spanning-tree problem, we were not concerned with a start or finish node. In this section we will be. Problems of this type are called *minimum-path problems* rather than minimum-spanning-tree problems.

Example 1 *Business.* A trucking company wishes to route its trucks from *A* to *G* along the path of minimum length. Assuming that all roads are two-way, find the minimum-length path. For later convenience, the node labels have been enclosed in circles, with double circles at the origin and destination. Note that the arc lengths represent distances and are thus nonnegative.

OBJECTIVES

You should be able to find the minimum paths from origin to destination for a network.

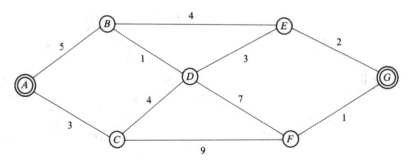

Solution This problem could be solved using a minimum-type linear program, but it is again easier to use network theory.

The basis for the algorithm we are about to consider is that, if an overall path is minimum, then any portion of the path is also minimum. We construct the minimum path by stepwise *tagging* the nodes.

The minimum-path algorithm is as follows:

1. **Tag the origin (here node A) with "0, –"; all other nodes being *untagged* for the present.**

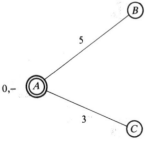

2. **Of all node tags not enclosed in a box* (none are at the outset), find those whose first number is a minimum and enclose those node tags in a box.**

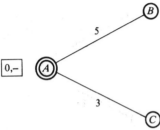

We have only the origin A tagged thus far, so we box the tag of the origin.

---

* Parentheses could also be used.

3. **Using the node(s) with tags** *just boxed* **(in this case it is just one, node A),** *tag* **all neighboring nodes (nodes that can be reached in one arc from the just-boxed node) with a pair of indices. The** *first* **index is the sum of the first number of the just-boxed node tags and the connecting arc length.**

The first index for B is $0 + 5$, and that for C is $0 + 3$.

The *second* **index is the label of the just-boxed node,** in this case A.

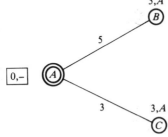

4. **Repeat Steps 2 and 3 until the destination is reached.**

Of all node tags not enclosed in a box, B and C, find those whose first index is a minimum and enclose those node tags in a box. The tags at C have the minimum first index.

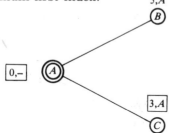

Using the node(s) with tags just boxed, tag all neighboring nodes with a pair of indices. The first index is the sum of the first index of the just-boxed node tags and the connecting arc length. The second index is the label of the just-boxed node. Thus, node D is tagged $7,C$ and node F is tagged $12,C$.

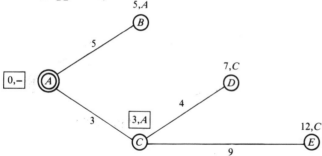

Repeating Steps 2 and 3, we find that node D can be retagged from node B with a *smaller* first tag index (distance) than from node C. *Note:* Nodes are never retagged with *larger* numbers since we are interested in *minimizing* overall arc length.

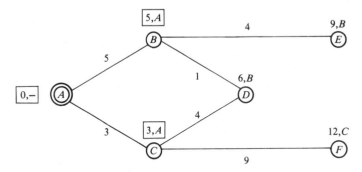

Repeating Steps 2 and 3, we find that node E can be tagged from node D with the *same* first tag index as from node B. Since the two tags are the same, there are two paths from A to E with the *same* overall arc length, and hence both are retained. Node F is *not* retagged since the distance through node D is greater than that through node C.

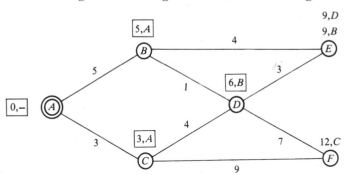

Repeating Steps 2 and 3, we box *both* tags of node E and tag node G.

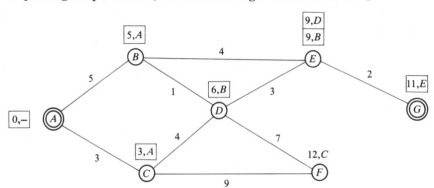

5. **The destination (node G) has been tagged but not boxed. Continue with Step 2 and, if necessary, Step 3 until the destination is boxed. (All nodes must then have been tagged, but not necessarily boxed.)**

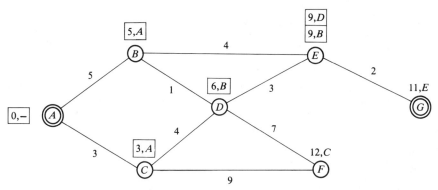

The destination has now been boxed. This terminates this part of the algorithm.

6. **To determine the minimum path, trace back from the destination to the origin the path indicated by following sequentially the *second index* of boxed tags.**

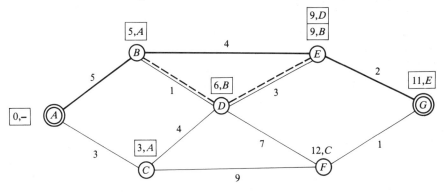

We start by noting the *second* tag index of the destination, node G. This leads us back to node E, which has two tags. Thus, we can be led back either to node B or to node D. Continuing, we obtain the two paths $ABEG$ and $ABDEG$, each with an overall path length of 11.

7. **The presence of more than one set of boxed tags at a node along the minimum path being constructed indicates more than one minimum path.**

Note that as you become more familiar with the algorithm, you may anticipate that certain nodes, if tagged, will have to be retagged later. Thus you can avoid some work by skipping the tagging until it becomes necessary. If in doubt, tag.

6. Find the minimum path from node A to node J through the following network.

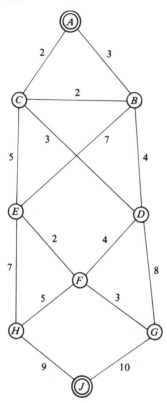

DO EXERCISE 6.

Digraphs

So far we have considered graphs which were *undirected*; that is, any arc can be traversed in either direction. We now consider *directed graphs*, or *digraphs*, in which the direction is indicated by arrows. Similarly, networks can also be directed. We shall assume that all arc lengths are nonnegative; this prevents having any cycles with negative path length. Also, the arc length between two nodes need *not* be the same in both directions. For example, if time on an air route is considered to be arc length, traveling east is usually much faster than traveling west because of the jetstream (a high-speed westerly wind).

The minimum path through a directed network can be found by the same algorithm as for an undirected network, that is, the algorithm described in Example 1.

Example 2 *Business, Operations.* At a certain time of the day, airline flight times along available routes of a flight network are indicated as arc lengths in the accompanying diagram. Find the minimum-time path from the origin to destination (each denoted by double-circled node labels).

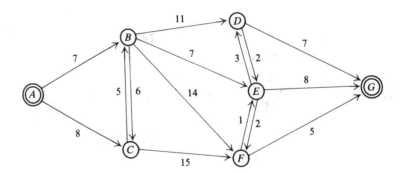

Note that the flight time from B to C is not the same as from C to B, due, for example, to the jetstream or to intermediate stops that do not connect further with this network.

Solution Carrying out the minimum-path algorithm described in Example 1, we reach the stage shown below, where B is the just-boxed node. Again, we are repeating Steps 2 and 3.

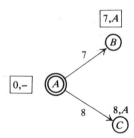

Using node B with tag just boxed, we draw arcs to all neighboring nodes, and tag. Note that D is tagged 18,B and E is tagged 14,B, and F is tagged 21,B. The tag at C is left as 8,A because the route from A to B to C would have a larger path length, 13. We find that the tags for node C can now be boxed, since the first index is minimum.

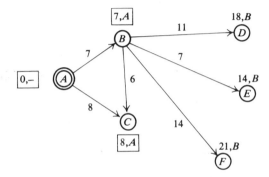

Note that we drew only the arc from B to C, since we are thinking of traveling *from* the just-boxed node.

Using node C, just boxed, we draw arcs to all neighboring nodes, and tag. Note that we now draw the arc from C to B. The tag for B is not retagged. A boxed tag never has to be retagged. The tag at F does not change. We find that the tags for node E can now be boxed.

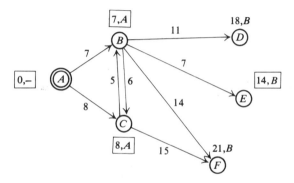

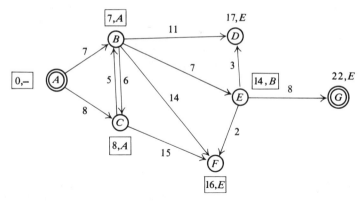

We draw arcs from E to all neighboring nodes and tag. Note that D is retagged $17,E$ and F is retagged $16,E$, and G is tagged $22,E$. We can now box the *tags* for F (see figure above).

We draw arcs from F to all neighboring nodes and tag. Note that G is retagged $21,F$. We can now box the tags for node D.

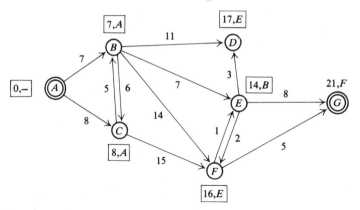

Now draw arcs from D to all neighboring nodes and tag. Note that no changes in tags occur. The tag for node G can now be boxed.

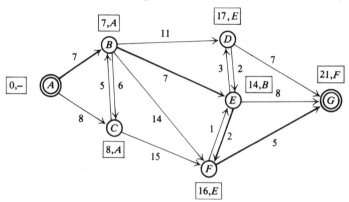

Since the destination has now been boxed, the algorithm is terminated. The minimum path is thus *ABEFG* with length 21.

DO EXERCISE 7.

7. Find the minimum path from origin to destination through the network.

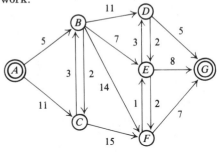

EXERCISE SET 6.2

1. *Business*. A portion of a highway network is indicated in the following diagram, where each arc length represents a distance in miles. The highway department wishes to increase the traffic flow from *A* to *G* by doubling the number of lanes along the path of minimum cost. Since the cost is proportional to the length, find the path of minimum length.

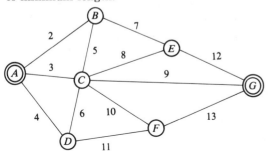

2. *Business*. A trucking company wants to make a refrigerated shipment in minimum travel time from *A* to *F* in the following network, where each arc length represents transit time in hours. Find the route with minimum transit time.

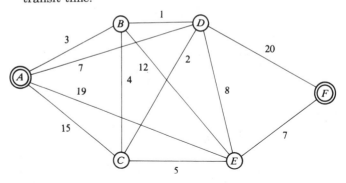

For each of the following networks find the minimum paths from origin to destination (each indicated by double circles about their node label).

3. (Network of Exercise 1, in reverse order.)

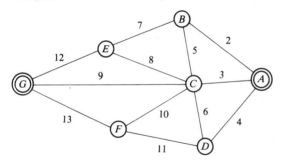

4. (Network of Exercise 2, in reverse order.)

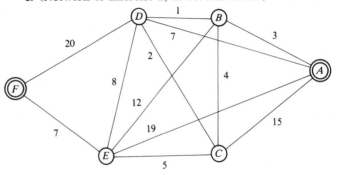

Business, Operations. In Exercises 5 through 7, assume that the networks describe airline flight routes at a certain time of the day, where the arc lengths represent flight times. Find the minimum-time path from origin to destination.

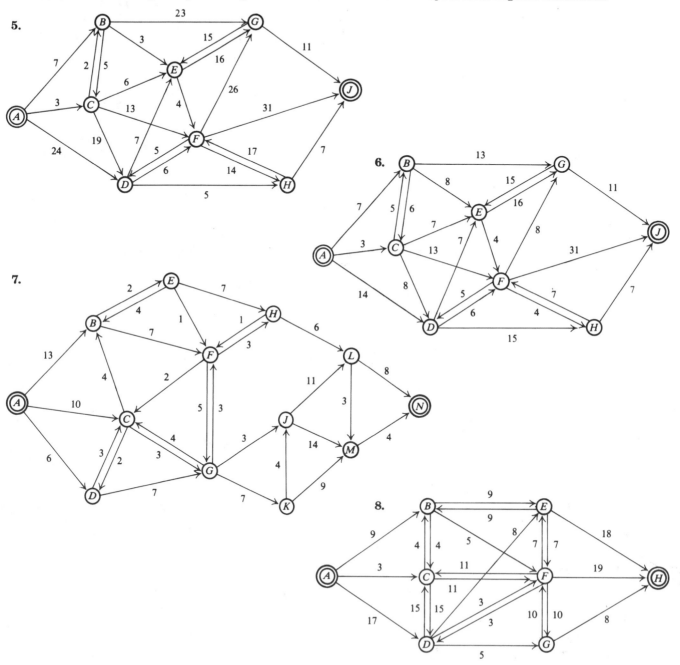

6.3 MAXIMUM FLOW THROUGH A NETWORK

Here we are concerned with a directed network where arc lengths represent capacities rather than distances or times. For example, the arc length on a highway might represent the capacity of a highway in thousands of cars per hour or, in a sewer, the flow of liquid in cubic feet (feet^3) per second. We seek to *maximize* the flow between a source (origin) and a sink (destination), where the amount of flow in any arc cannot exceed the capacity. For example, a certain stretch of highway may only be able to accommodate up to 8000 cars per hour, but there may be less in a given hour. Or a sewer may have a capacity of 10,000 cubic feet per second, but the volume flow may be less than that. We shall assume that there is unlimited flow at the source.

Example *Business, Operations.* The various routes between towns A and G can be represented by a directed network, where the arc length is the capacity of that link (in thousands of cars per hour). Find the maximum traffic flow that can pass through the following network.

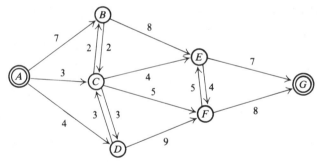

Solution We have a systematic procedure to find the maximum traffic flow. Called the *Uppermost Algorithm*, it involves tagging, and sometimes retagging, each arc with an ordered pair (c, x), where c represents the *capacity* of the arc and never changes, and x represents the *actual flow* in that arc. The actual flow must be nonnegative and no larger than the capacity of the arc, so that

$$0 \leqslant x \leqslant c.$$

Initially, $x = 0$ and we do not bother to write in the second coordinate. Thus, in the given network, the capacity of arc BE is 8 and is understood to be tagged $(8, 0)$. The fact that $x = 0$ means that initially we start with no flow in the network, so that the flow in each arc is 0.

OBJECTIVE

You should be able, given a network, to determine the maximum flow, using both the uppermost and bottommost algorithms; list the successive uppermost (bottommost) path, the flow in each path, and its saturating arc; and find the minimum cut.

THE *UPPERMOST ALGORITHM** IS AS FOLLOWS:

1. Starting at the origin, determine the *uppermost* available path to the destination along which some *positive* amount of flow can be sent. The uppermost path is as close to the top of the diagram as possible.

Initially, the uppermost available path is *ABEG*. The initial tags along this path are 7, 8, and 7, or, writing in the omitted 0's, these tags are $(7, 0)$, $(8, 0)$, and $(7, 0)$. Since the first tag coordinate is *greater* than the second tag coordinate all along this path, a positive amount of flow can be sent from *A* to *G* along *ABEG*.

2. Determine the maximum amount of flow that can be sent along the path found in (1).

The difference between the first and second tag coordinate, $(c - x)$, represents the unused arc capacity. Thus, arc *AB* can accommodate 7 units of flow, arc *BE* can accommodate 8 units of flow, while arc *EG* can accommodate 7 units of flow. The most flow that this path can accommodate is 7 units, which saturates arcs *AB* and *EG*. The flow in the network is increased (from 0) by 7 units.

3. Increase the second tag coordinate of the arcs found in Step 1 by the amount of flow that the flow in the network is increased, found in Step 2.

Since the flow in the network was increased 7 units along path *ABEG*, we increase the second tag coordinate of arcs *AB*, *BE*, and *EG* by 7 and obtain $(7, 7)$, $(8, 7)$, and $(7, 7)$.

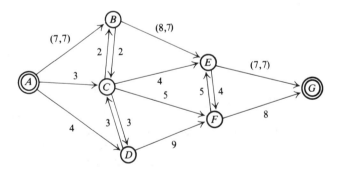

4. Continue Steps 1, 2, and 3 until no more flow can be routed through the network. Then go to Step 5.

Along the original *ABEG* arcs, *AB* and *EG* are saturated—that is, $c = x$—and they can accommodate *no* more flow, while arc *BE* has an

---

* This algorithm can be used for flows whose arcs do not intersect except at nodes. Otherwise, a modification *may* be necessary.

unused capacity $(c - x)$ of 1 unit, so that the flow in arc BE can be (but need not be) increased 1 unit. Thus, the next uppermost path is $ACBEFG$. The most flow that we can get through this path is 1 unit, which saturates arc BE. Thus, the flow in the network is increased by 1 unit, so the second coordinate of the tag along this path is increased by 1.

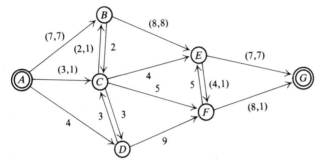

The next uppermost path is $ACEFG$, which can accommodate 2 units of flow before saturating arc AC. Thus, the actual flow in arcs AC, CE, EF, and FG has been increased by 2 units, and we have:

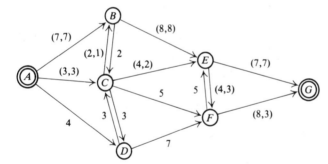

The next uppermost path is $ADCEFG$, which can accommodate 1 unit of flow before saturating arc EF. Thus, the actual flow in arcs AD, DC, CE, EF, and FG is increased by 1 unit of flow, and we have:

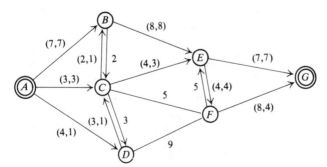

The next uppermost path is *ADCFG*, which can accommodate 2 units of flow before saturating arc *DC*. Thus, the actual flow in arcs *AD*, *DC*, *CF*, and *FG* is increased by 2 units, and we have:

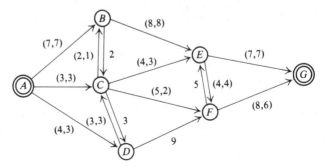

The last uppermost path is *ADFG*, which can accommodate 1 unit of flow before saturating arc *AD*. Thus, the actual flow in arcs *AD*, *DF*, and *FG* can be increased by 1 unit of flow, and we have:

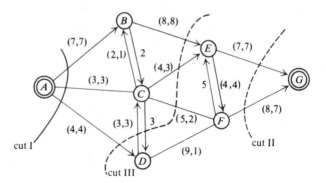

No more flow can be accommodated by the network, since all the arcs out of the source are saturated.

5. **The maximum flow through the network can be determined by making a *disconnecting cut*. This is a cut across arcs by some "line" such that *no* flow can pass from source to sink. The *maximum flow* is the sum of the second coordinates, the actual flows.**

Thus, along cut I, seen in the preceding diagram, the maximum flow is 7 + 3 + 4, or 14 units. Along cut II, we should get the same answer, which is 7 + 7, or 14 units. Along cut III, we should still get the same answer, 8 + 3 + 2 − 3 + 4, or 14 units. Note that, for cut II, there is excess capacity in arc *FG* while, for cut III, there is "back" flow, since the direction of flow in arc *DC* is opposite to that in the other arcs.

For cut I, there is *no excess capacity or any back flow.* Such a cut is called a *minimum cut,* and every network contains one someplace. The capacity of a cut is the sum of all the second-tag coordinates of its arcs.

The Duality Theorem for networks states that the amount of the maximum flow is equal to the capacity of the minimum cut.

This may be used to verify that a flow is indeed maximal.

Why did we start with the uppermost path through the network? In fact, we could have started with the bottommost path through the network. The maximum flow will still be 14. If there is time, this is a way to check your work; that is, carry out the uppermost algorithm to get the maximum flow—then carry out the *bottommost algorithm* to check your answer. In Example 1, we have:

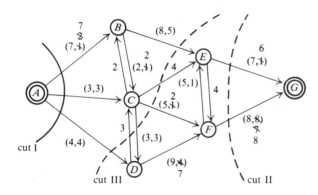

The successive bottommost paths are:

  i) *ADFG* with 4 units of flow and saturating arc *AD.*
  ii) *ACDFG* with 3 units of flow and saturating arcs *AC* and *CD.*
 iii) *ABCFG* with 1 unit of flow and saturating arc *FG.*
 iv) *ABCFEG* with 1 unit of flow and saturating arc *BC.*
  v) *ABEG* with 5 units of flow and saturating arc *AB.*

The maximum flow is $7 + 3 + 4$, or 14 units. The saturating cut is

$$\{AB, AC, AD\}.$$

The amount of flow in the various arcs differs from that obtained using the uppermost algorithm but the maximum flow is the same. The minimum cut is still cut I, but this need not have been so.

DO EXERCISE 8.

8. *Business.* Many commodities can be transported via pipes (for example, oil, gas, sewage, and so on) of various capacities. The layout of pipes between sites forms a directed network (directed, since pumps or gravity must direct the flow). Determine the maximum flow that can be put through the following network. First, use the uppermost algorithm and then the bottommost algorithm. List the successive uppermost (bottommost) paths, the flow in each, and its saturating arc.

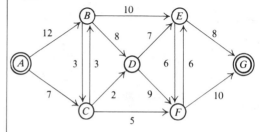

EXERCISE SET 6.3

For each of the following networks, determine the maximum flow using both the uppermost and bottommost algorithms. List the successive uppermost (bottommost) path, the flow in each, and its saturating arc. Find the minimum (that is, saturated) cut to verify your solution.

1.

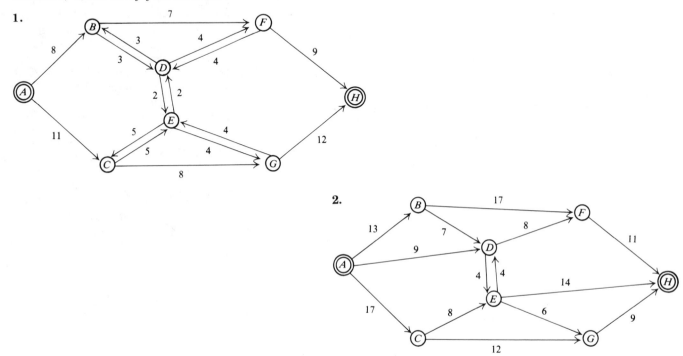

2.

3. *Business.* A portion of a sewer network is indicated in the following diagram, which is directed because the flow is induced by gravity. The arc length represents the capacity in feet³ per second. Find the maximum possible flow from A to J (as indicated just ahead of Exercise 1).

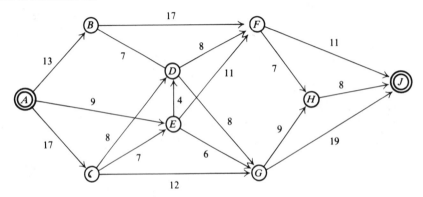

4. *Business.* Natural gas is pumped throughout a country to different cities by a network of pipes of varying capacities (in feet^3 per second) as indicated in the following diagram. Find the maximum possible flow from A to M (as indicated just ahead of Exercise 1).

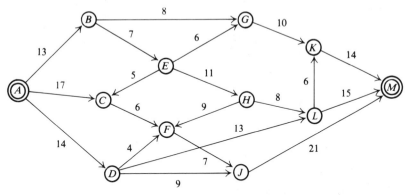

CHAPTER 6 TEST

1. a) Find the mimimum spanning tree of the following network.
   b) Find the maximum spanning tree of the following network.

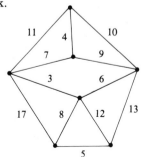

2. Find the minimum-length path from node A to node J:

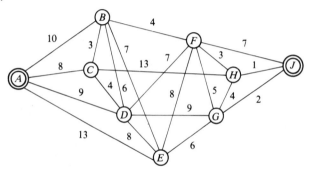

3. Find the maximum flow from node A to node J in the following network, and the saturating cut (show all work):

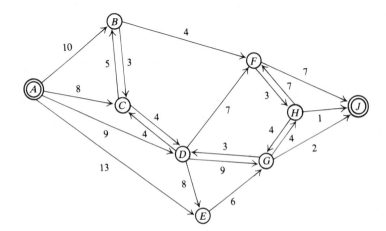

PART III

CHAPTER SEVEN

Sets and Counting Techniques

OBJECTIVES

You should be able to:

a) Determine whether an object is an element of a set.
b) Decide whether two sets are equal.
c) Name a set using the roster method or set-builder notation.
d) Decide whether one set is a subset of another.
e) Find the complement A^c of a given set.
f) Find the Cartesian Product of two sets.
g) Find the cardinality of a given set (finite).

7.1 SETS*

Sets form the foundation of our study of probability.

What do the following have in common?

> a *flock* of birds,   a *school* of fish,   a *crowd* of people,
> a *herd* of animals,   a *pod* of whales,   a *host* of angels.

In each case we are dealing with a collection of objects of a certain type. Rather than use a different word for each type of collection, it is convenient to denote them all by the one word "*set*."

A *set* is a collection of well-defined objects called *elements*.

One can talk about the set of all "employees" in a company since an "employee" is *well-defined*. On the other hand, one cannot talk about the set of all "good" employees unless one can provide an objective way to distinguish "good" employees.

We now develop the *notation* necessary to deal with sets.

The set of vowels can be written

$$V = \{a, e, i, o, u\}$$

where the *capital letter V* denotes the *set* as do the braces enclosing the *elements*. The elements are separated by commas, read "and." Elements of sets are usually denoted by lower-case letters. This way of describing a set is known as the "roster" method.

That "a is a vowel" or "a is an element of *V*" can be written

$$a \in V.$$

Similarly, that "b is not a vowel," or "b is not an element of *V*" can be written

$$b \notin V.$$

Two sets A and B are *equal*, written $A = B$, if and only if each element of either set is also an element of the other set. For example

a) $\{a, e, i, o, u, a\} = \{a, e, i, o, u\}$,   and
b) $\{u, o, i, e, a\} = \{a, e, i, o, u\}$,

where in (a) the *repetition* of an element does not change the set and in (b) the *order* in which we write the elements does not change the set.

---

* If the appendix on Logic is to be studied, it would be most advantageous to do it just prior to this chapter.

Consider the set of all students in a given class. The class is not changed if some student's name is listed twice. Furthermore, the class is not changed whether the student's names are listed alphabetically, or by Social Security number, or any other way.

Example 1 Given $A = \{1, 2, 3, 4, 5, 6\}$, $B = \{2, 3, 4, 5, 6\}$, and $C = \{2, 3, 4, 5, 6, 1\}$.

a) Is $1 \in A$?, $1 \in B$?
b) Is $A = B$?, $A = C$?

Solution
a) Yes, $1 \in A$; no, $1 \notin B$.
b) No, $A \neq B$, since $1 \in A$ but $1 \notin B$;
   yes, $A = C$, since order of elements is immaterial.

DO EXERCISE 1.

Other Set Notation

The set containing all the integers that are greater than 1 can be written

$$\{2, 3, 4, \ldots\}.$$

The dots are read "and so forth," and indicate that the pattern of the listed elements continues. This set can also be written using words as

"The set of all integers greater than 1."

or

"The set of all x such that x is an integer and $x > 1$."

or

$$\{x \mid x \text{ is an integer}, x > 1\}$$

In the latter notation, the first set brace "{" is read "The set of all," and the vertical line "|" is read "such that." The entire notation is read "The set of all x such that x is an integer and x is greater than 1." This is called "set builder" notation.

Example 2 Write $C = \{1, 4, 9, \ldots\}$ using set builder notation.

Solution

$$C = \{x \mid x \text{ is an integer and } x \text{ is a perfect square}\}$$

Answers may vary. We could have written this set

$$C = \{x \mid x = n^2, \quad n \text{ is an integer}, \quad n > 0\};$$

1. Given $A = \{a, b, c, d\}$,

$B = \{a, e, i, o, u\}$, $C = \{a, b, c, d, c\}$.

a) Is $b \in A$?, $e \in C$?

b) Is $A = B$?, $A = C$?

2. a) Write

$$D = \{1, 3, 5, \ldots\}$$

using set-builder notation. Answers may vary.

b) Write

$$F = \{x \mid x = 3y + 1, \; y \in I, \; y \geqslant 0\}$$

using the roster method.

or, if we let I represent the set of all integers,

$$I = \{\ldots, -3, -2, -1, 0, 1, 2, 3, \ldots\}$$

then we could use more abbreviated notation for C.

$$C = \{x \mid x = n^2, \quad n \in I, \quad n > 0\}$$

Example 3   Write $E = \{x \mid x = 4n - 1, \quad n \in I, \quad n \geqslant 0\}$ using the roster method.

Solution

$$E = \{-1, 3, 7, 11, 15, \ldots\}$$

DO EXERCISE 2.

The Empty Set

The set without elements is known as the *empty set*, or *null set*, and is denoted by the symbol $\emptyset$, or { }. The following is an example of the empty set:

The set of all odd numbers whose squares are even.

The notation {0} is not appropriate for the empty set, since this set has one element in it, namely the number 0. Also, $\{\emptyset\}$ is not notation for the empty set. If $\emptyset$ represents an empty set (empty paper sack), then $\{\emptyset\}$ would represent an empty sack in a sack.

Subsets

Set A is said to be a *subset* of B if and only if every element of A is an element of B. For example, if $B = \{a, b, c\}$, the subsets of B are

$$\{a\}, \quad \{b\}, \quad \{c\}, \quad \{a, b\}, \quad \{a, c\}, \quad \{b, c\}, \quad \{a, b, c\}, \quad \emptyset$$

That "A is a subset of B" is symbolized $A \subset B$, and is often read "A is *contained* in B." Thus

$$\{a\} \subset \{a, b, c\}$$
$$\{b, c\} \subset \{a, b, c\}$$
$$\{a, b, c\} \subset \{a, b, c\}$$
$$\emptyset \subset \{a, b, c\}$$

Note that, for any set A, $A \subset A$; that is, any set is a subset of itself.*

---

* A *proper* subset is a subset such that $A \subset B$, but $A \neq B$. Thus $\{a, b\}$ is a proper subset of $\{a, b, c\}$, but $\{a, b, c\}$ is not a proper subset of $\{a, b, c\}$. When A is a proper subset of B, there is at least one element in B which is not in A. Some texts use $\subseteq$ to indicate a subset, and reserve the symbol $\subset$ only for a proper subset, but this distinction is not useful to us in this text.

Also, for any set A, $\emptyset \subset A$; that is, the empty set is a subset of every set. For example, let A be the set of all fish in a lake and C be the subset of A consisting of all fish caught by a fisherman. Thus, $C \subset A$. Since it is possible that the fisherman catches *nothing*, we can have $C = \emptyset$, so that $\emptyset \subset A$.

DO EXERCISE 3.

Example 4 Let $V = \{a, e, i, o, u\}$, $A = \{a, e\}$, and $B = \{a, t, s\}$.
a) Is $A \subset V$?
b) Is $B \subset V$?

Solution
a) Yes, $A \subset V$, since each element of A is also in V.
b) No, $B \not\subset V$, since the elements t and s of B are not in V.

DO EXERCISE 4.

Cardinality of a Set

The *cardinality* of a set A is the *number of distinct elements* it contains and is written $\mathcal{N}(A)$. Thus, if

$$V = \{a, e, i, o, u\},$$

then

$$\mathcal{N}(V) = 5.$$

Also, $\mathcal{N}(\emptyset) = 0$; the cardinality of the empty set is 0.

DO EXERCISE 5.

Universal Sets and Complements

Any mathematical situation has a frame of reference called a *universal set*. For example, in elementary algebra the universal set is usually the set of real numbers. In plane geometry the universal set is the set of points in the plane. In probability the universal set might be, for example, the set of all possible outcomes for drawing a particular card from a well-shuffled deck. Usually it is clear what the universal set is, though it may have to be inferred from the context of a problem or application. We might think of the empty set as being on the low end of the frame of reference and the universal set as being on the high end. The universal set is usually denoted $\mathcal{U}$, and in any given application it must be true that for any set A, $A \subset \mathcal{U}$.

Example 5 A survey is to be made to determine the opinion of readers of the *National Observer* on a particular issue. What is the universal set?

3. List all the subsets of
$$C = \{a, b, c, d\}.$$

4. Let
$$A = \{1, 2, 3, 4, 5, 6, 7, 8, 9\},$$
$$B = \{2, 4, 6, 8\},$$
$$C = \{0, 1, 3, 5\}.$$

a) Is $B \subset A$?

b) Is $C \subset A$? Why?

5. Find:

a) $\mathcal{N}(\{a, b, c, d\})$

b) $\mathcal{N}$ (The set of all odd numbers whose squares are even)

6. A person enters a department store with the intention of spending $10. What is the universal set of his purchases?

Solution *Before* the survey is taken, the universal set consists of all readers of *National Observer*.

After the survey is taken, the universal set consists of those readers questioned in the survey. The set of readers surveyed is a subset of all the readers. The readers surveyed with opinions, pro, con, or otherwise, are all subsets of the readers surveyed.

DO EXERCISE 6.

The *absolute complement* or, simply, the *complement* of a set A, written A^c, is what is left in the universal set after the elements of A are removed.

7. Given $A = \{1, 3, 5, 7, 9\}$ and $\mathcal{U} = \{1, 2, 3, 4, 5, 6, 7, 8, 9\}$, find A^c.

Example 6 Given $A = \{a, e, i\}$ and $\mathcal{U} = \{a, e, i, o, u\}$, find A^c.

Solution $A^c = \{o, u\}$.

Note. $\mathcal{U}^c = \emptyset$ and $\emptyset^c = \mathcal{U}$.

DO EXERCISE 7.

Given two sets A and B, the *Cartesian product* of these sets, written $A \times B$, is the set of all ordered pairs (a, b) where $a \in A$ and $b \in B$. Formally

$$A \times B = \{(a, b) \mid a \in A, \quad b \in B\}.$$

Note that, as the name implies, the *order* of the elements in an *ordered pair* is important.

8. If $A = \{a, b, c\}$ and $B = \{1, 2, 3, 4\}$, find $A \times B$ and $\mathcal{N}(A \times B)$.

Example 7 If $A = \{2, 4\}$ and $B = \{1, 2, 3\}$, find $A \times B$, $\mathcal{N}(A)$, $\mathcal{N}(B)$, and $\mathcal{N}(A \times B)$.

Solution

$$A \times B = \{(2, 1), (4, 1), (2, 2), (4, 2), (2, 3), (4, 3)\}$$
$$\mathcal{N}(A) = 2 \quad \text{and} \quad \mathcal{N}(B) = 3$$
$$\mathcal{N}(A \times B) = 2 \cdot 3 = 6 \quad \text{(as can be verified by counting)}.$$

The cardinality of a Cartesian product of two sets can be found simply by counting to be

$$\mathcal{N}(A \times B) = \mathcal{N}(A) \cdot \mathcal{N}(B),$$

with the obvious extension to more than two sets.

DO EXERCISE 8.

EXERCISE SET 7.1

For Exercises 1 through 16, let $\mathcal{U} = \{x \mid x \in I, \;\; 0 \leq x \leq 12\}$, where I is the set of all integers, and

$$A = \{x \mid x \in \mathcal{U}, \;\; 0 \leq x \leq 10\}, \qquad D = \{x \mid x \in \mathcal{U}, \;\; x \in A, \;\; x \notin C\},$$

$$B = \{x \mid x \in \mathcal{U}, \;\; 0 < x < 10\}, \qquad E = \left\{x \mid x \in \mathcal{U}, \;\; \frac{x}{3} \in \mathcal{U}\right\},$$

$$C = \left\{x \mid x \in \mathcal{U}, \;\; \frac{x}{2} \in A\right\}, \qquad F = \{x \mid x \in \mathcal{U}, \;\; x \in E, \;\; x \leq 10\}.$$

1. Write the sets A through C using the roster method.

2. Write the sets D through F using the roster method.

3. Determine the complements of sets A through C.

4. Determine the complements of sets D through F.

State whether each of the following is true or false.

5. $5 \in A$

6. $7 \in C$

7. $A \subset B$

8. $F \subset B$

9. $\emptyset \subset D$

10. $\emptyset \subset A$

11. $F = \{0, 6, 3, 9\}$

12. $E = \{3, 3, 6, 9, 12\}$

13. $E = \{6, 3, 12, 9\}$

14. $F = \{0, 3, 6, 3, 9\}$

15. $E \subset F$

16. $D^c \subset \mathcal{U}$

For Exercises 17 through 34, let

$$\mathcal{U} = \{a, e, i, o, u, m, n, r, t, c\}$$

and

$$A = \{a, e, i, o, u\}, \qquad D = \{m, i, n, t\},$$
$$B = \{c, m, n, r, t\}, \qquad E = \{e, i\},$$
$$C = \{a, c, e\}, \qquad F = \{r, t\}.$$

17. Determine the complements of sets A through C.

18. Determine the complements of sets D through F.

State whether each of the following is true or false.

19. $E = \{c, e, i, o, u\}$

20. $D = \{t, i, m, e\}$

21. $\{c, e, i, o, u\} \subset \mathcal{U}$

22. $\{a, e, z\} \subset \mathcal{U}$

23. $C = \{c, a, s, e\}$

24. $F = \{t, r\}$

25. $\mathcal{N}(C) = 4$

26. $\mathcal{N}(E) = 2$

Answer the following.

27. $E \times F = ?$

28. $F \times E = ?$

29. Is $(E \times F) \subset (A \times B)$?

30. Is $(F \times E) \subset (A \times B)$?

31. $C \times E = ?$

32. $C \times F = ?$

33. $\mathcal{N}(C \times E) = ?$

34. $\mathcal{N}(C \times F) = ?$

35. a) Find $\mathcal{N}(\emptyset)$.
   b) List the subsets of $\emptyset$.
   c) How many subsets are there?

36. a) Find $\mathcal{N}(\{a\})$.
   b) List the subsets of $\{a\}$.
   c) How many subsets are there?

37. a) Find $\mathcal{N}(\{a, b\})$.
  b) List the subsets of $\{a, b\}$.
  c) How many subsets are there?

38. a) Find $\mathcal{N}(\{a, b, c\})$.
  b) List the subsets of $\{a, b, c\}$.
  c) How many subsets are there?

39. On the basis of Exercises 35 through 38, complete the following table.

Cardinality of a set	0	1	2	3	4	n
Number of subsets	1					

40. Complete, based on Exercise 39.

A set with n elements has _____ subsets.

OBJECTIVES

You should be able to:

a) Given two sets, A and B, find their union, $A \cup B$, and intersection, $A \cap B$.

b) Given two sets A and B, find their difference $A - B$.

c) Given a collection of subsets of a given set, determine whether it is a partition of that set.

d) Find several partitions of a given set.

e) Solve problems involving sets.

7.2 UNION AND INTERSECTION OF SETS AND VENN DIAGRAMS

Consider two sets A and B.

Let A be the set of all students in a Finite Mathematics class and B be the set of all students in an Economics class. The *union* of A and B, written $A \cup B$, is the set of all students taking Finite Mathematics *or* Economics *or both*.

Formally, the *union* of A and B is the set of all elements contained in *either A or B or both* and can be written

$$A \cup B = \{x \mid x \in A \quad or \quad x \in B\}.$$

$A \cup B$ is read "A union B."

Note. Throughout this book, "or" will be used in the *inclusive* sense of "either or both." If the *exclusive* "or"—meaning "either but not both"—is intended, it will be specifically stated.

The union of sets can be represented geometrically by means of *Venn* diagrams. If two sets A and B are each represented by the interior of some closed curve, then the union of A and B is represented by the shaded area of the following diagram.

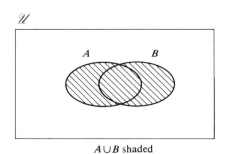

$A \cup B$ shaded

Consider again the two sets A and B, where A is the set of all students in a Finite Mathematics class, and B is the set of all students in an Economics class.

The *intersection* of A and B, written $A \cap B$, represents the set of all students taking *both* Finite Mathematics *and* Economics.

Formally, the *intersection* of A and B is the set of all elements contained in *both A and B* and can be written

$$A \cap B = \{x \mid x \in A \quad and \quad x \in B\}.$$

$A \cap B$ is read "A intersect B."

On a Venn diagram, the intersection of A and B is represented by the shaded area in the following diagram.

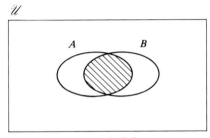

$A \cap B$ shaded

If two sets A and B have *no* elements in common, that is,

$$A \cap B = \emptyset$$

or

$$\mathcal{N}(A \cap B) = 0,$$

then A and B are said to be *disjoint*, and can be represented by the following Venn diagram:

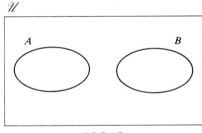

$A \cap B = \emptyset$

Note that $A \cap A^c = \emptyset$ and $A \cup A^c = \mathcal{U}$. (See the following diagram.)

9. Given $A = \{a, b, c, e, g\}$ and $B = \{b, c, e, f, s\}$. What are $A \cup B$ and $A \cap B$? Represent each of these sets by means of a Venn diagram.

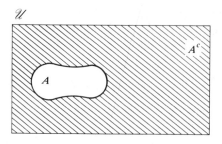

Example 1 Given $A = \{1, 2, 3, 5, 7\}$ and $B = \{0, 2, 3, 6, 9\}$. What are $A \cup B$ and $A \cap B$? Represent these sets by means of a Venn diagram.

Solution
$A \cup B = \{0, 1, 2, 3, 5, 6, 7, 9\}$
$A \cap B = \{2, 3\}$

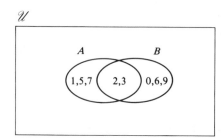

DO EXERCISE 9.

Consider the set of all students in a Finite Mathematics class. Divide the class into groups, or subsets, such that:

 i) each student in the class is in some group, and
ii) no student in the class is in more than one group.

These groups constitute a *partition* of the class.

Let A be the set of all molecules in a cookie. Break the cookie. Whatever way the cookie crumbles, the pieces are a *partition* of the cookie.

Formally, a *partition* of a set A is a division of A into subsets X_i $(i = 1, 2, \ldots, n)$ such that each element of A is contained in one and only one subset. Thus, the subsets X_i are disjoint (any two different sets have no elements in common):

$$X_i \cap X_j = \emptyset \qquad \text{for all } i \neq j,$$

and their union is A:

$$X_1 \cup X_2 \cup \cdots \cup X_n = A.$$

If these conditions are satisfied, then

$$\{X_1, X_2, \ldots, X_n\}$$

expresses the statement that the X_1's are a partition of A. For example, let

A = Set of natural numbers = $\{1, 2, 3, 4, \ldots\}$,

X_1 = Set of odd natural numbers = $\{1, 3, 5, \ldots\}$

X_2 = Set of even natural numbers = $\{2, 4, 6, \ldots\}$.

Then $\{X_1, X_2\}$ is a partition of A.

This partition can be represented by the following Venn diagram:

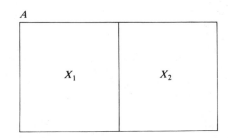

A set may be partitioned more than one way.

Example 2   Write several partitions of the set $A = \{a, e, r, s\}$ and draw the Venn diagram for each partition.

Solution

	Partitions		*Venn diagrams*
i)	$\{\{a\}, \{e\}, \{r\}, \{s\}\}$	i)	a | e | r | s
ii)	$\{\{a, e, r, s\}\}$	ii)	a, e, r, s
iii)	$\{\{a, e\}, \{r, s\}\}$	iii)	a, e | r, s
iv)	$\{\{a\}, \{e, r, s\}\}$	iv)	a | e, r, s

Note that $\{\{a, e\}, \{e, r, s\}\}$ is not a partition, because the sets are not disjoint. That is, $\{a, e\} \cap \{e, r, s\} = \{e\} \neq \emptyset$. Also, $\{\{a\}, \{r, s\}\}$ is not a partition because the union of the sets is not set A.

DO EXERCISE 10.

10. Write several partitions of the set $B = \{1, 2, 3, 4, 5\}$, and draw the Venn diagram for each. Answers may vary.

The *relative complement* of a set B with respect to set A, or, simply, the *difference* of A and B, written $A - B$, is the set of all elements contained *in A* but *not in B*. Formally,

$$A - B = \{x \mid x \in A, \quad x \notin B\}.$$

The *absolute complement* of a set A, or, simply, the *complement* of A, written A^c, can now be expressed as the relative complement of set A with respect to the universal set. Formally,

$$A^c = \mathcal{U} - A,$$

or

$$A^c = \{x \mid x \in \mathcal{U}, \quad x \notin A\}.$$

Using Venn diagrams, we can represent $A - B$, $A \cap B$, and $B - A$ by the following diagram:

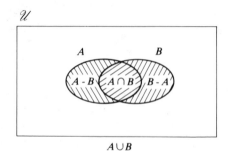

$A \cup B$

Thus, the union of two sets $A \cup B$ can always be partitioned

$$\{A - B, \quad A \cap B, \quad B - A\}.$$

Similarly, A and A^c can be represented by the Venn diagram:

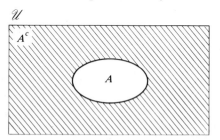

so that the universal set $\mathcal{U}$ can always be partitioned

$$\{A, A^c\}.$$

Example 3 Given $A = \{1, 2, 3, 5, 7\}$, $B = \{0, 2, 3, 6, 9\}$, and $\mathcal{U} = \{0, 1, 2, 3, 4, 5, 6, 7, 8, 9\}$. Find $A - B$, $B - A$, A^c, and $(A - B)^c$. Represent $(A - B)^c$ by a Venn diagram.

Solution

$$A - B = \{1, 5, 7\}, \qquad B - A = \{0, 6, 9\}, \qquad A^c = \{0, 4, 6, 8, 9\},$$
$$(A - B)^c = \{0, 2, 3, 4, 6, 8, 9\}$$

and is the shaded area in the figure.

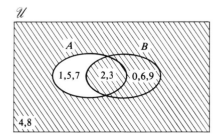

DO EXERCISE 11.

Since Venn diagrams embody set concepts, they can be used to *solve* problems involving sets. It is apparent from the following Venn diagram:

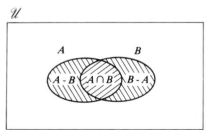

that

$$A = (A - B) \cup (A \cap B) \qquad \text{and} \qquad B = (B - A) \cup (A \cap B).$$

The cardinality of sets A and B is simply

$$\mathcal{N}(A) = \mathcal{N}(A - B) + \mathcal{N}(A \cap B) \qquad (1)$$

and

$$\mathcal{N}(B) = \mathcal{N}(B - A) + \mathcal{N}(A \cap B). \qquad (2)$$

Note that the union operation in the set relation has been replaced by the addition operation in the cardinality relation. This can be done when the components are disjoint.

Similarly, from the Venn diagram, we can write:

$$A \cup B = (A - B) \cup (A \cap B) \cup (B - A)$$

and

$$\mathcal{N}(A \cup B) = \mathcal{N}(A - B) + \mathcal{N}(A \cap B) + \mathcal{N}(B - A). \qquad (3)$$

11. $A = \{a, b, c, d, e, g\}$,

$\quad B = \{b, c, d, f, h\}$,

$\quad \mathcal{U} = \{a, b, c, \ldots, z\}$

Find $A - B$, $B - A$, A^c, B^c, $B^c - A^c$. Represent $B^c - A^c$ using a Venn diagram.

Solving for $\mathcal{N}(A - B)$ and $\mathcal{N}(B - A)$ in Eqs. (1) and (2) and substituting in Eq. (3), we obtain

$$\mathcal{N}(A \cup B) = \mathcal{N}(A) + \mathcal{N}(B) - \mathcal{N}(A \cap B).$$

This is a very important set property.

Thus, if $A = \{a, b, c, f\}$ and $B = \{a, b, c, d, e\}$, then

$$A \cup B = \{a, b, c, d, e, f\} \quad \text{and} \quad A \cap B = \{a, b, c\}.$$

Furthermore, $\mathcal{N}(A) = 4$, $\mathcal{N}(B) = 5$, $\mathcal{N}(A \cup B) = 6$, and $\mathcal{N}(A \cap B) = 3$. Substituting these quantities in

$$\mathcal{N}(A \cup B) = \mathcal{N}(A) + \mathcal{N}(B) - \mathcal{N}(A \cap B),$$

we have

$$6 = 4 + 5 - 3,$$

or

$$6 = 6,$$

which satisfies the above relation. Note that

$$\mathcal{N}(A \cup B) = \mathcal{N}(A) + \mathcal{N}(B) \quad \text{only if } A \text{ and } B \text{ are disjoint.}$$

An example more typical of the type usually solved using Venn diagrams is the following.

Example 4 Out of a sample of people surveyed,

> 43% smoked,
> 67% drank,
> 24% smoked and drank.

What percent neither smoked nor drank?

Solution Let

$$\mathcal{U} = \text{The set of all people surveyed,}$$
$$S = \text{The set of all smokers (surveyed),}$$

and

$$D = \text{The set of all drinkers (surveyed).}$$

Then

$$\mathcal{N}(\mathcal{U}) = 100 \quad \text{(dropping the \% sign, for simplicity),}$$
$$\mathcal{N}(S) = 43,$$
$$\mathcal{N}(D) = 67,$$

and the number of people who smoked *and* drank is given by

$$\mathcal{N}(S \cap D) = 24.$$

We think of the set symbol "∩" as corresponding to the word "and."

We want to find the number of people in the survey who neither smoked nor drank. We can express this with sets as

$$\mathscr{U} - S - D.$$

That is, we take out the smokers and drinkers and what is left are those people who neither smoke nor drink. Thus we want

$$\mathscr{N}(\mathscr{U} - S - D).$$

The appropriate Venn diagram is as follows:

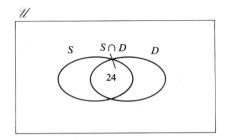

where $\mathscr{N}(S \cap D) = 24$ has been written in the area representing $S \cap D$. Now, as before, it is apparent from the Venn diagram that

$$\mathscr{N}(S - D) = \mathscr{N}(S) - \mathscr{N}(S \cap D)$$
$$= 43 - 24$$
$$= 19.$$

Similarly,

$$\mathscr{N}(D - S) = \mathscr{N}(D) - \mathscr{N}(S \cap D)$$
$$= 67 - 24$$
$$= 43.$$

Writing these two numbers, 19 and 43, into the areas representing $S - D$ and $D - S$ on the Venn diagram, we obtain:

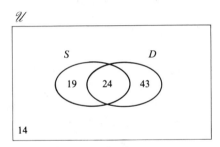

12. Out of a group of 240 students surveyed, 150 spent Friday night at a movie, 120 spent Saturday night at a basketball game, and 60 did both. How many students went to the basketball game, but not the movie?

Thus, the number of people who *either* smoked or drank is

$$\mathcal{N}(S \cup D) = \mathcal{N}(S - D) + \mathcal{N}(S \cap D) + \mathcal{N}(D - S)$$
$$= 19 + 24 + 43$$
$$= 86, \quad \text{or} \quad 86\%,$$

and the number of people who *neither* smoked nor drank is

$$\mathcal{N}(\mathcal{U} - S - D) = \mathcal{N}(\mathcal{U}) - \mathcal{N}(S \cup D)$$
$$= 100 - 86$$
$$= 14, \quad \text{or} \quad 14\%.$$

DO EXERCISE 12.

Here we used the Venn diagram both as a place to record the numbers representing the cardinality of various sets and also to indicate which differences to take to obtain new set cardinalities. The advantage of this approach may not be as apparent with the preceding example with two sets as it is in the next example with three sets.

Example 5   There are 220 students in a certain freshman class. Of these,

> 115 are taking Economics,
> 60 are taking Spanish,
> 95 are taking Mathematics,
> 20 are taking Economics and Spanish,
> 30 are taking Economics and Mathematics,
> 25 are taking Spanish and Mathematics,
> 15 are taking all three subjects.

How many students are taking only *one* of these three subjects?

Solution   Formally, this problem may be posed in this manner:

Let

> $\mathcal{U}$ = set of all students in the freshman class,
> E = set of all students taking Economics,
> S = set of all students taking Spanish, and
> M = set of all students taking Mathematics.

Then

$$\mathcal{N}(\mathcal{U}) = 220,$$
$$\mathcal{N}(E) = 115,$$
$$\mathcal{N}(S) = 60,$$
$$\mathcal{N}(M) = 95,$$
$$\mathcal{N}(E \cap S) = 20 \qquad \text{(those taking Economics } and \text{ Spanish),}$$
$$\mathcal{N}(E \cap M) = 30,$$
$$\mathcal{N}(S \cap M) = 25,$$
$$\mathcal{N}(E \cap S \cap M) = 15 \qquad \text{(those taking Economics } and \text{ Spanish}$$
$$and \text{ Mathematics).}$$

The question is then to determine

$$\mathcal{N}(E - S - M) + \mathcal{N}(S - E - M) + \mathcal{N}(M - S - E),$$

where $\mathcal{N}(E - S - M)$ is the number of students taking Economics, but not Spanish or Mathematics, and so on.

We start with information at the bottom of the data list:

$$\mathcal{N}(E \cap S \cap M) = 15 \quad \text{or} \quad (15)_1,$$

where the number 15 is being written in the Venn diagram as $(15)_1$. Here the subscript is being used to indicate the order in which the numbers are entered on the Venn diagram, to avoid redrawing it for each step. Thus, $(15)_1$ is the first entry in this diagram.

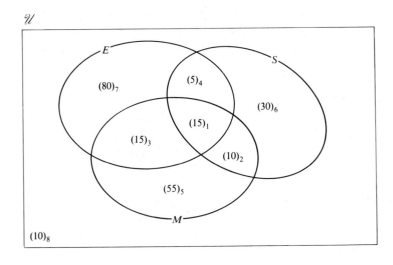

13. A survey of 195 people indicated that:

> 90 read *Time* (*T*),
> 45 read *National Review* (*R*),
> 15 read *Nation* (*N*),
> 3 read *T* and *R*,
> 5 read *T* and *N*,
> 1 read *R* and *N*,
> 1 read all three.

How many people read *none* of these three magazines? Use a Venn diagram.

Working *up* the data list, we have:

$\mathcal{N}[(S \cap M) - E]$

$\qquad = \mathcal{N}(S \cap M) - \mathcal{N}(S \cap M \cap E)$ (those taking Spanish *and* Math, *but not* Economics)

$\qquad = 25 - 15$

$\qquad = 10,$ or $(10)_2$;

$\mathcal{N}[(E \cap M) - S] = 30 - 15 = 15,$ or $(15)_3$;

$\mathcal{N}[(E \cap S) - M] = 20 - 15 = 5,$ or $(5)_4$.

Continuing, we have

$$\mathcal{N}(M - S - E) = 95 - 15 - 15 - 10 \qquad \text{(Why?)}$$
$$= 55, \quad \text{or} \quad (55)_5.$$

That is, 40 of the elements of *M* are in *S* or *E*, leaving 55 elements for $M - S - E$. Similarly,

$$\mathcal{N}(S - E - M) = 60 - 5 - 15 - 10 = 30, \quad \text{or} \quad (30)_6;$$

and

$$\mathcal{N}(E - S - M) = 115 - 5 - 15 - 15 = 80, \quad \text{or} \quad (80)_7.$$

Thus, the number of students taking *only one* subject is

$$\mathcal{N}(E - S - M) + \mathcal{N}(S - E - M) + \mathcal{N}(M - S - E)$$
$$= 80 + 30 + 55$$
$$= 165.$$

Furthermore, the number of students taking *exactly two* subjects is

$$\mathcal{N}[(E \cap S) - M] + \mathcal{N}[(E \cap M) - S] + \mathcal{N}[(S \cap M) - E]$$
$$= 5 + 15 + 10$$
$$= 30;$$

and the number of students taking *all three* subjects is

$$\mathcal{N}(E \cap S \cap M) = 15.$$

The number of students taking *at least one* subject is the sum of these numbers:

$$165 + 30 + 15, \quad \text{or} \quad 210.$$

Since $\mathcal{N}(\mathcal{U}) = 220$, there are $220 - 210 = 10$ who are *not taking any* of the courses mentioned. At this point the Venn diagram is completely filled out and the answer to many questions can be obtained by a simple sum.

DO EXERCISE 13.

EXERCISE SET 7.2

For Exercises 1 through 5 (as in Exercises 1 through 16, Exercise Set 7.1)

$$\mathcal{U} = \{x \mid x \in I, \ \ 0 \le x \le 12\}, \quad \text{where } I \text{ is the set of all integers,}$$

and

$$A = \{x \mid x \in \mathcal{U}, \ \ 0 \le x \le 10\}, \qquad D = \{x \mid x \in \mathcal{U}, \ \ x \in A, \ \ x \notin C\}$$

$$B = \{x \mid x \in \mathcal{U}, \ \ 0 < x < 10\}, \qquad E = \left\{x \mid x \in \mathcal{U}, \ \ \frac{x}{3} \in \mathcal{U}\right\},$$

$$C = \left\{x \mid x \in \mathcal{U}, \ \ \frac{x}{2} \in A\right\}, \qquad F = \{x \mid x \in \mathcal{U}, \ \ x \in E, \ \ x \le 10\}.$$

1. Determine the set indicated by each of the following:

a) $A \cup B$

b) $A \cap B$

c) $A - B$

d) $B - A$

e) $A - A^c$

f) $A - (D \cup E)$

g) $(C \cup E)^c$

h) $(A - B) \cup (C - F)$

i) $(A - B) \cap (C - F)$

j) $[(D - F)^c]^c$

2. Do any of the sets B, C, D, E, F partition set A? If yes, which ones?

3. Draw a Venn diagram illustrating the relationship among sets A, B, C, D, E, F.

For Exercises 4 through 6,

$$\mathcal{U} = \{a, e, i, o, u, c, m, n, r, t\};$$

$$A = \{a, e, i, o, u\}, \qquad D = \{m, i, n, t\},$$

$$B = \{c, m, n, r, t\}, \qquad E = \{e, i\},$$

$$C = \{a, c, e\}, \qquad F = \{r, t\}.$$

4. Answer the following:

a) Is $\{C, D, F\}$ a partition of $\mathcal{U}$?

b) Are B and C disjoint?

5. Determine the set indicated by each of the following:

a) $E \cup F$

b) $E \cap F$

c) $B - D$

d) $D - B$

e) $(A \cup B)^c$

f) $(D - E) \cup F$

g) $(D^c - E^c) \cap F$

h) $(A \cap C) - (C \cap D)$

6. Draw a Venn diagram illustrating the relationship among sets A, B, C, D, E, F.

7. Of 68 people surveyed, 33 smoked, 57 drank, and 27 did both. How many did neither?

8. *Marketing.* Of 87 people surveyed, 49 read *Time*, 21 read *Nation*, and 5 read both. How many read *only one* magazine? How many read *neither* magazine?

9. Of 73 men surveyed, 54 wore belts, 20 wore suspenders, and 3 wore neither. How many wore both?

10. Of 123 students, 79 could ride a bike, 53 could drive a car, and 15 could do both. How many could do neither?

11. *Political Science.* In a recent poll of 230 people, 50 thought only the Republicans could solve our problems, 70 thought only the Democrats could, and 25 thought neither could. How many thought one could do as well as the other (that is, both)?

12. *Medicine.* Blood can be typed as A (having Type A antigen), B (having Type B antigen), AB (having both), and O (having neither). Of 140 patients in a hospital, there were 53 with Type A blood, 47 with Type B, and 24 with Type AB. How many had Type O? *Note.* Having Type A blood implies *only* Type A antigens.

13. Of the students in a certain university,

> 55% took Finite Mathematics,
> 65% took English Composition,
> 35% took Spanish,
> 30% took Mathematics and English,
> 24% took Mathematics and Spanish,
> 18% took English and Spanish,
> 12% took all three.

How many students took only one of these three subjects? Only two of these three subjects? None of these courses?

14. *Marketing.* Of recent car buyers,

65% bought automatic transmissions,
20% bought air conditioning,
70% bought posh seats,
50% bought automatic transmissions and posh seats,
10% bought automatic transmissions and air
    conditioning,
10% bought posh seats and air conditioning,
5% bought all three.

How many bought posh seats alone? No option? Posh seats but not air conditioning?

OBJECTIVES

You should be able to

a) Draw a tree to represent the sample space in a counting problem.

b) Count the branches of a tree to determine the number of possible outcomes in a sample space.

7.3 EXPERIMENTS, TREES, AND DISTINGUISHABILITY OF OUTCOMES

Consider an "experiment" in coin-flipping. Assuming that the coin does not land standing on its edge, there are two possible outcomes: Heads (H) or Tails (T). The set S of all possible outcomes of some particular experiment is called the *sample space* in probability theory, and is the universal set. Thus, the sample space for a single flip of the coin is:

$$S = \{H, T\}.$$

Consider flipping a coin *twice*. The first coin flip is *distinguished* from the second in time. This sample space is

$$S = \{HH, HT, TH, TT\}.$$

Each element of the set S is an ordered pair written HH rather than (H, H) for simplicity. The *first* flip is indicated by the *first* element of the ordered pair and the *second* flip is indicated by the *second* element. The sample space for flipping two coins labelled 1 and 2 (for example, a penny and a dime) is the same, with the *time ordering* replaced by the *number* (or label) *ordering*. In either case, the outcomes of each *part* of the total experiment are *considered distinguishable*.

On the other hand, we might be interested only in *how many* heads and *how many* tails occur and not in how each individual coin lands. This sample space is

$$S = \{HH, HT, TT\}.$$

Here the element TH is *considered* identical to the element HT, and is thus omitted. The labelling on the coins is ignored and they are *considered indistinguishable*.

This concept of distinguishability applies as well to more than two trials. This is one of the first concerns in probability, even though it is only implicitly, not explicitly, stated in most problems.

Example 1 An election is being held with three candidates and 101 voters. The offices of president and vice-president go to the candidates with the highest and next highest number of votes. If there is a tie for president, then there is to be a runoff election between these two candidates for president and vice-president. If there is a tie for vice-president, then there is to be a runoff election for vice-president. What is the sample space of possible outcomes?

Solution Here we are not interested in which candidate a *particular* voter votes for, so that the *voters* are indistinguishable. Furthermore, the *number* of votes each candidate gets is relevant only in a ranking sense. Thus, the president is the winner whether by one vote or 100.

Let us denote the candidates by *A*, *B*, and *C*. Then the sample space is:

$$S = \{AB, AC, BA, BC, CA, CB\},$$

where the first letter of each pair denotes the president and the second denotes the vice-president. Thus, *AB* and *BA* are different outcomes.

DO EXERCISE 14.

Sample spaces can be represented geometrically by *trees*. Consider again coin-flipping. For a single coin, the sample space is $S = \{H, T\}$.

14. An old sea chest contains 10 bars of silver and 5 bars of gold. If 3 bars are drawn at random, what is the sample space of possible outcomes?

The corresponding tree is:

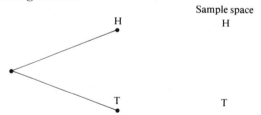

Leading out of a *vertex* (dot) representing the *flipping* of the coin, there are two *arcs* (line segments) representing the two *outcomes*, H *or* T. Thus, the tree can be *read* by saying that a coin is flipped and can land H *or* T (but *not* both H and T at the same time).

Each fork of a tree consists of a *vertex*, which represents an *experiment*, and a set of *arcs*, which represent all possible *mutually exclusive* outcomes.

Flipping a coin twice has the sample space $S = \{HH, HT, TH, TT\}$ and the tree:

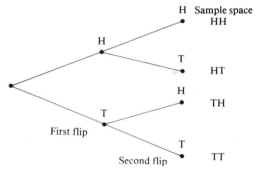

A *branch* of a tree is a sequence of connected arcs. Following along a branch of a tree from beginning to end indicates the succession of possible *individual* outcomes that constitute a single overall outcome. Thus, a tree with all its branches represents the sample space.

The upper branch of this tree indicates that the *first* coin landed H *and* the *second* coin landed H. The next lower branch indicates that the *first* coin landed H *and* the *second* coin landed T, and so forth.

The tree drawn for a given sample space may not be unique. For example, if we simultaneously flip two identical coins labelled "1" and "2," then either coin can be identified with the coin flipped first, so that two equivalent but different trees are possible.

The tree shown *distinguishes* between the two flips of the coin. If we wish to consider the coins *indistinguishable*, then we can ignore the labelling that distinguishes the coins.

Example 2   Draw a tree to represent the sample space of Example 1.

Solution

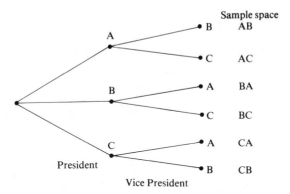

Here we are concerned only with who wins, not the actual vote. We have represented the presidential winner first on the tree, since it was represented first in the pair of the sample space.

DO EXERCISE 15.

15. Draw a tree to represent the sample space of Margin Exercise 14.

EXERCISE SET 7.3

1. A business manager has a problem that he discusses with each of his three assistants. Each comes up with a possible solution that looks equally acceptable. Hence, he resorts to his standard decision-maker: He tacks the solutions to the wall and throws darts to rank them I, II, and III. The first solution is tried. If it works, then no other solution is considered further. If the first solution doesn't work, the second solution is tried, and so forth. What is the sample space for possible outcomes? Draw a tree representing them. (Nothing has been said about what happens if III fails.)

2. Stockholders are being presented with several options on which to vote. If Option I is accepted, no further options need be considered. However, if Option I is rejected, then the voters are **asked** to vote on which option, II or III, to consider next. That is then done, with the losing option considered last. What is the sample space? Draw the tree.

3. On any particular day the weather can be classified by temperature (above normal, normal, below normal) and sky conditions (clear, cloudy, or precipitating (rain, snow, or whatever)). What is the sample space of possible weather conditions? Draw a tree.

4. *Marketing.* A survey is made to determine who buys a particular product. The people surveyed are classified by income (low, medium, or high) and education (grade school, high school, or college). What is the sample space of possible classifications? Draw a tree.

5. *Political Science.* A political survey is made and the responders are classified by sex (male or female), marital status (single or married), or age (<18 or ≥ 18). What is the sample space of the responders? Draw a tree.

6. The menu for a meal gives one the choice of (i) soup or fruit cup, (ii) fish, chicken, or meat, (iii) salad or dessert. What is the sample space of possible menus? Draw a tree.

7. Telephone lines exist between towns as indicated in the following diagram.

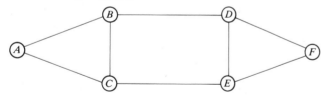

Draw a tree indicating all possible phone connections between *A* and *F*, assuming that no telephone *link* is used more than once.

9. *Sports.* In a single-elimination sports tournament consisting of *n* teams, a team is eliminated when it loses one game. How many games are required to complete the tournament?

8. As in Exercise 7, but with the phone lines as in the diagram below.

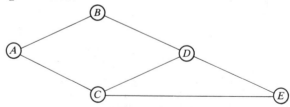

10. *Sports.* In a double-elimination softball tournament consisting of *n* teams, a team is eliminated when it loses two games. At most how many games are required to complete the tournament?

OBJECTIVES

You should be able to:

a) Express a factorial such as 6!, as a product $6 \cdot 5 \cdot 4 \cdot 3 \cdot 2 \cdot 1$, and evaluate.

b) Use the formula $P(n, k) = n(n - 1) \times (n - 2) \cdots [n - (k - 1)]$,

to find the number of permutations of *n* objects taken *k* at a time, without replacement or repetition.

c) Find the number of circular arrangements of *n* objects by evaluating $(n - 1)!$.

d) Find the number of ordered arrangements of *n* objects taken *k* at a time with replacement, or repetition, by evaluating n^k.

e) Use the Fundamental Counting Principle and the procedures of (b) through (d) to solve various types of counting problems.

7.4 COUNTING TECHNIQUES—PERMUTATIONS

Trees were used in the preceding section to show the possible outcomes of an experiment. We want to develop faster counting techniques. The following example leads up to the *Fundamental Counting Principle*.

Example 1   How many 4-letter words (not necessarily meaningful or pronounceable) can be formed using the letters P, D, Q, X *without* repetition?

Solution   Such a word would have the general form

Any of the 4 letters can be used for the first letter in the word. Once this letter has been selected, the second can be selected from the 3 remaining letters, and the third from the remaining 2 letters. The fourth letter is already determined, since only 1 possible letter remains. Thus there are

$$4 \cdot 3 \cdot 2 \cdot 1, \quad \text{or} \quad 24 \text{ words.}$$

We could, of course, have determined this by writing down all the possibilities; and even though this can be quite cumbersome in general, we list them below for reference.

PDQX   PDXQ   PQDX   PQXD   PXDQ   PXQD

DPQX   DPXQ   DQPX   DQXP   DXPQ   DXQP

QPDX   QPXD   QDPX   QDXP   QXDP   QXPD

XPDQ   XPQD   XDPQ   XDQP   XQPD   XQDP

DO EXERCISE 16.

FUNDAMENTAL COUNTING PRINCIPLE. Given a combined action, or event, in which the 1st action can be performed in n_1 ways, the 2nd action can be performed in n_2 ways, and so on, then the total number of ways the combined action can be performed is the product

$$n_1 \cdot n_2 \cdot n_3 \cdots n_k.$$

Let us demonstrate this for three actions with a tree. Assume the first action E_1 can be performed in 3 ways, the second E_2 can be performed in 4 ways, and the third E_3 in 2 ways.

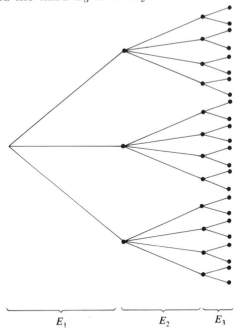

$$\underbrace{}_{E_1}\quad\underbrace{}_{E_2}\quad\underbrace{}_{E_3}$$

There are $3 \cdot 4 \cdot 2$, or 24 paths through the tree (count them), so there are 24 outcomes for all three actions.

DO EXERCISES 17 THROUGH 20.

16. How many 3-letter words can be formed using the letters P, D, Q *without* repetition?

17. How many 5-letter words can be formed using the letters P, D, Q, R, S *without* repetition?

18. How many 4-digit numbers can be formed using the digits 6, 7, 8, 9 without repetition?

19. An examination consists of ten true–false questions. How many possible different answer sheets can be turned in?

20. A woman is going out to eat. She will put on one of 5 pantsuits, one pair out of 40 pairs of shoes, and go to one of 8 restaurants. In how many ways can she dress and eat?

21. How many ways can 5 motorcycles be parked in a row?

22. How many permutations are there of a set of 5 objects? Consider a set

$$\{a, b, c, d, e\}.$$

23. Find $P(6, 6)$ and $P(5, 5)$.

Permutations

A *permutation* of a set of n objects is an ordered arrangement of all the objects. For example, consider the set of 4 objects,

$$\{P, D, Q, X\}$$

as in Example 1. There are 4 choices for the first letter, 3 choices for the second, 2 choices for the third, and 1 for the fourth. Thus, by the Fundamental Counting Principle, there are:

$$4 \cdot 3 \cdot 2 \cdot 1, \quad \text{or} \quad 24 \; permutations \text{ of a set of 4 objects.}$$

DO EXERCISES 21 AND 22.

In general, consider a set of n objects. There are n choices for the first object, $(n - 1)$ choices for the second, $(n - 2)$ choices for the third, and so on. The nth object is chosen in only 1 way.

The total number of permutations of a set of n objects, denoted $P(n, n)$, is given by

$$P(n, n) = \underbrace{n(n - 1)(n - 2) \cdots 3 \cdot 2 \cdot 1}_{n \textbf{ factors}}$$

Example 2 Find $P(7, 7)$ and $P(3, 3)$.

Solution

$$P(7, 7) = 7 \cdot 6 \cdot 5 \cdot 4 \cdot 3 \cdot 2 \cdot 1 = 5040,$$
$$P(3, 3) = 3 \cdot 2 \cdot 1 = 6$$

DO EXERCISE 23.

Factorial Notation

We will use products such as

$$6 \cdot 5 \cdot 4 \cdot 3 \cdot 2 \cdot 1 \quad \text{and} \quad 8 \cdot 7 \cdot 6 \cdot 5 \cdot 4 \cdot 3 \cdot 2 \cdot 1$$

so often that it is convenient to adopt a notation for them. We define

$$6 \cdot 5 \cdot 4 \cdot 3 \cdot 2 \cdot 1 = 6!, \quad \text{read "6 factorial".}$$

In general,

$$n! = n \text{ factorial} = \underbrace{n(n - 1)(n - 2) \cdots 3 \cdot 2 \cdot 1}_{n \textbf{ factors}}$$

For example,

$$n! = n \cdot (n - 1) \cdot (n - 2) \cdots 1$$

$$
\begin{aligned}
7! &= 7 \cdot 6 \cdot 5 \cdot 4 \cdot 3 \cdot 2 \cdot 1 = 5040 \\
6! &= 6 \cdot 5 \cdot 4 \cdot 3 \cdot 2 \cdot 1 = 720 \\
5! &= 5 \cdot 4 \cdot 3 \cdot 2 \cdot 1 = 120 \\
4! &= 4 \cdot 3 \cdot 2 \cdot 1 = 24 \\
3! &= 3 \cdot 2 \cdot 1 = 6 \\
2! &= 2 \cdot 1 = 2 \\
1! &= 1 = 1
\end{aligned}
$$

DO EXERCISES 24 AND 25.

We define

$$0! = 1$$

for consistency in formulas used later.

We can now restate the formula for the total number of permutations of n objects.

$$P(n, n) = n(n - 1)(n - 2) \cdots 3 \cdot 2 \cdot 1 = n!$$

DO EXERCISE 26.

We often use other notations for factorials. Note the following.

$$7! = 7 \cdot 6 \cdot 5 \cdot 4 \cdot 3 \cdot 2 \cdot 1 = 7 \cdot (6 \cdot 5 \cdot 4 \cdot 3 \cdot 2 \cdot 1) = 7 \cdot 6!$$

$$6! = 6 \cdot 5 \cdot 4 \cdot 3 \cdot 2 \cdot 1 = 6 \cdot (5 \cdot 4 \cdot 3 \cdot 2 \cdot 1) = 6 \cdot 5!$$

In general,

$$n! = n(n - 1)!$$

DO EXERCISE 27.

Note also that

$$
\begin{aligned}
7! &= 7 \cdot 6! \\
&= 7 \cdot 6 \cdot 5! \\
&= 7 \cdot 6 \cdot 5 \cdot 4!
\end{aligned}
$$

In general, for any $k < n$,

$$n! = \underbrace{\underbrace{n(n - 1)(n - 2) \cdots [n - (k - 1)]}_{k \text{ factors}} \underbrace{(n - k)!}_{(n - k) \text{ factors}}}_{n \text{ factors}}$$

24. Find 8!.

25. Find 9!.

26. Use factorial notation to represent the number of permutations of 6 objects.

27. Express each in the form $n(n - 1)!$.

a) 8!.

b) 38!.

Permutations of *n* Objects Taken *k* at a Time

Consider the set

$$\{P, D, Q, X, Y\}.$$

We have 5 objects. Suppose we wanted to determine the number of permutations of 3 objects taken from the set. There would be 5 choices for the first object. Then there would remain 4 choices for the second object, and 3 for the third selection. By the Fundamental Counting Principle, there would be

$5 \cdot 4 \cdot 3,$ or 60 permutations of a set of 5 objects taken 3 at a time.

In general, suppose we had a set of *n* objects and we wanted to determine the number of permutations of these *n* objects taken *k* at a time. There would be *n* choices for the first object. Then there would remain $(n - 1)$ choices for the second object, $(n - 2)$ for the third, and so on. We would make *k* choices in all, so there will be *k* factors in the product.

The total number of permutations of *n* objects taken *k* at a time, denoted *P(n, k)*, is given by

$$P(n, k) = \underbrace{n(n - 1)(n - 2) \cdots [n - (k - 1)]}_{k \text{ factors}}$$

An alternative symbol for $P(n, k)$ can be found by multiplying by 1, using $\dfrac{(n - k)!}{(n - k)!}$ as follows:

$$P(n, k) = n(n - 1)(n - 2) \cdots [n - (k - 1)] \cdot \frac{(n - k)!}{(n - k)!}$$

$$= \frac{n(n - 1)(n - 2) \cdots [n - (k - 1)](n - k)!}{(n - k)!}$$

$$= \frac{n!}{(n - k)!}$$

Thus,

The total number of permutations of *n* objects taken *k* at a time is given by

$$P(n, k) = n(n - 1)(n - 2) \cdots [n - (k - 1)] \qquad (1)$$

$$P(n, k) = \frac{n!}{(n - k)!}. \qquad (2)$$

Formula (1) is most useful in application, but formula (2) will be important in a development in Section 7.5.

Example 3   Evaluate $P(7, 3)$ using both formulas.

Solution   Using formula (1), we have:

$$P(7, 3) = \underbrace{7 \cdot 6 \cdot 5}_{} = 210$$

This number tells where to start.

This number tells how many factors.

Using formula (2) we have

$$P(7, 3) = \frac{7!}{(7 - 3)!} = \frac{7!}{4!} = \frac{7 \cdot 6 \cdot 5 \cdot 4 \cdot 3 \cdot 2 \cdot 1}{4 \cdot 3 \cdot 2 \cdot 1} = 7 \cdot 6 \cdot 5 = 210.$$

DO EXERCISES 28 AND 29.

Example 4   How many ways can the letters of "organize" be arranged

a) taking 8 at a time?   b) taking 6 at a time?
c) taking 4 at a time?   d) taking 2 at a time?

Solution

a) $P(8, 8) = 8 \cdot 7 \cdot 6 \cdot 5 \cdot 4 \cdot 3 \cdot 2 \cdot 1 = 40{,}320$
b) $P(8, 6) = 8 \cdot 7 \cdot 6 \cdot 5 \cdot 4 \cdot 3 \quad\quad = 20{,}160$
c) $P(8, 4) = 8 \cdot 7 \cdot 6 \cdot 5 \quad\quad\quad\quad\quad = \quad 1680$
d) $P(8, 2) = 8 \cdot 7 \quad\quad\quad\quad\quad\quad\quad\quad = \quad\quad 56$

DO EXERCISE 30.

Circular Permutations (Optional)

Consider arranging the 5 letters A, B, C, D, and E in a circular permutation.

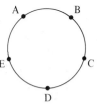

In a circle, the permutations ABCDE, BCDEA, CDEAB, DEABC, and EABCD are no longer distinguishable as they would be on a line. Therefore, for each circular permutation, there would be 5 distinguishable permutations on a line. (Think of cutting the circle open and bending it out to a line.)

Suppose we have P circular permutations. Each of these would yield $5 \cdot P$ permutations on a line, and we know there are 5! of these. Then

28. Evaluate $P(6, 4)$ using both formulas.

29. Evaluate, using formula (1):

a) $P(10, 4)$

b) $P(9, 9)$

30. How many ways can the letters of "soybean" be arranged,

a) taking 7 at a time?

b) taking 5 at a time?

c) taking 4 at a time?

d) taking 3 at a time?

31. How many ways can 6 different foods be arranged in 6 dishes *around* a Lazy Susan?

$5 \cdot P = 5!$, so

$$P = \frac{5!}{5} = 4! = 4 \cdot 3 \cdot 2 \cdot 1 = 24.$$

In general,

The number of circular permutations of n objects taken n at a time is $(n - 1)!$.

Example 5 How many ways can 10 college students sit around a campfire?

Solution

$$(10 - 1)! = 9! = 9 \cdot 8 \cdot 7 \cdot 6 \cdot 5 \cdot 4 \cdot 3 \cdot 2 \cdot 1 = 362,880.$$

DO EXERCISE 31.

Permutations "with Repetition" or "with Replacement"

32. How many 2-card permutations can be made by selecting 2 cards from a deck of 52,

a) without replacement?

b) with replacement?

Example 6 A standard deck of cards has 52 different cards. (For the exact makeup of the deck, see p. 291.) How many 3-card permutations can be made by selecting the 3 cards without replacement? with replacement?

Solution

a) The case "without replacement" is equivalent to "without repetition," as in Example 1. This is the number of permutations of 52 things taken 3 at a time.

$$P(52, 3) = 52 \cdot 51 \cdot 50 = 132,600$$

b) The case "with replacement" is considered by first making a selection. This can be done in 52 ways. Then the card is "replaced" and we make another selection. This can still be done 52 ways. Similarly, the third selection can be made 52 ways. Thus there are

$$52 \cdot 52 \cdot 52, \quad \text{or} \quad 140,608$$

possible permutations.

The number of permutations of n objects taken k at a time, with replacement, or with repetition, is n^k.

DO EXERCISE 32.

Mixed Counting

In the following example we carry out several types of counting.

Example 7 How many ways can 3 men and 3 women be seated in a row of **6** seats

a) with no seating restrictions?
b) if men and women must alternate?
c) if a particular couple (man–woman) must sit together?
d) if a particular couple must *not* sit together?

Solution

a) There are **6** people and **6** seats. Hence there are

$$P(6,6) = 6! = 6 \cdot 5 \cdot 4 \cdot 3 \cdot 2 \cdot 1 = 720 \text{ possible arrangements.}$$

b) They can sit either MWMWMW *or* WMWMWM. These yield two possibilities. Considering the possibility MWMWMW, we think of the number of ways the *men* can be arranged within this possibility.

This is equivalent to arranging 3 objects 3 at a time, so there are

$$P(3,3) = 3! = 3 \cdot 2 \cdot 1 = 6 \text{ ways.}$$

Considering the same possibility MWMWMW, we think of the number of ways the *women* can be arranged within this possibility.

MWMWMW
↑   ↑   ↑
① ② ③

In all there are **6 · 6**, or **36** ways within the possibility MWMWMW. Now consider the possibility WMWMWM. By a similar argument, there are **6·6**, or **36** ways within the possibility WMWMWM. Then the total number of arrangements is $2 \cdot 6 \cdot 6$, or 72.

c) i) The particular couple can sit MW or WM, yielding 2! or 2 possibilities.

   ii) Considering this particular couple as *one* object and the other four men and women as 4 objects, we have a total of 5 objects to be permuted (arranged 5 at a time). Then we have

$$P(5,5) = 5! = 5 \cdot 4 \cdot 3 \cdot 2 \cdot 1 = 120 \text{ ways.}$$

   iii) The total number of arrangements, from (i) and (ii), is $2 \cdot 120 = 240$.

33. The flags of five nations are to be raised on six flagpoles arranged in a row. How many ways can the flags be raised:

a) If there are no restrictions?

b) If all the flags must be together with no gaps between them?

c) If three particular nations want their flags all together with no separation either with an empty flagpole or another flag?

d) If three particular nations do *not* want their flags together, an empty flagpole being equivalent to a flag as separation?

d) If one action can be performed in n_1 ways and another disjoint action (see p. 259) can be performed in n_2 ways, then *either* action can be performed in:

$$n_1 + n_2 \quad \text{ways.}$$

This is in contrast to the Fundamental Counting Principle where we consider *both* actions occurring and would have multiplied.

Since the couple must *either* sit together or *not* sit together, the number of arrangements with the couple sitting together (c) plus the number of arrangements with the couple not sitting together (d) must equal the number of arrangements with no seating restrictions (a). Thus, the number of arrangements with the couple not sitting together is the result of (a) minus the result of (c):

$$720 - 240 = 480.$$

DO EXERCISE 33.

EXERCISE SET 7.4

Evaluate.

1. 5!   **2.** 6!   **3.** 0!   **4.** 1!   **5.** $P(6, 6)$ **6.** $P(5, 5)$

7. $P(20, 2)$ **8.** $P(30, 2)$ **9.** $P(7, 5)$ **10.** $P(6, 4)$ **11.** $P(n, 3)$ **12.** $P(n, 2)$

13. A person can get to the airport 3 ways, fly by any of 4 airlines, and get from the airport to his final destination in 2 ways. How many different ways can a person get to his destination? [*Hint.* Use the Fundamental Counting Principle.]

14. A person driving from Boston to Los Angeles wishes to go through Washington D.C., Atlanta, New Orleans, and Denver. If there are 6 main routes between Boston and Washington, 2 between Washington and Atlanta, 4 between Atlanta and New Orleans, 5 between New Orleans and Denver, and 7 between Denver and Los Angeles, how many possible routes can he take?

15. An office manager hires 4 secretaries, one for each of his assistants. If the secretaries are assigned at random, how many different assignments are possible? Give the solution in factorial and permutation notation. (*Hint.* See Example 1 and Example 2.)

16. An ice cream store has 6 different flavors of ice cream and room under the counter for 6 cartons of ice cream. How many different ways can the ice cream cartons be arranged under the counter? Give the solution in factorial and permutation notation.

17. As in Exercise 15, 4 secretaries apply for 2 different positions. How many ways can these two positions be filled if the secretaries are hired at random? Express the solution in permutation notation. (*Hint.* See Example 3 and Example 4.)

18. As in Exercise 16, if 8 ice cream flavors are available, in how many ways can 6 flavors be selected and arranged under the counter? Express the solution in permutation notation.

19. How many words (not necessarily meaningful or pronounceable) can be formed by rearranging the letters of the word "LOVE"? Express your answer in permutation notation. (*Hint.* See Example 4.)

20. How many 3-letter words can be formed from the letters of the word "ZORCH"? Express your answer in permutation notation.

21. How many ways can 4 people be seated at a circular bridge table? Would the answer change if the table were square? (*Hint.* See Example 5.)

22. How many ways can 8 people on a committee be seated at a circular table?

23. How many 4-number license plates can be made using the digits 1, 2, 3, 4, and 5 if repetitions are permitted? not permitted? (*Hint.* See Example 6.)

24. As in Exercise 23, but the number must be even.

25. How many 3-digit numbers can be formed from the numbers 1, 2, 3, 4, and 5, if repetitions are (a) allowed, (b) not allowed?

26. As in Exercise 25, if repetitions are not allowed, how many of these are (a) larger than 300? (b) less than 500?

27. How many ways can 4 different contracts be awarded to 7 different firms? if no firm gets more than one contract?

28. How many ways can 3 people be assigned to 5 offices?

29. As in Exercise 27, but two particular contracts must be awarded to the same firm with no other restrictions.

30. As in Exercise 28, but 2 particular people want adjacent offices.

31. How many ways can 4 executives from a given company and 3 visitors from another company be seated in a row? if they must alternate host, visitor, etc.? if a member of the host company sits at each end?

32. As in Exercise 31, but (a) two particular people wish to sit together, (b) two particular people should not be seated together.

▶

33. A car holds three people in the front and three people in the back. How many ways can 6 people be seated in the car? if a given couple must sit together?

34. As in Exercise 33, but two particular couples must sit together.

7.5 COUNTING TECHNIQUES—COMBINATIONS

We may sometimes make selections from a set without regard for order. Such selections are called *combinations*.

Example 1 How many combinations can be formed by taking elements 3 at a time from the set {A, B, C, D, E}?

Solution The combinations are

$$\{A, B, C\}, \quad \{A, B, D\}, \quad \{A, B, E\}, \quad \{A, C, D\}, \quad \{A, C, E\},$$
$$\{A, D, E\}, \quad \{B, C, D\}, \quad \{B, C, E\}, \quad \{B, D, E\}, \quad \{C, D, E\}.$$

Note that finding all the combinations of 5 objects taken 3 at a time is the same as forming all the 3-element subsets. Thus a combination is a subset. This is consistent with our earlier work regarding sets. That is, the set, or combination, {A, C, E} is the same as the set, or combination, {C, A, E}, because *order is not considered* when describing sets.

OBJECTIVES

You should be able to

a) Evaluate $C(n, k)$ and $\binom{n}{k}$ to find the number of combinations of n objects taken k at a time.

b) Solve counting problems involving repeated and/or mixed uses of $C(n, k)$ and $P(n, k)$.

c) Find the number of distinguishable arrangements of n objects of which n_1 are of one kind, n_2 are of a second kind, and so on, where finally there are n_k of a kth kind, by evaluating

$$\frac{n!}{n_1! n_2! \cdots n_k!}$$

34. Consider the set {A, B, C, D, E}. How many combinations taken:

a) 4 at a time are there?

b) 5 at a time?

c) 2 at a time?

d) 1 at a time?

e) 0 at a time?

Query: Why should a "combination" lock really be called a "permutation" lock?

A *combination* containing k objects is a subset containing k objects.

DO EXERCISE 34.

The number of combinations of n objects taken k at a time, denoted $C(n, k)$, is the number of different subsets of k elements.

In Example 1 and Margin Exercise 34, we see that

$$C(5, 5) = 1, \qquad C(5, 2) = 10,$$
$$C(5, 4) = 5, \qquad C(5, 1) = 5,$$
$$C(5, 3) = 10, \qquad C(5, 0) = 1.$$

We can derive some general results here. First, it is always true that $C(n, n) = 1$, because a set with n objects has only 1 subset with n objects. Second, $C(n, 1) = n$ because a set with n objects has n subsets with 1 element each. Finally, $C(n, 0) = 1$, because a set with n objects has only one subset, namely the empty set $\emptyset$, with 0 elements.

We now derive a general formula for $C(n, k)$ for any $k \leq n$. Let us return to Example 1 and compare the number of combinations with the number of permutations.

Combinations	Permutations
{A, B, C} ⟶	ABC BCA CAB CBA BAC ACB
{A, B, D} ⟶	ABD BDA DAB DBA BAD ADB
{A, B, E} ⟶	ABE BEA EAB EBA BAE AEB
{A, C, D} ⟶	ACD CDA DAC DCA CAD ADC
{A, C, E} ⟶	ACE CEA EAC ECA CAE AEC
{A, D, E} ⟶	ADE DEA EAD EDA DAE AED
{B, C, D} ⟶	BCD CDB DBC DCB CBD BDC
{B, C, E} ⟶	BCE CEB EBC ECB CBE BEC
{B, D, E} ⟶	BDE DEB EBD EDB DBE BED
{C, D, E} ⟶	CDE DEC ECD EDC DCE CED

Note that each combination of 3 objects, say {A, C, E} yields 3!, or 6, permutations, as shown above. It follows that

$$3! \cdot C(5, 3) = 60 = P(5, 3) = 5 \cdot 4 \cdot 3$$

so

$$C(5, 3) = \frac{P(5, 3)}{3!} = \frac{5 \cdot 4 \cdot 3}{3 \cdot 2 \cdot 1} = 10.$$

In general, the number of combinations of n objects taken k at a time, $C(n, k)$, times the number of permutations of these k objects, $k!$, must equal the number of permutations of n objects taken k at a time:

$$k! \cdot C(n, k) = P(n, k)$$

so

$$C(n, k) = \frac{P(n, k)}{k!} = \frac{1}{k!} \cdot P(n, k) = \frac{1}{k!} \cdot \frac{n!}{(n - k)!} = \frac{n!}{k!(n - k)!}. \quad (1)$$

We also have

$$C(n, k) = \frac{n(n - 1)(n - 2) \cdots [n - (k - 1)]}{k(k - 1)(k - 2) \cdots 3 \cdot 2 \cdot 1}. \quad (2)$$

Note that this expression for $C(n, k)$ is the quotient of two quantities, each of which has k factors.

An alternative notation, called a *Binomial Coefficient*, is also used for $C(n, k)$:

$$\binom{n}{k} = C(n, k).$$

You should be able to use either notation.

Example 2 Evaluate $\binom{7}{5}$, using expressions (1) and (2).

Solution

a) By (1),

$$\binom{7}{5} = \frac{7!}{5!2!} = \frac{7 \cdot 6 \cdot 5 \cdot 4 \cdot 3 \cdot 2 \cdot 1}{5 \cdot 4 \cdot 3 \cdot 2 \cdot 1 \cdot 2 \cdot 1} = \frac{7 \cdot 6 \cdot 5 \cdot 4 \cdot 3}{5 \cdot 4 \cdot 3 \cdot 2 \cdot 1} = \frac{7 \cdot 6}{2 \cdot 1} = 21.$$

b) By (2),   This tells where to start

$$\binom{7}{5} = \frac{7 \cdot 6 \cdot 5 \cdot 4 \cdot 3}{5 \cdot 4 \cdot 3 \cdot 2 \cdot 1} = \frac{7 \cdot 6}{2 \cdot 1} = 21.$$

This tells how many factors in numerator and denominator, and where to start the denominator.

The method in (b), using formula (2), is easiest to carry out, but in some situations formula (1) does become useful.

DO EXERCISES 35 AND 36.

35. Given the set of 4 letters

$$\{A, B, C, D\},$$

a) Determine the number of permutations of this set, taking 3 letters at a time.

b) List these permutations.

c) Determine the number of combinations of this set, taking 3 letters at a time.

d) List these combinations.

36. Evaluate.

a) $\binom{10}{3}$ b) $\binom{10}{7}$

c) $C(9, 4)$ d) $C(9, 5)$

37.

a) Evaluate $\binom{8}{5}$ and $\binom{8}{3}$.

b) Which seemed easier to compute?

Note that

$$\binom{7}{2} = \frac{7 \cdot 6}{2 \cdot 1} = 21$$

so that

$$\binom{7}{5} = \binom{7}{2}.$$

In general,

$$\binom{n}{k} = \binom{n}{n-k}.$$

This is because every set of k elements automatically determines the *set complement* with $(n-k)$ elements. So there are the same number of k-element subsets as there are $(n-k)$-element subsets. Try this with your fingers. Hold your hands up and bend down 3 fingers. This determines not only the 3-element set you turned down, but the 7-element set still up. Knowing this may ease some computation.

DO EXERCISE 37.

38. An examination consists of 10 questions. A student is required to answer any 8 of them. How many different ways can the student pick 8 questions to answer? (*Hint.* Is the *order* in which he answers the questions important, assuming the answers themselves are numbered?)

Example 3 For a psychology study, 4 people are chosen at random from a group of 10 people. In how many ways can this be done?

Solution No order is implied here, nor does it seem to be important, so the number of ways the 4 people can be selected is given by

$$\binom{10}{4} = \frac{\overset{3}{10 \cdot \cancel{9} \cdot \cancel{8} \cdot 7}}{\cancel{4} \cdot \cancel{3} \cdot \cancel{2} \cdot 1} = 10 \cdot 3 \cdot 7 = 210.$$

DO EXERCISE 38.

Some problems involve repeated and/or mixed uses of combination and permutation notation.

Example 4 A university offers 5 science courses, 6 humanity courses, and 3 literature courses. How many ways can a student choose 2 science courses, 3 humanity courses, and 1 literature course?

Solution Since the *order* of choosing the subjects is irrelevant, the student can choose 2 science courses $\binom{5}{2}$ ways, 3 humanity courses $\binom{6}{3}$ ways, and 1 literature course $\binom{3}{1}$ ways. Hence, by the Fundamental Counting Principle, the total number of choices available is the product

$$\binom{5}{2} \cdot \binom{6}{3} \cdot \binom{3}{1} = \frac{5 \cdot 4}{2 \cdot 1} \cdot \frac{6 \cdot 5 \cdot 4}{3 \cdot 2 \cdot 1} \cdot \frac{3}{1} = 600.$$

DO EXERCISE 39.

Not all solutions involve products. Some require additions.

Example 5 How many ways can 2 people be chosen out of 3 men and 4 women such that *at least* one is a man?

Solution   *Either* one or two men must be chosen to satisfy the constraints of the problem. If one man is chosen, then one woman must also be chosen. This can be done $\binom{3}{1} \cdot \binom{4}{1} = 3 \cdot 4 = 12$ ways. Two men can be chosen $\binom{3}{2} = 3$ ways. Since *either* is permitted, the total number is the sum $12 + 3 = 15$.

DO EXERCISE 40.

Some problems involve partitions.

Example 6   How many ways can 9 different books be distributed among three children so that the oldest gets 4, the middle child gets 3, and the youngest gets 2?

Solution   Here the 9 books are "partitioned" into 3 groups of 4, 3, and 2. The oldest child can "choose" his books $\binom{9}{4}$ ways. The next child can "choose" his 3 books out of the remaining $9 - 4 = 5$ books $\binom{5}{3}$ ways. The youngest child can "choose" his 2 books out of the remaining $5 - 3 = 2$ books $\binom{2}{2}$ ways. The total number of "choices" for all is the product $\binom{9}{4} \cdot \binom{5}{3} \cdot \binom{2}{2}$. Making use of the fact that

$$\binom{n}{k} = \frac{n!}{k!(n-k)!},$$

we have

$$\binom{9}{4}\binom{5}{3}\binom{2}{2} = \frac{9!}{4!5!} \cdot \frac{5!}{3!2!} \cdot \frac{2!}{2!0!}$$

$$= \frac{9!}{4!3!2!} = 9 \cdot 4 \cdot 7 \cdot 5 = 1260.$$

Verify that the same solution is obtained regardless of which child "chooses" first or second.

DO EXERCISE 41.

There is a related type of problem in which not all the elements are distinguishable in an arrangement.

39. How many ways can a committee of 4 men and 3 women be chosen out of a group of 7 men and 5 women?

40. How many ways can 2 people be chosen out of 3 men and 4 women such that at least one is a woman?

41. How many ways can 7 different toys be distributed to 4 children with the oldest getting one and the others 2 each?

42. Consider the word "TENNESSEE".

a) How many letters in all are there?

b) How many T's are there?

c) How many E's?

d) How many S's?

e) How many N's?

f) Evaluate

$$\frac{9!}{1! \cdot 4! \cdot 2! \cdot 2!}.$$

(*Hint.* Write out the factorials and simplify.)

g) How many distinguishable words can be made up of all the letters of the word "TENNESSEE"?

Example 7 How many *distinguishable* words can be made up of all the letters of the word "MISSISSIPPI"?

Solution The word "MISSISSIPPI" has 11 letters. Thus, if all the letters were *distinguishable*, there would be

$$11! = 39,916,800 \text{ possible arrangements.}$$

However, we actually have

$$1 \text{ M}, \quad 4 \text{ I's}, \quad 4 \text{ S's}, \quad 2 \text{ P's.}$$

The *different* letters are distinguishable, but the 4 I's are *indistinguishable* from each other, as are the 4 S's and the 2 P's. The 4 I's, 4 S's, and 2 P's can be made distinguishable by putting *tags* on them for the moment. Thus, we obtain "eleven" letters:

$$\text{M}, \quad \text{I}_1, \quad \text{I}_2, \quad \text{I}_3, \quad \text{I}_4, \quad \text{S}_1, \quad \text{S}_2, \quad \text{S}_3, \quad \text{S}_4, \quad \text{P}_1, \quad \text{P}_2.$$

Given *any* particular word, the 4 I's can be permuted $4! = 24$ ways, the 4 S's can be permuted $4! = 24$ ways, and the 2 P's can be permuted $2! = 2$ ways. Thus the I's, S's, and P's of *each* distinguishable word can be permuted $4! \cdot 4! \cdot 2! = 1152$ ways to make words distinguishable by tags, or indistinguishable when the tags are dropped. Hence, the number of *distinguishable* words is:

$$\frac{11!}{4!4!2!} = 34,650.$$

In general,

Given a set of *n* objects of which n_1 are of one kind, n_2 are of a second kind, n_3 are of a third kind, and so on, where finally there are n_k of a *k*th kind, then the number of distinguishable arrangements is

$$\frac{n!}{n_1! \cdot n_2! \cdot n_3! \cdots n_k!}.$$

DO EXERCISE 42.

EXERCISE SET 7.5

Evaluate.

1. $C(13, 2)$

2. $C(9, 6)$

3. $\binom{13}{11}$

4. $\binom{9}{3}$

5. $C(7, 1)$

6. $C(8, 8)$

7. $C(n, 2)$

8. $C(n, 3)$

9. An office manager interviews 6 secretaries. If 4 are hired at random, how many ways can they be selected? (*Hint.* See Example 3.)

10. If ice cream comes in 8 flavors, in how many ways can a particular store select 6 flavors at random?

11. On a test, a student must answer any 7 of the first 10 questions, and any 5 of the remaining 8 questions. In how many ways can this be done? (*Hint.* See Example 4.)

13. From a group consisting of 6 men and 8 women, 5 are to be hired at random as sales representatives of a company. How many ways can this be done if:

a) It does not matter how many are men and how many are women?
b) At least 3 must be women?
c) At least 4 must be women?
d) At least 1 must be a woman?

(*Hint.* See Example 4 and Example 5.)

15. How many ways can 11 different tools be distributed among four employees if the first gets 3, the second 4, and the remaining two employees get 2 each? (*Hint.* See Example 6.)

17. How many distinguishable words can be made up of all the letters in the word "CINCINNATI"? (*Hint.* See Example 7.)

19. A psychotic professor decides to grade his 20 students on a curve without regard to test performance. There will be 2 A's, 5 B's, 8 C's, 3 D's, and 2 F's. How many ways can this be done? (*Hint.* See Example 7.)

21. In how many ways can the expression $x^3y^2z^4$ be expressed without exponents?

23. From a group of 20 employees a delegation of 3 is to be selected at random.

a) How many ways can this be done?
b) How many ways can this be done if one of the 3 is selected at random to be spokesman?

25. How many ways can a group of 3 people be selected from a group of 5 with regard to order? without regard to order?

27. A folk-dance leader has 10 records for elementary dances and 20 records for more advanced dances.

a) How many ways can 6 dances be selected at random for a program?
b) How many ways can 4 elementary dances and 2 advanced dances be selected for the program?

12. How many committees can be formed from a set of 8 senators and 5 representatives if each committee contains 4 senators and 3 representatives?

14. From a group consisting of 10 smokers and 10 nonsmokers, 4 are to be chosen for a medical study. How many ways can this be done if:

a) It does not matter how many smokers or nonsmokers there are?
b) At least 2 must be smokers?
c) At least 3 must be smokers?
d) At least 1 must be a smoker?

16. How many ways can 8 different records be distributed among three students if the first gets 2, the second gets 5, and the third gets 1?

18. How many distinguishable words can be made up of all the letters in the word "ABRACADABRA"?

20. A psychotic professor decides to grade her 24 students on a curve without regard to test performance. There will be 3 A's, 5 B's, 9 C's, 4 D's, and 3 F's. How many ways can this be done?

22. In how many ways can the expression $p^3q^5r^2$ be expressed without exponents?

24. *Psychology.* For a psychological study, a group of 3 people is selected at random from a group of 4 men and 3 women.

a) How many ways can this be done?
b) How many ways can this be done if there must be at least 1 man and at least 1 woman in the group?

26. How many ways can a group of 4 people be selected from a group of 6 with regard to order? without regard to order?

28. How many ways can a bridge team of 4 be selected at random from 6 husband–wife pairs,

a) If husband–wife pairs cannot be broken?
b) If no husband or wife is on the same team as his or her spouse, but the team still consists of 2 men and 2 women?

29. How many ways can 12 work assignments be made at random, 4 to each of 3 machinists? (*Note.* Machinists are people; people are distinguishable.)

30. How many ways can 10 work assignments be made at random if one machinist gets 4 and the other two machinists 3 each?

31. As in Exercise 29, how many ways can these 12 work assignments be made on 3 work sheets *before* the work sheets are assigned to a particular machinist?

32. As in Exercise 30, how many ways can these 10 work assignments be made on 3 work sheets *before* the work sheets are assigned to particular machinists?

▶ ───

33. How many distinguishable words can be formed from the letters of ALGEBRA if each word is to contain 7 letters? 6 letters? 5 letters?

34. How many distinguishable words can be formed from the letters of PRECEDE if each word is to contain 7 letters? 6 letters? 5 letters?

35.* A class consists of 10 students of whom 4 are women of whom 3 are married to 3 of the 5 married men in the class. They all spend an evening at a motel which has 2 rooms with 2 single beds each (a double) and 2 rooms with 3 single beds each (a triple). Assuming that only one sex occupies any room except for married couples who *may* share a double, how many ways can the students distribute themselves into the rooms?

36. In Exercises 35 through 40 of Exercise Set 7.1, we found that a set with n elements has 2^n subsets. Show that

$$2^n = \binom{n}{0} + \binom{n}{1} + \binom{n}{2} + \cdots + \binom{n}{n} = \sum_{i=0}^{n} \binom{n}{i}.$$

37. Wendy's Hamburgers, a national firm, advertises "We Fix Hamburgers 256 Ways!" This is accomplished by various combinations of catsup, onion, mustard, pickle, mayonnaise, relish, tomato, or lettuce. Of course, one can also have a plain hamburger. Assume single portions.

a) Use combination notation to show the number of possible hamburgers.

b) Use the result of Exercise 36 to show how the expression in (a) can be evaluated quickly.

c) Interestingly, Wendy's excludes cheese from the possibilities. This may be to avoid false advertising, because then one would have a cheeseburger. Including cheese as a possibility, how many ways does Wendy's fix hamburgers?

───

(Optional) Counting Numbers of POKER Hands

In answering each problem *do not* just give the answer; provide a reasoned expression as well. Read all the problems before beginning.

38. How many different 5-card hands can be dealt from a standard 52-card deck? Note that no order is considered, even though one might "order" cards after receiving them.

39. A *royal flush* consists of a 5-card hand with A-K-Q-J-10 of the same suit. How many are there?

40. A *straight flush* consists of five cards in sequence in the same suit, but does not include royal flushes. How many are there? (Assume that an ace can be used at either the high or low end.)

41. *Four of a kind* is a five-card hand where 4 of the cards are the same denomination, such as 4 jacks, 4 aces, or 4 deuces. How many are there?

───

* This problem was assigned on a test by one of the authors (JCC) and subsequently submitted by the mother of one of his students to *TIME* magazine and printed with the cover letter stating that this was an example of "nonsexist mathematics," whatever that is.

A standard deck of 52 cards is made up as follows:

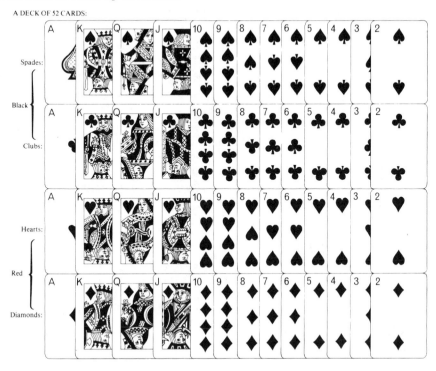

A DECK OF 52 CARDS:

42. A *full house* consists of a pair and three of a kind, such as K-K-K-7-7.

a) How many full houses are there consisting of kings and sevens?
b) How many full houses are there?

44. *Three of a kind* consists of a five-card hand where 3 of the cards are of the same denomination and the other two cards are not the same denomination, such as K-K-K-10-3. How many are there?

46. *Two pairs* is just what it says—a hand such as Q-Q-8-8-A. How many are there?

43. A *pair* is a five-card hand where just 2 of the cards are of the same denomination, such as Q-Q-8-A-3. How many are there?

45. A *flush* is a five-card hand where all the cards are the same suit, but not all in sequence (not a straight flush or royal flush). How many are there?

47. A *straight* is any five cards in sequence but not in the same suit. How many are there?

Exercises 48 through 71 provide a mixture of permutation and combination problems.

48. There are 27 people at a party. How many possible handshakes are there?

50. How many ways can 4 cars be placed in 9 garages, one car to a garage?

49. How many ways can 6 novels be assigned to 10 students if each gets at most one novel?

51. How many softball games are played in a league with 8 teams if each team plays each other team once? twice?

52. How many games are played in a league with n teams if each team plays each other team once? twice?

53. How many distinguishable words can be formed from all the letters of the word MATH? BUSINESS? PHILOSOPHICAL?

54. There are 8 points on a circle. How many triangles (inscribed) can be drawn with these points as vertices?

55. A money clip contains one each of the following bills: $1, $2, $5, $10, $20, $50, $100. How many different sums of money can be formed using the bills?

56. How many words can be formed using 4 out of 5 letters of A, B, C, D, E if the letters

a) are not repeated?
b) can be repeated?
c) are not repeated but must begin with D?
d) are not repeated but must end with DE?

57. A state forms its license plates by first listing a number which corresponds to the county the car owner dwells in (the names of the counties are alphabetized and the number is its location in the order). Then the plate lists a letter of the alphabet and this is followed by a number from 1 to 9999. How many such plates are possible if there are 80 counties?

58. How many diagonals does a hexagon have?

59. How many diagonals does an n-agon have?

60. How many distinguishable words can be formed from the letters of the word ORANGE? BIOLOGY? MATHEMATICS?

61. How many words can be formed using 5 out of 6 of the letters of G, H, I, J, K, L if the letters

a) are not repeated?
b) can be repeated?
c) are not repeated but must begin with K?
d) are not repeated but must end with IGH?

62. A set of 5 parallel lines crosses another set of 8 parallel lines. How many parallelograms are formed?

63. There are n points on a circle. How many quadrilaterals can be drawn with these points as vertices?

Solve for n.

64. $P(n, 5) = 7 \cdot P(n, 4)$ **65.** $C(n + 1, 3) = 2 \cdot C(n, 2)$ **66.** $\binom{n + 2}{4} = 6 \cdot \binom{n}{2}$ **67.** $P(n, 4) = 8 \cdot P(n - 1, 3)$

68. $C(n, n - 2) = 6$ **69.** $\binom{n}{3} = 2 \cdot \binom{n - 1}{2}$ **70.** $P(n, 5) = 9 \cdot P(n - 1, 4)$ **71.** $P(n, 4) = 8 \cdot P(n, 3)$

OBJECTIVES

You should be able to

a) Find the rth term of a binomial expansion of $(a + b)^n$.

b) Evaluate expressions like $\binom{5}{3}$ to
$$\frac{5!}{3!2!}.$$

c) Use the binomial theorem to expand expressions like $(x^2 + 3y)^9$ and 2^n.

*7.6 (OPTIONAL) THE BINOMIAL THEOREM

Consider the following expanded powers of $(a + b)^n$, where $a + b$ is any binomial. Look for patterns.

$$(a + b)^0 = \qquad\qquad\qquad\qquad\qquad 1$$
$$(a + b)^1 = \qquad\qquad\qquad\qquad a \quad + \quad b$$
$$(a + b)^2 = \qquad\qquad\quad a^2 \quad + \quad 2ab \quad + \quad b^2$$
$$(a + b)^3 = \qquad\quad a^3 \quad + \quad 3ab^2b \quad + \quad 3ab^2 \quad + \quad b^3$$
$$(a + b)^4 = \quad a^4 \quad + \quad 4a^3b \quad + \quad 6a^2b^2 \quad + \quad 4ab^3 \quad + \quad b^4$$
$$(a + b)^5 = a^5 \quad + \quad 5a^4b \quad + \quad 10a^3b^2 \quad + \quad 10a^2b^3 \quad + \quad 5ab^4 \quad + \quad b^5$$

There are some patterns to be noted in the expansions.

1. In each term, the sum of the exponents is n.

2. The exponents of a start with n and decrease. The last term has no factor of a. The first term has no factor of b. The exponents of b start in the second term with 1 and increase to n.

3. There is one more term than the degree of the polynomial. The expansion of $(a + b)^n$ has $n + 1$ terms.

We now find a way to determine the coefficients. Let us consider the nth power of a binomial $(a + b)$:

$$(a + b)^n = \underbrace{(a + b)(a + b)(a + b) \cdots (a + b)}_{n \text{ factors}}$$

When we multiply, we will find all possible products of a's and b's. For example, when we multiply all the first terms we will get n factors of a, or a^n. Thus the first term in the expansion is a^n, and the first coefficient is 1. Similarly, the coefficient of the last term is 1.

To get a term such as the $a^{n-r}b^r$ term, we will take a's from $n - r$ factors and b's from r factors. Thus we take n objects, $n - r$ of them a's and r of them b's. The number of ways we can do this is

$$\frac{n!}{(n - r)!r!}.$$

This is $\binom{n}{r}$. Thus the $(r + 1)$st term in the expansion is

$$\binom{n}{r}a^{n-r}b^r.$$

We now have a theorem.

THE BINOMIAL THEOREM. For any binomial $(a + b)$ and any natural number n,

$$(a + b)^n = \binom{n}{0}a^n + \binom{n}{1}a^{n-1}b + \binom{n}{2}a^{n-1}b^2 + \cdots + \binom{n}{n}b^n.$$

Sigma notation for a binomial **expansion is as follows.**

$$(a + b)^n = \sum_{r=0}^{n} \binom{n}{r}a^{n-r}b^r.$$

Because of the Binomial Theorem, $\binom{n}{r}$ is called a *binomial coefficient*. It can now be made apparent why 0! is defined to be 1. In the binomial expansion we want $\binom{n}{0}$ to equal 1 and we also want the

43. Find the 4th term of $(x - 3)^8$.

44. Find the 5th term of $(x - 3)^8$.

45. Find the 6th term of $(y + 2)^{10}$.

46. Find the 1st term of $(3x + 5)^5$.

definition

$$\binom{n}{r} = \frac{n!}{(n - r)!r!}$$

to hold for all whole numbers n and r. Thus we must have

$$\binom{n}{0} = \frac{n!}{(n - 0)!0!} = \frac{n!}{n!0!} = 1.$$

This will be satisfied if 0! is defined to be 1.

Example 1   Find the 7th term of $(4x - y^2)^9$.

Solution   We let $r = 6$, $n = 9$, $a = 4x$, and $b = -y^2$ in the formula $\binom{n}{r}a^{n-r}b^r$. Then

$$\binom{9}{6}(4x)^3(-y^2)^6 = \frac{9!}{6!3!}(4x)^3(-y^2)^6$$

$$= \frac{9 \cdot 8 \cdot 7 \cdot 6!}{3! \cdot 6!}64x^3y^{12}$$

$$= 5376x^3y^{12}.$$

DO EXERCISES 43 THROUGH 46.

Example 2   Expand $(x^2 - 2y)^5$.

Solution   Note that $a = x^2$, $b = -2y$, and $n = 5$. Then, using the binomial theorem, we have

$$(x^2 - 2y)^5 = \binom{5}{0}(x^2)^5 + \binom{5}{1}(x^2)^4(-2y)$$

$$+ \binom{5}{2}(x^2)^3(-2y)^2 + \binom{5}{3}(x^2)^2(-2y)^3$$

$$+ \binom{5}{4}x^2(-2y)^4 + \binom{5}{5}(-2y)^5$$

$$= \frac{5!}{0!5!}x^{10} + \frac{5!}{1!4!}x^8(-2y)$$

$$+ \frac{5!}{2!3!}x^6(-2y)^2 + \frac{5!}{3!2!}x^4(-2y)^3$$

$$+ \frac{5!}{4!1!}x^2(-2y)^4 + \frac{5!}{5!1!}(-2y)^5$$

$$= x^{10} - 10x^8y + 40x^6y^2 - 80x^4y^3 + 80x^2y^4 - 32y^5.$$

Example 3 Expand $\left(\dfrac{2}{x} + 3\sqrt{x}\right)^4$.

Solution Note that $a = 2/x$, $b = 3\sqrt{x}$, and $n = 4$. Then, using the binomial theorem, we have

$$\left(\dfrac{2}{x} + 3\sqrt{x}\right)^4 = \binom{4}{0}\cdot\left(\dfrac{2}{x}\right)^4 + \binom{4}{1}\cdot\left(\dfrac{2}{x}\right)^3(3\sqrt{x})$$

$$+ \binom{4}{2}\cdot\left(\dfrac{2}{x}\right)^2(3\sqrt{x})^2$$

$$+ \binom{4}{3}\left(\dfrac{2}{x}\right)(3\sqrt{x})^3 + \binom{4}{4}(3\sqrt{x})^4$$

$$= \dfrac{4!}{0!4!}\cdot\dfrac{16}{x^4} + \dfrac{4!}{1!3!}\cdot\dfrac{8}{x^3}\,3\sqrt{x}$$

$$+ \dfrac{4!}{2!2!}\cdot\dfrac{4}{x^2}\cdot 9x$$

$$+ \dfrac{4!}{3!1!}\cdot\dfrac{2}{x}\cdot 27x^{3/2} + \dfrac{4!}{4!0!}\cdot 81x^2$$

$$= \dfrac{16}{x^4} + \dfrac{96}{x^{5/2}} + \dfrac{216}{x} + 216\sqrt{x} + 81x^2.$$

DO EXERCISES 47 THROUGH 49.

Example 3 Use $(1 + 1)^n$ for 2^n to find a binomial expansion for 2^n.

Solution

$$2^n = (1 + 1)^n = \binom{n}{0} + \binom{n}{1} + \binom{n}{2} + \cdots + \binom{n}{n}$$

The right side is an expression for the number of subsets of a set of n objects. Thus we have a proof of the following.

A set with n objects has 2^n subsets.

DO EXERCISE 50.

Pascal's Triangle

When the coefficients of the binomial expansions of $(a + b)^n$ in a triangular array are arranged as follows, we get what is known as

47. Expand $(x^2 - 1)^5$.

48. Expand $\left(2x + \dfrac{1}{y}\right)^4$.

49. Expand $(x - \sqrt{2})^6$.

50. Use $(2 + 1)^n$ for 3^n to find a binomial expansion of 3^n.

51. Try to write the next row of numbers in Pascal's Triangle.

Pascal's Triangle:

$$
\begin{array}{cccccccccccc}
(a + b)^0 & & & & & & 1 & & & & & \\
(a + b)^1 & & & & & 1 & & 1 & & & & \\
(a + b)^2 & & & & 1 & & 2 & & 1 & & & \\
(a + b)^3 & & & 1 & & 3 & & 3 & & 1 & & \\
(a + b)^4 & & 1 & & 4 & & 6 & & 4 & & 1 & \\
(a + b)^5 & 1 & & 5 & & 10 & & 10 & & 5 & & 1 \\
\end{array}
$$

There are many patterns in the triangle. Find as many as you can.

DO EXERCISE 51.

Perhaps you have discovered a way to write the next row of numbers, given the numbers in the row above it. There are always 1's on the outside. Each remaining coefficient is found by adding the two numbers above. This is shown as follows:

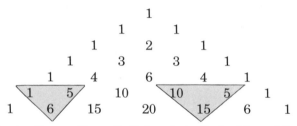

One could actually use Pascal's Triangle to find a binomial expansion. The triangle is extended until the row is obtained which corresponds to the given power. For example, using the row we just found,

$$(a + b)^6 = a^6 + 6a^5b + 15a^4b^2 + 20a^3b^3 + 15a^2b^4 + 6ab^5 + b^5.$$

EXERCISE SET 7.6

Find the indicated term of the binomial expansion.

1. 3rd, $(a + b)^6$

2. 6th, $(x + y)^7$

3. 12th, $(a - 2)^{14}$

4. 11th, $(x - 3)^{12}$

5. 5th, $(2x^3 - \sqrt{y})^8$

6. 4th, $\left(\dfrac{1}{b^2} + \dfrac{b}{3}\right)^7$

7. Middle, $(2u - 3v^2)^{10}$

8. Middle two, $(\sqrt{x} + \sqrt{3})^5$

Expand.

9. $(m + n)^5$

10. $(a - b)^4$

11. $(x^2 - 3y)^5$

12. $(3c - d)^6$

13. $(x^{-2} + x^2)^4$

14. $\left(\dfrac{1}{\sqrt{x}} - \sqrt{x}\right)^6$

15. $(1 - 1)^n$

16. $(1 + 3)^n$

17. $(\sqrt{2} + 1)^6 - (\sqrt{2} - 1)^6$

18. $(1 - \sqrt{2})^4 + (1 + \sqrt{2})^4$

19. $(\sqrt{3} - t)^4$

20. $(\sqrt{5} + t)^6$

▶

Solve for x.

21. $\displaystyle\sum_{r=0}^{8} \binom{8}{r} x^{8-r} 3^r = 0$

22. $\displaystyle\sum_{r=0}^{5} \binom{5}{r} (-1)^r x^{5-r} 3^r = 32$

23. $\displaystyle\sum_{r=0}^{4} \binom{4}{r} 5^{4-r} x^r = 64$

CHAPTER 7 TEST

Consider these sets for Questions 1 through 6:

$$A = \{a, b, c, d, e\}, \qquad B = \{a, b, c, d, e, f, g\},$$
$$C = \{a, b, c\}, \qquad D = \{d, e\}.$$

Find:

1. $A \cup B$ **2.** $A \cap B$ **3.** $C \times D$ **4.** $A - D$

5. Do sets C and D form a partition of A?

6. Do sets C and D form a partition of B?

7. $P(7, 4)$ **8.** $\binom{7}{4}$ **9.** 6!   **10.** 1!

11. If 3 dice of different colors are rolled simultaneously, how many ways can they land?

12. How many ways can 2 cards be drawn at random (without regard for order) from a deck of 52?

13. How many ways can 6 people be seated at a circular table?

14. How many 3-digit numbers can be formed using the digits 2, 3, 5, and 7, with repetition? without repetition?

15. How many ways can 4 men and 4 women be seated in a row if men and women occupy alternate seats?

16. How many different words can be formed from all the letters of the word JENNIFER?

17. From a group of 12 Democrats and 9 Republicans, how many committees of 3 can be formed consisting of at least 2 Democrats?

18. There are 240 students in a certain freshman class. Of these

120 are taking Economics,
65 are taking Russian,
84 are taking Mathematics,
34 are taking Economics
and Russian,
22 are taking Economics
and Mathematics,
18 are taking Russian
and Mathematics,
10 are taking all three
subjects.

How many are not taking any of these subjects?

19. Expand: $(x^2 + 3y)^4$.

CHAPTER EIGHT

Probability

OBJECTIVES

You should be able to:

a) Compute the probability of a simple event.

b) Given $p(E)$, find $p(E^c)$ as $1 - p(E)$.

1. A die, used in craps and other games, is a cube with one of the numbers 1 through 6 on each side.

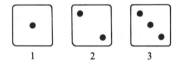

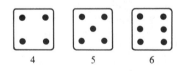

a) What would you reason to be the probability of getting a 5 on a roll of the die?

b) Roll the die 18 times and keep track of the results. What fraction of the rolls resulted in a 5?

c) Compare your answers to (a) and (b).

8.1 INTRODUCTION TO PROBABILITY

Suppose we toss a nickel 100 times and it comes up heads 53 times. We might say that the

Probability of getting a head is $\frac{53}{100}$, or 0.53.

Suppose we toss the same nickel another 100 times and it comes up heads 49 times. We might say that the

Probability of getting a head is $\frac{49}{100}$, or 0.49.

We might also reason about the probability of getting a head. There are 2 outcomes of tossing the nickel, heads and tails. If the coin is "fair," we might reason that the chances are 1 out of 2 of getting a head, so the probability of getting a head is $\frac{1}{2}$, or 0.50.

We call 0.53 and 0.49 *experimental probabilities*, and 0.50 *theoretical probability*. Which is correct? We really never know this. Such is the nature of probability. For example, to determine whether a coin is indeed fair, we may carry out an experiment: toss it a thousand times—or a million times. The information gathered may lead us to reject or accept the fairness of the coin. On the other hand, it may be quite cumbersome and time-consuming to determine probabilities experimentally, so we attempt to determine them theoretically. This will be our main objective in this chapter.

DO EXERCISE 1.

We need some terminology before we continue. Suppose we perform an *experiment* such as flipping a coin, drawing a card from a deck, or checking an item off an assembly line for quality. The results of an experiment are called the *outcomes*. The set of all possible outcomes is called the *sample space*. An *event* is a set of outcomes; that is, a subset of the sample space. For example, for the experiment "flipping a coin," an *event* is "getting a head," from the *sample space* consisting of "head, tail."

We will denote the probability that an event can occur as $p(E)$, or p_E. For example, "getting a head" may be denoted by H. Then $p(H)$, or p_H, represents the probability of getting a head. When the outcomes of an experiment all have the same probability of occurring, we say that they are *equally likely*, or *equiprobable*. To see the distinction between events that are equiprobable and those that are not, consider these dartboards (exicutive decision-makers).

For dart board A, the events, hitting "Yes," hitting "No," hitting "Maybe", are equally likely, but for board B they are not. A sample

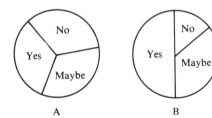

A

B

space that can be expressed as a union of equiprobable events can allow us to calculate probabilities of other events.

BASIC PROBABILITY PRINCIPLE. **If an event *E* can occur *m* ways out of *n* possible *equiprobable* outcomes of a sample space *S*, the probability of that event is given by**

$$p(E) = p_E = \frac{m}{n}.$$

That is,

$$p(E) = \frac{\mathcal{N}(E)}{\mathcal{N}(S)}.$$

We will give many examples related to a standard bridge deck of 52 cards. Such a deck is made up as follows:

A DECK OF 52 CARDS:

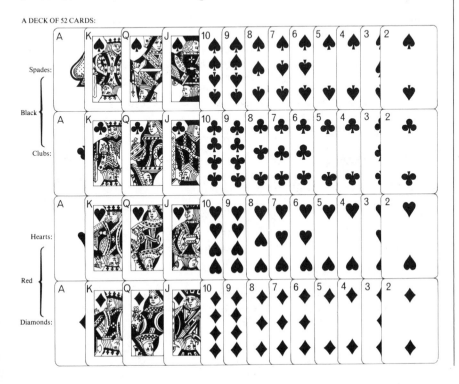

2. A card is drawn at random out of a standard deck of cards. What is the probability of drawing

a) a king?

b) a red ace?

c) a black card?

d) a face card (a king, queen, or jack)?

3. In Example 2, what is the probability of selecting a green marble?

4. A card is drawn at random out of a *standard* deck of cards. What is the probability of getting a *blue* 18?

5. For flipping a fair coin, what is the probability of getting a head or a tail?

Example 1 What is the probability of drawing an ace at random out of a standard deck of cards?

Solution There are 52 equally likely outcomes and there are 4 ways to get an ace, so

$$p(\text{Drawing an ace}) = \tfrac{4}{52}, \quad \text{or } \tfrac{1}{13}.$$

The wording "at random" is a way of implying that each card has the same probability of being drawn as any other card; that is, is equally likely to be drawn.

DO EXERCISE 2.

Example 2 Suppose we select, without looking, one marble from a sack containing 5 green marbles and 3 yellow marbles. What is the probability of selecting a yellow marble?

Solution There are 8 equally likely outcomes and 3 ways to get a yellow marble, so

$$p(\text{Selecting a yellow marble}) = \tfrac{3}{8}.$$

DO EXERCISE 3.

Example 3 In Example 2, what is the probability of selecting a red marble?

Solution There are 0 ways of selecting a red marble since there are no red marbles in the sack, so

$$p(\text{Selecting a red marble}) = \tfrac{0}{8}, \quad \text{or } 0.$$

For an event which cannot occur, $p(E) = 0$. It follows that $p(\emptyset) = 0$.

DO EXERCISE 4.

Example 4 In Example 2, what is the probability of selecting either a green marble *or* a yellow marble?

Solution Since the sack contains only green and yellow marbles, there are 8 ways of selecting either one, so

$$p(\text{Selecting a green or yellow marble}) = \tfrac{8}{8}, \quad \text{or } 1.$$

For an event which is *certain* to occur, or in the case of repeated trials, will *always* occur, $p(E) = 1$. Since the sample space contains all possible events, $p(S) = 1$.

DO EXERCISE 5.

The previous examples lead us to:

The probability that an event will occur is a number from 0 to 1; that is, for any event E,

$$0 \leq p(E) \leq 1.$$

Example 5 Suppose 3 cards are drawn at random from a well-shuffled deck of cards. What is the probability that all 3 are diamonds?

Solution The number of ways of drawing 3 cards out of a deck of 52 is

$$\binom{52}{3}.$$

Now 13 of the 52 cards are diamonds, so the number of ways of drawing the 3 diamonds is

$$\binom{13}{3}$$

Thus,

$$p(\text{All 3 diamonds}) = \frac{\binom{13}{3}}{\binom{52}{3}} = \frac{\dfrac{13 \cdot 12 \cdot 11}{3 \cdot 2 \cdot 1}}{\dfrac{52 \cdot 51 \cdot 50}{3 \cdot 2 \cdot 1}} = \frac{13 \cdot 12 \cdot 11}{52 \cdot 51 \cdot 50} = \frac{11}{850}.$$

DO EXERCISE 6.

Example 6 *Psychology.* For a psychology study, 2 people are selected at random from a group consisting of 8 men and 6 women. What is the probability that both are women?

Solution The number of ways of selecting 2 people from the group of 14 is $\binom{14}{2}$. The number of ways of selecting 2 women from a group of 6 is $\binom{6}{2}$. Thus,

$$p(\text{Both are women}) = \frac{\binom{6}{2}}{\binom{14}{2}} = \frac{\dfrac{6 \cdot 5}{2 \cdot 1}}{\dfrac{14 \cdot 13}{2 \cdot 1}} = \frac{6 \cdot 5}{14 \cdot 13} = \frac{3 \cdot 5}{7 \cdot 13} = \frac{15}{91}.$$

DO EXERCISE 7.

6. Suppose 2 cards are drawn at random from a well-shuffled deck of cards. What is the probability that both are clubs?

7. In Example 6, what is the probability that both are men?

8. In Example 7, what is the probability that 2 men and 1 woman are selected?

Example 7 *Psychology.* For a psychology study, 3 people are selected at random from a group consisting of 8 men and 6 women. What is the probability that 1 man and 2 women are selected?

Solution The number of ways of selecting 3 people out of a group of 14 is $\binom{14}{3}$. The number of ways of selecting 1 man out of a group of 8 is $\binom{8}{1}$. The number of ways of selecting 2 women out of a group of 6 is $\binom{6}{2}$. Then, by the Fundamental Counting Principle, we know that the number of ways of selecting 1 man and 2 women is the product

$$\binom{8}{1} \cdot \binom{6}{2}$$

Thus,

$$p(1 \text{ man and 2 women}) = \frac{\binom{8}{1} \cdot \binom{6}{2}}{\binom{14}{3}} = \frac{\dfrac{8}{1} \cdot \dfrac{6 \cdot 5}{2 \cdot 1}}{\dfrac{14 \cdot 13 \cdot 12}{3 \cdot 2 \cdot 1}} = \frac{30}{91}.$$

DO EXERCISE 8.

9. On a roll of a pair of dice, what is the probability of getting a total of

a) 8?

b) 7?

c) 11?

d) 14?

Example 8 What is the probability of getting a total of 6 on a roll of a pair of dice?

Solution We assume the dice are different, say one white and one black. There are 6 possible outcomes on one die (singular of "dice") so, by the Fundamental Counting Principle, there are $6 \cdot 6$, or 36 possible equiprobable outcomes in the sample space for rolling two dice. We show this as follows.

White

6	(1,6)	(2,6)	(3,6)	(4,6)	(5,6)	(6,6)
5	(1,5)	(2,5)	(3,5)	(4,5)	(5,5)	(6,5)
4	(1,4)	(2,4)	(3,4)	(4,4)	(5,4)	(6,4)
3	(1,3)	(2,3)	(3,3)	(4,3)	(5,3)	(6,3)
2	(1,2)	(2,2)	(3,2)	(4,2)	(5,2)	(6,2)
1	(1,1)	(2,1)	(3,1)	(4,1)	(5,1)	(6,1)
	1	2	3	4	5	6

Black

The pairs that total 6 are enclosed in the diagram. There are 5 such pairs, so the probability of getting a total of 6 is $\frac{5}{36}$.

DO EXERCISE 9.

Complementary Events

Drawing an ace from a deck of cards is an event. *Not* drawing an ace is also an event. If E is an event, the nonoccurrence of E is expressed by the symbol E^c, read "not E." Thus, if E is the event of drawing an ace, then E^c is the event of *not* drawing an ace. E and E^c are called *complementary* events.

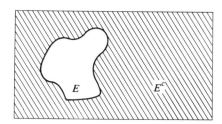

The probability that an event E will *not* occur is given by

$$p(E^c) = 1 - p(E).$$

We can demonstrate this with a Venn diagram. The total number of ways E can occur is m. Then, if there are n total outcomes, it follows that there are $(n - m)$ ways in which E does not occur, so

$$p(E^c) = \frac{n - m}{n} = \frac{n}{n} - \frac{m}{n} = 1 - p(E).$$

Example 9 Suppose $p(E) = \frac{8}{25}$. Find $p(E^c)$.

Solution $p(E^c) = 1 - \frac{8}{25} = \frac{17}{25}$.

DO EXERCISE 10.

Example 10 One card is drawn at random from a well-shuffled deck. What is the probability that it is *not* an ace?

Solution Let $p(E)$ = probability of drawing an ace. Then $p(E^c)$ = probability that the card is not an ace. From Example 9, $p(E) = \frac{1}{13}$, so

$$p(E^c) = 1 - \frac{1}{13} = \frac{12}{13}.$$

DO EXERCISE 11.

Odds

If p is the probability for an event to occur, then, from the preceding

10. Find $p(E^c)$ if

a) $p(E) = \frac{11}{34}$

b) $p(E) = 0.63$

11. One card is drawn at random from a well-shuffled deck. What is the probability that it is

a) a red ace?

b) not a red ace?

c) a club?

d) not a club?

12. One card is drawn at random from a well-shuffled deck.

a) What are the odds for a red ace to occur?

b) What are the odds against a red ace occurring?

result, we know that $(1 - p)$ is the probability for the event not to occur. The ratio

$$p:(1 - p)$$

is the *odds for* the event to occur. The ratio

$$(1 - p):p$$

is the *odds against* the occurrence.

Example 11 One card is drawn from a well-shuffled deck.

a) What are the odds for an ace to occur?
b) What are the odds against the occurrence of an ace?

Solution From Example 10, we let $p = p(\text{getting an ace}) = \frac{1}{13}$, and $1 - p = \frac{12}{13}$. Thus,

a) The odds are $1:12$ (read "1 to 12") *for* drawing an ace.
b) The odds are $12:1$ *against* drawing an ace.

DO EXERCISE 12.

Probability theory came about historically as a way to calculate odds in games of chance. Today it has ever-expanding application to fields such as business, social science, behavioral science, and physics.

EXERCISE SET 8.1

One card is drawn at random from a well-shuffled deck of 52. What is the probability of drawing

1. a queen?	**2.** a jack?	**3.** a spade?	**4.** a diamond?	**5.** an 8?
6. a 10?	**7.** a black card?	**8.** a red club?	**9.** a 7 or a jack?	**10.** an 8 or 10?

Suppose we select one billiard ball from a bag containing 6 red billiard balls and 10 white ones. What is the probability of selecting:

11. a red ball?   **12.** a white ball?   **13.** a chartreuse ball?   **14.** a red or white ball?

15. For a sociological study, a group of 4 people are chosen from a group containing 7 men and 8 women. What is the probability that 2 men and 2 women are chosen?

16. Suppose 4 pens are selected at random from a box containing 9 yellow pens and 6 blue pens. What is the probability that 2 will be yellow and 2 will be blue?

17. What is the probability of getting a total of 9 on a roll of a pair of dice?

18. What is the probability of getting a total of 10 on a roll of a pair of dice?

19. What is the probability of getting a total of 12 ("boxcars") on a roll of a pair of dice?

20. what is the probability of getting a total of 2 ("snake eyes") on a roll of a pair of dice?

21. What is the probability of getting a total of 1 on a roll of a pair of dice?

22. What is the probability of getting a total of 13 on a roll of a pair of dice?

23. From a bag containing 7 dimes, 8 nickels, and 10 quarters, 7 coins are drawn at random. What is the probability of getting 3 dimes, 2 nickels, and 2 quarters?

24. From a sack containing 7 dimes, 5 nickels, and 10 quarters, 8 coins are drawn at random. What is the probability of getting 4 dimes, 3 nickels, and 1 quarter?

Suppose 5 cards are dealt from a well-shuffled deck of 52. What is the probability of dealing:

25. 3 sevens and 2 kings? This is a type of *full house.*

26. 2 jacks and 3 aces?

27. 4 aces and 1 five?   **28.** 4 kings and 1 queen?

29. 5 aces?   **30.** 5 kings?

31. The sales force of a business consists of 10 men and 10 women. A production unit of 4 people is set up at random. What is the probability that 2 men and 2 women are chosen?

32. A union is made up of 14 women and 7 men. A bargaining unit of 3 is chosen at random. What is the probability that 2 women and 1 man are chosen?

33. At a personnel office 5 men and 3 women apply for a job. If 2 are hired at random, what is the probability that:

a) 1 is a man and 1 is a woman?
b) both are men?
c) both are women?
d) both are men *or* both are women?

34. Repeat Exercise 33, but assume that 5 men and 6 women apply for the job.

35. a) Find $p(E^c)$ if $p(E) = \frac{17}{45}$.
    b) What are the odds *for* the event E to occur?
    c) What are the odds *against* the event E occurring?

36. a) Find $p(E^c)$ if $p(E) = \frac{29}{63}$.
    b) What are the odds *for* the event E to occur?
    c) What are the odds *against* the event E occurring?

37. *Business, Advertising.* In many state-run lotteries the odds *against* winning a

1) \$20 prize are 200:1,
2) \$500 prize are 250,000:3, and
3) \$1000 prize are 500,000:3.

What is the probability to win:

a) \$20 on one lottery ticket?
b) \$500 on one lottery ticket?
c) \$1000 on one lottery ticket?
d) \$20 on two lottery tickets?

38. *Business, Advertising.* The following is an odds chart for an actual giveway game from a national food chain.

Prize value	No. of prizes	Odds for one store visit	Odds for 13 store visits	Odds for 26 store visits
\$1000	42	147,619:1	11,355:1	5678:1
\$100	450	13,778:1	1,060:1	530:1
\$20	895	6,927:1	533:1	267:1
\$5	2,385	2,600:1	200:1	100:1
\$2	7,450	832:1	64:1	32:1
\$1	59,600	104:1	8:1	4:1
Total	70,822	88:1	7:1	3.5:1

a) What is the probability to win \$1 in one store visit?
b) What is the probability to win \$5 in 13 store visits?
c) What is the probability to win something in one store visit?
d) What is the probability to win something in 13 store visits?

OBJECTIVES

You should be able to:

a) Given $p(E_1)$, $p(E_2)$, and $p(E_1 \cap E_2)$, find $p(E_1 \cup E_2)$ as

$$p(E_1) + p(E_2) - p(E_1 \cap E_2).$$

b) Given $p(E_1)$ and $p(E_2)$ and that E_1 and E_2 are *mutually exclusive*, find

$$p(E_1 \cup E_2) \quad \text{as} \quad p(E_1) + p(E_2).$$

c) Compute probabilities involving complementary and mutually exclusive events.

8.2 COMPOUND EVENTS

Let us consider an experiment where two of the outcomes are E_1 and E_2. Suppose we draw one card from a well-shuffled deck. Let

$$E_1 = \text{Event of drawing an ace}$$

and

$$E_2 = \text{Event of drawing a king.}$$

The

Event of drawing an ace or a king, or both,

is denoted

$$E_1 \cup E_2, \quad \text{or} \quad E_1 \text{ or } E_2,$$

and is called a *disjunction*. The

Event of drawing an ace and a king

is denoted

$$E_1 \cap E_2, \quad \text{or} \quad E_1 \text{ and } E_2,$$

and is called a *conjunction*. The events $E_1 \cup E_2$ and $E_1 \cap E_2$ are examples of *compound events*.

Since in *one* draw of a card it is impossible to draw an ace *and* a king at the same time, the probability of drawing an ace and a king is

$$p(E_1 \cap E_2) = p(\emptyset) = 0.$$

The probability of drawing an ace or a king is

$$p(E_1 \cup E_2) = \tfrac{8}{52} = \tfrac{2}{13}.$$

If we know the probabilities $p(E_1)$, $p(E_2)$, and $p(E_1 \cap E_2)$, we can compute $p(E_1 \cup E_2)$ using the following result.

For any events E_1 and E_2.

$$\mathbf{p(E_1 \cup E_2) = p(E_1) + p(E_2) - p(E_1 \cap E_2).}$$

This follows from a result of Chapter 6:

$$\mathcal{N}(E_1 \cup E_2) = \mathcal{N}(E_1) + \mathcal{N}(E_2) - \mathcal{N}(E_1 \cap E_2).$$

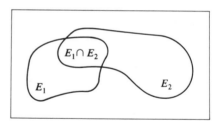

Let n = the number of elements in the sample space. Then

$$p(E_1 \cup E_2) = \frac{\mathcal{N}(E_1 \cup E_2)}{n}$$

$$= \frac{\mathcal{N}(E_1) + \mathcal{N}(E_2) - \mathcal{N}(E_1 \cap E_2)}{n}$$

$$= \frac{\mathcal{N}(E_1)}{n} + \frac{\mathcal{N}(E_2)}{n} - \frac{\mathcal{N}(E_1 \cap E_2)}{n}$$

$$= p(E_1) + p(E_2) - p(E_1 \cap E_2).$$

This result is called the *addition theorem* and will be considered in more detail later in this chapter. For now we will be more interested in the following consequence of this result.

DO EXERCISE 13.

For the event E_1 = Drawing an ace in one draw of a card and the event E_2 = Drawing a king in one draw of a card, we say that they are *mutually exclusive*, meaning that they cannot both happen at the same time. That is, $p(E_1 \cap E_2) = p(\emptyset) = 0$. Then

$$p(E_1 \cup E_2) = p(E_1) + p(E_2) - p(E_1 \cap E_2)$$

$$= p(E_1) + p(E_2) - 0$$

$$= p(E_1) + p(E_2).$$

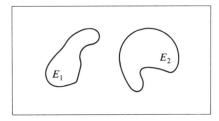

This leads us to the following.

For any events E_1 and E_2 which are mutually exclusive,

$$p(E_1 \cup E_2) = p(E_1) + p(E_2).$$

Example 1 Suppose E_1 and E_2 are mutually exclusive, and $p(E_1) = 0.45$ and $p(E_2) = 0.22$. Find $p(E_1 \cup E_2)$.

Solution $p(E_1 \cup E_2) = p(E_1) + p(E_2) = 0.45 + 0.22 = 0.67$

DO EXERCISE 14.

13. Suppose

$$p(E_1) = 0.34,$$
$$p(E_2) = 0.42,$$

and

$$p(E_1 \cap E_2) = 0.13.$$

Find $p(E_1 \cup E_2)$.

14. Suppose

$$p(E_1) = 0.34,$$
$$p(E_2) = 0.42,$$

and that E_1 and E_2 are mutually exclusive. Find $p(E_1 \cup E_2)$.

Two events which are mutually exclusive can be represented as *branches* of a tree.

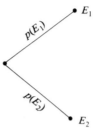

For coin-flipping, a coin cannot land both "heads up" and "tails up" at the same time, assuming the coin does not land on edge. Thus,

$$p(H \cap T) = p(\text{Heads and Tails}) = p(\emptyset) = 0,$$

so the events are mutually exclusive and

$$p(H \cup T) = p(H) + p(T) = \tfrac{1}{2} + \tfrac{1}{2} = 1.$$

Note that these events *partition* the sample space, that is, are mutually exclusive and fill the sample space.

For any events E_1 and E_2 which partition the sample space,

$$p(E_1 \cup E_2) = p(E_1) + p(E_2) = 1.$$

For any two events which partition the sample space, one event is the *complement* of the other, so

$$E_2 = E_1^c,$$

and

$$p(E_1) + p(E_2) = p(E_1) + p(E_1^c) = 1.$$

This follows from the fact that $p(E_1^c) = 1 - p(E_1)$.

Consider the example of drawing a card from a well-shuffled deck, and these 13 events:

$$E_1 = \text{drawing an ace},$$
$$E_2 = \text{drawing a king},$$
$$E_3 = \text{drawing a queen},$$
$$\cdot \qquad \cdot$$
$$\cdot \qquad \cdot$$
$$\cdot \qquad \cdot$$
$$E_{12} = \text{drawing a three},$$
$$E_{13} = \text{drawing a two}.$$

These events partition the sample space into 13 equiprobable subsets.

That is, the sample space S consisting of all possible outcomes, is given by

$$S = E_1 \cup E_2 \cup E_3 \cup \cdots \cup E_{12} \cup E_{13},$$

and any two pairs of events are mutually exclusive. We express this as:

$$E_i \cap E_j = \emptyset, \qquad \text{for all } i \neq j,$$

or

$$p(E_i \cap E_j) = 0, \qquad \text{for all } i \neq j.$$

We can represent the events as branches of a tree:

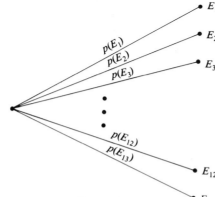

Since the events are mutually exclusive,

$$p(S) = p(E_1 \cup E_2 \cup E_3 \cup \cdots \cup E_{13})$$
$$= p(E_1) + p(E_2) + p(E_3) + \cdots + p(E_{13}).$$

The sample space S includes *all* events, so in any trial, *one* of them must occur, so that $p(S) = 1$. Thus

$$p(E_1) + p(E_2) + p(E_3) + \cdots + p(E_{13}) = 1,$$

where for each event $p(E_i)$, $0 \leq p(E_i) \leq 1$.

Let us use this partitioning idea in an example.

Example 2 *Manufacturing, Quality Control.* A box contains 20 transistors, 5 of which are defective. Three transistors are taken out at random. What is the probability that at least one is defective?

Solution Let E_i be the *event* of drawing exactly i defective transistors and $p_i = p(E_i)$. Drawing different numbers of transistors are mutually exclusive events, so that

$$p_0 + p_1 + p_2 + p_3 = 1.$$

Note that no more than 3 defective transistors can be drawn. Thus,

15. An old sea chest contains 10 bars of silver and 5 bars of gold. If 3 bars are drawn at random, what is the probability that exactly one is gold?

the probability of drawing at least one defective transistor, $p_{i \geqslant 1}$, is given by

$$p_{i \geqslant 1} = p_1 + p_2 + p_3 = 1 - p_0.$$

We have a choice of computing the three probabilities p_1, p_2, p_3, or the one probability p_0 and subtracting from 1. Since computing p_0 not only involves less work, but less work in turn involves less opportunity for error, we shall compute p_0.

Remember that this is a combination (unordered) rather than a permutation (ordered), since the order in which the transistors is drawn is not relevant. Thus, the number of ways 3 transistors can be drawn out of a total of 20 is $\binom{20}{3}$, and the number of ways 3 nondefective transistors can be drawn is $\binom{15}{3}$, since there are $20 - 5$ or 15 nondefective transistors. Thus, the probability p_0 is:

$$p_0 = \frac{\binom{15}{3}}{\binom{20}{3}} = \frac{\dfrac{15 \cdot 14 \cdot 13}{3 \cdot 2 \cdot 1}}{\dfrac{20 \cdot 19 \cdot 18}{3 \cdot 2 \cdot 1}} = \frac{15 \cdot 14 \cdot 13}{20 \cdot 19 \cdot 18} = \frac{91}{228}$$

and

$$p_{i \geqslant 1} = 1 - p_0 = 1 - \frac{91}{228} = \frac{137}{228}.$$

Even though it is more work, consider obtaining $p_{i \geqslant 1}$ from

$$p_{i \geqslant 1} = p_1 + p_2 + p_3.$$

Now, p_1 is the ratio of the number of ways one defective and two nondefective transistors can be drawn, or

$$p_1 = \frac{\binom{5}{1}\binom{15}{2}}{\binom{20}{3}} = \frac{105}{228}.$$

Similarly,

$$p_2 = \frac{\binom{5}{2}\binom{15}{1}}{\binom{20}{3}} = \frac{30}{228}, \quad \text{and} \quad p_3 = \frac{\binom{5}{3}\binom{15}{0}}{\binom{20}{3}} = \frac{2}{228}.$$

Thus,

$$p_{i \geqslant 1} = \frac{105}{228} + \frac{30}{228} + \frac{2}{228} = \frac{137}{228}, \quad \text{as before.}$$

DO EXERCISE 15.

EXERCISE SET 8.2

1. a) Suppose $p(E_1) = 0.73$, $p(E_2) = 0.24$, and $p(E_1 \cap E_2) = 0.20$. Find $p(E_1 \cup E_2)$.
  b) Suppose $p(E_1) = \frac{2}{7}$ and $p(E_2) = \frac{5}{14}$ and E_1 and E_2 are mutually exclusive. Find $p(E_1 \cup E_2)$.

2. a) Suppose $p(E_1) = 0.46$, $p(E_2) = 0.50$, and $p(E_1 \cap E_2) = 0.35$. Find $p(E_1 \cup E_2)$.
  b) Suppose $p(E_1) = \frac{3}{8}$ and $p(E_2) = \frac{5}{15}$ and E_1 and E_2 are mutually exclusive. Find $p(E_1 \cup E_2)$.

3. Two fair dice are rolled and the sum of the numbers showing is noted. What is the probability that the sum is 7? (Express this as the union of mutually exclusive events).

4. As in Exercise 3, what is the probability that the sum is even?

5. *Manufacturing, Quality Control.* A crate of 20 machine parts contains 3 defective parts. Two parts are drawn at random. What is the probability that:

a) Neither is defective?
b) One is defective (the other could be either defective or not defective)?
c) Only one is defective?
d) Both are defective?

6. *Public Health.* If 5 workers in a group of 25 have mononucleosis and 3 workers are chosen at random, what is the probability that:

a) None have mono?
b) All have mono?
c) Only one has mono? (1 has mono, 2 do not)
d) At least one has mono? (See Example 5 for a hint.)

7. There are 5 married couples in a room. If two people are chosen at random, what is the probability that:

a) One is a man and the other a woman?
b) They are of the same sex? (2 men or 2 women)
c) They are married (to each other)?

8. There are 3 married couples and 3 unmarried couples in a room. If two people are chosen at random, what is the probability that:

a) They are of the opposite sex?
b) They are of the same sex?
c) They are married?
d) They are married to each other?

9. Given a standard deck of cards. Two cards are dealt from a shuffled deck. What is the probability that:

a) Both are aces?
b) They are a pair?
c) Both are the same suit?
d) They are neither a pair nor the same suit?

10. As in Exercise 9, what is the probability that:

a) they are a pair but not aces?
b) One is an ace and the other is of the same suit?
c) Both are the same color?
d) They are a pair of different colors?

11. Two couples (four people) go to a theatre and find a row of six seats. If all are seated at random, what is the probability that a given couple will sit together?

*12. Two couples (four people) go to a restaurant and are seated at a round table with six seats. If all are seated at random, what is the probability that a given couple will sit together? [*Hint.* Look up circular permutations, Section 7.4, p. 279.]

13. If the letters of the word "hooch" are scrambled and reassembled by a chimpanzee (that is, randomly), what is the probability that they spell "hooch" correctly? [*Hint.* See Section 7.4, p. 280, and/or Section 7.5, p. 288. Compare methods.]

14. A little red wagon consists of a base with four wheels and a handle. Each wheel is held on the axle with a cotter pin. The handle is attached with a larger cotter pin. If the wheels and handle are removed, what is the probability that a chimpanzee without instructions (that is, at random) will put the wagon together correctly?

OBJECTIVES

You should be able to:

a) Given that E_1 and E_2 are independent, and $p(E_1)$ and $p(E_2)$, find $p(E_1 \cap E_2)$ as $p(E_1) \cdot p(E_2)$.

b) Compute probabilities involving independent events.

8.3 INDEPENDENT EVENTS—MULTIPLICATION THEOREM

In Section 8.2 we considered, among other things, probabilities of disjunctions

$$p(E_1 \cup E_2),$$

where the joint probability $p(E_1 \cap E_2) = 0$. Such events were *mutually exclusive*. That is, we could compute the probability by adding the respective probabilities:

$$p(E_1 \cup E_2) = p(E_1) + p(E_2).$$

Now we want to consider probabilities of conjunctions

$$p(E_1 \cap E_2).$$

We shall see that, under certain conditions, we can compute these probabilities by multiplying the respective probabilities:

$$p(E_1 \cap E_2) = p(E_1) \cdot p(E_2).$$

Let us consider some examples.

Example 1 A black die and a white die are rolled. What is the probability that a 5 is obtained on the black die and an odd number is obtained on the white die?

Solution Let

$$E_1 = \text{the event of a 5 on the black die and}$$
$$E_2 = \text{the event of an odd number on the white die.}$$

a) A 5 is obtained on the black die in 1 out of 6 outcomes, so

$$p(5 \text{ on black}) = p(E_1) = \tfrac{1}{6}.$$

b) An odd number is obtained on the white die in 3 of the 6 outcomes, so

$$p(\text{Odd on white}) = p(E_2) = \tfrac{3}{6} = \tfrac{1}{2}.$$

c) Look at the sample space in Example 8 of Section 8.1. there are 3 outcomes where a 5 on black and an odd number on white occur, so

$$p(5 \text{ on black and odd on white}) = p(E_1 \cap E_2) = \tfrac{3}{36} = \tfrac{1}{12}.$$

Note that this same result can be obtained from the product of the individual probabilities:

$$p(5 \text{ on black } and \text{ odd on white}) = p(5 \text{ on black}) \cdot p(\text{Odd on white})$$

or

$$p(E_1 \cap E_2) = p(E_1) \cdot p(E_2)$$
$$= \tfrac{1}{6} \cdot \tfrac{1}{2} = \tfrac{1}{12}.$$

DO EXERCISE 16.

In Example 1 E_1 (the event of a 5 on the black die) is *independent* of E_2 (the event of an odd number on the white die.)

Not all events are independent, however. Consider the probability of rain on either of two particular days. Weather patterns being what they are, the weather on the second of two successive days is very much influenced by the weather on the preceding day.* Thus, the weather on the second day is *dependent* (to some extent) on the weather on the first day. On the other hand if the two days are sufficiently separated, say Easter and Christmas, then the weather on the second day is not likely to be influenced by the weather on the first day. Thus, in this case the weather on the second day is (essentially) *independent* of the weather on the first day. Dependent events are considered further later.

The independence of two events usually has to be inferred from the nature of the problem.

For two events E_1 and E_2 which do *not* affect each other, we have

MULTIPLICATION THEOREM FOR INDEPENDENT EVENTS. If neither of the events E_1 and E_2 affects the other, we say that the events are *independent*. Then

$$p(E_1 \cap E_2) = p(E_1) \cdot p(E_2).$$

The probability for both E_1 and E_2 to occur is the product of their individual probabilities.

DO EXERCISES 17 AND 18.

Example 2 A fair coin is flipped and two fair dice are rolled. What is the probability for a tail to show on the coin *and* the total, or sum, on the dice to be 7?

Solution The flipping of the coin does not affect the sum on the dice nor does the sum on the dice affect the flipping of the coin (unless there was some unstated strange condition, such as that the dice and coins were glued together). Thus the events

$$E_1 = \text{Getting a tail}, \qquad E_2 = \text{The total is 7}$$

are independent.

a) $p(\text{Getting a tail}) = p(E_1) = \frac{1}{2}$.

---

* In fact, if you used today's weather as a prediction of tomorrow's, you would be correct about 80% of the time.

16. A black die and a white die are rolled.

a) What is the probability that a 2 is obtained on the black die?

b) What is the probability that an even number is obtained on the white die?

c) What is the probability that a 2 is obtained on the black die *and* an even number on the white die?

d) Does

$p(2 \text{ on black and even on white}) =$
$p(2 \text{ on black}) \cdot p(\text{Even on white})$?

17. A sack contains 4 red and 4 yellow marbles. One marble is drawn and replaced, and a second is drawn. What is the probability that the

a) first is red?

b) second is yellow?

c) first is red *and* the second is yellow?

18. Given that E_1 and E_2 are independent, find $p(E_1 \cap E_2)$ where

a) $p(E_1) = \frac{3}{4}$, $p(E_2) = \frac{2}{9}$;

b) $p(E_1) = 0.44$, $p(E_2) = 0.3$.

19. A fair dime and nickel are flipped and one fair die is rolled. What is the probability that the coins show a head on the dime and a tail on the nickel and the die shows a 4?

b) $p(\text{Total is 7}) = p(E_2) = \frac{1}{6}$. See the sample space in Example 8 of Section 8.1.

c) Thus,

$$p(\text{Getting a tail } and \text{ the total is 7}) = p(\text{Getting a tail}) \cdot p(\text{Total is 7})$$

or

$$p(E_1 \cap E_2) = p(E_1) \cdot p(E_2)$$
$$= \frac{1}{2} \cdot \frac{1}{6}$$
$$= \frac{1}{12}.$$

DO EXERCISE 19

Sometimes we have to infer the "and" quality of an event from the wording of a problem.

Example 3 A die is rolled four times. What is the probability of getting a 5 on all four rolls?

Solution The event

$$E = \text{Getting a 5 on all four rolls}$$

can be reexpressed with "and" as

$$E = (\text{1st roll 5}) \text{ and } (\text{2nd roll 5}) \text{ and } (\text{3rd roll 5}) \text{ and } (\text{4th roll 5}).$$

20. A die is rolled three times. What is the probability of getting a 2 on all three rolls?

Each of these four events is independent of the others, so that from the Multiplication Theorem

$$p(E) = \frac{1}{6} \cdot \frac{1}{6} \cdot \frac{1}{6} \cdot \frac{1}{6} = \frac{1}{1296}.$$

DO EXERCISE 20.

Example 4 A coin is flipped five times. What is the probability that the flips come out in the order H, T, T, H, T?

Solution The event

$$E = \text{flips come out in the order H, T, T, H, T}$$

can be reexpressed with "and" as

$$E = (\text{1st flip H}) \text{ and } (\text{2nd flip T}) \text{ and } (\text{3rd flip T}) \text{ and }$$
$$(\text{4th flip H}) \text{ and } (\text{5th flip T}).$$

21. A coin is flipped four times. What is the probability that the flips come out in the order H, H, T, H?

Each of these five events is independent of the others, so that, from the Multiplication Theorem

$$p(E) = \frac{1}{2} \cdot \frac{1}{2} \cdot \frac{1}{2} \cdot \frac{1}{2} \cdot \frac{1}{2} = \frac{1}{32}.$$

DO EXERCISE 21.

Example 5   The probability that a man will live another 20 years is $\frac{1}{5}$ and the probability that his wife will live another 20 years is $\frac{1}{4}$. What is the probability that:

a) Both will live another 20 years?
b) Neither will live another 20 years?
c) Only one will live another 20 years?

Solution   Let

$$E_1 = \text{Event that the man lives another 20 years,}$$
$$E_2 = \text{Event that the woman lives another 20 years.}$$

a) Assuming that the longevity of each is independent (this assumption might be questioned, especially in the case of older people, where the death of one mate sometimes affects the death of the other), the joint probability that *both* will be alive in 20 years can be expressed as

$$p_{\text{Both}} = p(E_1 \text{ and } E_2) = p(E_1 \cap E_2),$$

and from the Multiplication Theorem is given by

$$p_{\text{Both}} = p(E_1 \cap E_2) = p(E_1) \cdot p(E_2) = \tfrac{1}{5} \cdot \tfrac{1}{4} = \tfrac{1}{20}.$$

b) That *neither* lives another 20 years can be expressed

(The man does not live another 20 years)

and

(The woman does not live another 20 years)

or, in set language,

$$E_1^c \cap E_2^c.$$

Assuming these events independent, we have from the Multiplication Theorem

$$
\begin{aligned}
p_{\text{Neither}} = p(E_1^c \cap E_2^c) &= p(E_1^c) \cdot p(E_2^c) \\
&= [1 - p(E_1)] \cdot [1 - p(E_2)] \\
&= (1 - \tfrac{1}{5})(1 - \tfrac{1}{4}) = \tfrac{4}{5} \cdot \tfrac{3}{4} \\
&= \tfrac{3}{5}.
\end{aligned}
$$

c) We look at this two ways. That only one (either) will live another 20 years can be expressed as

(The man is alive and the woman is not)

or

(The man is not alive and the woman is),

or, in set language,

$$(E_1 \cap E_2^c) \cup (E_1^c \cap E_2).$$

22. The probability that a student passes Economics is $\frac{4}{5}$ and the probability that the student passes Finite Mathematics is $\frac{2}{3}$. Assuming no dependence between passing or failing one course and passing or failing the other course, what is the probability that the student:

a) Passes both courses?
b) Passes neither course?
c) Passes one of the courses?

Do this two ways.

Since the events $E_1 \cap E_2^c$ and $E_1^c \cap E_2$ are mutually exclusive, both cannot occur at the same time. Thus,

$$p_{\text{Either}} = p(E_1 \cap E_2^c) + p(E_1^c \cap E_2).$$

Now we can assume E_1 and E_2^c to be independent, and E_1^c and E_2 to be independent, so

$$
\begin{aligned}
p_{\text{Either}} &= p(E_1 \cap E_2^c) + p(E_1^c \cap E_2) \\
&= p(E_1) \cdot p(E_2^c) + p(E_1^c) \cdot p(E_2) \\
&= \frac{1}{5} \cdot \frac{3}{4} + \frac{4}{5} \cdot \frac{1}{4} = \frac{3}{20} + \frac{4}{20} = \frac{7}{20}.
\end{aligned}
$$

Considering this another way, the probability that *either* will be alive in 20 years, plus the probability that *both* will be alive, plus the probability that *neither* will be alive exhausts the possibilities, so that

$$p_{\text{Either}} + p_{\text{Both}} + p_{\text{Neither}} = 1,$$

and

$$
\begin{aligned}
p_{\text{Either}} &= 1 - p_{\text{Both}} - p_{\text{Neither}} \\
&= 1 - \frac{1}{20} - \frac{3}{5} \\
&= \frac{7}{20}.
\end{aligned}
$$

DO EXERCISE 22.

Let us solve the problem in Example 5 using trees. Since either the man lives another 20 years, or he does not, the events E_1 and E_1^c are *mutually exclusive*, so that they can be represented along different branches of a tree, as in Section 8.2.

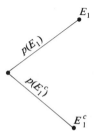

Since the events E_2 and E_2^c are also mutually exclusive whether or not the woman lives another 20 years, we can represent these events as *different branches* starting at the end of the previous branches,

that is

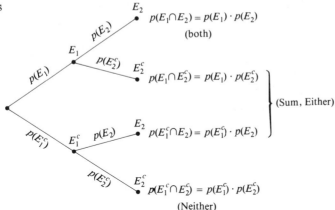

$$p(E_1 \cap E_2) = p(E_1) \cdot p(E_2) \quad \text{(both)}$$

$$p(E_1 \cap E_2^c) = p(E_1) \cdot p(E_2^c)$$

$$p(E_1^c \cap E_2) = p(E_1^c) \cdot p(E_2) \quad \Bigg\} \text{(Sum, Either)}$$

$$p(E_1^c \cap E_2^c) = p(E_1^c) \cdot p(E_2^c) \quad \text{(Neither)}$$

23. Solve Margin Exercise 22 using trees.

Note that since events E_1 and E_2 are independent, either event could have been written first on the tree.

From the Multiplication Theorem, the joint probability for two events (along the *same branch* of the tree) to both happen is the *product* of their individual probabilities as indicated to the right of the tree.

Using numerical values, we have:

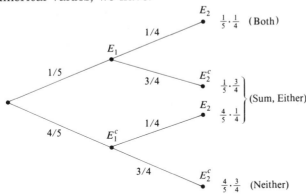

Thus, as before

$$p_{\text{Both}} = p(E_1 \cap E_2) = \tfrac{1}{5} \cdot \tfrac{1}{4} = \tfrac{1}{20},$$

$$p_{\text{Neither}} = p(E_1^c \cap E_2^c) = \tfrac{4}{5} \cdot \tfrac{3}{4} = \tfrac{3}{5},$$

and

$$p_{\text{Either}} = p(E_1 \cap E_2^c) + p(E_1^c \cap E_2) = \tfrac{1}{5} \cdot \tfrac{3}{4} + \tfrac{4}{5} \cdot \tfrac{1}{4} = \tfrac{7}{20}.$$

DO EXERCISE 23.

Example 6 *Manufacturing, Quality Control.* A box contains 20 transistors, 5 of which are defective. An inspector takes out 1 transistor at random, examines it for defects, and replaces it. After it has been replaced, another inspector does the same thing, and then so does a third inspector. What is the probability that at least one of the inspectors finds a defective transistor? (Compare with Example 2, Section 8.2.)

Solution The probability that at least one of the inspectors finds a defective transistor, denoted $p_{i \geqslant 1}$, plus the probability that none of them finds a defective transistor, denoted p_0, sums to 1 because the events partition the sample space, so that

$$p_{i \geqslant 1} = 1 - p_0.$$

Suppose we use this fact and try to compute p_0 to then get $p_{i \geqslant 1}$. This time we must allow for *replacement*. The probability that a defective transistor is not drawn in the first trial is $\frac{15}{20}$. Since each trial is *independent* and *identical*, the probability that no defective transistors are drawn in three trials is the product

$$p_0 = \tfrac{15}{20} \cdot \tfrac{15}{20} \cdot \tfrac{15}{20} = (\tfrac{3}{4})^3 = \tfrac{27}{64},$$

so that $p_{i \geqslant 1} = 1 - \tfrac{27}{64} = \tfrac{37}{64}$.

This problem can also be solved using a tree. If D_1 is the event that the first transistor drawn is defective and D_1^c is the event that it is not defective, then (as in Example 5) we can draw the following tree:

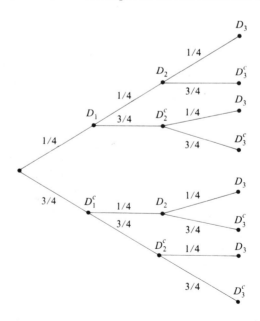

The probability that at least one defective transistor is drawn is the probability that:

(The first is defective) plus the probability that
(The first is not defective but the second is) plus the probability that
(The first and second are not defective but the third is). That is,

$$p_{i \geq 1} = p(D_1) + p(D_1^c \cap D_2) + p(D_1^c \cap D_2^c \cap D_3)$$
$$= \quad \tfrac{1}{4} \quad + \quad \tfrac{3}{4} \cdot \tfrac{1}{4} \quad + \quad \tfrac{3}{4} \cdot \tfrac{3}{4} \cdot \tfrac{1}{4}$$
$$= \tfrac{37}{64}, \quad \text{as before.}$$

DO EXERCISE 24.

(Optional) The Birthday Problem

THE BIRTHDAY PROBLEM. Of n people in a group, what is the probability that at least two of them have the same birthday (day and month, but not necessarily the same year)?

If $p(E)$ is the probability that at least two people have the same birthday,

then $p(E^c)$ is the probability that no two people have the same birthday and

$$p(E) = 1 - p(E^c).$$

The probability $p(E^c)$ will be evaluated two ways.

i) From Section 7.1,

$$p(E^c) = \frac{\mathcal{N}(E^c)}{\mathcal{N}(S)}.$$

One person can have a birthday on any of 365 days (ignoring Leap Year possibilities). Two people can have birthdays on 365^2 days. Thus, in general, n people can have birthdays on 365^n days, so that

$$\mathcal{N}(S) = 365^n.$$

Now if the second person has a birthday other than that of the first person, he has 364 possibilities. The third person, has then 363 possibilities. In general, the nth person has $[365 - (n - 1)]$ possibilities, so that the n people have $\mathcal{N}(E^c)$ possibilities, where

$$\mathcal{N}(E^c) = 365 \cdot 364 \cdot 363 \cdots [365 - (n - 1)].$$

Thus,

$$p(E^c) = \frac{365 \cdot 364 \cdot 363 \cdots [365 - (n - 1)]}{365^n}.$$

24. In Example 6, what is the probability that at least two of the inspectors find a defective transistor?

25. From Table 8.1, determine the probability that two or more students in your class have the same birthday. If your class is large (more than 40 students), divide the class into two groups and examine the results for each half as well as for the whole class.

ii) Alternately, let E_n be the event that the nth person does not have a birthday in common with the preceding $(n - 1)$ people. Thus,

$$p(E_n) = \frac{[365 - (n - 1)]}{365}.$$

(Note that $p(E_1) = 365/365 = 1$ is the probability that the first person has no common birthday with his predecessors, of which he has none.) Since each person's birthday is independent of another person's birthday, the events E_n are also independent, so that from *this* section

$$\begin{aligned}
p(E^c) &= p(E_1 \cap E_2 \cap \cdots \cap E_n) \\
&= p(E_1) \cdot p(E_2) \cdots p(E_n) \\
&= \frac{365}{365} \cdot \frac{364}{365} \cdots \frac{[365 - (n - 1)]}{365},
\end{aligned}$$

which is equivalent to the first value for $p(E^c)$.

Evaluating $p(E) = 1 - p(E^c)$, we obtain Table 8.1.

Table 8.1.
Probability for two or more people to have the same birthday.

n	$p(E)$	n	$p(E)$	n	$p(E)$	n	$p(E)$	n	$p(E)$
2	0.00274	16	0.284	24	0.538	32	0.753	40	0.891
5	0.0271	17	0.315	25	0.569	33	0.775	50	0.970
10	0.117	18	0.347	26	0.598	34	0.795	60	0.9951
11	0.141	19	0.379	27	0.627	35	0.814	70	0.99916
12	0.167	20	0.411	28	0.654	36	0.832	80	0.999914
13	0.194	21	0.444	29	0.681	37	0.849	90	0.999994
14	0.223	22	0.476	30	0.706	38	0.864	100	0.9999997
15	0.253	23	0.507	31	0.732	39	0.878		

For 23 or more people, the probability that two or more people will have the same birthday is greater than $\frac{1}{2}$, that is, it is more likely to occur than not!

DO EXERCISE 25.

In situations where class sizes are large, it may be of interest to consider:

THE EXTENDED BIRTHDAY PROBLEM. Of n people in a group,

what is the probability that at least k of them have the same birthday (day and month, but not necessarily the same year)?

We have just solved this for $k = 2$.

Let

$$p_n(k) = \text{the probability that of } n \text{ people in a group, there is at least one } k\text{-tuplet but no higher multiplets,}$$

and

$$p_n(\geqslant k) = \text{the probability that of } n \text{ people in a group, there is at least one } k\text{-tuplet.}$$

Thus,

$$p_n(1) = \text{what we previously called } p(E^c)$$

and

$$p_n(\geqslant 2) = \text{what we previously called } p(E);$$

so that

$$p_n(\geqslant 2) = 1 - p_n(1).$$

The Birthday Problem can be restated as seeking an expression for $p_n(\geqslant k)$.

We can express $p_n(\geqslant k)$ as

$$p_n(\geqslant k) = p_n[\geqslant (k - 1)] - p_n(k - 1).$$

Thus, for $k = 3$

$$p_n(\geqslant 3) = p_n(\geqslant 2) - p_n(2).$$

We have $p_n(\geqslant 2)$ and need only calculate $p_n(2)$ to solve the Birthday Problem for $k = 3$.

Similarly, for $k = 4$

$$p_n(\geqslant 4) = p_n(\geqslant 3) - p_n(3).$$

Once we have $p_n(\geqslant 3)$, we need calculate only $p_n(3)$ to be able to solve the Birthday Problem for $k = 4$, and so on for any value of k.

Consider now the Birthday Problem for $k = 3$.

Let

$$p_n(2, m) = \text{the probability that of } n \text{ people in a group, there are } m \text{ pairs of people with the same birthday and no more than two people with the same birthday.}$$

The probability $p_n(2, m)$ can be obtained by seeking the probability for m pairs with the same birthday in one configuration (for example, no pairs in the first $n - m$ people and the last matches the first and the next-to-last matches the second, and so on) times the number of possible configurations (how many ways m pairs can be distributed among n people):

$$p_n(2, m) = \frac{p_{n-m}(1)}{365^m} \cdot \frac{\binom{n}{2m}}{k!(2!)^k} .$$

The probability $p_n(2)$ is then the sum over all possible values of m:

$$p_n(2) = \sum_{m=1}^{m_L} p_n(2, m),$$

where the upper limit of the sum is

$$m_L = \text{the maximum possible number of pairs}$$

$$= \text{the largest integer} \leqslant \frac{n}{2} .$$

This computation is rather awkward in the form given. However, let

$$A_n(m) = \frac{(n - 2m + 2)(n - 2m + 1)}{2m(365 + m - n)} ;$$

then after some algebra, we can obtain

$$p_n(2) = p_n(1)A_n(1)[1 + A_n(2)(1 + A_n(3)(1 + \cdots (1 + A_n(m_L)))) \cdots].$$

In this form the computations were performed by one of the authors (JCC) on an HP–25. This handheld calculator has 49 programmable steps and for large values of n the program ran 2 to 3 minutes.

Table 8.2, showing values of $p_n(\geqslant 3)$, was obtained.

Table 8.2.
Probability that three or more people have the same birthday.

n	$p_n(\geqslant 3)$	n	$p_n(\geqslant 3)$	n	$p_n(\geqslant 3)$
20	0.00824	88	0.51107	160	0.98216
30	0.02853	90	0.53420	170	0.99169
40	0.06689	100	0.64586	180	0.99645
50	0.12638	110	0.74553	190	0.99862
60	0.20723	120	0.82796	200	0.999512
70	0.30649	130	0.89108	210	0.999844
80	0.41817	140	0.93570	220	0.9999552
87	0.49945	150	0.96477	240	0.99999999

Thus, for 88 or more people it is more likely than not for *three* or more people in a group to have the same birthday.

In a similar manner (see Exercise 30), we can obtain $p_n(3)$ and the following table of $p_n(\geq 4)$:

Table 8.3.
Probability that four or more people have the same birthday.

n	$p_n(\geq 4)$	n	$p_n(\geq 4)$	n	$p_n(\geq 4)$
50	0.00428	140	0.21109	210	0.65703
60	0.00884	150	0.26463	220	0.71873
70	0.01621	160	0.32390	230	0.77493
80	0.02723	170	0.38777	240	0.82737
90	0.04275	180	0.45480	250	0.88904
100	0.06358	186	0.49583	260	0.94889
110	0.09041	187	0.50269	270	0.98618
120	0.12380	190	0.52327	280	0.99751
130	0.16403	200	0.59130	290	0.99966

Thus, for 187 or more people it is more likely than not for *four* or more people to have the same birthday.

Effect of Leap Years

So far we have assumed 365 days in a year. Actually, 3 out of 4 years have 365 days and 1 out of 4 years has 366 days (to a good approximation).

Let

$p_{n,365} = p_n$ based on a 365-day year (these values are the ones previously obtained),

$p_{n,366} = p_n$ based on a 366-day year (these values can be obtained from the preceding formulas by replacing 365 by 366),

and

$$p_{n,L} = p_n \text{ allowing for leap years.}$$

Then, to this approximation, we have

$$p_{n,L} = \frac{3}{4} p_{n,365} + \frac{1}{4} p_{n,366}.$$

Consider $k = 3$. Then

$$p_{87,365}(\geqslant 3) = 0.49945 \qquad \text{(See Table 8.2)}$$

and

$$p_{87,\text{L}}(\geqslant 3) = 0.49901.$$

Thus, the leap-year correction does not change the results appreciably nor the break-even point of 88.

Similarly, for $k = 2$ and $k = 4$ the leap-year correction does not change the results appreciably nor the break-even points of 23 and 187.

Effect of Twins

Twins are born with a frequency of one pair in approximately 89 pregnancies; thus the probability of twins per pregnancy is

$$p^{(2)} \approx \frac{1}{89}.$$

If $p^{(k)}$ is the probability of k-tuplets per pregnancy, then Hellin's law* states that

$$p^{(3)} \approx [p^{(2)}]^2, \qquad p^{(4)} \approx [p^{(2)}]^3,$$

and, in general,

$$p^{(k)} \approx [p^{(2)}]^{k-1} \quad \text{for } k \geqslant 3.$$

The twinning rate is sufficiently small that for determining the break-even point of the Birthday Problem, we need not consider multiple births other than twins and a single occurrence of twins at that.

If twins are born once in about 89 pregnancies, or

$$p^{(2)} \approx \frac{1}{89},$$

then one pair of twins will be found in about 90 people, or

$$p_{\text{T}} \approx \frac{1}{90}.$$

The probability for two or more people to have the same birthday *allowing for twins* is given by

$$\begin{aligned} \bar{p}(\geqslant 2) &= (1 - p_{\text{T}})p_n(\geqslant 2) + p_{\text{T}} \\ &= p_n(\geqslant 2) + p_{\text{T}}[1 - p_n(\geqslant 2)]. \end{aligned}$$

---

*Encyclopedia Americana (Americana Corp.) 1977, Vol. 19, p. 558.

Thus, we have
$$p_{22}(\geqslant 2) = 0.47570$$
and obtain
$$\tilde{p}_{22}(\geqslant 2) = 0.48153;$$

so that, even though the presence of twins increases the probability for two or more people to have the same birthday, the break-even point of 23 is not changed.

The probability for *three* or more people to have the same birthday *allowing for twins* is given by

$$\tilde{p}_n(\geqslant 3) = (1 - p_T)p_n(\geqslant 3) + p_T p_n^T(\geqslant 3)$$
$$= p_n(\geqslant 3) + p_T[p_n^T(\geqslant 3) - p_n(\geqslant 3)],$$

where the superscript "T" indicates that the n people include one pair of twins, and

$$p_n^T(\geqslant 3) = 1 - p_n^T(2).$$

Here $p_n^T(2)$ can be obtained in a manner similar to that used for $p_n(2)$. (See Exercise 31.)

Thus, we obtain

$$p_{87}(\geqslant 3) = 0.49945,$$
$$p_{87}^T(\geqslant 3) = 0.58653,$$

and

$$\tilde{p}_{87}(\geqslant 3) = 0.50042;$$

so that the break-even point of 88, for three or more people to have the same birthday, drops to 87 with allowance for twins.

The break-even point of 187, for four or more people to have the same birthday, allowing for twins is considered in Exercise 32.

EXERCISE SET 8.3

Given that E_1 and E_2 are independent, find $p(E_1 \cap E_2)$:

1. $p(E_1) = \frac{7}{9}$, $p(E_2) = \frac{11}{14}$.

2. $p(E_1) = \frac{4}{5}$, $p(E_2) = \frac{3}{8}$.

3. $p(E_1) = 0.48$, $p(E_2) = 0.33$.

4. $p(E_1) = 0.77$, $p(E_2) = 0.101$.

5. One card is drawn from a well-shuffled deck of 52 and replaced, and a second card is drawn. What is the probability that:

a) The first is a spade?
b) The second is an ace?
c) The first is a spade and the second is an ace?

6. As in Exercise 5, what is the probability that

a) The first is a face card?
b) The second is a king?
c) The first is a face card and the second is a king?

7. One card is drawn from a well-shuffled deck of 52, but not replaced, and a second card is drawn.

a) What is the probability that the first card is a spade?

b) What is the conditional probability that the second is a diamond, given that the first is a spade?

c) What is the probability that the first is a spade and the second a diamond?

9. For an unfair coin, $p(H) = \frac{2}{3}$ and $p(T) = \frac{1}{3}$. The coin is flipped five times. What is the probability that the flips come out in the order T, H, H, T, H?

11. A student entering college has a $\frac{1}{3}$ probability for getting married while in college and a 0.6 probability of graduating. What is the probability that he will graduate married?

13. *Quality Control.* Candles are molded on a production line such that 10% are defective. What is the probability that a box of six are all good? All defective?

15. A class is one-third women and two-thirds men. Also, 60% are blond and 40% have dark hair. What is the probability that a person chosen at random is a blond woman?

17. Suppose your probability of passing a test over this chapter is $\frac{3}{4}$, and the probability of your passing a psychology test the same day is $\frac{5}{8}$. Assuming the events are independent, what is the probability that you:

a) Pass both tests?

b) Pass one test, but not the other?

c) Fail both tests?

19. *Quality Control.* A box contains 24 transistors, 6 of which are defective. An inspector takes out 1 transistor at random, examines it for defects, and replaces it. After it has been replaced, another inspector does the same thing, and then so does the third inspector. What is the probability that at least one of the examiners finds a defective transistor?

*21. *On the Birthday Problem.* Check an almanac for the birthdays of all the presidents of the U.S. How many presidents have there been? Do any two have the same birthday? Does this seem reasonable based on the table?

8. As in Exercise 7,

a) What is the probability that the first card is a face card?

b) What is the conditional probability that the second card is a four, given that the first card is a face card?

c) What is the probability that the first card is a face card and the second is a four?

10. For an unfair coin, $p(H) = \frac{4}{5}$ and $p(T) = \frac{1}{5}$. the coin is flipped three times. What is the probability that the flips come out in the order T, H, T?

12. Two pilots are trying to communicate with each other during a severe thunderstorm. If the probability for a malfunction of either transceiver is 25%, what is the probability that they **can** communicate?

14. *Quality Control.* A vintner is making three separate batches of wine. Due to circumstances beyond his control, the probability for success is 0.9 for any one batch. What is the probability all three are successful?

16. In a lake there are several kinds of fish of which 20% are pike. Of each kind 10% is tagged. What is the probability that a fish caught at random will be a tagged pike?

18. Suppose your probability of passing a test over this chapter is $\frac{5}{6}$, and the probability of your passing a sociology test the same day is $\frac{2}{3}$. Assuming the events are independent, what is the probability that you:

a) Pass both tests?

b) Pass one test, but not the other?

c) Fail both tests?

20. As in Exercise 19 but the box has 45 transistors of which 9 are defective.

*22. *On the Birthday Problem.* The probabilities regarding birthdays apply also to death dates. Check an almanac for the death dates of the U.S. presidents. Do any two have the same death date (excluding year)? Does this seem reasonable based on the table?

23. Democrats and Republicans are running for Congress (both Senate and House of Representatives). If the odds for the Democrat to win the Senate seat are 3:2 and the odds for the Republican to win the House seat are 5:4, what is the probability that both Democrats win? one Democrat and one Republican? Assume that one campaign does not affect the other. (*Hint.* Convert the odds to probabilities. For example, 3:2 converts to a probability of $\frac{3}{5}$.)

24. A baseball team is playing a doubleheader. If the odds are 4:3 to win the first game and 2:3 to win the second game, what is the probability that the team will win both games? only one game?

25. Five cards are dealt at random from a deck of 52 cards. What is the probability for:

a) A pair?
b) Three of a kind?
c) Two pairs?
d) A full house (a pair and three of a kind)?
e) Four of a kind?

(*Hint.* See Exercises 38 through 47 of Exercise Set 7.5.)

26. As in Exercise 25 (read the whole problem for definitions), what is the probability for:

a) A royal flush (an ace-high straight flush)?
b) A straight flush (not including an ace as the high card)?
c) A flush (five cards of the same suit, not all in sequence)?
d) A straight (five cards in sequence, not all the same suit)?

(*Hint.* See Exercises 38 through 47 of Exercise Set 7.5.)

27. There are three identical urns containing purple and yellow balls. The first has 7 purple and 2 yellow, the second, 3 purple and 3 yellow, and the third, 2 purple and 4 yellow. An urn is chosen at random and a ball selected at random. What is the probability it is purple? Draw a tree.

28. As in Exercise 27, a second ball is drawn at random from a random choice of one of the two urns not selected in the first choice. What is the probability the ball is yellow?

29. A coin weighted so that heads is twice as likely as tails is to be flipped three times. What is the probability for at least two heads? Solve using a tree diagram.

*30. *On the Birthday Problem.* Obtain an expression for $p_n(3)$.

*31. *On the Birthday Problem.* Obtain an expression for $p_n^T(2)$.

*32. *On the Birthday Problem.* Is the break-even point of 187 for four or more people to have the same birthday changed by allowing for twins?

8.4. THE ADDITION THEOREM

Addition Theorem—Two Events

So far we have studied

i) *Mutually exclusive events*: Two events E_1 and E_2 which cannot both happen at the same time, that is, $p(E_1 \cap E_2) = 0$;

OBJECTIVE

You should be able to calculate probabilities using the Addition Theorem (where events are independent or mutually exclusive, or neither) and trees.

ii) *Independent events:* The probability of *both* occurring is the *product* of their individual probabilities (Multiplication Theorem); that is,

$$p(E_1 \cap E_2) = p(E_1) \cdot p(E_2).$$

For mutually exclusive events, the probability for *either* to happen is the *sum* of the individual probabilities; that is,

$$p(E_1 \cup E_2) = p(E_1) + p(E_2).$$

We also considered briefly $p(E_1 \cup E_2)$ for the general case. Let us reconsider such probabilities.

In Section 8.2 we developed the following result.

THE ADDITION THEOREM.   For any events E_1 and E_2,

$$\boldsymbol{p(E_1 \cup E_2) = p(E_1) + p(E_2) - p(E_1 \cap E_2).}$$

The probability of either E_1 or E_2 is the probability of E_1 plus the probability of E_2 minus the probability of both E_1 and E_2.

Example 1   Suppose the probability that a student will pass Finite Mathematics is 68%, or 0.68, and the probability that the student will pass an Accounting course is 70%, or 0.70. The probability that the student will pass both courses is 64%, or 0.64. What is the probability that the student will pass either Finite Mathematics or Accounting, or both?

Solution   Let

$$E_1 = \text{Event of passing Finite Mathematics,}$$
$$E_2 = \text{Event of passing Accounting.}$$

Then

$$E_1 \cap E_2 = \text{Event of passing both courses,}$$

and

$$E_1 \cup E_2 = \text{Event of passing Finite Mathematics} \\ \textit{or} \text{ Accounting (or both.)}$$

We are seeking $p(E_1 \cup E_2)$, and we already know $p(E_1)$, $p(E_2)$, and $p(E_1 \cap E_2)$. Thus, by the Addition Theorem, we have

$$\begin{aligned} p(E_1 \cup E_2) &= p(E_1) + p(E_2) - p(E_1 \cap E_2) \\ &= 0.68 + 0.70 - 0.64 \\ &= 0.74. \end{aligned}$$

DO EXERCISE 26.

Note in Example 1 that the events E_1 and E_2 were not independent. That is,

$$(0.68)(0.70) = 0.476 \neq 0.64,$$

or

$$p(E_1) \cdot p(E_2) \neq p(E_1 \cap E_2).$$

Since E_1 and E_2 were *not* independent, the extra information about $p(E_1 \cap E_2)$ had to be supplied.

Now let us consider a problem where events are independent.

Example 2   If the probability for rain is 0.4 on April 1, and 0.3 on November 29, what is the probability for rain on either day?

Solution   Let

$$E_1 = \text{The event of rain on April 1}$$

and

$$E_2 = \text{The event of rain on November 29.}$$

Then

$$E_1 \cup E_2 = \text{The event of rain on April 1 or November 29.}$$

From the wording of the problem, we may infer that E_1 and E_2 are independent. Thus, we determine $p(E_1 \cap E_2)$ from the Multiplication Theorem and $p(E_1 \cup E_2)$ from the Addition Theorem:

$$p(E_1 \cup E_2) = p(E_1) + p(E_2) - p(E_1 \cap E_2)$$
$$= p(E_1) + p(E_2) - p(E_1) \cdot p(E_2)$$
$$= 0.4 + 0.3 - (0.4)(0.3)$$
$$= 0.58.$$

DO EXERCISE 27.

Using Trees to Solve Probability Problems

Let us see how we can use trees to facilitate solving problems.

In Example 3, we reconsider the Addition Theorem for the case where events are mutually exclusive, but now the events are themselves compound events.

26. Suppose the probability that a woman lives to be 70 is 0.81 and that she will go bald is 0.37. The probability that she will live to be 70 and also go bald is 0.28. What is the probability that she will live to be 70 or go bald?

27. A farmer faces a season of drought with a probability of 0.1 and an insect invasion with probability of 0.3.

a) What is the probability that either or both events will occur?

b) What is the probability that neither event will occur? [*Hint.*

$$p(E_1^c \cap E_2^c) = p(E_1^c) \cdot p(E_2^c).]$$

Example 3 Experience shows that, with fatal accidents involving two cars, the probability that neither driver is drunk is $\frac{1}{4}$, and the probability that both are drunk is $\frac{1}{8}$. What is the probability that only one driver is drunk?

Solution   Let

$$E_1 = \text{Event that the } \textit{first} \text{ driver is drunk,}$$

and

$$E_2 = \text{Event that the } \textit{second} \text{ driver is drunk.}$$

Now the event that only one driver is drunk translates to:

(First drunk *and* the second not drunk)

or

(First not drunk *and* the second drunk),

or, using set language, we have

$$(E_1 \cap E_2^c) \cup (E_1^c \cap E_2).$$

This is a union of mutually exclusive events $E_1 \cap E_2^c$ and $E_1^c \cap E_2$. We are seeking the probability, p_1, of this event:

$$p_1 = p[(E_1 \cap E_2^c) \cup (E_1^c \cap E_2)] = p(E_1 \cap E_2^c) + p(E_1^c \cap E_2).$$

To find this, consider a tree.

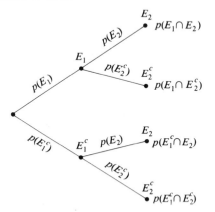

From the fact that $E_1 \cap E_2^c$ and $E_1^c \cap E_2$ represent *different* branches of the tree, we could have deduced that they are mutually exclusive events. But we also get more information from the tree, by noting that:

$$E_1 \cap E_2 = \text{The event that both are drunk,}$$
$$E_1^c \cap E_2^c = \text{The event that neither is drunk.}$$

Then since $E_1 \cap E_2$, $E_1 \cap E_2^c$, $E_1^c \cap E_2$, and $E_1^c \cap E_2^c$ represent *all* branches of the tree, they fill the sample space, so:

$$p(E_1 \cap E_2) + p(E_1 \cap E_2^c) + p(E_1^c \cap E_2) + p(E_1^c \cap E_2^c) = 1$$

or

$$p(E_1 \cap E_2) + p_1 + p(E_1^c \cap E_2^c) = 1,$$

so

$$p_1 = 1 - [p(E_1 \cap E_2) + p(E_1^c \cap E_2^c)]$$
$$= 1 - (\tfrac{1}{8} + \tfrac{1}{4})$$
$$= \tfrac{5}{8}.$$

Note that we never actually determined $p(E_1)$ or $p(E_2)$, nor can we from the data given (unless we assume that $p(E_1) = p(E_2)$).

We can also solve this problem using a Venn diagram.

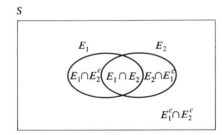

S

Putting in the available information we obtain:

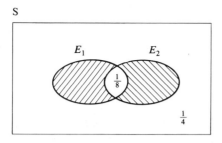

S

The probability that only one driver is drunk corresponds to the shaded area, and is equal to

$$p_1 = 1 - (\tfrac{1}{8} + \tfrac{1}{4})$$
$$= \tfrac{5}{8}, \quad \text{as before.}$$

DO EXERCISE 28.

28. *Public Health.* A country is afflicted with two kinds of flu during one winter. The probability that a person will get both kinds of flu is 0.43. The probability that a person will get only one kind of flu is 0.51. What is the probability that a person will get neither kind of flu?

(Optional) The Addition Theorem—Three or More Events

The Addition Theorem for two events E_1 and E_2 is:

$$p(E_1 \cup E_2) = p(E_1) + p(E_2) - p(E_1 \cap E_2).$$

This theorem can be extended to three events, E_1, E_2, and E_3, by replacing E_2 in the above relation by $(E_2 \cup E_3)$, yielding

$$p(E_1 \cup E_2 \cup E_3) = p(E_1) + p(E_2 \cup E_3) - p[E_1 \cap (E_2 \cup E_3)].$$

From a Venn diagram, we can verify the relation

$$p[E_1 \cap (E_2 \cup E_3)] = p[(E_1 \cap E_2) \cup (E_1 \cap E_3)].$$

Applying the Addition Theorem for two events, we obtain the Addition Theorem for three events:

THE ADDITION THEOREM FOR THREE EVENTS. For any events E_1, E_2 and E_3,

$$p(E_1 \cup E_2 \cup E_3) = p(E_1) + p(E_2) + p(E_3)$$
$$- p(E_1 \cap E_2) - p(E_1 \cap E_3) - p(E_2 \cap E_3)$$
$$+ p(E_1 \cap E_2 \cap E_3).$$

For more than three events, the Addition Theorem has the same pattern as the preceding with appropriate alternating changes of sign.

Example 4 a group of n men, each with a different type of car, would like to swap cars. So they toss their keys into a pile. Each man takes out a key at random. What is the probability that at least one man gets his own car key?

Solution Let E_i be the event that the ith man gets his own car back. Then the probability that at least one man out of n gets his own car back is

$$p_n = p(E_1 \cup E_2 \cup \cdots \cup E_n).$$

Using the Addition Theorem, this can be expanded into a series of terms representing events corresponding to one or more men getting their own cars back.

For $n = 1$, it is certain that the man will get his car back. That is, $p_1 = 1$.

Consider $n = 2$.

$$p_2 = p(E_1 \cup E_2) = p(E_1) + p(E_2) - p(E_1 \cap E_2).$$

Each man has a $\frac{1}{2}$ probability of getting his own car back; that is,

$$p(E_1) = p(E_2) = \tfrac{1}{2}.$$

But if one man gets his own car back, so must the other man, so that

$$p(E_1 \cap E_2) = \tfrac{1}{2}.$$

Thus,

$$p_2 = \tfrac{1}{2} + \tfrac{1}{2} - \tfrac{1}{2} = 1 - \tfrac{1}{2} = \tfrac{1}{2},$$

or

$$p_2 = p_1 - \tfrac{1}{2}.$$

For $n = 3$, we use the Addition Theorem for three events. The probability that any one man gets his own car back is:

$$p(E_i) = \tfrac{1}{3},$$

so that

$$p(E_1) + p(E_2) + p(E_3) = 1 = p_1.$$

Two men out of three can be chosen to get their cars back $\binom{3}{2}$ ways. A given pair can get their cars back with probability $1/3!$ Thus,

$$p(E_1 \cup E_2) + p(E_1 \cup E_3) + p(E_2 \cup E_3) = \binom{3}{2} \cdot \frac{1}{3!} = \frac{1}{2!}.$$

All three men can get their own cars back with probability $1/3!$, so that we have

$$p_3 = 1 - \frac{1}{2!} + \frac{1}{3!} = \frac{2}{3},$$

or

$$p_3 = p_2 + \frac{1}{3!}.$$

If we keep up this process, we find that:

$$p_n = p_{n-1} + (-1)^{n-1} \cdot \frac{1}{n!},$$

or

$$p_n = 1 - \frac{1}{2!} + \frac{1}{3!} - \frac{1}{4!} + \cdots + (-1)^{n-1} \cdot \frac{1}{n!}.$$

DO EXERCISE 29.

29.

a) Evaluate p_n for several increasing values of n. (*Note.* As n becomes increasingly large, p_n approaches the quantity

$$(1 - 1/e) = 0.632121\ldots,$$

e being $2.71828\ldots$, an important number in mathematics.

b) Pair off with someone, each person having a shuffled deck of cards. Each person turns up a first card, and the two cards are compared for a match. Then each turns up a second card, and these cards are compared for a match. This is continued until both have exhausted their cards. What approximate odds should you bet for at least one match? Did you get a match?

EXERCISE SET 8.4

Find $p(E_1 \cup E_2)$, where:

1. $p(E_1) = 0.67$, $p(E_2) = 0.65$, and $p(E_1 \cap E_2) = 0.61$.

2. $p(E_1) = 0.76$, $p(E_2) = 0.57$, and $p(E_1 \cap E_2) = 0.46$.

3. *Biology.* Examination of fruitflies indicates that 30% have an eye defect, 60% have a color variation, and 10% have both. What is the probability that a random fly will have either an eye defect or a color variation? [*Hint.* 30% have an eye defect, so the probability of an eye defect is 0.3.]

4. *Agriculture.* A farmer finds that 24% of his corn crop has a blight, and 30% has received insufficient rain, and 8% has both. What part of the crop had either blight or insufficient rain?

Find $p(E_1 \cup E_2)$, assuming E_1 and E_2 are independent.

5. $p(E_1) = \frac{3}{5}$, $p(E_2) = \frac{1}{3}$.

6. $p(E_1) = \frac{5}{8}$, $p(E_2) = \frac{2}{5}$.

7. If the probability for snow is 0.7 and the probability of having a fire in your home is 0.0006, what is the probability of snow *or* a fire in your home?

8. If the probability for a flood is 0.0004 and the probability of passing math is 0.73, what is the probability of a flood or passing math?

9. *Business.* The manager of a company is faced with the prospect of a snowstorm, which would keep 40% of the employees out, and a flu epidemic, which would keep 15% out. What percent of his employees should be expected to show up for work?

10. *Business.* A manufacturer is coming out with two new products. The first has a 70% chance of being successful and the second 80%. What is the probability that *either* will be successful?

11. *Public Affairs.* A survey indicates that 40% of car drivers do not wear seat belts. That is, the probability that a driver is wearing a seat belt is 0.6. In an accident between two cars what is the probability that neither driver was wearing a seat belt? just one driver? [*Hint.* Assume events independent, and draw a tree.]

12. If the probability for divorce in a marriage is $\frac{1}{3}$, what is the probability that two given couples will both stay married or both get divorced? [*Hint.* Draw a tree, assuming events independent.]

*__13.__ (This problem can be solved using the Addition Theorem for Three Events. It can also be solved—and more simply—using a Venn diagram.) Cars can be bought with any of the following options:

A) An engine package (higher horsepower, etc.)
B) A suspension package (heavy duty suspension, etc.)
C) An appointment package (vinyl seats, etc.)

If

65% buy option A,	30% buy options A and B
40% buy option B,	20% buy options A and C
40% buy option C,	10% buy options B and C
	10% buy options A, B, and C.

What percent of the car buyers buy no options at all?

OBJECTIVES

You should be able to

a) Compute conditional probabilities using the expression

$$p(E_2 \mid E_1) = \frac{p(E_1 \cap E_2)}{p(E_1)}$$

8.5 CONDITIONAL PROBABILITY--MULTIPLICATION THEOREM—DEPENDENT AND INDEPENDENT EVENTS

In the preceding sections we considered events E_1 and E_2 as *independent*. Now let us consider events E_1 and E_2 as *dependent*. The probability that E_1 occurs *provided* that E_2 occurs is written

$$p(E_1 \text{ provided } E_2) = p(E_1 \mid E_2).$$

Similarly, the probability that E_2 occurs *provided* that E_1 occurs is written

$$p(E_2 \text{ provided } E_1) = p(E_2 \mid E_1).$$

Example 1   A box contains 3 red billiard balls and 2 white billiard balls. One ball is selected, but not replaced, and a second is selected. What is the probability that the second is white given that the first is red?

Solution   Let

E_1 = The event that the first ball drawn is red,

and

E_2 = The event that the second ball drawn is white.

Then drawing a tree and letting the ordered pair (R, W) represent the number of red and white balls remaining in the box, we obtain

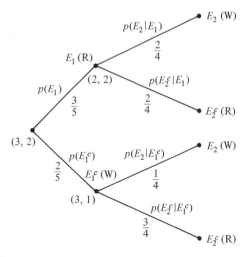

Thus, we find

$$p(\text{First red}) = p(E_1) = \tfrac{3}{5},$$

and

$$p(\text{Second white, provided first red}) = p(E_2 \mid E_1) = \tfrac{1}{2}.$$

Now consider the sample space as follows, where we label the red balls R_1, R_2, and R_3 and the white balls W_1, W_2. The first element in a pair represents the first ball selected and the second element, the second ball selected.

b) Given a probability problem, determine whether two events are independent.

c) Compute probabilities using the Multiplication Theorem for n events.

(R_1, R_2)	(R_1, R_3)	(R_1, W_1)	(R_1, W_2)
(R_2, R_1)	(R_2, R_3)	(R_2, W_1)	(R_2, W_2)
(R_3, R_1)	(R_3, R_2)	(R_3, W_1)	(R_3, W_2)

(W_1, R_1)	(W_1, R_2)	(W_1, R_3)	(W_1, W_2)
(W_2, R_1)	(W_2, R_2)	(W_2, R_3)	(W_2, W_1)

There are 20 equiprobable outcomes in all. Note that

$$p(\text{First red and second white}) = p(E_1 \cap E_2) = \frac{6}{20} = \frac{3}{10}.$$

Thus, we find that

$$p(\text{First red and second white})$$
$$= p(\text{First red}) \cdot p(\text{Second white, provided first red})$$

or

$$p(E_1 \cap E_2) = p(E_1) \cdot p(E_2 \mid E_1)$$

since

$$\frac{3}{10} = \frac{3}{5} \cdot \frac{1}{2}.$$

This result can be formalized as the

MULTIPLICATION THEOREM FOR ANY TWO EVENTS. If the occurrence of event E_2 depends on the occurrence of event E_1, then the probability for their joint occurrence is given by

$$p(E_1 \cap E_2) = p(E_1) \cdot p(E_2 \mid E_1).$$

Actually, this equation is valid for any two events, dependent or independent.

Assuming $p(E_1) \neq 0$, we can divide and obtain the following alternative form of the Multiplication Theorem:

$$p(E_2 \mid E_1) = \frac{p(E_1 \cap E_2)}{p(E_1)}.$$

Suppose S is the sample space. Then the preceding expression can be expressed as

$$p(E_2 \mid E_1) = \frac{p(E_1 \cap E_2)}{p(E_1)} = \frac{\dfrac{\mathcal{N}(E_1 \cap E_2)}{\mathcal{N}(S)}}{\dfrac{\mathcal{N}(E_1)}{\mathcal{N}(S)}} = \frac{\mathcal{N}(E_1 \cap E_2)}{\mathcal{N}(E_1)}.$$

We can interpret this meaningfully by considering a Venn diagram.

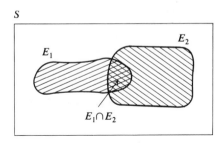

We can compute $p(E_2 \mid E_1)$ by first determining how many elements are in the *reduced sample space* E_1. This is $\mathcal{N}(E_1)$. Of those elements in E_1, how many are in E_2? This is $\mathcal{N}(E_1 \cap E_2)$. The division of $\mathcal{N}(E_1 \cap E_2)$ by $\mathcal{N}(E_1)$ give us $p(E_2 \mid E_1)$.

Returning to Example 1, we note that

$$p(\text{First red}) = p(E_1) = \tfrac{12}{20}$$

and

$$p(\text{First red and second white}) = p(E_1 \cap E_2) = \tfrac{6}{20}.$$

It then follows that the *conditional* probability $p(E_2 \mid E_1)$ can be obtained from the Multiplication Theorem as

$$p(\text{Second white, provided first red}) = p(E_2 \mid E_1)$$

$$= \frac{p(E_1 \cap E_2)}{p(E_1)}$$

$$= \frac{\frac{6}{20}}{\frac{12}{20}} = \frac{6}{12} = \frac{1}{2}.$$

Note that we can also obtain this probability by considering the reduced sample space, enclosed in the diagram. It has 12 equiprobable outcomes, of which 6 have the second ball selected as white; that is,

$$\mathcal{N}(E_1) = 12 \quad \text{and} \quad \mathcal{N}(E_1 \cap E_2) = 6.$$

Thus the conditional probability is

$$p(E_2 \mid E_1) = \frac{\mathcal{N}(E_1 \cap E_2)}{\mathcal{N}(E_1)} = \frac{6}{12} = \frac{1}{2}.$$

DO EXERCISES 30 AND 31.

30. Given

$$p(E_1 \cap E_2) = 0.125,$$

and

$$p(E_1) = 0.625,$$

find $p(E_2 \mid E_1)$.

31. In Example 2, what is the probability that the second ball is white, given that the first ball is white?

Suppose the events E_1 and E_2 are independent. Then

$$p(E_2 \mid E_1) = \frac{p(E_1 \cap E_2)}{p(E_1)} = \frac{p(E_1) \cdot p(E_2)}{p(E_1)} = p(E_2),$$

which is consistent with an earlier statement regarding independent events. That is, the occurrence of E_1 does not affect E_2.

Determining Whether Events are Independent

Let us now consider an example where we try to determine whether two events are independent. That is, we will try to decide whether either $p(E_2 \mid E_1) = p(E_2)$ or $p(E_1 \cap E_2) = p(E_1) \cdot p(E_2)$ is true.

Example 2 *Public Health.* A medical survey of 1000 people over the age of fifty-five was made to investigate the dependence of smoking on lung cancer. Of those surveyed 500 were steady smokers. Among the smokers, 200 had some form of lung cancer, while among the nonsmokers, only 120 had lung cancer.

a) What is the probability that a person smokes?
b) What is the probability that a person has lung cancer?
c) What is the probability that a person has lung cancer provided that person smokes?
d) Are smoking and lung cancer independent events?
e) What is the probability that a person has lung cancer given that person is a nonsmoker?

Solution Let

$$S = \text{Event of smoking,}$$
$$C = \text{Event of having lung cancer.}$$

To compute the various probabilities, we first organize the data into a table in terms of the number of people involved.

	C	C^c	Total
S	200	300	500
S^c	120	380	500
Total	320	680	1000

This data can also be organized, or represented, on a Venn diagram, as follows.

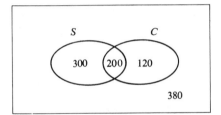

Either representation leads to the desired probabilities.

a) The probability that a person smokes is given by

$$p(S) = \frac{500}{1000} = 0.5.$$

b) The probability that a person has lung cancer (whether or not he smokes) is given by

$$p(C) = \frac{320}{1000} = 0.32.$$

c) The (conditional) probability that a person has lung cancer given that the person smokes is given by

$$p(C \mid S) = \frac{p(S \cap C)}{p(S)}.$$

From the table, we see that $p(S \cap C) = \dfrac{200}{1000} = 0.2$, so

$$p(C \mid S) = \frac{0.2}{0.5} = 0.4.$$

d) Now $p(C \mid S) = 0.4$ and $p(C) = 0.32$, and since

$$p(C \mid S) \neq p(C),$$

the events are dependent (not independent).

e) The probability that a person is a nonsmoker is given by

$$p(S^c) = \frac{500}{1000} = 0.50;$$

and the probability of the person being a nonsmoker and having lung cancer is given by

$$p(S^c \cap C) = \frac{120}{1000} = 0.12.$$

Thus, the probability that a person has lung cancer given that the

32. A sociological survey of 1000 people who were married at one time was made to investigate the dependence of age of marriage upon divorce. Of those surveyed 500 were divorced. Among those divorced, 280 were married as teenagers, while among those never divorced, only 190 were married as teenagers.

a) What is the probability that a person gets divorced?

b) What is the probability that a person was married as a teenager?

c) What is the probability that a person was married as a teenager given that the person gets divorced?

d) Are teenage marriages and divorce independent events?

e) What is the probability that a person was married as a teenager given that person does not get divorced?

33. The probability that a coed passes Economics is $\frac{4}{5}$, that she passes Mathematics is $\frac{2}{3}$, and that she passes both is $\frac{3}{5}$. Are the events of passing these two courses independent? That is, does

$$p(E_1 \cap E_2) = p(E_1) \cdot p(E_2)?$$

Compare this with Margin Exercise 22.

person is a nonsmoker is given by

$$p(C \mid S^c) = \frac{p(S^c \cap C)}{p(S^c)} = \frac{0.12}{0.50} = 0.24.$$

Note that

$$p(C \mid S^c) = 0.24 < 0.40 = p(C \mid S).$$

This confirms that (from the given data) nonsmokers will have less lung cancer than will smokers.

DO EXERCISES 32 AND 33.

Consider two *dependent* events E_1 and E_2. E_1 and E_1^c are mutually exclusive, as are E_2 and E_2^c, so that we can draw a tree:

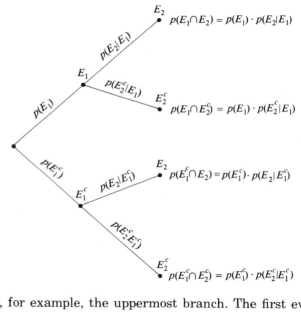

Consider, for example, the uppermost branch. The first event is E_1, and the probability that E_1 occurs is written in along the arc to E_1. If E_1 does happen, then E_2 may or may not happen. For the uppermost branch, E_2 happens, and the quantity written along the second arc represents the probability that E_2 happens provided E_1 happened, that is $p(E_2 \mid E_1)$. From the Multiplication Theorem, the joint probability for both E_1 and E_2 to happen is the product of the two probabilities written along the arcs of the branch. Thus, trees can be used for dependent events (as here), as well as for independent events (as in the preceding section).

Trees need not be limited to two branches or two arcs. Each *branch* of a tree can consist of any number of connected arcs. If some event E_1

precedes another event E_2, then the arc corresponding to E_1 must be closer to the vertex of the tree than is that of E_2. The *joint* probability for all events represented by the connected arcs of a branch of a tree is given by the *product* of the *conditional* probabilities written in along arcs. Thus, if $E_1, E_2, E_3, \ldots, E_n$ are events corresponding to the connected arcs of a branch of a tree, then

$$p(E_1 \cap E_2 \cap E_3 \cap \cdots \cap E_n) = p(E_1) \cdot p(E_2 \,|\, E_1) \cdot p(E_3 \,|\, E_1 \cap E_2)$$
$$\cdots p(E_n \,|\, E_1 \cap E_2 \cap \cdots \cap E_{n-1}).$$

This is the *Multiplication Theorem* for *n* events.

Example 3 A box contains 20 transistors, 5 of which are defective. Three transistors are taken out at random. What is the probability that at least one is defective? (This is Example 2 of Section 8.2 reconsidered, using the concepts of conditional probability.)

Solution Let D_i be the event that the *i*th transistor is defective. Then we obtain the following tree, where the ordered pair (d, g) represents the numbers of defective and good transistors, respectively, remaining in the box.

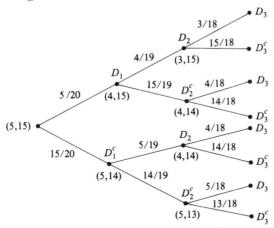

The probability for drawing at least one defective transistor is

$$p_{i \,\geqslant\, 1} = 1 - p_0,$$

where p_0 is the probability that none of the three transistors is defective, $p(D_i^c)$ or

$$p_0 = p(D_1^c \cap D_2^c \cap D_3^c).$$

Using the Multiplication Theorem, we obtain

$$p_0 = p(D_1^c) \cdot p(D_2^c \,|\, D_1^c) \cdot p(D_3^c \,|\, D_1^c \cap D_2^c).$$

34. An old sea chest contains 10 bars of silver and 5 bars of gold. If 3 bars are drawn at random, what is the probability that exactly one is gold? Solve using a tree and conditional-probability concepts. (This is also Margin Exercise 15.)

Since there are 5 defective transistors out of a total of 20, there must be 15 nondefective transistors, so that the probability that the first transistor is not defective is

$$p(D_1^c) = \tfrac{15}{20}.$$

If the first transistor is nondefective, then there are still 5 defective transistors remaining out of a reduced total of 19. Thus, there are 14 nondefective transistors and the probability that the second transistor is not defective *provided* the first transistor is not defective is

$$p(D_2^c \mid D_1^c) = \tfrac{14}{19}.$$

If both the first and second transistors are not defective, then there are 13 remaining nondefective transistors out of 18, so that the conditional probability that the third is not defective, given that the first and second were not defective, is

$$p(D_3^c \mid D_1^c \cap D_2^c) = \tfrac{13}{18}.$$

These three probabilities

$$p(D_1^c), \qquad p(D_2^c \mid D_1^c), \qquad p(D_3^c \mid D_1^c \cap D_2^c)$$

are written in along successive arcs of the lowest branch of the tree. Their *product* is their *joint* probability,

$$p_0 = \tfrac{15}{20} \cdot \tfrac{14}{19} \cdot \tfrac{13}{18} = \tfrac{91}{228}.$$

Then

$$p_{i \geqslant 1} = 1 - p_0 = \tfrac{137}{228}, \quad \text{as before.}$$

From the tree, it can be seen that the actual calculations involved here are quite simple, although the notation required to describe them can become quite lengthy.

DO EXERCISE 34.

Each *fork* of a tree represents a partition of the possible outcomes. That is, the arcs out of that vertex represent mutually exclusive events which fill that sample space. Thus, all *branches* represent mutually exclusive events, so that the probability for the occurrence of *either* of several events is the *sum* of the relevant probabilities.

Example 4 Using the data and tree from Example 3 and conditional-probability concepts, what is the probability for drawing:

a) Exactly one defective transistor?
b) Either one or two defective transistors?

Solution

a) The probability for drawing one defective transistor is

$$p_1 = p(D_1 \cap D_2^c \cap D_3^c) + p(D_1^c \cap D_2 \cap D_3^c) + p(D_1^c \cap D_2^c \cap D_3).$$

Each of these *joint* probabilities is, from the Multiplication Theorem, a *product* of the conditional probabilities written in along the arcs of appropriate *branches* of the tree. Since each joint event is mutually exclusive of the other, the probability for any *one* of the alternatives to occur is the *sum* of the various alternatives, as indicated by the appropriate branches,

$$p_1 = (\tfrac{5}{20} \cdot \tfrac{15}{19} \cdot \tfrac{14}{18}) + (\tfrac{15}{20} \cdot \tfrac{5}{19} \cdot \tfrac{14}{18}) + (\tfrac{15}{20} \cdot \tfrac{14}{19} \cdot \tfrac{5}{18}) = \tfrac{105}{228}, \quad \text{as before.}$$

b) Similarly, the probability for drawing two defective transistors is the sum

$$p_2 = p(D_1 \cap D_2 \cap D_3^c) + p(D_1 \cap D_2^c \cap D_3) + p(D_1^c \cap D_2 \cap D_3).$$

Evaluating, using the Multiplication Theorem, we obtain:

$$p_2 = (\tfrac{5}{20} \cdot \tfrac{4}{19} \cdot \tfrac{15}{18}) + (\tfrac{5}{20} \cdot \tfrac{15}{19} \cdot \tfrac{4}{18}) + (\tfrac{15}{20} \cdot \tfrac{5}{19} \cdot \tfrac{4}{18}) = \tfrac{30}{228}, \quad \text{as before.}$$

If E_i is the probability of drawing i defective transistors, then E_1 and E_2 are mutually exclusive events. The probability for *either* is the *sum*

$$p(E_1 \cup E_2) = p_1 + p_2 = \tfrac{105}{228} + \tfrac{30}{228} = \tfrac{135}{228},$$

and represents the sum of the probabilities corresponding to the branches of the tree representing one or two defective transistors.

DO EXERCISE 35.

35. Using the data and tree from Margin Exercise 34 and conditional-probability concepts, what is the probability for drawing 3 silver bars *or* 3 gold bars?

EXERCISE SET 8.5

1. Given $p(E_1 \cap E_2) = 0.0625$ and $p(E_1) = 0.125$, find $p(E_2 \mid E_1)$.

2. Given $p(E_1 \cap E_2) = 0.16$ and $p(E_1) = 0.64$, find $p(E_2 \mid E_1)$.

3. Given $p(E_1 \cap E_2) = \tfrac{12}{45}$ and $p(E_1) = \tfrac{13}{45}$, find $p(E_2 \mid E_1)$.

4. Given $p(E_1 \cap E_2) = \tfrac{16}{35}$ and $p(E_1) = \tfrac{19}{35}$, find $p(E_2 \mid E_1)$.

5. One card is drawn from a well-shuffled deck. What is the probability that it is an ace given that it is a red card?

6. One card is drawn from a well-shuffled deck. What is the probability that it is a king given that it is a black card?

7. A sack contains 5 blue marbles and 3 yellow marbles. One marble is selected, but not replaced, and a second is selected. What is the probability that

a) The second is yellow given that the first was blue?
b) The second is blue given that the first was blue?

8. Repeat Exercise 7, where the sack contains 5 blue marbles and 2 yellow marbles.

9. *Political Science.* The probability that Democrat A wins in the coming election is $\frac{3}{5}$; that of Democrat B winning the election for a separate seat is $\frac{2}{3}$; that for both is $\frac{1}{2}$. What is the probability that Democrat A wins provided Democrat B wins? Are their elections independent?

11. *Business.* The manager of a company is faced with a snowstorm, which usually keeps 40% of his employees out, and a flu epidemic, which usually keeps 15% out. If 45% of his employees show up, is the absentee rate what should be expected if the events were independent? Or is it possible that the employees were taking advantage of the situations? If 50% show up?

13. *Business.* A restaurant buyer orders some food. Due to a failure of refrigeration, 50% of the food was spoiled, and due to a fuel shortage, only 80% of the food could be transported. The buyer says she has received only 40% of her order. Is this reasonable if the two events are independent?

15. A box contains 10 candles, of which 7 are green and 3 purple. Three candles are taken out at random, one at a time. What is the probability that they alternate in color?

17. Statistics indicate that 4% of men are colorblind and 0.3% of women are colorblind. Assuming that a population is half male and half female, what is the probability that a person selected at random is colorblind? Draw a tree.

19. *Business, Public Health.* A stockyard gets cattle from three ranches. The first ranch supplies 300 cattle, the second 500, and the third 200. Due to an outbreak of hoof-and-mouth disease, 10%, 15%, and 20% of the cattle, respectively, have the disease. What percent of the combined stock from the first and third are infected? Overall? [*Hint.* Draw two tree diagrams.]

21. A person holds a two-tailed coin in one hand and a fair coin in the other. If a hand is chosen at random and that coin is flipped, what is the probability that *tails* shows? Draw a tree.

10. *Business.* A manufacturer is introducing two new products. From a survey, the first has 70% chance of success, the second 80%, and both 65%. What is the probability that the first product is successful given that the second is? Is the success of one product dependent on that of the other?

12. *Business, Sports.* The manager of a sports arena knows that bad weather will reduce attendance by 30% and another game on TV will reduce attendance by 25%. Both events occur (bad weather and another game on TV) and the attendance is 60% below normal. What drop in attendance would be expected if the events were independent? Or is it possible people are using bad weather as an excuse to stay home and watch TV?

14. *Business.* The manager of a shop has ordered some stock. Due to a severe rainstorm, 60% of the order was ruined; and due to a bridge washout, only 70% of the order arrived. The foreman claims that only 10% is usable. Is this reasonable if the rain damage and bridge washout are independent events? If only 30% is usable?

16. A class contains 6 men and 12 women. Four people are selected at random, one at a time. What is the probability that the first two are of the same sex and opposite to that of the last two?

18. The employees of a company are one-third women and two-thirds men. If men are twice as likely as women to have a car and a third of the women have cars, what is the probability that an employee selected at random has a car? Draw a tree.

20. *Business.* Three manufacturers, A, B, and C, supply respectively, 5, 5, and 6 cases of lightbulbs. Each case contains 24 lightbulbs. Manufacturer A makes lightbulbs which are 1% defective, B 2% defective, and C 5% defective. What percent of the combined order of A and B should be defective? Of the total order? [*Hint.* Draw two tree diagrams.]

22. In an urn are three coins, of which two are fair and one is two-tailed. A person reaches in and takes out at random one in his right hand and one in his left hand. What is the probability that the two-tailed coin is in either hand? Draw a tree.

23. In an urn are three coins of which two are fair and one two-tailed. A person reaches in and takes one out in each hand at random. (See Exercise 22.) Someone selects a hand at random. The coin in the hand selected is flipped. What is the probability that tails shows? Draw a tree.

25. One box contains 3 brass washers and 2 steel washers. Another box contains 4 brass washers and 3 steel ones. A box is chosen at random and a washer is taken out at random. The washer is then put into the other box (the one not chosen). A box is again chosen at random and a washer taken out at random. What is the probability it is brass? Draw a tree. What is the probability the same washer is picked both times?

27. A multiple-choice exam is being given. If a student knows the answer, he gets it right. If the student doesn't know the answer, he picks at random any of the four possible answers. It is also possible he "isn't sure" but has it narrowed down to one of two answers. If a student knows 80% of the answers and does not know 10% at all, what is the probability that he will get an arbitrary question correct? Draw a tree.

29. As in Exercise 28: If a voter is selected at random, what is the probability he or she will switch-vote (that is, a Republican voting Democratic or a Democrat voting Republican)?

24. Two urns contain 3 and 4 fair coins, respectively. A two-tailed coin is dropped at random into one of these urns. An urn is then selected at random and a coin is taken out at random and flipped. What is the probability for tails to show? [*Hint.* How many stages does the tree have?]

26. *Business, Banking.* A savings-and-loan institution classifies borrowers as AAA, AA, or A risks. AAA risks constitute 10% of the borrowers and default 5% of the time. AA risks constitute 25% and default 10% of the time. A risks constitute 65% and default 20% of the time. If a borrower is selected at random, what is the probability he will default? Draw a tree.

28. *Political Science.* In a certain city, registered voters are 40% Republican, 35% Democrat, and 25% independent. A Republican and a Democratic candidate are running for office. From a survey, 70% of the Republicans and 80% of the Democrats will vote for their party's candidate, while 75% of the independents will vote Democrat and the rest Republican. Which candidate has the better chance of winning, and by what odds? Draw a tree.

8.6 CONDITIONAL PROBABILITY—BAYES' THEOREM

In order to illustrate Bayes' theorem, consider first:

Example 1 Three manufacturers, I, II, and III, supply all the calculators to a particular store. I supplies 50, with 4% defective, II supplies 60 with 1% defective, and III supplies 30 with 2% defective. If a calculator is purchased at random, what is the probability that it is defective?

Solution   Let

$$D = \text{event that the calculator is defective,}$$
$$E_1 = \text{event that it came from I,}$$
$$E_2 = \text{event that it came from II,}$$
$$E_3 = \text{event that it came from III.}$$

OBJECTIVE

You should be able to use Bayes' Theorem to determine conditional probabilities.

We first draw a tree noting that the events E_1, E_2, and E_3 are mutually exclusive. Here we use the Multiplication Theorem (Section 8.5) to rewrite the joint probabilities.

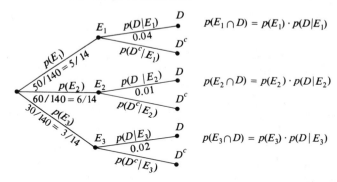

Since the outcome can be either D or D^c, the set of outcomes D is actually a *reduced sample space* which can be expressed as a union of mutually exclusive events:

$$D = (E_1 \cap D) \cup (E_2 \cap D) \cup (E_3 \cap D)$$

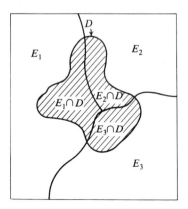

Then the probability that the calculator is defective, $p(D)$, is the sum:

$$
\begin{aligned}
p(D) &= p(E_1 \cap D) + p(E_2 \cap D) + p(E_3 \cap D) \\
&= p(E_1) \cdot p(D \,|\, E_1) + p(E_2) \cdot p(D \,|\, E_2) + p(E_3) \cdot p(D \,|\, E_3) \\
&= \tfrac{5}{14}(0.04) + \tfrac{6}{14}(0.01) + \tfrac{3}{14}(0.02) \\
&= \tfrac{5}{14} \cdot \tfrac{4}{100} + \tfrac{6}{14} \cdot \tfrac{1}{100} + \tfrac{3}{14} \cdot \tfrac{2}{100} \\
&= \tfrac{32}{1400} = \tfrac{4}{175}.
\end{aligned}
$$

Now with Example 1 in mind, consider:

Example 2 Given the data of Example 1 with a calculator purchased at random. What is the probability that, if it is defective, it came from manufacturer I?, II?, III?

Solution The probability that if a calculator is defective it came from manufacturer I is $p(E_1 \mid D)$; from II, $p(E_2 \mid D)$; from III, $p(E_3 \mid D)$.

To determine any of these conditional probabilities, we use the reduced sample space D with probability $p(D)$. Of these, $E_1 \cap D$ are defective and came from I. From the Multiplication Theorem (Section 8.5), we find $p(E_1 \mid D)$:

$$p(E_1 \mid D) = \frac{p(E_1 \cap D)}{p(D)} = \frac{p(E_1) \cdot p(D \mid E_1)}{p(D)} .$$

Using probabilities already computed in Example 1, we obtain:

$$p(E_1 \mid D) = \frac{\frac{5}{14} \cdot \frac{4}{100}}{\frac{32}{1400}} = \frac{20}{32} = \frac{5}{8} .$$

Similarly, the probability that if the calculator is defective it came from manufacturer II, $p(E_2 \mid D)$, is given by:

$$p(E_2 \mid D) = \frac{p(E_2) \cdot p(D \mid E_2)}{p(D)} = \frac{\frac{6}{14} \cdot \frac{1}{100}}{\frac{32}{1400}} = \frac{6}{32} = \frac{3}{16} .$$

And the probability that if the calculator is defective it came from manufacturer III, $p(E_3 \mid D)$, is given by

$$p(E_3 \mid D) = \frac{p(E_3) \cdot p(D \mid E_3)}{p(D)} = \frac{\frac{3}{14} \cdot \frac{2}{100}}{\frac{32}{1400}} = \frac{6}{32} = \frac{3}{16} .$$

Alternately, using the expression for $p(D)$ from Example 1, we can write:

$$p(E_1 \mid D) = \frac{p(E_1 \cap D)}{p(E_1 \cap D) + p(E_2 \cap D) + p(E_3 \cap D)} .$$

Each of these joint probabilities can be obtained from the Multiplication Theorem, so that we can write

$$p(E_1 \mid D) = \frac{p(E_1) \cdot p(D \mid E_1)}{p(E_1) \cdot p(D \mid E_1) + p(E_2) \cdot p(D \mid E_2) + p(E_3) \cdot p(D \mid E_3)} .$$

This is Bayes' Theorem for three events.

Note that here we first calculated $p(D)$ and then $p(E_1 \mid D)$ using the concept of reduced sample space. On the other hand Bayes' Theorem incorporates this concept implicitly so that one calculates $p(E_1 \mid D)$

36. Three manufacturers A, B, and C supply all the fire alarms to a group of residences. A supplies 200 with 3% defective, B supplies 150 with 4% defective, and C supplies 100 with 5% defective. If an alarm is selected at random, what is the probability

a) It is defective?

b) If it is defective, it came from A? from B? from C?

directly. Either way the result is the same but the first method may be easier to remember.

Note that the probabilities

$$p(D \mid E_1) = 0.04, \qquad p(D \mid E_2) = 0.01, \qquad \text{and} \qquad p(D \mid E_3) = 0.02$$

represent probabilities "before" the calculator is purchased. These are sometimes called *a priori* probabilities. The probabilities, given that a calculator has been purchased,

$$p(E_1 \mid D) = \tfrac{5}{8}, \qquad p(E_2 \mid D) = \tfrac{3}{16}, \qquad p(E_3 \mid D) = \tfrac{3}{16},$$

can be thought of as "after," or *a posteriori*, probabilities. That is, information known beforehand allows one to compute probabilities of what will later occur.

DO EXERCISE 36

The general Bayes' Theorem for n events as follows.

BAYES' THEOREM. **For any events $E_1, E_2, \ldots, E_n$ which partition a sample space, if the probability of each event is greater than 0 and if the events are conditional on some event C with $p(C) > 0$, then for each value of i ($i = 1, 2, \ldots, n$),**

$$p(E_i \mid C) = \frac{p(E_i \cap C)}{p(E_1 \cap C) + p(E_2 \cap C) + \cdots + p(E_n \cap C)}$$

$$= \frac{p(E_i) \cdot p(C \mid E_i)}{p(E_1) \cdot p(C \mid E_1) + p(E_2) \cdot p(C \mid E_2) + \cdots + p(E_n) \cdot p(C \mid E_n)}.$$

Example 3 *Political Science.* The public relations agent for a small political party wants to convince people how many people support his party even though his party received only 10% of the vote for the past three elections. The agent chooses for a survey an issue which is supported by 75% of his party and opposed by 75% of the other parties. If a voter is selected at random, what is the probability that:

a) The voter supports the issue?
b) If the voter supports the issue, the voter also supports the party?

Solution   Let

P represent voters of the party of the PR agent,

P^c represent voters of the opposition,

S represent the voters who support the issue,

S^c represent the voters who oppose the issue.

Then we can draw a tree to represent this situation and incorporate the data:

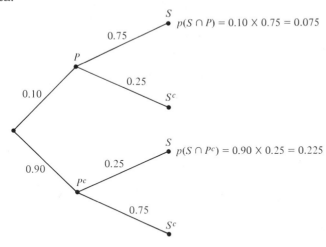

a) The probability that a voter supports the issue is

$$p(S) = p(S \cap P) + p(S \cap P^c)$$
$$= 0.075 + 0.225 = 0.30.$$

b) The probability that, if a voter supports the issue, he also supports the party is

$$p(P \mid S) = \frac{p(S \cap P)}{p(S)} = \frac{0.075}{0.30} = \frac{1}{4} = 0.25.$$

Note how different this probability is from the probability that a voter supports the party, $p(P) = 0.10$. The results of this calculation can be used quite effectively (albeit deceptively) to make support for the party seem greater than it is.

EXERCISE SET 8.6

1. *Public Health.* Statistics indicate that 4% of men are colorblind and 0.3% of women are colorblind. Assume that a population is half male and half female. If a person selected at random is colorblind, what is the probability the person is a woman? (See Exercise 17, Exercise Set 8.5.)

2. The employees of a company are one-third women and two-thirds men. Men are twice as likely as women to have a car and a third of the women have cars. If an employee selected at random has a car, what is the probability that the employee is a woman? (See Exercise 18, Exercise Set 8.5.)

3. *Business, Public Health.* A stockyard gets cattle from three ranches. The first supplies 300 cattle, the second 500, and the third 200. Due to an outbreak of hoof-and-mouth-disease 10%, 15%, and 20% of the cattle, respectively, have the disease. If an animal selected at random from those shipped by the first and third ranches is infected, what is the probability it came from the third ranch? If an animal selected at random from all three ranches is infected, what is the probability it came from the third ranch? (See Exercise 19, Exercise Set 8.5.)

5. A person holds a two-tailed coin in one hand and a fair coin in the other. A hand is chosen at random and the coin is flipped. If tails shows, what is the probability that the coin is two-tailed? (See Exercise 21, Exercise Set 8.5.)

7. Two urns contain 3 and 4 fair coins, respectively. A two-tailed coin is dropped at random into one of these urns. An urn is then selected at random and a coin is taken out at random and flipped. If tails shows, what is the probability that the coin came from the urn with the two-tailed coin? (See Exercise 24, Exercise Set 8.5.)

9. *Business, Banking.* A savings-and-loan institution classifies borrowers as AAA, AA, or A risks. AAA risks constitute 10% of their borrowers and default 5% of the time. AA risks constitute 25% and default 10%. A risks constitute 65% and default 20%. If a borrower defaults, what is the probability he was a AAA risk? (See Exercise 26, Exercise Set 8.5.)

11. *Business.* In drilling for oil, it is known that terrain with geological characteristics A yields oil once in 10 strikes and constitutes 50% of available land. Terrain with characteristics B yields oil once in 5 strikes and constitutes 30% of the available land. Terrain with characteristics C yields oil once in 4 strikes and constitutes 20% of the available land. If a site is selected at random, what is the probability for an oil strike? If oil is struck, what is the probability it came from land with characteristics A?

4. *Business.* Three manufacturers A, B, and C, supply, respectively, 5, 5, and 6 cases of lightbulbs. Each case contains 24 lightbulbs. Manufacturer A makes lightbulbs that are 1% defective, B's are 2% defective, and C's are 5% defective. If a lightbulb selected at random from those provided by manufacturers A and B is defective, what is the probability it came from B? selected from all three? (See Exercise 20, Exercise Set 8.5.)

6. In an urn are three coins of which two are fair and one is two-tailed. A person reaches in and takes out at random one coin in each hand. A hand is then selected at random and the coin in that hand is flipped. If tails shows, what is the probability the coin is fair? (See Exercise 22, Exercise Set 8.5.)

8. One box contains 3 brass washers and 2 steel washers. Another box contains 4 brass washers and 3 steel ones. A box is chosen at random and a washer is taken out at random and put into the other box (the one not chosen). A box is again chosen at random and a washer taken out at random. If the washer selected is brass, what is the probability that it came from the box from which a washer had been removed? (See Exercise 25, Exercise Set 8.5.)

10. A multiple-choice exam is being given. If a student knows the answer, he gets it right. If he doesn't know the answer, he picks at random any of the four given choices. It is also possible he "isn't sure" but has it narrowed down to one of two answers. A student knows 80% of the answers and does not know 10% at all. If he gets a question correct what is the probability he knew the answer? (See Exercise 27, Exercise Set 8.5.)

12. *Public Health, Biomedicine.* Patients entering a clinic have one of three (mutually exclusive) diseases A, B, or C. They also have one or more of the symptoms of fever, sore throat, or faintness. From the records of the clinic, 25% of the entering patients have disease A with symptoms of fever and sore throat, 35% disease B with symptoms of fever and faintness, and 40% disease C with all three symptoms. What is the probability that an undiagnosed patient has disease C given that he has a sore throat?

13. *Public Health, Biomedicine.* A test for mononucleosis administered to students among whom 10% have the disease is 90% accurate; that is, 90% of those with the disease will have a positive reaction and vice versa. If a student has a positive reaction, what is the probability he does not have mononucleosis?

14. As in Exercise 13, how accurate must the test be such that if a student has a positive reaction, the probability he does not have mononucleosis is 0.1?

15. *Pyschology, Biology.* In a laboratory there are 3 boxes (numbered I, II, and III). In the first are 6 white mice, in the second 3 white and 3 black, and in the third 6 black mice. While cleaning up, an assistant bumps into one box (at random) and a mouse (at random) escapes. This mouse is then caught but the assistant doesn't know which box to put the mouse back in. Therefore, he decides to take a mouse out of each box in turn (I, II, then III). If the two mice match in color, he puts both into that box; otherwise, he goes on to the next box to try for a match. What is the probability that the mouse is put back into the box from which it escaped? Draw a tree. If the mouse is put into the third box, what is the probability it came from that box?

16. *Psychology, Biology.* As in Exercise 15, but two mice (instead of one) escape from the same box. If they are both white, he puts them in the first box. If they are one of each color, he puts them in the second box. If they are both black, he puts them in the third box. What is the probability that he puts them back in the correct box? Draw a tree. If he puts them in the third box, what is the probability that they came from that box?

17. *Business, Quality Control.* A box of 6 clocks contains 3 defective ones. If 3 are chosen at random what is the probability that more than one is defective? In an effort to minimize the apparent number of defective clocks, if *one* or *no* clock chosen is defective, the sample is displayed. However, if more than one clock is defective, the defective clocks are replaced and that many are again drawn at random. Draw the tree. What is the probability that only one clock in the sample displayed is defective? If only one clock in the sample displayed is defective, what is the probability that some clocks had been redrawn?

18. As in Exercise 17, but the new clocks are redrawn *before* the old ones are replaced?

19. *Sports.* The winner of the National League playoffs is the first team to win three out of five games. The two teams A and B are evenly matched. What is the probability team A will win if:

a) Team A loses the first game?
b) Team A loses one of the first two games?
c) Team A loses the first two games?
d) Check a baseball almanac to compare these theoretical probabilities with the experimental probabilities you can determine from the almanac.

CHAPTER 8 TEST

1. a) Find $p(E^c)$ if $p(E) = \frac{23}{39}$.

b) What are the odds *for* the event E?

c) What are the odds *against* the event E?

3. Find $p(E_1 \cap E_2)$, where E_1 and E_2 are independent, $p(E_1) = \frac{11}{17}$ and $p(E_2) = \frac{34}{55}$.

5. a) Find $p(E_2 \mid E_1)$, where

$$p(E_1 \cap E_2) = 0.0043$$

and $p(E_1) = 0.125$.

b) Are the events E_1 and E_2 of (a) independent given that $p(E_2) = 0.1$?

7. A shipment of 125 stereos contains 5 defective stereos. Two stereos are selected at random. What is the probability that:

a) Both are defective?

b) Exactly one is defective?

c) Neither is defective?

9. A city is afflicted with two kinds of disease during one summer. The probability that a person gets both diseases is 0.28. the probability that a person gets neither of the diseases is 0.41. What is the probability that a person gets exactly one of the diseases?

11. Three manufacturers supply all the stereos to a particular music store. A supplies 20, with 5% defective, B supplies 70 with 3% defective, and C supplies 10 with 8% defective. If a stereo is purchased at random, what is the probability that:

a) It is defective?

b) If it is defective, it came from A? from B? from C?

2. Find $p(E_1 \cup E_2)$, where E_1 and E_2 are mutually exclusive, $p(E_1) = 0.34$ and $p(E_2) = 0.56$.

4. Find $p(E_1 \cup E_2)$, where E_1 and E_2 are independent, $p(E_1) = 0.11$ and $p(E_2) = 0.42$.

6. What is the probability of getting a 2 or 3 on a single roll of a die?

8. For an unfair coin, $p(H) = \frac{3}{5}$ and $p(T) = \frac{2}{5}$. The coin is flipped five times. What is the probability that the flips come out in the order H, T, T, T, H?

10. A manufacturer is introducing two new products. From a marketing survey, the first has an 80% chance of success, the second 60%, and both 48%. What is the probability that the first product is successful, given that the second is? Is the success of one product dependent on that of the other?

CHAPTER NINE

Random Variables—Statistics

Given a probability problem, you should be able to determine what the random variable is, its values, and the corresponding probability function.

9.1 RANDOM VARIABLES AND PROBABILITY FUNCTIONS

It is convenient in the study of probability and statistics to introduce the concepts of "random variable" and "probability function." We shall define them formally after illustrating them.

Example 1 Two fair dice are rolled. The *sum* of the numbers showing is noted. What is the random variable? What is the probability function for this random variable?

Solution A fair die has 6 sides each of which has the same probability to show (face up). The following is a list of the $6 \cdot 6$ or 36 possible equiprobable outcomes, the sample space, as developed in Example 8 of Section 8.1:

$$
\begin{array}{cccccc}
(1,1) & (1,2) & (1,3) & (1,4) & (1,5) & (1,6) \\
(2,1) & (2,2) & (2,3) & (2,4) & (2,5) & (2,6) \\
(3,1) & (3,2) & (3,3) & (3,4) & (3,5) & (3,6) \\
(4,1) & (4,2) & (4,3) & (4,4) & (4,5) & (4,6) \\
(5,1) & (5,2) & (5,3) & (5,4) & (5,5) & (5,6) \\
(6,1) & (6,2) & (6,3) & (6,4) & (6,5) & (6,6)
\end{array}
$$

Now let us form a table listing those outcomes (a, b) whose sum $a + b$ is a particular value x:

x	2	3	4	5	6	7	8	9	10	11	12
(a, b)	(1, 1)	(1, 2) (2, 1)	(1, 3) (2, 2) (3, 1)	(1, 4) (2, 3) (3, 2) (4, 1)	(1, 5) (2, 4) (3, 3) (4, 2) (5, 1)	(1, 6) (2, 5) (3, 4) (4, 3) (5, 2) (6, 1)	(2, 6) (3, 5) (4, 4) (5, 3) (6, 2)	(3, 6) (4, 5) (5, 4) (6, 3)	(4, 6) (5, 5) (6, 4)	(5, 6) (6, 5)	(6, 6)

Counting the number of times a given sum is obtained, we obtain a table of frequencies f:

x	2	3	4	5	6	7	8	9	10	11	12
f	1	2	3	4	5	6	5	4	3	2	1

Converting these frequencies into probabilities p as in Section 8.1, we obtain the table:

x	2	3	4	5	6	7	8	9	10	11	12
p	$\frac{1}{36}$	$\frac{2}{36}$	$\frac{3}{36}$	$\frac{4}{36}$	$\frac{5}{36}$	$\frac{6}{36}$	$\frac{5}{36}$	$\frac{4}{36}$	$\frac{3}{36}$	$\frac{2}{36}$	$\frac{1}{36}$

Here we have taken the *random variable X* (capital letter) to be the "sum of the numbers showing" and x (small letter) to be the *value* of the random variable. A *particular* value of the random variable is denoted by x_i.

Note that $p_1 + p_2 + \cdots + p_n = 1$, or simply,

$$\sum_{i=1}^{n} p_i = 1,$$

for probability functions.

In general, consider an experiment with n outcomes.

A *random variable X* is a *rule* (function) which assigns a numerical value x_i to each outcome.

As in the preceding example, the outcomes of an experiment need not be equiprobable.

To each outcome we assign not only a number x_i but also a probability p_i. This set of all ordered pairs $\{(x_i, p_i) \mid i = 1, 2, \ldots, n\}$, for example, as displayed in the preceding table, is called the *probability frequency function* of the random variable X, or simply, the *probability function*.

DO EXERCISE 1.

Many of the problems in the preceding chapter involved random drawings

 i) of objects restricted to *two* kinds,
ii) simultaneously (in sequence and *without* replacement).

Specifically, consider the following example.

Example 2   *Business, Quality Control.* Given a box of 20 transistors of which 5 are defective. Three are drawn at random and the number defective is noted. What is the random variable? What is the probability function for this random variable?

Solution   The random variable X here is the number of defective transistors drawn. It takes on the values $x = 0, 1, 2, 3$. The probability of drawing a particular number x_i of defective transistors is $p(x_i)$ or p_{x_i}.

Thus, to obtain p_0 we calculate the probability of drawing 0 defective transistors out of 5 available and 3 nondefective transistors out of 15

1. Two fair dice are rolled and the *difference* (in magnitude) between the numbers showing is noted. What is the probability function?

2. A sea chest contains 10 silver bars and 5 gold ones. Three are drawn at random and the number of gold bars is noted. What is the probability function?

available; that is,

$$p_0 = \frac{\binom{5}{0}\binom{15}{3}}{\binom{20}{3}} = \frac{91}{228}.$$

Similarly, we obtain:

$$p_1 = \frac{\binom{5}{1}\binom{15}{2}}{\binom{20}{3}} = \frac{105}{228},$$

$$p_2 = \frac{\binom{5}{2}\binom{15}{1}}{\binom{20}{3}} = \frac{30}{228},$$

$$p_3 = \frac{\binom{5}{3}\binom{15}{0}}{\binom{20}{3}} = \frac{2}{228}.$$

Thus, we obtain the probability function given in the following table:

x	0	1	2	3
p	$\dfrac{91}{228}$	$\dfrac{105}{228}$	$\dfrac{30}{228}$	$\dfrac{2}{228}$

DO EXERCISE 2.

Example 2 and Margin Exercise 2 just considered have certain characteristics in common. Specifically, such problems have

s objects,
m of which have a particular characteristic and
$s - m$ do not have this characteristic;
n objects are drawn at random and
r of those drawn have this characteristic.

If we take as the random variable R the number of those drawn that have this characteristic, then the values that this random variable can assume are $r = 0, 1, 2, \ldots, n$ and the probability that r have this characteristic is:

$$p_r = H(s, m; n, r) = \frac{\binom{m}{r}\binom{s - m}{n - r}}{\binom{s}{n}},$$

where

$$0 \leqslant r \leqslant n \leqslant s \quad \text{and} \quad 0 \leqslant r \leqslant m \leqslant s.$$

We call p_r the *hypergeometric* probability. We have already used it in Chapter 8. All we have done here is give it a name.

EXERCISE SET 9.1

1. The sales force of a business consists of 20 people, half of whom are men and the other half women. Four people are chosen at random. What is the probability function for the number of women chosen? (See Exercise 31, Set 8.1.)

2. A union has 21 members, 14 of whom are women and the other 7 are men. Three people are chosen at random. What is the probability function for the number of women chosen? men? (See Exercise 32, Set 8.1.)

3. Eight people apply for a job, 5 men and 3 women. Four are hired at random. What is the probability function for the number of women hired?

4. Eight people apply for a job, 4 men and 4 women. Four are hired at random. What is the probability function for the number of women hired?

5. *Business, Quality Control.* A crate of 20 machine parts contains 6 defective parts. Five parts are drawn at random. What is the probability function for the number of defective parts drawn?

6. *Public Health.* Five workers in a group of 25 have mononucleosis. Three workers are chosen at random. What is the probability distribution for the number who have mono? (See Exercise 6, Set 8.2.)

7. *Political Science.* If a party fields 3 candidates for office, each opposed by an equiprobable candidate, what is the probability function for the number of winners for the party in an election?

8. As in Exercise 7, but the first candidate has a probability of $\frac{1}{3}$ to win, the second $\frac{1}{2}$, and the third $\frac{2}{3}$?

9. *Business, Quality Control.* A wine rack contains 7 bottles of red wine and 2 of white wine. If 3 bottles are taken out at random, what is the probability function for the number of bottles of white wine? red wine?

10. *Business, Quality Control.* A case of 12 bottles of wine contains 3 which have spoiled. If 3 bottles are taken out at random, what is the distribution function for the number of bottles of spoiled wine among those drawn? If the case contains 4 spoiled bottles?

11. *Psychology, Biology.* One cage contains 3 white mice and 2 black ones and another cage contains 2 white mice and 3 black ones. A cage is chosen at random and 3 mice are taken out at random. What is the probability function for the number of white mice in the sample?

12. *Psychology, Biology.* There are 2 cages of mice as in Exercise 11. A cage is chosen at random. A mouse of unidentified color escapes from the chosen cage. Then 3 mice are taken out at random. What is the probability function for the number of white mice in the sample?

13. *Psychology, Biology.* There are 2 cages of mice as in Exercise 11. A white mouse escapes from an unidentified cage. Then a cage is chosen at random and 3 mice are taken out at random. What is the probability function for the number of white mice in the sample?

14. *Psychology, Biology.* There are 2 cages of mice as in Exercise 11. All the mice escape and are put back at random, 5 to each cage. A cage is then selected at random and 3 mice are taken out at random. What is the probability function for the number of white mice in the sample?

15. *Sports.* The winner in a World Series is the first team to win 4 out of seven games. What is the probability function for the number of games in the series if the two teams are evenly matched?

16. *Sports.* As in Exercise 15, but the winning team must win by two games. After 7 games, one extra game is to be played if one team is winning by only one game but the series ends if there is a 4-4 tie?

17. Two gamblers toss fair coins. One wins the toss if they match, the other if they don't. The winner of the game is the first to win two tosses in a row. What is the probability function for the number of tosses required for someone to win the game?

18. Three gamblers each toss a fair coin. The winner of a toss is the odd man, if there is one. The winner of the game is the first one to win two tosses in a row. What is the probability function for the number of tosses required for someone to win the game?

OBJECTIVE

You should be able to calculate the expected value of a random variable.

3. ▦ If the test scores for a class are 69, 72, 83, 74, 89, 67, 77, 82, 84, 93, 68, and 79, what is their average value?

9.2 AVERAGE AND EXPECTED VALUE

Frequently we have some data and would like some way of determining a "center" point. This is usually done by computing the *mean* or *average* value.

Example 1 ▦ The test scores for a particular class are 76, 72, 88, 90, 74, 83, 52, 79, 81, 84, and 69. What is the average score?

Solution The mean or average value $\bar{x}$ is simply the sum of the various test scores divided by the number of test scores; that is,

$$\bar{x} = \tfrac{1}{11}(76 + 72 + 88 + 90 + 74 + 83 + 52 + 79 + 81 + 84 + 69)$$
$$= \tfrac{1}{11}(848)$$
$$= 77.09.$$

In general, if there are n data points $x_1, x_2, \ldots, x_n$, then their *average* value is:

$$\bar{x} = \frac{1}{n}(x_1 + x_2 \cdots x_n),$$

or, using summation notation,

$$\bar{x} = \frac{1}{n} \sum_{i=1}^{n} x_i.$$

DO EXERCISE 3.

Sometimes a given data point is present more than once.

Example 2 ▦ *Sports.* On the fourteenth hole of the 1976 Andy Williams San Diego Open Golf Tournament, scores were obtained as given in the following table. What was the average score on this hole?

Score	Number with score
3 (Eagle)	1
4 (Birdie)	29
5 (Par)	176
6 (Bogie)	30
7 (Double bogie)	2

Solution Here each score x_i occurs with frequency f_i (number with a particular score). The total number of scores is:

$$N = 1 + 29 + 176 + 30 + 2 = 238,$$

so that the average score is

$$\bar{x} = \tfrac{1}{238}(1 \cdot 3 + 29 \cdot 4 + 176 \cdot 5 + 30 \cdot 6 + 2 \cdot 7)$$

$$= \tfrac{1193}{238}$$

$$= 5.0126 \quad \text{(to four decimal places).}$$

Note that the average score is quite close to par (5).

In general, if each data point x_i occurs with frequency f_i, then the total number of data points is:

$$N = f_1 + f_2 + \cdots + f_n = \sum_{i=1}^{n} f_i,$$

and the average value is given by:

$$\bar{x} = \frac{1}{N} \sum_{i=1}^{n} f_i x_i.$$

DO EXERCISE 4.

Example 3 ▦ Two fair dice are rolled. The sum of the numbers showing is noted. The random variable X corresponds to the sum.

a) Suppose we roll the dice 144 times and obtain the following frequency table.

x	2	3	4	5	6	7	8	9	10	11	12
f	4	9	12	17	21	23	19	16	13	7	3

4. ▦ *Sports.* On the sixteenth hole of the 1976 Andy Williams San Diego Open Golf Tournament, scores were obtained as given in the following table.

Score	Number with score
1 (Hole-in-one)	3
2 (Birdie)	164
3 (Par)	61
4 (Bogies)	3

a) What was the average score on this hole?

b) Based on your answer to part (a), was this an easy or difficult hole?

What is the average value?

b) Suppose we roll the dice 1440 times and obtain the following frequency table.

x	2	3	4	5	6	7	8	9	10	11	12
f	38	84	119	163	207	239	193	159	122	75	41

What is the average value?

Solution

a) The average value of x is given by:

$$\bar{x} = \tfrac{1}{144} \sum_{i=1}^{11} x_i f_i = \tfrac{1}{144}(994) = 6.90278.$$

b) The average value of x is given by:

$$\bar{x} = \tfrac{1}{1440} \sum_{i=1}^{11} x_i f_i = \tfrac{1}{1440}(10,046) = 6.97639.$$

We might call the average values calculated in Example 3 *experimental estimates* of the "center" point. Suppose we had no data, or did not want to bother to obtain any and we wanted to determine a theoretical "center" point. Then we start with the probability distribution function, if such is available. It is. We considered it in Example 1 of the preceding section:

x	2	3	4	5	6	7	8	9	10	11	12
p	$\frac{1}{36}$	$\frac{2}{36}$	$\frac{3}{36}$	$\frac{4}{36}$	$\frac{5}{36}$	$\frac{6}{36}$	$\frac{5}{36}$	$\frac{4}{36}$	$\frac{3}{36}$	$\frac{2}{36}$	$\frac{1}{36}$

Note that the average value can be expressed as

$$\bar{x} = \frac{1}{N} \sum_{i=1}^{n} x_i f_i = \sum_{i=1}^{n} \frac{f_i}{N} x_i.$$

In the long run, after many rolls of the dice, we would expect that the probability p_i for a given outcome would be quite close to the quotient of the frequency f_i for that outcome and the total number of trials N:

$$p_i \approx \frac{f_i}{N}.$$

Thus,

$$\sum_{i=1}^{n} \frac{f_i}{N} x_i \approx \sum_{i=1}^{n} p_i x_i.$$

Then, rather than compute an *average value* from the data, we can compute an *expected value* from the probability function. This is given by:

$$E(X) = \mu = \sum_{i=1}^{n} p_i x_i,$$

where μ is the Greek letter "mu."

Here x_i are now the values of the random variable. Note that we use $\bar{x}$ to denote the average value computed from experimental data and μ to denote the expected value computed theoretically from the probability function.

The expected value of the sum of the dice is given by

$$E(X) = 2 \cdot \tfrac{1}{36} + 3 \cdot \tfrac{2}{36} + 4 \cdot \tfrac{3}{36} + 5 \cdot \tfrac{4}{36} + 6 \cdot \tfrac{5}{36}$$
$$+ 7 \cdot \tfrac{6}{36} + 8 \cdot \tfrac{5}{36} + 9 \cdot \tfrac{4}{36} + 10 \cdot \tfrac{3}{36} + 11 \cdot \tfrac{2}{36} + 12 \cdot \tfrac{1}{36},$$
$$E(X) = 7.$$

In the long run, the more we roll the dice, the closer we "expect" the average values to be to the expected value. Note in Example 3 that the average values 6.90278 and 6.97639 are getting closer to the expected value 7.

DO EXERCISE 5.

For the *hypergeometric frequency function* (see preceding section), it can be shown that the expected value is given by

$$E(X) = n \cdot p$$

where

$$p = \frac{m}{s}.$$

As before, n is the number of trials to draw a random number of objects of which m out of s have a particular characteristic, so that p represents the probability of drawing one with this characteristic *initially*.

Example 4 *Business, Quality Control.* Given a box of 20 transistors, of which 5 are defective. Three are drawn at random and the number defective is noted. What is the expected number of defective transistors in the sample? Use both the general formula and the special one for hypergeometric probabilities. [See Example 2, Section 9.1.]

5. Two fair dice are rolled and the *difference* (in magnitude) between the numbers showing is noted. What is the expected value of this difference? See Margin Exercise 1.

6. A sea chest contains 10 silver bars and 5 gold ones. Three are drawn at random and the number of gold bars is noted. Determine the expected number of gold bars to be drawn using both the general formula and that for hypergeometric frequency functions. See Margin Exercise 2.

Solution   From Example 2, Section 9.1, we have the probability function

x	0	1	2	3
p	$\frac{91}{228}$	$\frac{105}{228}$	$\frac{30}{228}$	$\frac{2}{228}$

Using the general formula for expected value and the probability function previously obtained, we have

$$E(X) = 0 \cdot \tfrac{91}{228} + 1 \cdot \tfrac{105}{228} + 2 \cdot \tfrac{30}{228} + 3 \cdot \tfrac{2}{228} = \tfrac{171}{228} = \tfrac{3}{4}.$$

Using the special formula, we have

$$E(X) = 3 \cdot \tfrac{5}{20} = \tfrac{3}{4}, \qquad \text{as above.}$$

Note that the expected value $E(X)$ need not be a possible value of the random variable; that is, no value of the random variable is $\tfrac{3}{4}$.

DO EXERCISE 6.

In a game of chance, the game is said to be *favorable* or *unfavorable* to the player as the expected value is positive or negative. The game is considered *fair* if the expected value is zero.

Example 5   Consider a lottery in which 10,000 tickets are sold at $1 each. Five tickets are drawn at random. The first-place winner gets a $5000 car, the second-place winner gets a $700 stereo, and the next three winners get $100 each. What is the expected value of a ticket? Is the game fair?

Solution   We let the amount of winnings per ticket be the random variable. The probability function is as follows:

x	0	100	700	5000
p	$\frac{9995}{10000}$	$\frac{3}{10000}$	$\frac{1}{10000}$	$\frac{1}{10000}$

The expected value of a *ticket* is then

$$E_{\mathrm{T}} = 0 \cdot \tfrac{9995}{10000} + 100 \cdot \tfrac{3}{10000} + 700 \cdot \tfrac{1}{10000} + 5000 \cdot \tfrac{1}{10000} = \tfrac{60}{100}, \quad \text{or } \$0.60.$$

Since one is paying $1 for a ticket worth $0.60, one would be suspicious that the game is not fair. The expected value of the *game* E_{G} can be obtained by identifying a new random variable X_i' with the *net* winnings; that is, total winnings minus the cost. Thus, $x_i' =$

$x_i - 1$, so that

$$E_G = E_T - 1 = 0.60 - 1.00 = -0.40.$$

Since $E_G \neq 0$, the game is not fair. Now let c be the cost of a ticket for a fair game. The expected value of a *ticket* E_T is still $0.60, but the expected value of the *game* is now

$$E_G = E_T - c = 0.60 - c,$$

so that $E_G = 0$ for $c = 0.60$ or $0.60. Thus the game is fair when the cost of a ticket equals the expected value.

DO EXERCISE 7.

Example 6 You have a choice between buying one chance for $1 in the lottery of Example 5 or 4 chances for $0.25 each in the following lottery. There are 1000 chances being sold for $0.25 each with a first prize of a $100 TV set. Assuming that the prize can always be exchanged for some other article of equal value, which lottery is the better buy?

Solution The probability function for this lottery is:

x	0	100
p	$\frac{999}{1000}$	$\frac{1}{1000}$

where the random variable is the winnings per ticket. Thus, the expected value per ticket is

$$E_T = 0 \cdot \tfrac{999}{1000} + 100 \cdot \tfrac{1}{1000} = 0.1, \quad \text{or } \$0.10.$$

The expected value of 4 tickets is

$$E_{4T} = 4 \cdot E_T' = 4(0.1) = 0.4, \quad \text{or } \$0.40.$$

The same answer would have been obtained had we taken the random variable as the winnings for 4 tickets. In that case:

x	0	100
p	$\frac{996}{1000}$	$\frac{4}{1000}$

and

$$E_{4T} = 0 \cdot \tfrac{996}{1000} + 100 \cdot \tfrac{4}{1000} = 0.4.$$

Thus $1 would buy four tickets in this lottery with an expected value of $0.40 compared to one ticket in the lottery of Example 5 with an

7. A raffle is being held to raise money for a charity. There are 1000 tickets to be sold for $10 each, with a first prize of a three-week vacation in Europe worth $1500, a second prize of one week in lovely downtown Burbank worth $250, and 5 third prizes of $2 tickets to a movie travelogue. What is the expected value of a ticket? How much of the cost of each ticket goes to charity if the printing expenses are $40 and all other labor is volunteer?

8. Consider a comparison between the lotteries of Examples 5 and 6. If the random variable is taken as winning something (rather than a given amount), which lottery is the better buy

expected value of $0.60, so that the better buy is from the lottery of Example 5.

DO EXERCISE 8.

Craps (Optional)

Example 7 *Business, Casinos.* Craps is a dice game with many variations of the basic rules. The rules used in casinos are the following: A shooter rolls two dice. If the sum of the numbers showing totals 7 or 11, he wins; if the sum totals 2, 3, or 12, he loses. If the sum is anything else (that is, 4, 5, 6, 7, 8, 9, or 10), this becomes his "point." To win, he must roll his point before he rolls a 7, in which case he loses. What is the probability of winning? If even money is bet, what is the expected value?

Solution First let us draw a tree and indicate for each branch the *conditional* probability of occurrence using the probabilities determined in Example 1 of Section 9.1.

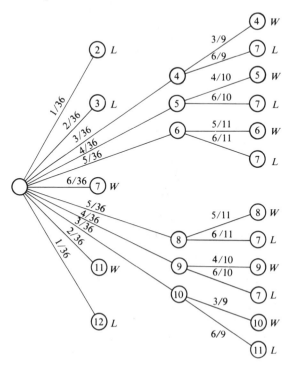

The tree has been simplified in the following respect. If, for example, the "point" is 4, then the branch of the tree for this point would be

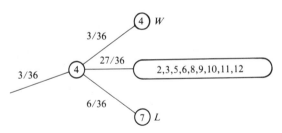

If the point 4 is obtained, the player wins and the game ends. If a 7 is obtained, the player loses and the game ends. If anything but a 4 or a 7 is obtained, the game continues until a 4 or 7 is obtained.

Using Bayes' Theorem, we can determine the probability of obtaining the point 4 provided the game ends, that is, the player gets a 4 or 7:

$$p(4 \mid 4 \text{ or } 7) = \frac{\frac{3}{36}}{\frac{3}{36} + \frac{6}{36}} = \frac{3}{9}.$$

Similarly, we can determine the probability of obtaining a 7 provided the game ends:

$$p(7 \mid 4 \text{ or } 7) = \frac{\frac{6}{36}}{\frac{3}{36} + \frac{6}{36}} = \frac{6}{9}.$$

Using only these two options, we can simplify this branch of the tree to

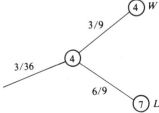

"Points" other than 4 are considered in the same manner. Thus, the probability to win p_W is:

$$p_W = \tfrac{3}{36} \cdot \tfrac{3}{9} + \tfrac{4}{36} \cdot \tfrac{4}{10} + \tfrac{5}{36} \cdot \tfrac{5}{11} + \tfrac{6}{36} + \tfrac{5}{36} \cdot \tfrac{5}{11}$$
$$+ \tfrac{4}{36} \cdot \tfrac{4}{10} + \tfrac{3}{36} \cdot \tfrac{3}{9} + \tfrac{2}{36} = \tfrac{244}{495}.$$

The probability to lose, p_L, is

$$p_L = 1 - p_W = \tfrac{251}{495}.$$

Thus the odds to win are 244:251. The expected value is:

$$E(X) = 1 \cdot \tfrac{244}{495} + (-1) \cdot \tfrac{251}{495} = -\tfrac{7}{495},$$

indicating that, in the long run, the *shooter* will lose.

DO EXERCISE 9.

9. *Business, Casinos.* From Example 7, it is apparent that one can win in the long run by betting *against* the shooter. Since this is the traditional role of the "house," in the long run a casino would lose if it accepted such bets. Since gambling casinos cannot afford to lose, they modify the rules as follows: If one bets *with* the shooter, then the same rules prevail. However, if one bets *against* the shooter, then a standoff feature is added; namely, if the initial roll is 2, the game ends with no win and no loss. What is the expectation of winning by betting against the shooter? Is it better to bet for or against the shooter?

EXERCISE SET 9.2

1. ▦ *Business, Racing.* The lap speeds in the speed trials of a stock-car race were

90.791, 89.237, 89.108, 87.926, 86.401, 85.858, 83.271, and 79.669 mph.

What is the average lap speed?

2. ▦ *Business, Racing.* The lap speeds for a set of speed trials were

91.101, 90.973, 89.257, 86.118, 85.879, 82.438, 81.962, 78.113, and 78.104 mph.

What is the average lap speed?

3. ▦ *Business.* In the course of an evening, a restaurant served 23 lobster dinners at $7.95, 47 steak dinners at $6.95, 53 roast beef dinners at $6.45, 33 shrimp dinners at $5.95, 29 Salisbury steak dinners at $4.95, 37 fried chicken dinners at $4.45, and 23 fish dinners at $3.75. What was the average price of a dinner?

4. ▦ *Business.* A theater has 320 seats for which tickets cost $6.90, 480 seats at $5.90, 624 seats at $4.70, and 484 seats at $3.60. What is the average cost of a ticket?

5. A union has 21 members, 14 of whom are women and the other 7 are men. Three people are chosen at random. What is the expected number of women chosen? men? Use both formulas. [See Exercise 2, Set 9.1.]

6. The sales force of a business consists of 20 people, half of whom are men and the other half women. Four people are chosen at random. What is the expected number of women chosen? Use both formulas. [See Exercise 1, Set 9.1.]

7. Eight people apply for a job, 4 men and 4 women. Four are hired at random. What is the expected number of men hired? Use both formulas. [See Exercise 4, Set 9.1.]

8. Eight people apply for a job, 5 men and 3 women. Four are hired at random. What is the expected number of women hired? Use both formulas. [See Exercise 3, Set 9.1.]

9. *Public Health.* Five workers in a group of 25 have mononucleosis. Three workers are chosen at random. What is the expected number of those chosen who have mono? [See Exercise 6, Set 9.1.]

10. *Business.* A crate of 20 machine parts contains 6 defective parts. Five parts are taken out at random. What is the expected number of defective parts taken out? [See Exercise 5, Set 9.1.]

11. *Political Science.* If a party fields 3 candidates for office, each opposed by an equiprobable candidate, what is the expected number of winners for the party? [See Exercise 7, Set 9.1.]

12. *Political Science.* As in Exercise 11, but the first candidate has a probability of $\frac{1}{3}$ to win, the second $\frac{1}{2}$, and the third $\frac{2}{3}$? [See Exercise 8, Set 9.1.]

13. *Business.* A case of 12 bottles of wine contains 3 which have spoiled. Three bottles are taken out at random. What is the expected number of bottles of spoiled wine among those drawn? if the case contains 4 spoiled bottles? [See Exercise 10, Set 9.1.]

14. *Business.* A wine rack contains 7 bottles of red wine and 2 of white wine. If 3 bottles are taken out at random, what is the expected number of bottles of white wine drawn? red wine? [See Exercise 9, Set 9.1.]

15. *Psychology, Biology.* One cage contains 3 white mice and 2 black ones and another cage contains 2 white mice and 3 black ones. A cage is chosen at random and 3 mice are taken out at random. What is the expected number of white mice in the sample? [See Exercise 11, Set 9.1.]

16. *Psychology, Biology.* There are two cages of mice, as in Exercise 15. A cage is chosen at random. A mouse of unidentified color escapes from the chosen cage. Then 3 mice are taken out at random. What is the expected number of white mice in the sample? [See Exercise 12, Set 9.1.]

17. *Psychology, Biology.* There are 2 cages of mice, as in Exercise 15. A white mouse escapes from an unidentified cage. Then a cage is chosen at random and 3 mice are taken out at random. What is the expected number of white mice in the sample? [See Exercise 13, Set 9.1.]

18. *Psychology, Biology.* There are 2 cages of mice, as in Exercise 11. All the mice escape and are put back at random, 5 to each cage. A cage is then selected at random and 3 mice are taken out at random. What is the expected number of white mice in the sample? [See Exercise 14, Set 9.1.]

19. *Sports.* The winner in a World Series is the first team to win 4 out of 7 games. If the two teams are evenly matched, what is the expected number of games required for one team to win the series? [See Exercise 15, Set 9.1.]

20. *Sports.* As in Exercise 19, but the winning team must win by two games. After 7 games one extra game is to be played if one team is winning by only one game but the series ends if there is a 4-4 tie [See Exercise 16, Set 9.1.]

9.3 VARIANCE AND STANDARD DEVIATION

In addition to wanting to know the mean or average value of the data or the expected value of a probability-distribution function for some random variable, we may want to know about the "spread" of the data. To do this, we use the *variance* and the *standard deviation.*

Let us consider the test scores from Example 1 of Section 9.2:

$$76, \quad 72, \quad 88, \quad 90, \quad 74, \quad 83, \quad 52, \quad 79, \quad 81, \quad 84, \quad 69.$$

In that example we found that the average value was

$$\bar{x} = 77.09.$$

We want to consider how the data *varies* from the average value. How can we do this? One way might be to consider the differences between a score and the average value

$$x_i - \bar{x}.$$

For example, $76 - 77.09 = -1.09$ represents the deviation of the score 76 from the mean. If we add all of these deviations and average them, we would have

$$\frac{1}{n} \sum_{i=1}^{n} (x_i - \bar{x}),$$

which is called the *average deviation from the mean.* Unfortunately, this quantity always adds to 0. Thus, it would not yield any results which would vary between different sets of data to yield information about the spread of the data. (The reader should verify that for the above set of data, the average deviation from the mean is 0.) To avoid this difficulty, we consider the *deviation squared:*

$$(x_i - \bar{x})^2.$$

OBJECTIVES

You should be able to calculate the variance and standard deviation of:

a) Given data
b) A random variable, given its probability function.

10. ▦ Determine the variance of the test scores of Margin Exercise 3: 69, 72, 83, 74, 89, 67, 77, 82, 84, 93, 68, and 79.

The average of the deviation squared, called the sample *variance*, is given by*

$$s^2 = \frac{1}{n} \sum_{i=1}^{n} (x_i - \bar{x})^2.$$

Now calculating the variance, we obtain

$$s^2 = \tfrac{1}{11}[(76 - 77.09)^2 + (72 - 77.09)^2 + \cdots + (69 - 77.09)^2]$$

or

$$s^2 = 101.72.$$

DO EXERCISE 10.

When a probability function is known, we can find a "theoretical" variance defined as the *expected value* of the *deviation squared*:

$$\sigma^2 = E[(X - \mu)^2] = \sum_{i=1}^{n} p_i(x_i - \mu)^2.$$

11. ▦ A sea chest contains 10 silver bars and 5 gold ones. Three are drawn at random. What is the variance of the number of gold bars drawn? (See Margin Exercise 2, Section 9.1, and Margin Exercise 6, Section 9.2.)

Example 1 ▦ Given a box of 20 transistors, of which 5 are defective. What is the variance of the number of defective transistors?

Solution From Example 2, Section 9.1, we have the probability function:

x	0	1	2	3
p	$\frac{91}{228}$	$\frac{105}{228}$	$\frac{30}{228}$	$\frac{2}{228}$

From Example 4, Section 9.2, we have the expected value

$$\mu = E(X) = \tfrac{3}{4}.$$

The variance is given by:

$$\sigma^2 = \tfrac{91}{228}(0 - \tfrac{3}{4})^2 + \tfrac{105}{228}(1 - \tfrac{3}{4})^2 + \tfrac{30}{228}(2 - \tfrac{3}{4})^2 + \tfrac{2}{228}(3 - \tfrac{3}{4})^2$$

$$= \tfrac{91}{228} \cdot \tfrac{9}{16} + \tfrac{105}{228} \cdot \tfrac{1}{16} + \tfrac{30}{228} \cdot \tfrac{25}{16} + \tfrac{2}{228} \cdot \tfrac{81}{16}$$

$$= 0.503.$$

DO EXERCISE 11.

Note that we started with certain units. For example, in Example 1 the test scores were measured in percentile "points." So is the average

---

* Here we consider *all* the data. If we were using only a *sample* of the data, then the n in the denominator should be replaced by $(n - 1)$, for statistical reasons which we cannot go into in this text. We need the definition of variance using n for later use.

value. On the other hand, the variance is measured in the *square* of these units, or "points-squared." We can obtain a measure of the "spread" of the data in the same units as the data by using the *standard deviation* which is the square root of the variance.

Example 2 ▦ Obtain the standard deviation of the test scores of Example 1, Section 9.2.

Solution The standard deviation is the positive square root of the variance:

$$s = +\sqrt{s^2} = \sqrt{\frac{1}{n} \sum_{i=1}^{n} (x_i - \bar{x})^2}.$$

From earlier work in this section, we know that $s^2 = 101.72$, so that $s = \sqrt{101.72} = 10.09$. (Use a square-root table or hand calculator.)

DO EXERCISE 12.

In general, a small standard deviation relative to the mean indicates that the data has little spread about the mean, while a large standard deviation relative to the mean indicates a large spread about the mean.

Example 3 ▦ Find the standard deviation for the number of defective transistors of Example 1.

Solution The standard deviation is

$$\sigma = \sqrt{\sigma^2} = \sqrt{\sum_{i=1}^{n} p_i (x_i - \mu)^2}.$$

In Example 1, we found that $\sigma^2 = 0.503$, so that

$$\sigma = \sqrt{0.503} = 0.709.$$

DO EXERCISE 13.

12. ▦ Find the standard deviation of the test scores of Margin Exercise 10.

13. ▦ Find the standard deviation of the number of gold bars of Margin Exercise 11.

EXERCISE SET 9.3

1. ▦ *Business, Racing.* Find the variance and standard deviation for the lap speeds of Exercise 1, Set 9.2:

  90.791, 89.237, 89.108, 87.926, 86.401,

  85.858, 83.271, and 79.669 mph.

2. ▦ *Business, Racing.* Find the variance and standard deviation for the lap speeds of Exercise 2, Set 9.2:

  91.101, 90.973, 89.257, 86.118, 85.879,

  82.438, 81.962, 78.113, and 78.104 mph.

3. ▥ *Business.* Find the variance and standard deviation for the dinners of Exercise 3, Set 9.2:

23 at \$7.95, 47 at \$6.95, 53 at \$6.45, 33 at \$5.95, 29 at \$4.95, 37 at \$4.45, and 23 at \$3.75.

4. ▥ *Business.* Find the variance and standard deviation for the tickets of Exercise 4, Set 9.2:

320 at \$6.90, 480 at \$5.90, 624 at \$4.70, and 484 at \$3.60.

5. A union has 21 members, of whom 14 are women and 7 are men. Three people are chosen at random. What is the variance and standard deviation of the number of women chosen? [See Exercise 5, Set 9.2.]

6. The sales force of a business consists of 20 people, half of whom are men and the other half women. Four people are chosen at random. What is the variance and standard deviation of the number of women chosen? [See Exercise 6, Set 9.2.]

7. *Public Health.* Five workers in a group of 25 have mononucleosis. Three workers are chosen at random. What is the variance and standard deviation of the number with mono? [See Exercise 9, Set 9.2.]

8. *Business.* A crate of 20 machine parts contains 6 defective parts. Five parts are taken out at random. What is the variance and standard deviation of the number of defective parts taken out? [See Exercise 10, Set 9.2.]

OBJECTIVES

You should be able to:

a) Calculate the expected value of a random variable (two ways) in a binomial-probability problem.

b) Solve binomial-probability problems.

9.4 BERNOULLI TRIALS—BINOMIAL PROBABILITY

Many experiments have outcomes which fall naturally into two disjoint sets designated simply "success" or "failure." For example,

i) The flipping of a coin and its landing "heads" or "tails" (barring its standing on edge);

ii) The winning or losing of an election (allowing for the ultimate resolution of a tie);

iii) The passing or failing of a manufactured article to a particular tolerance for quality control.

Experiments or trials with two possible outcomes are called *Bernoulli trials.*

Most of the problems considered so far involved repeated trials in which each trial changed the conditions of subsequent trials. In particular for hypergeometric probabilities, an object was drawn at random from a group of objects and *not* replaced before the next trial.

Now we consider trials such that whatever is removed in one trial is *replaced* before the next trial. Each trial is the same and hence *independent* of the others. Thus, we consider *repeated independent Bernoulli trials.*

Example 1 A fair coin is flipped repeatedly. What is the probability for heads to show 3 times out of 5 flips?

Solution   Here a "success" can be identified with a coin showing "heads." Since the coin is fair, the probability for either heads or tails to show is $\frac{1}{2}$. One way for heads to show 3 times out of 5 is for the first 3 flips to show heads and the last two to show tails, that is

HHHTT.

Each particular outcome (H or T) occurs with probability $\frac{1}{2}$. In the outcome HHHTT, each individual flip is independent of the others. Thus the combined outcome HHHTT will happen with a probability that is the product of the individual probabilities (Multiplication Theorem):

$$\left(\tfrac{1}{2}\right)^3\left(\tfrac{1}{2}\right)^2 = \tfrac{1}{32}.$$

There are 10 configurations in which we can get 3 heads out of 5 flips. They are

HHHTT	TTHHH
HHTHT	THTHH
HTHHT	THHTH
HHTTH	HTTHH
HTHTH	THHHT.

Each of these configurations is equally probable, since it is calculated exactly the same way as the previous calculation for HHHTT.

Using techniques from Chapter 8, we can compute directly the number of ways 3 heads can show in 5 flips:

$$C(5, 3) = \binom{5}{3} = \frac{5 \cdot 4 \cdot 3}{1 \cdot 2 \cdot 3} = 10.$$

The probability for heads to show 3 times out of five is the product of the probability for heads to show 3 times out of five in a particular configuration times the number of different ways heads can show 3 times out of five, that is

$$p_{3H} = \tfrac{1}{32} \cdot 10 = \tfrac{10}{32}.$$

Note that the answer is the same whether *one* coin is flipped *5* times in a row or *5* coins are flipped *once* at the same time. In either case each trial is identical.

DO EXERCISE 14.

14. What is the probability for tails to show twice in 6 flips of a fair coin? List the configurations.

Problems of this type have a *binomial* probability. This means that for a *binomial* random variable,

  i) There are two outcomes (Bernoulli trials);
 ii) Each trial is independent of preceding trials;
iii) Each trial is identical to preceding trials.

Binomial probability is similar to *hypergeometric* probability in that objects in random drawings are restricted to *two* kinds, but differs in that *binomial* probability assumes *replacement* or identical trials while *hypergeometric* probability assumes *no replacement* or that successive trials differ.

Example 2 *Business, Quality Control.* Electrical switches are manufactured with 10% being defective. Five switches are drawn at random and without replacement. What is the probability that two of these are defective?

Solution If this problem had specified that 5 switches were to be drawn from some *fixed* number of switches, then the probability would be hypergeometric. However, a fixed number is *not* specified. Rather, switches are being manufactured, as on a production line, and continually fed into some container from which the sample of 5 is taken. Thus, within the limits of the information available to us, the trials are independent and hence the probability is binomial.

Let the *event* be the drawing of a *defective* switch. This event has a probability $p = 0.10$. The probability for drawing a nondefective switch is $q = 1 - p = 0.90$.

The probability for getting the *first two* switches defective and the *last three* switches nondefective is

$$(0.10)^2(0.90)^3.$$

The number of ways two defective switches can be drawn out of a sample of five is $\binom{5}{2}$.

Thus, the probability that there will be two defective switches in a sample of five is

$$\binom{5}{2}(0.10)^2(0.90)^3 = 10(0.10)(0.729) = 0.0729.$$

In general, if the probability is p that some event *will* happen in one trial and $q = 1 - p$ that it will *not*, then the binomial probability p_k for the event to happen k times out of n trials is given by

$$p_k = B(n, k, p) = \binom{n}{k}p^k q^{n-k}.$$

In Example 2, we have

$$p = 0.10, \quad q = 0.90, \quad n = 5, \quad k = 2,$$

so that

$$p_2 = \binom{5}{2}(0.10)^2(0.90)^3 = 0.0729, \quad \text{as before.}$$

Alternately, we may take the *event* as the drawing of a *nondefective* switch. Then we seek the probability of drawing *three nondefective* switches out of a sample of five. Thus

$$p = 0.90, \quad q = 0.10, \quad n = 5, \quad k = 3,$$

and

$$p_3 = \binom{5}{3}(0.90)^3(0.10)^2 = 0.0729.$$

This is the same answer as before but with different notation. Recall that

$$\binom{n}{k} = \binom{n}{n-k}.$$

DO EXERCISE 15.

Example 3 What is the probability function for the problem of Example 2?

Solution The probability in this case is binomial. Thus, taking the random variable K to be the number of defective switches observed in the sample, we have

k	p_k
0	$\binom{5}{0}(0.1)^0(0.9)^5 = 0.59049$
1	$\binom{5}{1}(0.1)^1(0.9)^4 = 0.32805$
2	$\binom{5}{2}(0.1)^2(0.9)^3 = 0.07290$
3	$\binom{5}{3}(0.1)^3(0.9)^2 = 0.00810$
4	$\binom{5}{4}(0.1)^4(0.9)^1 = 0.00045$
5	$\binom{5}{5}(0.1)^5(0.9)^0 = 0.00001$

15. *Public Health.* Treatment for a certain disease is effective 80% of the time. If six treated patients are surveyed, what is the probability that four of them will be cured?

16. What is the probability function for the problem of Margin Exercise 15?

DO EXERCISE 16.

In Section 9.2 the expected value of a random variable X was defined by

$$\mu = E(X) = \sum_{i=1}^{n} x_i p_i.$$

For binomial probability, the random variable is K, the number of successes in a series of trials. It takes on the values $k = 0, 1, \ldots, n$. Thus, we write

$$\mu = E(K) = \sum_{k=1}^{n} k p_k.$$

It can be shown that the expected value for the random variable with binomial probability is given by

$$\mu = E(K) = n \cdot p,$$

where n is the number of trials and p is the probability for the event to occur in one trial.

Note that the expected value for the *hypergeometric* probability distribution is the same as that for the *binomial* probability *provided* that, in the former case, p is taken as the probability of an *initial* success; that is, $p = m/s$, as in Section 9.2.

17. Determine the expected value for the problem of Margin Exercise 15 using both formulas.

Example 4 Determine the expected value for the problem of Example 2 using both formulas.

Solution Using

$$E(K) = \sum_{k=1}^{n} k p_k,$$

we have

$$E(K) = 0(0.59049) + 1(0.32805) + 2(0.07290) + 3(0.00810)$$
$$+ 4(0.00045) + 5(0.00001) = 0.50000.$$

Using $E(K) = n \cdot p$, we have $n = 5$ and $p = 0.1$, so that

$$E(K) = 5(0.1) = 0.5, \quad \text{as above.}$$

DO EXERCISE 17.

We can also determine the variance and standard deviation for a binomial probability distribution.

Example 5 ▦ Find the variance and standard deviation for the problem of Example 2.

Solution   The expected value for this problem was found in Example 4 to be
$$\mu = 0.5.$$

Thus, the variance is
$$\sigma^2 = 0.59049(0 - 0.5)^2 + 0.32805(1 - 0.5)^2 + 0.07290(2 - 0.5)^2$$
$$+ 0.00810(3 - 0.5)^2 + 0.00045(4 - 0.5)^2 + 0.00001(5 - 0.5)^2$$

or
$$\sigma^2 = 0.45000,$$

so that the standard deviation is
$$\sigma = \sqrt{0.45000} = 0.67082.$$

The variance and standard deviation for a binomial probability-distribution function can be obtained simply from the formula
$$\sigma^2 = np(1 - p).$$

In this case, $n = 5$ and $p = 0.1$ (see **Example 2**), so that
$$\sigma^2 = 5 \cdot 0.1(1 - 0.1), \quad \text{or} \quad \sigma^2 = 0.45, \quad \text{as before.}$$

DO EXERCISE 18.

Example 6   *Business, Quality Control.* Given a box of 20 transistors of which 5 are defective. Three transistors are drawn at random, the number defective is noted, and the transistors are replaced. This is repeated 5 times. What is the probability that at least one is defective in at least 4 trials?

Solution   The probability p that at least one transistor is defective is the *hypergeometric* probability given by the quantity $p_{i \geqslant 1}$ in Example 2 of Section 8.2; that is,
$$p = \tfrac{137}{228} \quad \text{and} \quad q = 1 - p = \tfrac{91}{228}.$$

This corresponds to a "success" (that is, the event happening) in the second part of the problem. The probability for at least 4 successes is
$$p_{i \geqslant 4} = p_4 + p_5$$

where p_4 and p_5 are the *binomial* probabilities
$$p_4 = \binom{5}{4}(\tfrac{137}{228})^4(\tfrac{91}{228})^1 \quad \text{and} \quad p_5 = \binom{5}{5}(\tfrac{137}{228})^5(\tfrac{91}{228})^0.$$

Note that while in Example 2 the probability was given, in this Example it had to be computed. Furthermore, one problem may involve more than one type of probability, in this case both hypergeometric and binomial.

DO EXERCISE 19.

18. ▦ Determine the variance and standard deviation for the problem of Margin Exercise 15 using both formulas.

19. An old sea chest contains 10 bars of silver and 5 bars of gold. Three are drawn out at random, the number of gold bars is noted, and the bars are replaced. This is repeated four times. What is the probability for drawing two gold bars three times? *Caution.* Interpret subscripts carefully.

EXERCISE SET 9.4

1. Five fair coins are tossed. What is the probability function for the number of tails showing? What is the expected number of tails? What is the variance and standard deviation? Use both formulas.

3. *Political Science.* Half the people in a community favor a certain political stand and half oppose it. Of 6 people selected at random, what is the expected number to favor the stand? to oppose the stand? What is the probability that of these 6 people 4 or more will favor the stand or oppose it?

5. *Business.* An impostor applies for a job as a wine taster. As a test he is given 5 wines to taste to determine whether they are *vin ordinaire* or a great vintage wine. What is the probability that he gets at least 4 out of 5 correct by guessing? What is the expected number of correct evaluations? What is the variance and standard deviation?

7. *Science.* A complex experiment consists of 6 components each with a reliability of 0.9 (that is, the probability for the component to work is 0.9). If the experiment is so constructed that it can be run if no more than one out of the 6 components fails to function properly, what is the probability that the experiment can be run? What is the expected number of failures? What is the variance and standard deviation?

9. *Public Health.* Treatment for a certain disease is effective 80% of the time. If 5 patients are sampled, what is the probability function for the number of effective treatments? What is the expected value? What is the probability that the treatment is ineffective for at least one patient?

11. *Business, Quality Control.* Sparkplugs are manufactured and pass along a conveyor belt for inspection. A sample of 5 is taken at random. What is the probability function for the number of defective plugs if the defective rate is 10%? if the defective rate is 20%? if the defective rate is 30%? Which defective rate is most probable if *no* defective plugs are found in the sample? if one defective plug is found in the sample? if two are found? if three are found?

2. *Public Health.* If the birth rate for boys and girls were equal, what would be the distribution of girls in a four-child family? What is the expected number of girls? What is the variance and standard deviation? Use both formulas.

4. *Political Science.* One-third of the people in a community favor a certain political stand and two thirds oppose it. Of 6 people selected at random, what is the expected number to favor the stand? to oppose the stand? What is the probability that of these 6 people at least half will oppose the issue?

6. *Business.* As in Exercise 5, but the applicant is genuine and can distinguish the two wines 4 times out of 5. What is the probability that he fails the test (that is, fails to get at least 4 out of 5 correct)? What is the expected number of correct evaluations? What is the variance and standard deviation?

8. *Science.* A successful flight of an exploratory space rocket requires that no more than two of the 10 components fail to function properly (due to use of interlocking failsafe circuits). If each component has a reliability of 0.98, what is the probability for a successful flight? What is the expected number of component failures? What is the variance and standard deviation?

10. *Demographics.* If 30% of marriages end in divorce by the fifth year of marriage, what is the probability function for the number of couples out of a sample of 6 who have been divorced after no more than 5 years of marriage? What is the probability that half or more of the sample has been divorced? What is the expected number of divorced couples? the variance? the standard deviation?

12. *Business, Quality Control.* Machine A makes ballpoint pens with a deficiency rate of 10% and machine B makes pens with a deficiency rate of 30%. Two pens are taken from each machine. What is the probability that two of the four are defective? If the pens from the two machines are mixed half and half (that is, with a deficiency rate of 20%) before a sample of 4 is taken, what is the probability that there will be two defective pens in the sample?

13. *Business.* A company manufactures a type of mousetrap which is 50% effective. They would like to claim that it is at least 80% effective. A sample of 5 traps is tested for effectiveness. What is the probability that the sample is at least 80% effective? If the testing of a sample of 5 traps is repeated 5 times, what is the expected number of trials for which the samples tested are at least 80% effective?

14. *Business.* A company manufactures thermometers with a deficiency rate of 20%. A sample of 5 thermometers is tested. What is the probability that at least one thermometer is defective? If this test is repeated 5 times, what is the probability that there is at least one defective thermometer in each trial? in at least 4 out of 5 trials?

15. As in Exercise 14, what is the probability that at least two thermometers are defective? If the test is repeated 5 times, what is the probability that at least two defective thermometers are found in at least 2 trials out of the five?

---

*9.5 (OPTIONAL) BERNOUILLI TRIALS—GEOMETRIC AND NEGATIVE BINOMIAL PROBABILITIES

Bernoulli trials lead not only to binomial probability but to geometric and other related probabilities.

Example 1 *Business.* An oil company claims to drill 5 dry holes for each 1 that produces oil. That is, they hit oil 1 time in 6. What is the probability that the first producing well is obtained on the 4th try?

Solution The probability for a success (a producing well) in one trial is $p = \frac{1}{6}$. The probability for a failure (a dry hole) is $q = 1 - p = \frac{5}{6}$. Since each trial is *independent*, the probability of 3 failures *and then* 1 success on the 4th try is the *product*

$$\frac{5}{6} \cdot \frac{5}{6} \cdot \frac{5}{6} \cdot \frac{1}{6} \quad \text{or} \quad \left(\frac{5}{6}\right)^3\left(\frac{1}{6}\right)^1.$$

If we now take the random variable to be the number of trials for the first success, then the resulting probability is called *geometric*. Thus,

$$p_k = (1 - p)^{k-1} \cdot p,$$

where the subscript k now refers to the trial on which the first success occurs.

DO EXERCISE 20.

Example 2 *Business.* As in Example 1, how many holes should be drilled to be at least 70% certain, that is, $p_k = 0.70$, of hitting oil at least once?

Solution In this case, we need the (geometric) probability function where the random variable can take on the values $k = 1, 2, 3, \ldots$

OBJECTIVE

You should be able to solve problems involving geometric and related probabilities.

20. *Public Health.* Treatment for a certain disease is effective 80% of the time. What is the probability that the first success occurs on the fourth try? (Compare with Margin Exercise 15.)

21. As in Margin Exercise 20, how many patients should be treated to have a 99% probability of at least one effective treatment?

Note that there is no *upper* limiting value to k. Thus, there is no *guarantee* for success at any stage. Since these trials are independent, the probability for success *by* the kth trial, p_k^*, is the probability for success *on* the first trial *or on* the second trial, etc., so that p_k^* is given by the *sum*

$$p_k^* = p_1 + p_2 + \cdots + p_k.$$

Tabulating the probability function, we have:

k	p_k	p_k^*	
1	$\frac{1}{6}$	$\frac{1}{6}$	
2	$\frac{5}{6} \cdot \frac{1}{6}$	$\frac{11}{36}$	
3	$(\frac{5}{6})^2 \cdot \frac{1}{6}$	$\frac{91}{216}$	
4	$(\frac{5}{6})^3 \cdot \frac{1}{6}$	$\frac{671}{1296}$	
5	$(\frac{5}{6})^4 \cdot \frac{1}{6}$	$\frac{4651}{7776}$	< 0.6
6	$(\frac{5}{6})^5 \cdot \frac{1}{6}$	$\frac{31031}{46656}$	$< \frac{2}{3}$
7	$(\frac{5}{6})^6 \cdot \frac{1}{6}$	$\frac{201811}{279936}$	> 0.7
.	.	.	.
.	.	.	.
.	.	.	.

Thus, 7 holes must be drilled in order to have at least a 70% probability of hitting oil at least once. Note that not all of the values of the probability function need be determined to obtain the solution.

DO EXERCISE 21.

22. As in Margin Exercise 20, what is the expected number of patients that must be treated to achieve the first successful treatment? Note that expected value need not be an integer.

Example 3 In Example 1, what is the expected number of trials to achieve the first success?

Solution The expected value of a geometric random variable is

$$\mu = E(K) = \sum_{k=1}^{\infty} k p_k = \sum_{k=1}^{\infty} k p (1 - p)^{k-1},$$

where the sum to infinity (∞) means that we keep adding terms for larger and larger values of k until the terms no longer effect $E(K)$.

The expected value for a *geometric* probability distribution (that is, for the first success) is represented by an infinite *geometric* series whose sum is

$$\mu = E(K) = \frac{1}{p}.$$

In this example, $p = \frac{1}{6}$. Thus, the expected number of holes to be drilled to find one producing well is $1/(\frac{1}{6})$, or 6.

DO EXERCISE 22.

Example 4 ▦ Determine the variance and standard deviation of the problem of Example 1.

Solution   In the case of geometric probability the general formula for variance gives us a "sum to infinity" as for the expected value:

$$\sigma^2 = \sum_{k=1}^{\infty} p_k (k - \mu)^2.$$

However, this sum can be represented by

$$\sigma^2 = \frac{(1 - p)}{p^2}.$$

Thus, since $p = \frac{1}{6}$ in this Example, we obtain for the variance

$$\sigma^2 = \frac{(1 - \frac{1}{6})}{(\frac{1}{6})^2} = 30,$$

and for the standard deviation

$$\sigma = \sqrt{30} = 5.47723.$$

DO EXERCISE 23.

Geometric probabilities can be generalized to *negative binomial probabilities*. Since there are still two outcomes, the trials are Bernoulli.

Example 5   *Business.* (Compare with Example 1) An oil company claims to drill five dry holes for each one that produces oil. What is the probability that it will require 4 trials to produce 2 producing wells? What is the probability that in the first 4 trials 2 producing wells are obtained?

Solution   The probability that the company will need 4 trials to produce 2 producing wells means that one of the two producing wells must be struck on the fourth try. The other may be struck on any of the first three. Thus, the probability that four trials are required to obtain two good wells is the product of the probability to be successful on one of the first 3 trials *times* the probability to be successful on the fourth trial:

$$[\tbinom{3}{1}(\tfrac{5}{6})^2(\tfrac{1}{6})] \cdot (\tfrac{1}{6}) = \tbinom{3}{1}(\tfrac{5}{6})^2(\tfrac{1}{6})^2 = \tfrac{75}{1296}.$$

The probability that of the first four trials, two are successful can be obtained as the binomial probability (as in the previous section):

$$p_2 = \tbinom{4}{2} \cdot (\tfrac{5}{6})^2 \cdot (\tfrac{1}{6})^2 = \tfrac{150}{1296}.$$

In general, if p is the probability for success on one trial and we seek the probability $p_{s,k}$ that k trials will be required for s successes, then

23. ▦ Find the variance and standard deviation for the problem of Margin Exercise 20.

24. *Public Health.* Treatment for a certain disease is 80% effective. What is the probability that five trials will be required for four successful treatments? What is the probability that four of the first five treatments will be successful?

the kth trial must be a success and the remaining $s - 1$ successes can be achieved in the first $(k - 1)$ trials; thus

$$P_{s,k} = \left[\binom{k-1}{s-1}(1 - p)^{[(k-1)-(s-1)]}p^{s-1} \right] \cdot p$$

or

$$P_{s,k} = \binom{k-1}{s-1}(1 - p)^{k-s}p^{s}.$$

Geometric probability p_k corresponds here to the negative binomial probability $p_{1,k}$, that is, for $s = 1$.

DO EXERCISE 24.

Example 6 Find the expected value for the problem of Example 5.

Solution As for geometric probability, the formula for expected value for negative binomial probability is a "sum to infinity." It can be represented by

$$\mu = E(K) = \frac{s}{p},$$

where s is the number of successes. This reduces to that for geometric probability for $s = 1$.

In this example, we have $s = 2$ and $p = \frac{1}{6}$, so that

$$\mu = \frac{2}{\frac{1}{6}} = 12.$$

DO EXERCISE 25.

25. Find the expected value for the problem of Exercise 24.

Example 7 ▦ Find the variance and standard deviation for the problem of Example 5.

Solution As for expected value, the variance for negative binomial probabilities is a "sum to infinity" which can be represented by

$$\sigma^2 = \frac{s(1 - p)}{p^2}.$$

The standard deviation is obtained by taking the square root of the variance:

$$\sigma = \sqrt{\sigma^2}.$$

Thus, in this example, the variance is

$$\sigma^2 = \frac{2(1 - \frac{1}{6})}{(\frac{1}{6})^2} = 60$$

and the standard deviation is

$$\sigma = \sqrt{60} = 7.74597.$$

DO EXERCISE 26.

26. ▦ Find the variance and standard deviation for the problem of Exercise 24.

EXERCISE SET 9.5

1. A fair coin is flipped. What is the probability that the third head shows on the fifth flip?

2. A fair coin is flipped. What is the probability that the fourth tail shows on the seventh flip?

3. A salesman selling encyclopedias makes a sale at 1 house in 10. What is the probability that he makes a sale *on* the 3rd try? *by* the 3rd try?

4. *Business.* A new process for manufacturing diodes makes them at one-tenth of the previous cost but with a 20% defect rate. What is the probability that the first defective diode is the fifth tested? is one of the first five tested?

5. As in Exercise 3, how many houses must be visited for the probability of a sale to be at least 0.5?

6. As in Exercise 4, how many diodes must be sampled for the probability of finding the first defective diode to be 0.4? to be 0.5?

7. As in Exercise 3, what is the expected number of trials for the first sale? What is the standard deviation?

8. As in Exercise 4, what is the expected number of diodes to be sampled for the first defective one to be found? What is the standard deviation?

9. An automobile salesman sells one car for each seven customers he talks to. What is the probability that his first sale is to his seventh customer?

10. As in Exercise 9, what is the expected number of customers for the first sale? What is the standard deviation?

11. As in Exercise 9, but his *second* sale is to his seventh customer?

12. As in Exercise 11, but his *first* sale is to his first customer?

13. A fisherman has a one-in-four chance at catching a trout on a cast. What is the probability that on the sixth cast he will have caught his second trout? that in six casts he will have caught just two trout?

14. As in Exercise 13, what is the expected number of trout to be caught in 6 casts? What is the expected number of casts for the first trout to be caught?

▶

15. A man flips a fair coin until he gets two tails in a row. What is the distribution function for the number of flips required?

16. As in Exercise 15, but the coin is flipped until there are either two tails in a row or two heads in a row?

17. *Business, Casinos.* A gambler with $1 enters a casino having unlimited funds. The gambler makes $1 bets in a game where the probability of winning is $\frac{1}{2}$ and the game is fair. What is the probability function for the number of games that the gambler must play to be wiped out? [*Note.* If the gambler wins the first game, he has $2 with which to bet.]

18. *Business, Casinos.* As in Exercise 17, but the gambler enters with $2.

19. *Business, Casinos.* As in Exercise 17, if the gambler plays indefinitely, what is the probability that he gets wiped out?

21. *Business, Casinos.* As in Exercise 17, but the gambler quits as soon as he is ahead $1. What is the probability that he is ahead $1 before he is wiped out?

23. *Business, Casinos.* As in Exercise 22, but the gambler quits as soon as he is $2 ahead? (*Note.* The bets are still $1.) Compare answer with that of Exercise 21.

20. *Business, Casinos.* As in Exercise 18, if the gambler plays indefinitely, what is the probability that he gets wiped out?

22. *Business, Casinos.* As in Exercise 21, but the gambler enters with $2?

OBJECTIVES

You should be able to

a) put a random variable in standardized form;

b) solve problems, using the standardized random variable and the normal probability distribution.

27. ▦ *Psychology.* The test scores for a class are (Margin Exercise 3, Section 9.2): 69, 72, 83, 74, 89, 67, 77, 82, 84, 93, 68, and 79. The mean of these test scores is $\bar{x} = 78.08$ and the standard deviation is $\sigma = 8.06$. Convert a test score of 85 to standardized form.

9.6 STANDARDIZED RANDOM VARIABLES AND THE NORMAL PROBABILITY DISTRIBUTION

The smaller the standard deviation, the less spread there is in the data and the larger the standard deviation, the more spread there is. We shall pursue this concept further by converting our random variable to *standardized form* with 0 mean and a standard deviation of 1. This we do by calculating the standardized random variable Z with numerical values z:

$$z = \frac{x - \bar{x}}{s} \quad \text{or} \quad z = \frac{x - \mu}{\sigma},$$

depending on whether one is using sample or theoretical data.

Example 1 ▦ The test scores of Example 1, Section 9.2, were 76, 72, 88, 90, 74, 83, 52, 79, 81, 84, and 69. The mean of these test scores is $\bar{x} = 77.09$ and the standard deviation is $\sigma = 10.09$. Convert a test score of 85 to standardized form.

Solution Substituting into the preceding formula, we obtain:

$$z = \frac{85 - 77.09}{10.09} = \frac{7.91}{10.09} = 0.78.$$

DO EXERCISE 27.

Example 2 ▦ To what test scores do $z = 1$ and $z = -1$ correspond for the data of Example 1? to what test scores do $z = 2$ and $z = -2$ correspond?

Solution   Using the standardized random variable, we have

$$z = \frac{x - 77.09}{10.09} = 1.$$

Solving for x, we obtain a test score $x = 87.18$.

Similarly,

$$z = \frac{x - 77.09}{10.09} = -1$$

yields a test score $x = 67.00$.

Continuing,

$$z = \frac{x - 77.09}{10.09} = 2 \quad \text{yields } x = 97.27$$

and

$$z = \frac{x - 77.09}{10.09} = -2 \quad \text{yields } x = 56.91.$$

DO EXERCISE 28.

Example 3   What percent of the students of Example 1 and 2 have test scores that lie within one standard deviation of the mean, that is, that lie between $z = 1$ and $z = -1$?

Solution   Out of 11 students, 8 had test scores between 67.00 ($z = -1$) and 87.18 ($z = 1$). This corresponds to $100 \cdot \frac{8}{11}$ or 73%.

DO EXERCISE 29.

Example 4   *Psychology.* What percent of the students of a "normal" class would be expected to have test scores within one standard deviation of the mean? Compare the result with the actual class of Example 3.

Solution   Starting with the binomial probability distribution and assuming a large enough sample that inclusion of more data would not affect the results, one can obtain the "normal" probability distribution. While not all probability distributions are "normal," their use often yields a good approximation with simpler calculations. Whether there is little or much spread in the data, the use of the standardized random variable permits us to use a *single* normal probability distribution. This is a *bell-shaped curve*, illustrated in Fig. 9.1, where "up" is to the "right."

28. ▦ *Psychology.* To what test scores do $z = 1$ and $z = -1$ correspond for the data of Margin Exercise 27? $z = 2$ and $z = -2$?

29. *Psychology.* What percent of the students of Margin Exercises 27 and 28 have test scores within one standard deviation of the mean, that is, lie between $z = -1$ and $z = 1$?

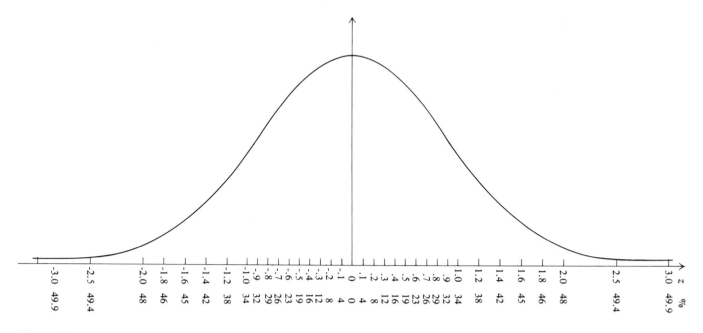

Figure 9.1

30. *Psychology.* What percent of the students of a "normal" class would be expected to have test scores within two standard deviations of the mean? Compare with the actual class of Exercise 29.

From the figure, we see that there is as much of the curve above the center ($z = 0$, the standardized mean) as below the center. The curve is symmetrical about $z = 0$. Between the center and one standard deviation *above* the center (between $z = 0$ and $z = 1$), 34% of the data (test scores) should lie. Similarly, between the center and one standard deviation *below* the center (between $z = -1$ and $z = 0$), another 34% of the data should lie. Thus, within one standard deviation of the mean (that is between $z = -1$ and $z = 1$) should lie 34% + 34%, or 68% of the data.

The class of Example 3 has 73% of the students with test scores within one standard deviation of the mean. A "normal" class would have 68%. The difference is due to the fact that the actual class has a small number of students. If more students were added to the class, the percent within one standard deviation of the mean would become closer to that for a "normal" class.

DO EXERCISE 30.

Example 5 Using the mean and standard deviation from Example 1, what percent of a normal class would be expected to have test scores of 85 or more? Compare results with the actual test scores of Example 1.

Solution   As calculated in Example 1, a test score of 85 corresponds to $z = 0.78$. Rounding this value to the nearest value in Fig. 9.1, we obtain $z = 0.8$ and find that 29% of the test scores will be *between* $z = 0$ and $z = 0.8$. Since 50% of the test scores are *above* $z = 0$, the difference,

$$50\% - 29\% = 21\%,$$

will be above $z = 0.8$ or a test score of 85. Actually, out of 11 students, 2 students, or 18%, had test scores of 85 or more.

Frequently, we use a normal probability distribution to estimate certain quantities when we lack sufficient data.

Example 6   *Public Health.* Assume that the heights of male college students are distributed according to a normal distribution with a mean of 5′9″ and a standard deviation of 3″. How many students in a freshman class of 1800 (males) may be expected to be under 5′4″ tall?

Solution   A height of 5′4″ corresponds to

$$z = \frac{5'4'' - 5'9''}{3''} = \frac{-5''}{3''} = -1.67.$$

From the normal distribution, we see that 45% ($z = -1.6$) of the students will have a height between 5′4″ and 5′9″ and

$$50\% - 45\% = 5\%$$

will be shorter than 5′4″. Then 5% of 1800 students is

$$0.05 \cdot 1800 = 90 \text{ students.}$$

DO EXERCISE 31.

31. *Public Health.* Assuming that the heights of female college students have a normal probability distribution with a mean of 5′3″ and a standard deviation of 2.5″, how many students out of a freshman class of 2000 (female) students can be expected to be under 5′ tall?

EXERCISE SET 9.6

1. ▦ *Business, Racing.* Given the lap speeds:

   90.791,   89.237,   89.108,   87.926,   86.401,
   85.858,   83.271,   and   79.669 mph.

The mean is 86.533 mph (Exercise 1, Set 9.2) and the standard deviation is 3.396 mph (see Exercise 1, Set 9.3).

a) What value of the standardized random variable corresponds to 87.533 mph? to 85.533 mph?

b) What speed corresponds to $z = 0.5$? to $z = -0.5$?

c) What percent of the cars had lap speeds within 0.5 standard deviation of the mean?

d) What percent of the cars, assuming a normal-probability distribution, would be expected to have lap speeds within 0.5 standard deviation of the mean?

2. ▦ *Business, Racing.* Given the lap speeds:

91.101, 90.973, 89.257, 86.118, 85.879,

82.438, 81.962, 78.113, and 78.104 mph.

a) What is the mean speed? (Cf. Exercise 2, Set 9.2.)
b) What is the standard deviation? (See Exercise 2, Set 9.3.)
c) What value of the standardized random variable corresponds to 84 mph? to 85 mph?
d) What speed corresponds to $z = 0.5$? to $z = -0.5$?
e) What percent of the cars had lap speeds within 0.5 standard deviation of the mean?
f) What percent of the cars, assuming a normal probability distribution, would be expected to have lap speeds within 0.5 standard deviation of the mean?

4. *Business.* If transistors have a mean life of 525 days with a standard deviation of 90 days, what percent would be expected to last less than 360 days? If one started with 2000 transistors, how many would be expected to last less than 360 days?

6. *Business.* A movie theater that can seat 720 has a mean attendance of 490 and a standard deviation of 170.

a) If an attendance of 300 is required to cover expenses, what percent of the time should they expect to cover expenses?
b) What percent of the time should they expect more to come than they can seat?

3. *Business.* If lightbulbs have a mean life of 270 days with a standard deviation of 30 days, what percent would be expected to last 360 days? If one started with 1000 lightbulbs, how many would be expected to last more than 360 days?

5. *Business.* A lecture series has a mean attendance of 670 and a standard deviation of 110.

a) If an attendance of at least 500 is necessary to pay expenses, what percent of the time would you expect them not to cover expenses?
b) If the hall can seat 825, what percent of the time would you expect not to be able to seat all those who come?

7. *Business, Agriculture.* A farmer's mean crop of soybeans is 1150 bushels with a standard deviation of 240 bushels.

a) What percent of the time should the farmer expect a crop of less than 1000 bushels?
b) What percent of the time should the farmer expect a crop of more than 1250 bushels?

CHAPTER 9 TEST

1. A multiple-choice test consists of 5 questions, each with a choice of 4 answers. If a student guesses answers at random, what is the probability function for the number of correct answers?

3. In question 2, what is the number of questions the student should expect to get correct?

5. Using the data of question 4, find the variance.

7. Using the data of question 4, what is the standardized random variable corresponding to $x = 1$?

2. A multiple-choice test consists of 5 questions each with a choice of 4 answers. If the probability is $\frac{3}{4}$ that a student knows an answer to a question, what is the probability that he gets at least 4 correct?

4. Given the following probability function, find the expected value.

x	0	1	2	3	4
p	$\frac{1}{9}$	$\frac{2}{9}$	$\frac{3}{9}$	$\frac{2}{9}$	$\frac{1}{9}$

6. Using the data of question 5, find the standard deviation.

8. A drunk has 5 keys on his key ring and is trying to unlock his front door. He tries a key at random, forgetting which keys he may have tried previously. What is the probability he opens the door on the third try?

PART IV

CHAPTER TEN

Markov Chains

OBJECTIVES

You should be able to:

a) Given a problem involving a Markov chain, draw the transition diagram, and find the transition matrix;
b) Given a matrix, decide whether it qualifies as a transition matrix;
c) Given an initial probability vector P_0 and a transition matrix T, find P_n either as

$$P_n = P_{n-1}T,$$

or

$$P_n = P_0 T^n.$$

10.1 TRANSITION MATRICES AND PROBABILITY VECTORS

A *Markov chain* is a sequence of experiments with certain features which we shall illustrate before presenting a formal definition.

Example 1 *Business, Marketing Surveys.* A child, looking back over the many ice cream cones he has eaten through the years, recalls that:

a) After he had eaten a vanilla cone, the probability was:

   i) 0 that he would pick vanilla next time,
   ii) $\frac{1}{2}$ that he would pick chocolate next time,
   iii) $\frac{1}{2}$ that he would pick strawberry next time;

b) After he had eaten a chocolate cone, the probability was:

   i) $\frac{1}{5}$ that he would pick vanilla next time,
   ii) $\frac{2}{5}$ that he would pick chocolate next time,
   iii) $\frac{2}{5}$ that he would pick strawberry next time;

c) After he had eaten a strawberry cone, the probability was:

   i) $\frac{1}{3}$ that he would pick vanilla next time,
   ii) 0 that he would pick chocolate next time,
   iii) $\frac{2}{3}$ that he would pick strawberry next time.

Assuming that the child's first ice cream cone is vanilla, draw the tree describing possible outcomes through his third ice cream cone.

Solution   The tree can be drawn in a straightforward manner.

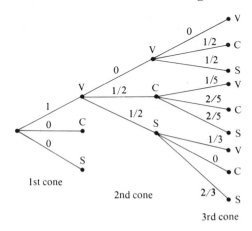

In view of the *repetitive* nature of the example, continuing the tree through further cycles (that is, the fourth and following ice cream cones) becomes increasingly awkward. Hence, it is useful to adopt an alternate way of representing these trials—that is, by means of a *transition diagram.*

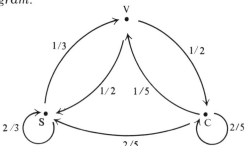

Here the directed line represents a *transition* from one *state* to another (here the state corresponds to a flavor). The number along the line corresponds to the probability that, if one starts in one state, next time he will be in the other state; that is, the probability of transition from one state to the other. The "transition" from one state at one stage to the *same* state at the next stage is represented by a "self" loop. The line is omitted where the transition probability is zero.

This transition diagram illustrates the general features of a *Markov chain:*

1. **The outcome of each experiment (or process, or choice) is one of a *set of discrete states* (a state is another name for an outcome).**
2. **The probability for transition from one state to another depends only on the present state (that is, the state one is in and is leaving).**

In the present example the states correspond to flavors. And since the next choice of flavor depends only on the previous choice, the whole process is a Markov chain.

DO EXERCISE 1.

Transition *diagrams* provide a *graphical* way of representing Markov chains. However, for computational purposes, *transition matrices* are more convenient.

We can define a transition matrix *T* by

$$T = [t_{ij}]_{n \times n},$$

a square matrix, where there are *n* states and t_{ij} represents the transition probability from state *i* to state *j*, so that $0 \le t_{ij} \le 1$. Note that the order of the indices *is* important. Since the object in question must be in one of the *n* states,

$$\sum_{j=1}^{n} t_{ij} = 1 \qquad \text{for all } i = 1, \ldots, n;$$

1. *Political Science.* Of voters sampled, 60% of the Democrats (that is, who voted Democrat in the last election) will vote Democrat in the next election, 20% will vote Republican, and 20% will vote Independent.

Of the Republicans, 40% will vote Democrat and 60% will vote Republican. Of the Independents, 40% will vote Democrat, 20% will vote Republican, and 40% will vote Independent.

a) Assuming that voters are split evenly among the three parties, draw a tree indicating voting patterns through one election.

b) Draw the transition diagram.

2. What is the transition matrix for the problem of Margin Exercise 1?

that is, **the sum of the *row* entries must be 1. There is no corresponding restriction for *column* entries.**

Example 2   What is the transition matrix for the problem of Example 1?

Solution   Let vanilla be state 1, chocolate be state 2, and strawberry be state 3. Then

$$T = \begin{matrix} & \begin{matrix} \text{V} & \text{C} & \text{S} \end{matrix} \\ \begin{bmatrix} 0 & \frac{1}{2} & \frac{1}{2} \\ \frac{1}{5} & \frac{2}{5} & \frac{2}{5} \\ \frac{1}{3} & 0 & \frac{2}{3} \end{bmatrix} & \begin{matrix} \text{V} \\ \text{C} \\ \text{S} \end{matrix} \end{matrix}$$

Note that the *row* elements do sum to 1.

DO EXERCISE 2.

Example 3   Given the following matrix, determine whether it qualifies as a transition matrix. If it does, draw the corresponding transition diagram.

$$T = \begin{matrix} & \begin{matrix} 1 & 2 & 3 & 4 \end{matrix} \\ \begin{bmatrix} \frac{1}{2} & \frac{1}{2} & 0 & 0 \\ \frac{1}{2} & 0 & \frac{1}{2} & 0 \\ 0 & \frac{1}{3} & \frac{1}{3} & \frac{1}{3} \\ \frac{1}{2} & 0 & 0 & \frac{1}{2} \end{bmatrix} & \begin{matrix} 1 \\ 2 \\ 3 \\ 4 \end{matrix} \end{matrix}$$

Solution

i) Since the elements are all nonnegative and the row elements all sum to one, we have a transition matrix and can proceed.

ii) Labelling the states 1, 2, 3, and 4, we have:

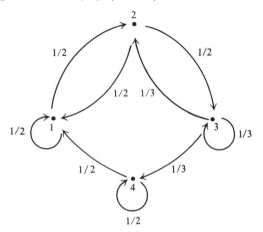

Note that the various states can be located where convenient or advantageous, so that pictorially the transition diagram may not be unique.

DO EXERCISE 3.

Transition matrices are useful in determining the probability of being in various states at later stages.

Example 4 For Example 1, determine the probability that the child will choose each of the different flavors for the second cone; for the third cone.

Solution Let us solve this problem first by using the tree diagram, then using the transition matrix.

Method 1 It is convenient to describe the initial state of the system by a probability *vector* (here a row matrix),

$$P_0 = [p_1 \quad p_2 \cdots p_n],$$

where p_i $(i = 1, \ldots, n)$ is the probability of being in state i at that stage. Thus, $0 \leq p_i \leq 1$, because p_i is a probability and $\sum_{i=1}^{n} p_i = 1$, since the n states exhaust the possibilities. In the present case, the initial probability vector is

$$P_0 = [1 \quad 0 \quad 0],$$

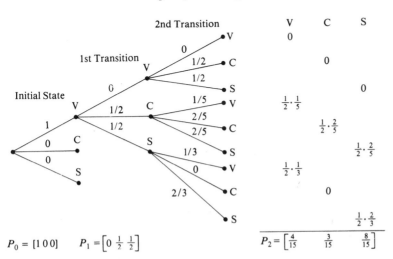

$$P_0 = [1\,0\,0] \qquad P_1 = \left[0 \ \tfrac{1}{2} \ \tfrac{1}{2}\right] \qquad P_2 = \left[\tfrac{4}{15} \quad \tfrac{3}{15} \quad \tfrac{8}{15}\right]$$

since the child's first ice cream cone was vanilla (the first state) and there were three flavors. If the child were equally likely to have

3. For each matrix, determine whether it qualifies as a transition matrix. If it does, draw the corresponding transition diagram.

a) $T = \begin{bmatrix} \frac{1}{3} & \frac{1}{3} & \frac{1}{3} & 0 \\ \frac{1}{2} & \frac{1}{2} & 0 & 0 \\ \frac{1}{3} & 0 & \frac{1}{3} & \frac{1}{3} \\ 0 & 0 & \frac{1}{2} & \frac{1}{2} \end{bmatrix}$

b) $T = \begin{bmatrix} \frac{1}{8} & \frac{2}{8} & 0 & \frac{5}{8} \\ \frac{1}{9} & \frac{2}{9} & 0 & \frac{5}{9} \\ \frac{1}{2} & 0 & 0 & \frac{1}{2} \\ \frac{1}{4} & \frac{3}{4} & 0 & 0 \end{bmatrix}$

picked any flavor, then the initial probability vector would have been

$$P_0 = [\tfrac{1}{3} \quad \tfrac{1}{3} \quad \tfrac{1}{3}].$$

Multiplying along branches of the tree and adding among the branches, we obtain the results of the *first* choice as the probability vector

$$P_1 = [0 \quad \tfrac{1}{2} \quad \tfrac{1}{2}]. \qquad \text{(See figure on page 397.)}$$

The components represent the probability that vanilla, chocolate, or strawberry will be chosen for the *second* cone, given that the *first* cone was vanilla.

Continuing along the branches of the tree for the *second* choice (that is, the third cone), we obtain

$$P_2 = [\tfrac{4}{15} \quad \tfrac{3}{15} \quad \tfrac{8}{15}].$$

The components represent the probability that vanilla, chocolate, or strawberry will be chosen for the *third* cone, given that the *first* cone was vanilla.

Method 2 As can be seen from the transition *diagram*, the same results can be obtained from the transition *matrix* in the following manner:

$$P_1 = P_0 T,$$
$$P_2 = P_1 T,$$
$$\cdot \qquad \cdot$$
$$\cdot \qquad \cdot$$
$$\cdot \qquad \cdot$$

Here,

$$P_1 = [1 \quad 0 \quad 0] \cdot \begin{bmatrix} 0 & \tfrac{1}{2} & \tfrac{1}{2} \\ \tfrac{1}{5} & \tfrac{2}{5} & \tfrac{2}{5} \\ \tfrac{1}{3} & 0 & \tfrac{2}{3} \end{bmatrix} = [0 \quad \tfrac{1}{2} \quad \tfrac{1}{2}],$$

$$P_2 = [0 \quad \tfrac{1}{2} \quad \tfrac{1}{2}] \cdot \begin{bmatrix} 0 & \tfrac{1}{2} & \tfrac{1}{2} \\ \tfrac{1}{5} & \tfrac{2}{5} & \tfrac{2}{5} \\ \tfrac{1}{3} & 0 & \tfrac{2}{3} \end{bmatrix} = [\tfrac{4}{15} \quad \tfrac{3}{15} \quad \tfrac{8}{15}],$$

$$\cdot$$
$$\cdot$$
$$\cdot$$

The numerical computation is the same in both cases, but the transition matrix facilitates the computation.

Alternately, we could substitute the expression for P_1 into the expression for P_2, obtaining:

$$P_2 = (P_0 T)T = P_0 T^2.$$

Thus, here we have

$$P_2 = \begin{bmatrix} 1 & 0 & 0 \end{bmatrix} \cdot \begin{bmatrix} 0 & \frac{1}{2} & \frac{1}{2} \\ \frac{1}{5} & \frac{2}{5} & \frac{2}{5} \\ \frac{1}{3} & 0 & \frac{2}{3} \end{bmatrix}^2 = \begin{bmatrix} 1 & 0 & 0 \end{bmatrix} \cdot \begin{bmatrix} \frac{4}{15} & \frac{3}{15} & \frac{8}{15} \\ \frac{32}{150} & \frac{39}{150} & \frac{79}{150} \\ \frac{4}{18} & \frac{3}{18} & \frac{11}{18} \end{bmatrix}$$

or

$$P_2 = \begin{bmatrix} \frac{4}{15} & \frac{3}{15} & \frac{8}{15} \end{bmatrix}, \qquad \text{as before.}$$

In general, we can obtain each probability vector from its predecessor,

$$P_n = P_{n-1}T;$$

or directly from the initial probability vector,

$$P_n = P_0 T^n.$$

DO EXERCISE 4.

4. For Margin Exercise 1, determine the probability vector for the first election using both a tree diagram and the transition matrix. Assuming the transition matrix does not change for the subsequent election, determine the probability vector for the second election from the transition matrix, in two ways.

EXERCISE SET 10.1

1. *Business.* A taxi company in a certain town has set up three zones. Taxis picking up a passenger in the first zone have 50% probability of delivering the passenger to that zone and are twice as likely to deliver a passenger to the second zone as to the third zone. A passenger picked up in the second zone will be let off there with a probability equal to that for being delivered to either other zone. A passenger picked up in the third zone is twice as likely to go to the first zone as either to go to the second zone or to stay in the third zone. Draw the transition diagram and find the transition matrix.

2. *Business, Marketing Surveys.* Of car owners surveyed, 60% of the VW owners would buy a VW for their next car while 20% each would buy a Ford or Chevy. Of the Ford owners, 30% would buy a Ford next time, 30% would buy VW, and 40% a Chevy. Of the Chevy owners, 40% would buy a Chevy, 40% a VW, and 20% a Ford. Draw the transition diagram and find the transition matrix.

For each matrix of Exercises 3 through 14, determine whether it qualifies as a transition matrix. If not, state why. If so, draw the transition diagram.

3. $\begin{bmatrix} \frac{1}{2} & -\frac{1}{8} & \frac{5}{8} \\ \frac{1}{3} & \frac{1}{3} & \frac{1}{3} \\ \frac{1}{5} & \frac{2}{5} & \frac{2}{5} \end{bmatrix}$

4. $\begin{bmatrix} \frac{2}{5} & \frac{2}{5} & \frac{1}{5} \\ 0 & 1 & 0 \\ 1 & 0 & 0 \end{bmatrix}$

5. $\begin{bmatrix} \frac{1}{2} & \frac{1}{2} & 0 & 0 \\ \frac{1}{2} & \frac{1}{2} & 0 & 0 \\ 0 & 0 & \frac{2}{3} & \frac{1}{3} \\ 0 & 0 & \frac{1}{3} & \frac{2}{3} \end{bmatrix}$

6. $\begin{bmatrix} \frac{1}{3} & \frac{2}{3} & 0 \\ \frac{1}{2} & \frac{3}{8} & \frac{3}{8} \\ 0 & \frac{2}{3} & \frac{1}{3} \end{bmatrix}$

7. $\begin{bmatrix} 1 & 0 \\ 0 & 1 \end{bmatrix}$

8. $\begin{bmatrix} 0 & 1 \\ 1 & 0 \end{bmatrix}$

9. $\begin{bmatrix} 1 & 0 & 0 \\ 0 & 1 & 0 \\ \frac{1}{2} & \frac{1}{2} & 0 \end{bmatrix}$

10. $\begin{bmatrix} 0 & 1 & 0 \\ 0 & 0 & 1 \\ 1 & 0 & 0 \end{bmatrix}$

11. $\begin{bmatrix} 0 & 1 & 0 \\ 0 & 0 & 1 \\ 0 & 0 & 1 \end{bmatrix}$

12. $\begin{bmatrix} \frac{1}{2} & \frac{1}{2} & 0 & 0 \\ 0 & \frac{1}{2} & \frac{1}{2} & 0 \\ 0 & 0 & \frac{1}{2} & \frac{1}{2} \\ 0 & 0 & \frac{1}{2} & \frac{1}{2} \end{bmatrix}$

13. $\begin{bmatrix} \frac{1}{2} & \frac{1}{2} & 0 & 0 \\ 0 & 1 & 0 & 0 \\ 0 & \frac{1}{3} & \frac{1}{3} & \frac{1}{3} \\ 0 & 0 & 0 & 1 \end{bmatrix}$

14. $\begin{bmatrix} 0 & 1 & 0 & 0 \\ \frac{1}{3} & \frac{1}{3} & \frac{1}{3} & 0 \\ 0 & \frac{1}{3} & \frac{1}{3} & \frac{1}{3} \\ 0 & 0 & 1 & 0 \end{bmatrix}$

15. *Business.* As in Exercise 1, a taxi starts in the second zone. Using a tree, determine its probable location after discharging its second passenger. What is the initial probability vector? Determine the probability vector after the second passenger, two ways.

16. *Business, Marketing Surveys.* As in Exercise 2, assume that initially car ownership is equally divided among VW, Ford, and Chevy. Using a tree, determine the probable ownership distribution for the second car; for the third car. What is the initial probability vector? Determine the probability vector for the second car, and for the third car, two ways.

17. As in Exercise 15, but the taxi starts in the *third* zone. Which method is less work?

18. As in Exercise 16, but the initial distribution is all VW's; is all Fords; is all Chevies.

In each of Exercises 19 through 35, given P_0 and T, determine P_n.

19. $P_0 = [1 \quad 0 \quad 0]$, T from Exercise 4, $P_2 = ?$

20. $P_0 = [1 \quad 0 \quad 0]$, T from Exercise 4, $P_3 = ?$

21. $P_0 = [\frac{1}{4} \quad \frac{1}{4} \quad \frac{1}{4} \quad \frac{1}{4}]$, T from Exercise 5, $P_2 = ?$

22. $P_0 = [\frac{1}{5} \quad \frac{4}{5}]$, T from Exercise 7, $P_5 = ?$

23. $P_0 = [\frac{1}{2} \quad \frac{1}{2}]$, T from Exercise 8, $P_1 = ?$

24. $P_0 = [1 \quad 0]$, T from Exercise 8, $P_1 = ?$

25. $P_0 = [1 \quad 0]$, T from Exercise 8, $P_2 = ?$

26. $P_0 = [0 \quad 0 \quad 1]$, T from Exercise 9, $P_1 = ?$

27. $P_0 = [0 \quad 0 \quad 1]$, T from Exercise 9, $P_2 = ?$

28. $P_0 = [\frac{1}{3} \quad \frac{1}{3} \quad \frac{1}{3}]$, T from Exercise 11, $P_2 = ?$

29. $P_0 = [\frac{1}{3} \quad \frac{1}{3} \quad \frac{1}{3}]$, T from Exercise 11, $P_3 = ?$

30. $P_0 = [\frac{1}{4} \quad \frac{1}{4} \quad \frac{1}{4} \quad \frac{1}{4}]$, T from Exercise 12, $P_1 = ?$

31. $P_0 = [\frac{1}{4} \quad \frac{1}{4} \quad \frac{1}{4} \quad \frac{1}{4}]$, T from Exercise 12, $P_2 = ?$

32. $P_0 = [\frac{1}{4} \quad \frac{1}{4} \quad \frac{1}{4} \quad \frac{1}{4}]$, T from Exercise 13, $P_1 = ?$

33. $P_0 = [\frac{1}{4} \quad \frac{1}{4} \quad \frac{1}{4} \quad \frac{1}{4}]$, T from Exercise 13, $P_2 = ?$

34. $P_0 = [\frac{1}{4} \quad \frac{1}{4} \quad \frac{1}{4} \quad \frac{1}{4}]$, T from Exercise 14, $P_1 = ?$

35. $P_0 = [\frac{1}{4} \quad \frac{1}{4} \quad \frac{1}{4} \quad \frac{1}{4}]$, T from Exercise 14, $P_2 = ?$

OBJECTIVES

You should be able to

a) Determine whether a transition matrix is regular.
b) Determine whether an ergodic transition matrix is regular.

10.2 REGULAR AND IRREGULAR MARKOV CHAINS

In order to determine the long-range characteristics of a Markov chain, we shall be interested in whether or not its transition matrix is *regular*.

A Markov chain is *regular* if its transition matrix is regular. A transition matrix

$$T = [t_{ij}]_{n \times n}$$

is regular if some power k of T has all positive elements; that is,

$$T^k = [t_{ij}^{(k)}]_{n \times n}$$

where

$$t_{ij}^{(k)} > 0 \qquad \text{for all } i, j.$$

Example 1 Determine whether the following transition matrix is regular and draw its transition diagram:

$$T = \begin{bmatrix} \frac{1}{3} & \frac{2}{3} & 0 \\ \frac{1}{2} & 0 & \frac{1}{2} \\ 0 & \frac{1}{3} & \frac{2}{3} \end{bmatrix}$$

Solution   The transition diagram is

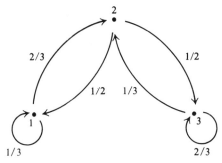

Since the first power of T contains zero elements, we proceed to square T and obtain:

$$T^2 = \begin{bmatrix} \frac{8}{18} & \frac{4}{18} & \frac{6}{18} \\ \frac{3}{18} & \frac{9}{18} & \frac{6}{18} \\ \frac{3}{18} & \frac{4}{18} & \frac{11}{18} \end{bmatrix}.$$

The elements of T^2 are all positive. Thus T is regular. If regularity had not been determined with T^2, then it is usually faster to obtain T^4, T^8, and so on, rather than T^3, T^4, and so on.

DO EXERCISE 5.

A Markov chain is *ergodic* if it is possible to go from each state to each other state. This might mean going to a state by first going to an intermediate state. If it is *not* possible to go from each state to each other state, then the Markov chain and its transition diagram are *not* ergodic and *not* regular. This frequently permits *irregularity* to be determined by inspection of the transition diagram.

Example 2 Determine from its transition diagram whether or not the following transition matrix is regular:

$$T = \begin{bmatrix} 1 & 0 & 0 \\ 0 & 1 & 0 \\ \frac{1}{3} & \frac{1}{3} & \frac{1}{3} \end{bmatrix}.$$

5. Determine whether the following transition matrix is regular and draw its transition diagram:

$$T = \begin{bmatrix} 0 & 1 & 0 \\ \frac{1}{3} & \frac{1}{3} & \frac{1}{3} \\ 0 & 1 & 0 \end{bmatrix}$$

6. Determine whether or not the following transition matrix is regular, using both the transition matrix and its transition diagram:

$$T = \begin{bmatrix} \frac{2}{5} & \frac{2}{5} & \frac{1}{5} \\ 0 & 1 & 0 \\ \frac{1}{4} & \frac{1}{4} & \frac{1}{2} \end{bmatrix}$$

Solution The transition diagram is

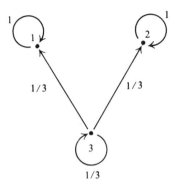

Since there is no arc leading either out of state 1 or out of state 2, it is not possible to leave either of these states once they are entered. Hence the chain is not ergodic and hence not regular.

Successive powers of T are:

$$T^2 = \begin{bmatrix} 1 & 0 & 0 \\ 0 & 1 & 0 \\ \frac{4}{9} & \frac{4}{9} & \frac{1}{9} \end{bmatrix}, \quad T^4 = \begin{bmatrix} 1 & 0 & 0 \\ 0 & 1 & 0 \\ \frac{40}{81} & \frac{40}{81} & \frac{1}{81} \end{bmatrix}, \quad \dots,$$

and further squaring will not eliminate the zeros, so T is not regular, as already determined from the transition diagram.

States that cannot be exited once entered are called *absorbing states*, and can be detected by a "1" in the corresponding location in the main diagonal (upper left to lower right) of the transition matrix.

In Example 2 there are 1's as the first two elements in the main diagonal of the transition matrix. Thus, states 1 and 2 are absorbing. Markov chains with absorbing states are not regular (since they are not ergodic).

DO EXERCISE 6.

If a Markov chain is ergodic (that is, it is possible to go from each state to each other state), then it is regular if there is at least one nonzero element on the main diagonal of the transition matrix.

Example 3 Determine whether the transition matrix of Example 1 is regular without using the power test.

Solution   From the transition diagram it can be seen that this Markov chain is ergodic. Since there is at least one (in this case, two) nonzero element(s) on the main diagonal of the transition matrix, the Markov chain is regular.

DO EXERCISE 7.

If a Markov chain is ergodic, then it is either regular or *periodic* (cyclic). In a periodic Markov chain, some sequence of transitions repeats regularly, as do some powers of the transition matrix. Thus, zeroes cannot be eliminated, so that periodic Markov chains are *not* regular. The distinction between regular and periodic Markov chains can be determined using the power test given at the beginning of this section.

If the main diagonal of an ergodic Markov chain contains *at least one* nonzero element, then it is regular. However, if the main diagonal of an ergodic Markov chain contains *no* zero element, it *may* still be regular. To determine whether it is regular or periodic, consider successive powers or squarings of the transition matrix. If this yields any *ones* on the main diagonal, the transition matrix is periodic but if there are no *ones* any nonzero elements less than one on the main diagonal, the transition matrix is regular.

Example 4   Determine whether the following transition matrix is regular and draw its transition diagram:

$$T = \begin{bmatrix} 0 & 1 & 0 \\ \frac{1}{2} & 0 & \frac{1}{2} \\ 0 & 1 & 0 \end{bmatrix}.$$

Solution   The transition diagram is:

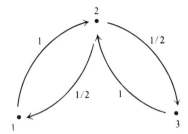

Here it is possible to go from each state to each other state, so that the transition matrix is ergodic. However, is *not* apparent whether or

7. Determine whether the transition matrix of Margin Exercise 5 is regular without using the power test.

8. Determine whether the following transition matrix is regular and draw its transition diagram:

$$T = \begin{bmatrix} 0 & \frac{1}{2} & \frac{1}{2} \\ \frac{1}{2} & 0 & \frac{1}{2} \\ \frac{1}{2} & \frac{1}{2} & 0 \end{bmatrix}.$$

not it is periodic. Hence we must resort to the power test. Squaring T, we obtain

$$T^2 = \begin{bmatrix} \frac{1}{2} & 0 & \frac{1}{2} \\ 0 & 1 & 0 \\ \frac{1}{2} & 0 & \frac{1}{2} \end{bmatrix}.$$

Further squaring yields

$$T^4 = \begin{bmatrix} \frac{1}{2} & 0 & \frac{1}{2} \\ 0 & 1 & 0 \\ \frac{1}{2} & 0 & \frac{1}{2} \end{bmatrix} = T^2.$$

Thus, the zeros cannot be eliminated and T is *not* regular. Also the presence of a one on the main diagonal of T^2 indicates that T is periodic rather than regular.

DO EXERCISE 8.

EXERCISE SET 10.2

For each of the transition matrices of Exercises 1 through 10, determine which are regular using, if possible, information along the main diagonal and the transition diagrams.

1. $\begin{bmatrix} \frac{2}{5} & \frac{2}{5} & \frac{1}{5} \\ 0 & 1 & 0 \\ 1 & 0 & 0 \end{bmatrix}$
2. $\begin{bmatrix} \frac{1}{2} & \frac{1}{2} & 0 & 0 \\ \frac{1}{2} & \frac{1}{2} & 0 & 0 \\ 0 & 0 & \frac{2}{3} & \frac{1}{3} \\ 0 & 0 & \frac{1}{3} & \frac{2}{3} \end{bmatrix}$
3. $\begin{bmatrix} 1 & 0 \\ 0 & 1 \end{bmatrix}$
4. $\begin{bmatrix} 0 & 1 \\ 1 & 0 \end{bmatrix}$
5. $\begin{bmatrix} 1 & 0 & 0 \\ 0 & 1 & 0 \\ \frac{1}{2} & \frac{1}{2} & 0 \end{bmatrix}$

6. $\begin{bmatrix} 0 & 1 & 0 \\ 0 & 0 & 1 \\ 1 & 0 & 0 \end{bmatrix}$
7. $\begin{bmatrix} 0 & 1 & 0 \\ 0 & 0 & 1 \\ 0 & 0 & 1 \end{bmatrix}$
8. $\begin{bmatrix} \frac{1}{2} & \frac{1}{2} & 0 & 0 \\ 0 & \frac{1}{2} & \frac{1}{2} & 0 \\ 0 & 0 & \frac{1}{2} & \frac{1}{2} \\ 0 & 0 & \frac{1}{2} & \frac{1}{2} \end{bmatrix}$
9. $\begin{bmatrix} \frac{1}{2} & \frac{1}{2} & 0 & 0 \\ 0 & 1 & 0 & 0 \\ 0 & \frac{1}{3} & \frac{1}{3} & \frac{1}{3} \\ 0 & 0 & 0 & 1 \end{bmatrix}$
10. $\begin{bmatrix} 0 & 1 & 0 & 0 \\ \frac{1}{3} & \frac{1}{3} & \frac{1}{3} & 0 \\ 0 & \frac{1}{3} & \frac{1}{3} & \frac{1}{3} \\ 0 & 0 & 1 & 0 \end{bmatrix}$

Determine which of the following ergodic transition matrices are regular using the power test if necessary.

11. $\begin{bmatrix} 0 & 1 \\ \frac{1}{2} & \frac{1}{2} \end{bmatrix}$
12. $\begin{bmatrix} 0 & 1 & 0 \\ \frac{2}{3} & 0 & \frac{1}{3} \\ \frac{1}{3} & \frac{2}{3} & 0 \end{bmatrix}$
13. $T = \begin{bmatrix} 0 & 1 & 0 \\ 0 & 0 & 1 \\ \frac{2}{5} & \frac{3}{5} & 0 \end{bmatrix}$
14. $T = \begin{bmatrix} 0 & 1 & 0 \\ 0 & \frac{2}{5} & \frac{3}{5} \\ 1 & 0 & 0 \end{bmatrix}$

15. $T = \begin{bmatrix} 0 & 0 & 1 \\ 0 & 0 & 1 \\ \frac{2}{5} & \frac{3}{5} & 0 \end{bmatrix}$
16. $T = \begin{bmatrix} \frac{1}{5} & \frac{2}{5} & \frac{2}{5} \\ \frac{1}{2} & 0 & \frac{1}{2} \\ \frac{1}{4} & \frac{1}{4} & \frac{1}{2} \end{bmatrix}$
17.* $T = \begin{bmatrix} 0 & \frac{2}{5} & 0 & \frac{3}{5} \\ 0 & 0 & 0 & 1 \\ 1 & 0 & 0 & 0 \\ 0 & 0 & 1 & 0 \end{bmatrix}$
18. $T = \begin{bmatrix} 0 & \frac{1}{3} & \frac{2}{3} & 0 \\ 0 & 0 & 1 & 0 \\ 0 & 0 & 0 & 1 \\ \frac{2}{3} & \frac{1}{3} & 0 & 0 \end{bmatrix}$

---

* This problem is of special interest because there are four states and only five nonzero elements.

19. $T = \begin{bmatrix} 0 & 0 & 0 & 1 \\ \frac{1}{2} & 0 & \frac{1}{2} & 0 \\ 0 & \frac{1}{2} & 0 & \frac{1}{2} \\ 0 & 0 & 1 & 0 \end{bmatrix}$ **20.** $\begin{bmatrix} 0 & 0 & \frac{3}{4} & \frac{1}{4} \\ \frac{1}{2} & 0 & 0 & \frac{1}{2} \\ \frac{1}{4} & \frac{3}{4} & 0 & 0 \\ 0 & \frac{1}{2} & \frac{1}{2} & 0 \end{bmatrix}$

10.3 FIXED POINTS AND "THE LONG RUN"

In Section 10.2, we sought to determine the regularity of transition matrices. Here we make use of their regularity to determine what happens in the long run.

Example 1 ▨ *Business, Marketing Surveys.* A travel agency recommends three vacation resorts: Acapulco, Bermuda, and the Caribbean. The probability that a vacationer who has been to one place will go to the same or another place is given by the following transition matrix (Example 1, Section 10.2):

$$T = \begin{matrix} & \begin{matrix} A & B & C \end{matrix} & \\ & \begin{bmatrix} \frac{1}{3} & \frac{2}{3} & 0 \\ \frac{1}{2} & 0 & \frac{1}{2} \\ 0 & \frac{1}{3} & \frac{2}{3} \end{bmatrix} & \begin{matrix} A \\ B \\ C \end{matrix} \end{matrix}$$

Assuming that one starts by going to Acapulco, what is the probability that one will be in any one of the three places after *many* visits?

Solution If one starts in Acapulco, then the initial probability vector is

$$P_0 = [1 \quad 0 \quad 0].$$

Successive probability vectors can be determined as in Section 10.1:

$$P_1 = P_0 T = [\tfrac{1}{3} \quad \tfrac{2}{3} \quad 0],$$
$$P_2 = P_1 T = [\tfrac{4}{9} \quad \tfrac{2}{9} \quad \tfrac{3}{9}],$$
$$P_3 = P_2 T = [\tfrac{7}{27} \quad \tfrac{11}{27} \quad \tfrac{9}{27}],$$
$$P_4 = P_3 T = [\tfrac{47}{162} \quad \tfrac{46}{162} \quad \tfrac{69}{162}], \quad \text{and so forth.}$$

Although it may not be apparent, this sequence of vectors is approaching some limit vector, although rather slowly. To speed up the calculation, let us consider *successive squarings* of T.

From Section 10.1,

$$P_n = P_0 T^n.$$

OBJECTIVES

You should be able to

a) Determine the fixed probability vector of a transition matrix;

b) Solve probability problems involving Markov Chains and fixed probability vectors.

Thus,

$$T^2 = \begin{bmatrix} \frac{8}{18} & \frac{4}{18} & \frac{6}{18} \\ \frac{3}{18} & \frac{9}{18} & \frac{6}{18} \\ \frac{3}{18} & \frac{4}{18} & \frac{11}{18} \end{bmatrix},$$

and

$$P_2 = P_0 T^2 = [\tfrac{4}{9} \quad \tfrac{2}{9} \quad \tfrac{3}{9}], \quad \text{as before.}$$

Continuing,

$$T^4 = \frac{1}{324} \begin{bmatrix} 94 & 92 & 138 \\ 69 & 117 & 138 \\ 69 & 92 & 163 \end{bmatrix},$$

so that $P_4 = P_0 T^4 = [\tfrac{47}{162} \quad \tfrac{46}{162} \quad \tfrac{69}{162}], \quad$ as before; and

$$T^8 = \frac{1}{104976} \begin{bmatrix} 24706 & 32108 & 48162 \\ 24081 & 32733 & 48162 \\ 24081 & 32108 & 48787 \end{bmatrix},$$

so that $P_8 = [24706 \quad 32108 \quad 48162]/104976$, and so forth.

This calculation is quite tedious, so let us seek an alternate method.

Note that, if we started with the initial probability vector

$$P_0 = [0 \quad 1 \quad 0],$$

then we would have obtained

$$P_8 = [24081 \quad 32733 \quad 48162]/104976.$$

Furthermore, if we had started with

$$P_0 = [0 \quad 0 \quad 1],$$

then

$$P_8 = [24081 \quad 32108 \quad 48787]/104976.$$

By comparing these three values of P_8, we find that, regardless where we start, we seem to be approaching the same probability vector.[*]

If these sequences tend to approach some limit vector, than after a while there should be little change from one transition to the next. This leads us to ask what some probability vector $\bar{P}$ must be such that it is the same after a transition as before. That is, we want to find $\bar{P}$ such that

$$\bar{P} = \bar{P}T.$$

---

[*] This is also true for any convex combination of these states.

Such a probability vector is called a *fixed probability vector,* or *fixed point.*

The equation $\bar{P} = \bar{P}T$ can be written $O = \bar{P}T - \bar{P}$, or, since matrix multiplication is distributive,

$$\bar{P}(T - I) = O$$

where I is the identity matrix of appropriate order.

This system of equations, $\bar{P}(T - I) = O$, is homogeneous (which we cannot pursue here in detail, but see Section 3.2) and hence does not have a unique (nontrivial) solution. The solution can be made unique, however, since the probability vector

$$\bar{P} = [\bar{p}_1 \quad \bar{p}_2 \quad \cdots \quad \bar{p}_n]$$

must have components which sum to 1; that is,

$$\sum_{i=1}^{n} \bar{p}_i = 1.$$

The system can now be solved using the echelon method of Chapter 3.

For Example 1, we have

$$\sum_{i=1}^{3} \bar{p}_i = \bar{p}_1 + \bar{p}_2 + \bar{p}_3 = 1,$$

and

$$\bar{P}(T - I) = O$$

or

$$[\bar{p}_1 \quad \bar{p}_2 \quad \bar{p}_3]\left(\begin{bmatrix} \frac{1}{3} & \frac{2}{3} & 0 \\ \frac{1}{2} & 0 & \frac{1}{2} \\ 0 & \frac{1}{3} & \frac{2}{3} \end{bmatrix} - \begin{bmatrix} 1 & 0 & 0 \\ 0 & 1 & 0 \\ 0 & 0 & 1 \end{bmatrix}\right) = [0 \quad 0 \quad 0],$$

so that

$$[\bar{p}_1 \quad \bar{p}_2 \quad \bar{p}_3]\begin{bmatrix} -\frac{2}{3} & \frac{2}{3} & 0 \\ \frac{1}{2} & -1 & \frac{1}{2} \\ 0 & \frac{1}{3} & -\frac{1}{3} \end{bmatrix} = [0 \quad 0 \quad 0].$$

Thus, the initial echelon tableau is

$\bar{p}_1$	$\bar{p}_2$	$\bar{p}_3$	1
1	1	1	1
$-\frac{2}{3}$	$\frac{1}{2}$	0	0
$\frac{2}{3}$	-1	$\frac{1}{3}$	0
0	$\frac{1}{2}$	$-\frac{1}{3}$	0

9. Convert P_8 and $\bar{P}$ to decimals, and compare their numerical values.

Note that the *columns* of $(T - I)$ become the *rows* of the echelon tableau. Solving, we obtain the final tableau.

$\bar{p}_1$	$\bar{p}_2$	$\bar{p}_3$	1
1	0	0	$\frac{3}{13}$
0	1	0	$\frac{4}{13}$
0	0	1	$\frac{6}{13}$
0	0	0	0

The last row of zeros is evidence of the linear dependence of the system $\bar{P}(T - I) = O$. The fixed probability vector is:

$$\bar{P} = \begin{bmatrix} \frac{3}{13} & \frac{4}{13} & \frac{6}{13} \end{bmatrix},$$

which does satisfy the fixed-point equation. Note that $\bar{P}$ is fairly close to P_8 previously obtained. (See Margin Exercise 9.)

DO EXERCISE 9.

10. Determine the fixed probability vector and long-range properties of the transition matrix (Margin Exercise 5 Section 10.2):

$$T = \begin{bmatrix} 0 & 1 & 0 \\ \frac{1}{3} & \frac{1}{3} & \frac{1}{3} \\ 0 & 1 & 0 \end{bmatrix}.$$

The calculation of the fixed probability vector $\bar{P}$ does not depend on the initial state. Thus, regardless of the initial probability vector, in the long run, after sufficient transitions of a *regular* matrix chain have occurred, the probability vector will approach a unique fixed probability vector.

Thus, the significance of *regularity* is that where one ends up in the long run does *not* depend upon where one started.

DO EXERCISE 10.

Example 2 Determine the fixed probability vector and long-range properties of the transition matrix (Example 4, Section 10.2):

$$T = \begin{bmatrix} 0 & 1 & 0 \\ \frac{1}{2} & 0 & \frac{1}{2} \\ 0 & 1 & 0 \end{bmatrix}.$$

Solution This transition matrix is ergodic and hence has a unique fixed probability vector, which we find to be

$$\bar{P} = \begin{bmatrix} \frac{1}{4} & \frac{1}{2} & \frac{1}{4} \end{bmatrix}.$$

Consider now successive powers of the transition matrix T:

$$T = T^3 = T^5 = \cdots = \begin{bmatrix} 0 & 1 & 0 \\ \frac{1}{2} & 0 & \frac{1}{2} \\ 0 & 1 & 0 \end{bmatrix}$$

and

$$T^2 = T^4 = T^6 = \cdots = \begin{bmatrix} \frac{1}{2} & 0 & \frac{1}{2} \\ 0 & 1 & 0 \\ \frac{1}{2} & 0 & \frac{1}{2} \end{bmatrix}$$

This cyclic transition matrix does not exhibit the same long-range characteristics as does the *regular* transition matrix of Example 1.

Since an ergodic transition matrix which is not regular must be *periodic*, there is no "*long-run*" probability vector to end up with since the transitions keep on cycling even though a *fixed* probability vector exists.

DO EXERCISE 11.

Example 3   Determine the fixed probability vector and long-range properties of the transition matrix (Example 2, Section 10.2):

$$T = \begin{bmatrix} 1 & 0 & 0 \\ 0 & 1 & 0 \\ \frac{1}{3} & \frac{1}{3} & \frac{1}{3} \end{bmatrix}.$$

Solution   This transition matrix is neither regular nor ergodic. Since states 1 and 2 are absorbing states, once one enters either of these states, one cannot exit. Thus, by inspection,

$$\bar{P}_1 = [1 \quad 0 \quad 0] \quad \text{and} \quad \bar{P}_2 = [0 \quad 1 \quad 0]$$

must each be fixed probability vectors, as can be verified by evaluating the product $\bar{P}_1 T$ obtaining $\bar{P}_1$ and by evaluating the product $\bar{P}_2 T$ obtaining $\bar{P}_2$.*

For nonergodic transition matrixes, the fixed probability vector may *not* be *unique* and hence where one ends up in the "long run" *may* depend on where one started.

***Regularity* of the transition matrix ensures that the "long run" *will* correspond to the fixed probability vector.**

11. Attempt to determine the fixed probability vector and long range properties of the (ergodic) transition matrix

$$T = \begin{bmatrix} 0 & \frac{2}{3} & \frac{1}{3} \\ 1 & 0 & 0 \\ 1 & 0 & 0 \end{bmatrix}$$

Why can no "long-run" probability vector be found?

EXERCISE SET 10.3

Determine the fixed probability vector for each transition matrix of Exercises 1 through 10 (Exercises 11 through 20, Set 10.2).

1. $T = \begin{bmatrix} 0 & 1 \\ \frac{1}{2} & \frac{1}{2} \end{bmatrix}$ **2.** $T = \begin{bmatrix} 0 & 1 & 0 \\ \frac{2}{3} & 0 & \frac{1}{3} \\ \frac{1}{3} & \frac{2}{3} & 0 \end{bmatrix}$ **3.** $T = \begin{bmatrix} 0 & 1 & 0 \\ 0 & 0 & 1 \\ \frac{2}{5} & \frac{3}{5} & 0 \end{bmatrix}$ **4.** $T = \begin{bmatrix} 0 & 1 & 0 \\ 0 & \frac{2}{3} & \frac{3}{5} \\ 1 & 0 & 0 \end{bmatrix}$ **5.** $T = \begin{bmatrix} 0 & 0 & 1 \\ 0 & 0 & 1 \\ \frac{2}{5} & \frac{3}{5} & 0 \end{bmatrix}$

---

* Furthermore, any convex combination of $\bar{P}_1$ and $\bar{P}_2$ is also a fixed probability vector.

6. $T = \begin{bmatrix} \frac{1}{5} & \frac{2}{5} & \frac{2}{5} \\ \frac{1}{2} & 0 & \frac{1}{2} \\ \frac{1}{4} & \frac{1}{4} & \frac{1}{2} \end{bmatrix}$ **7.** $T = \begin{bmatrix} 0 & \frac{2}{5} & 0 & \frac{3}{5} \\ 0 & 0 & 0 & 1 \\ 1 & 0 & 0 & 0 \\ 0 & 0 & 1 & 0 \end{bmatrix}$ **8.** $T = \begin{bmatrix} 0 & \frac{1}{3} & \frac{2}{3} & 0 \\ 0 & 0 & 1 & 0 \\ 0 & 0 & 0 & 1 \\ \frac{2}{3} & \frac{1}{3} & 0 & 0 \end{bmatrix}$ **9.** $T = \begin{bmatrix} 0 & 0 & 0 & 1 \\ \frac{1}{2} & 0 & \frac{1}{2} & 0 \\ 0 & \frac{1}{2} & 0 & \frac{1}{2} \\ 0 & 0 & 1 & 0 \end{bmatrix}$ **10.** $T = \begin{bmatrix} 0 & 0 & \frac{3}{4} & \frac{1}{4} \\ \frac{1}{2} & 0 & 0 & \frac{1}{2} \\ \frac{1}{4} & \frac{3}{4} & 0 & 0 \\ 0 & \frac{1}{2} & \frac{1}{2} & 0 \end{bmatrix}$

11. Using the data from Exercise 1, Set 10.1, determine the "long-run" distribution (fixed-point probability vector) of taxi location. By successive squaring, show that each row of T^m is approaching the fixed point.

12. Using the data from Exercise 2, Set 10.1, determine the "long-run" distribution of car ownership. By successive squaring, show that each row of T^m is approaching the fixed point.

13. Consider the "main" street of a town with cross streets numbered 1, 2, 3, 4, and 5. A drunk is standing at the intersection of 3rd and Main. He flips a fair coin. If heads shows, he walks one block toward his home at 5th and Main, while if tails shows, he walks one block toward the bar at 1st and Main. At each intersection he repeats this procedure. If he reaches either his home or the bar he stays there.

Find the transition matrix and the transition diagram. By inspection of these, determine *two* fixed points of the transition matrix. What is the initial probability vector? Determine successive values of the probability vector and find how this is related to the fixed points. What is the probability that the drunk eventually (that is, in the long run) reaches home?

15. As in Exercise 14, but the bar is closed, so that, after reaching the bar, the next step is *always* to walk a block toward home (this corresponds to a *reflecting* barrier).

14. As in Exercise 13, but:

i) if the drunk gets home, he stays there, as before (this corresponds to an *absorbing* barrier);
ii) but if the drunk gets to the bar, he again flips a coin: Heads, he walks one block toward home, and tails, he stays a while and has a drink (this corresponds to a *retaining* barrier).

Find the transition matrix and transition diagram. By inspection of these, determine the fixed point. Determine successive values of the probability vector and find how this is related to the fixed point. What is the probability that the drunk gets home?

16. Two gamblers, one with \$2 and the other with \$3, flip a fair coin. If tails shows, the first pays the second \$1 and if heads shows, the second pays the first \$1. The game ends when either becomes broke.

Set up the transition matrix and draw the transition diagram. Is the transition matrix regular? Explain. How much should the first gambler expect to have after 3 flips of the coin? after 4 flips? Note that the probability is 60% that the first gambler will go broke and 40% that the second one will go broke.

17. *Psychology.* A mouse is placed in the maze in the accompanying figure. There is an equal probability that it will pass through each door to an adjoining room. Find the transition matrix and the fixed point. What happens in the long run?

CHAPTER 10 TEST

1. In a certain city a review of electoral records indicated that in an election in which the incumbent was:

a) Democratic, the probability was:

    i) 0.4 that the next mayor would be Democratic,
    ii) 0.3 that the next mayor would be Republican,
    iii) 0.3 that the next mayor would be Independent;

b) Republican, the probability was:

    i) 0.4 that the next mayor would be Democratic,
    ii) 0.6 that the next mayor would be Republican,
    iii)   0 that the next mayor would be Independent;

c) Independent, the probability was:

    i) 0.3 that the next mayor would be Democratic,
    ii)   0 that the next mayor would be Republican,
    iii) 0.7 that the next mayor would be Independent.

Assuming that the mayor is originally a Republican, draw the tree describing the possible outcomes. Draw the transition diagram and set up the transition matrix.

3. Given

$$P_0 = [1 \quad 0 \quad 0] \quad \text{and} \quad T = \begin{bmatrix} \frac{1}{2} & \frac{1}{2} & 0 \\ 0 & \frac{2}{3} & \frac{1}{3} \\ \frac{1}{4} & 0 & \frac{3}{4} \end{bmatrix},$$

find P_2 and the fixed probability vector.

2. Determine which of the following transition matrices are regular. Explain why.

a) $\begin{bmatrix} 0 & 0 & 1 \\ 0 & 1 & 0 \\ 1 & 0 & 0 \end{bmatrix}$

b) $\begin{bmatrix} 0 & 1 & 0 & 0 \\ 0 & 0 & 1 & 0 \\ 0 & 0 & 0 & 1 \\ 1 & 0 & 0 & 0 \end{bmatrix}$

c) $\begin{bmatrix} \frac{1}{2} & \frac{1}{2} & 0 & 0 \\ \frac{1}{3} & \frac{2}{3} & 0 & 0 \\ 0 & 0 & \frac{2}{5} & \frac{3}{5} \\ 0 & 0 & \frac{1}{4} & \frac{3}{4} \end{bmatrix}$

d) $\begin{bmatrix} 0 & 1 & 0 \\ 0 & 0 & 1 \\ \frac{1}{2} & \frac{1}{2} & 0 \end{bmatrix}$

CHAPTER ELEVEN

Games and Decisions

OBJECTIVE

You should be able to find the pure-strategy solution of a matrix game.

1. Partition your class into student pairs with each student of a pair representing a prisoner. Then, *without any discussion*, let each student write his decision on a slip of paper. The instructor can then compile the results.

11.1 PURE STRATEGIES AND MATRIX GAMES

Many situations involving two or more people can be represented as *games* whose solutions aid in *decision*-making.

Formulation of Games as Matrices

Example 1   *The Prisoner's Dilemma. (Psychology)* A district attorney has two prisoners whom he suspects of a big robbery. Lacking evidence, he puts each in a separate isolation cell and makes each the following proposition: "If neither of you confess, I will see that you both get 1 year on trumped up charges. If you both confess, you will both get 10 years. However, if just one of you confesses, that one will get off with just probation (0 years) while the other one will get 20 years. So you better confess before he does." Draw the tree for this game and convert to matrix form.

Before discussing this problem,

DO EXERCISE 1.

Solution   Let us first represent this situation using a tree diagram. Starting at node A, Prisoner I can decide to confess (C) or not to confess (C^c):

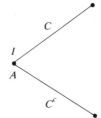

Prisoner II has the same choices. Thus, we write:

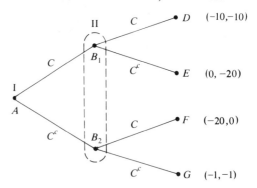

Here the dashed line about nodes B_1 and B_2 indicates that Prisoner II does *not* know Prisoner I's decision; that is, he doesn't know whether he is at node B_1 or B_2.

The nodes that a player cannot distinguish belong to the same *information set*.

The choices available to Prisoner II at either node B_1 or node B_2 (referred to simply as B) are to confess or not to confess. Prisoner I has the same two choices at node A. All of I's choices are represented by information set A, $\mathcal{I}_A = \{C, C^c\}$, and all of II's choices are represented by information set B, $\mathcal{I}_B = \{C, C^c\}$. The *pure strategies* for I are $S_I = \mathcal{I}_A$ and the *pure strategies* for II are $S_{II} = \mathcal{I}_B$.

In general, the pure strategies for I can be represented by $S_I = \{\alpha_1, \alpha_2, \ldots, \alpha_m\}$ and for II by $S_{II} = \{\beta_1, \beta_2, \ldots, \beta_n\}$. Corresponding to each pair of pure strategies $\alpha_i \beta_j$, there is a resultant payoff to each, denoted r_{ij}.

A *game* is a situation involving two or more players with rules leading to choices at various information sets and a payoff for each possible outcome.

So far we have represented a game by a tree. Now, in order to facilitate the solution, we represent the game by a matrix using pure strategies. Let I's pure strategy α_i correspond to the ith row of a matrix and II's pure strategy β_j correspond to the jth column. The payoff r_{ij} then corresponds to the ij-element of the payoff matrix R.

Thus, for the Prisoner's Dilemma example, we have

$$
\begin{array}{c}
\\
\\
I \quad \begin{matrix} \alpha_1 = C \\ \alpha_2 = C^c \end{matrix}
\end{array}
\overset{\displaystyle \text{II}}{
\overset{\displaystyle \beta_1 = C \qquad \beta_2 = C^c}{
\begin{bmatrix} (-10, -10) & (0, -20) \\ (-20, 0) & (-1, -1) \end{bmatrix}}}
$$

Here the resultant payoff is written as an ordered pair, each component (coordinate) being the payoff to a particular player. For example, $r_{11} = r_D = (-10, -10)$.

We will solve this game later in Example 4.

Pure strategies play a very important role in game theory. The Prisoner's Dilemma problem illustrates the determination of the pure strategies and their use in obtaining the matrix form of a game when each player has only *one* information set.

The following example illustrates the determination of the pure strategies and their use in obtaining the matrix form of the game when a player has *two or more* information sets.

Example 2 (Optional) *The Prisoner's Dilemma Compounded.* (*Psychology*). Starting with the situation of the Prisoner's Dilemma (Example 1), the assistant district attorney learns that at most one of the prisoners has confessed. Wanting to make a name for himself by getting *both* to confess, he says to each separately: "Your partner has confessed and you will receive a lighter sentence if you confess also. Consequently, I am giving you a chance to *change* your decision." Actually, the assistant district attorney does *not* know whether a particular prisoner has confessed or not and is thus giving each prisoner a chance to retract his confession. Each prisoner does know that the offer to change decisions would not have been made if *both* had confessed.

Draw the game tree and convert to matrix form.

Solution For the second decision, let H stand for "*hold* decision" and H^c stand for "do *not hold* decision" or "change decision." As in Example 1, we obtain the tree:

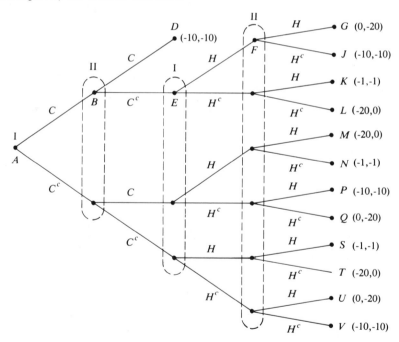

There are two steps in the conversion of a game to matrix form:

1. For each player, list all pure strategies:

$$S_I = \{\alpha_1, \alpha_2, \ldots, \alpha_m\},$$
$$S_{II} = \{\beta_1, \beta_2, \ldots, \beta_n\}.$$

2. For each pair of pure strategies $\alpha_i \beta_j$, determine the resultant payoff r_{ij}.

To determine the pure strategies, we first list the information sets of each:

Prisoner I's information sets are

$$\mathcal{I}_A = \{C, C^c\} \quad \text{and} \quad \mathcal{I}_E = \{H, H^c\}.$$

Similarly, Prisoner II's information sets are

$$\mathcal{I}_B = \{C, C^c\} \quad \text{and} \quad \mathcal{I}_F = \{H, H^c\}.$$

The set of pure strategies is a list of what each player should do in all conceivable situations.

All a player knows is the *choices available* at each of his information sets and the *final payoffs*. Thus, each pure strategy for a particular player can be specified by taking one choice from each of his information sets, so that all of his pure strategies are contained in the *Cartesian product* of his information sets.

Thus, for Prisoner I, the set of pure strategies is:

$$S_I = \mathcal{I}_A \times \mathcal{I}_E = \{C, C^c\} \times \{H, H^c\}$$
$$= \{(C, H), (C, H^c), (C^c, H), (C^c, H^c)\}$$
$$= \{\alpha_1, \alpha_2, \alpha_3, \alpha_4\};$$

and for prisoner II, the set of pure strategies is:

$$S_{II} = \mathcal{I}_B \times \mathcal{I}_F = \{C, C^c\} \times \{H, H^c\}$$
$$= \{(C, H), (C, H^c), (C^c, H), (C^c, H^c)\}$$
$$= \{\beta_1, \beta_2, \beta_3, \beta_4\}.$$

Next, using the tree, we identify the payoff r_{ij} corresponding to each $\alpha_i \beta_j$.

Consider the pure-strategy pair $\alpha_1 \beta_1 = (C, H), (C, H)$:

1. We start at node A, where Prisoner I has choices C or C^c.

2. $\alpha_1 = (C, H)$ says I chooses C, which leads to node B.

3. At node B, Prisoner II has choices C and C^c.

4. $\beta_1 = (C, H)$ says II chooses C, which leads to node D.

5. Node D is terminal, so that $\alpha_1 \beta_1$ leads to payoff $r_{11} = r_D = (-10, -10)$.

Consider now the pure-strategy pair $\alpha_1 \beta_3 = (C, H), (C^c, H)$:

1. We start at node A, where Prisoner I has choices C or C^c.

2. $\alpha_1 = (C, H)$ says I chooses C, which leads to node B.

3. At node B, Prisoner II has choices C or C^c.

4. $\beta_3 = (C^c, H)$ says II chooses C^c, which leads to node E.

5. At node E, Prisoner I has choices H or H^c.

6. $\alpha_1 = (C, H)$ says I chooses H, which leads to node F.

7. At node F, Prisoner II has choices H or H^c.

8. $\beta_3 = (C^c, H)$ says II chooses H, which leads to node G.

9. Node G is terminal, so that $\alpha_1 \beta_3$ leads to payoff $r_{13} = r_G = (0, -20)$.

Continuing in this manner until all pure-strategy pairs $\alpha_i \beta_j$ have been considered, we determine all the payoffs r_{ij} and obtain:

$$
\text{I} \quad
\begin{array}{c}
 \\
\alpha_1 = (C, H) \\
\alpha_2 = (C, H^c) \\
\alpha_3 = C^c, H) \\
\alpha_4 = (C^c, H^c)
\end{array}
\begin{array}{cccc}
\beta_1 = (C, H) & \beta_2 = (C, H^c) & \beta_3 = (C^c, H^c) & \beta_4 = (C^c, H^c) \\
\left[r_D = (-10, -10) \right. & r_D = (-10, -10) & r_G = (0, -20) & \left. r_J = (-10, -10) \right] \\
\left[r_D = (-10, -10) \right. & r_D = (-10, -10) & r_K = (-1, -1) & \left. r_L = (-20, 0) \right] \\
\left[r_M = (-20, 0) \right. & r_N = (-1, -1) & r_S = (-1, -1) & \left. r_T = (-20, 0) \right] \\
\left[r_P = (-10, -10) \right. & r_Q = (0, -20) & r_U = (0, -20) & \left. r_V = (-10, -10) \right]
\end{array}
$$

We will solve this game later in Example 6.

We will limit our consideration in this chapter mainly to *zero-sum* games. Then what one player wins, the other loses, and vice versa, so that the sum of their combined winnings is zero. Thus, for a zero-sum game, the payoff is simply the amount Player I *wins* from Player II. A negative payoff indicates the amount Player I *loses* to Player II. Note that the games in Examples 1 and 2 are *not* zero-sum.

Solution of Matrix Games

The basis for solution of zero-sum games is the principle of *individual rationality*. This states that each player will try to *maximize* the payoff to *himself*.

Example 3 Solve the following zero-sum game.

$$
\text{I} \quad
\begin{array}{c}
 \\
\alpha_1 \\
\alpha_2 \\
\alpha_3
\end{array}
\begin{array}{cccc}
\beta_1 & \beta_2 & \beta_3 & \beta_4 \\
\left[\begin{array}{cccc} 3 & 2 & 4 & 1 \\ 1 & 2 & 2 & 3 \\ 7 & 3 & 3 & 5 \end{array} \right]
\end{array}
$$

Solution   The first step is to try to identify those pure strategies that are unprofitable to a player. Positive payoffs represent amounts that player II must pay player I. Thus, player I wants to maximize what he wins and player II wants to minimize what he loses.

After a little searching, it is apparent that player II will *always* choose pure strategy β_2 in preference to pure strategy β_3, since whatever player I chooses (α_1, α_2, or α_3), player II's loss is less or no greater. That is, if player II chooses β_2 rather than β_3, then,

  i) if player I chooses α_1, then player II loses 2 rather than 4;
 ii) if player I chooses α_2, then player II loses 2 rather than 2;
iii) if player I chooses α_3, then player II loses 3 rather than 3.

Thus, the column pure strategy β_2 *dominates* the column pure strategy β_3.

A *column* strategy β_j dominates another *column* strategy β_i if:

a) Each element of the jth column is less than or equal to the rowwise corresponding element of the ith column and
b) At least one such element is less (rather than less than *or equal*).

Since β_2 dominates β_3, we can eliminate the β_3 column from consideration and obtain:

$$
\begin{array}{c}
 & & \text{II} \\
 & \begin{array}{cccc} \beta_1 & \beta_2 & \beta_3 & \beta_4 \end{array} \\
\text{I} \begin{array}{c} \alpha_1 \\ \alpha_2 \\ \alpha_3 \end{array} &
\left[\begin{array}{cccc}
3 & 2 & 4 & 1 \\
1 & 2 & 2 & 3 \\
7 & 3 & 3 & 5
\end{array} \right]
\end{array}
$$

Having done this, we now see that player I will *always* choose pure strategy α_3 in preference to pure strategy α_1, since whatever player II chooses (β_1, β_2, or β_4), player I's winnings are equal or greater. That is, if player I chooses α_3 rather than α_1, then,

  i) if player II chooses β_1, then player I wins 7 rather than 3,
 ii) if player II chooses β_2, then player I wins 3 rather than 2,
iii) if player II chooses β_4, then player I wins 5 rather than 1.

Thus, the row pure strategy α_3 *dominates* the row pure strategy α_1.

A *row* strategy α_j dominates another *row* strategy α_i if:

a) Each element of the jth row is greater than or equal to the columnwise corresponding element of the ith row and
b) At least one such element is greater (rather than greater than *or equal*).

2. Solve the following game.

$$
\text{I}\quad
\begin{array}{c}
 \\ \alpha_1 \\ \alpha_2 \\ \alpha_3 \\ \alpha_4
\end{array}
\begin{array}{c}
\text{II} \\
\begin{array}{ccc}
\beta_1 & \beta_2 & \beta_3
\end{array} \\
\left[\begin{array}{ccc}
4 & 2 & 8 \\
1 & 2 & 4 \\
5 & 4 & 5 \\
2 & 4 & 7
\end{array}\right]
\end{array}
$$

Since α_3 dominates α_1, we can eliminate the α_1 row from consideration and obtain:

In a similar manner, we find that α_3 dominates α_2, so that we obtain:

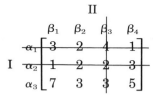

Player I is now reduced to the single pure strategy α_3. Knowing this, player II can now see that pure strategy β_2 dominates β_1 and β_4. Thus, we obtain:

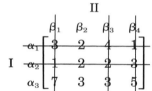

The game is now reduced to a single pure strategy α_3 for player I and a single pure strategy β_2 for player II.

The payoff that player I obtains from player II with this resulting optimum pair of pure strategies is called the *value* of the game, denoted by v; that is, $v = 3$ here.

The solution to this game is thus:

a) the optimum pure strategies, $\alpha_3\beta_2$;
b) the value of the game, $v = 3$.

DO EXERCISE 2.

The concept of dominance can sometimes be used to solve problems which are *not* zero sum.

Example 4 Solve the Prisoner's Dilemma problem (Example 1) which in matrix form is:

$$
\text{I}\quad
\begin{array}{c}
 \\ \alpha = C \\ \alpha_2 = C^c
\end{array}
\begin{array}{c}
\text{II} \\
\begin{array}{cc}
\beta_1 = C & \beta_2 = C^c
\end{array} \\
\left[\begin{array}{cc}
(-10, -10) & (0, -20) \\
(-20, 0) & (-1, -1)
\end{array}\right]
\end{array}
$$

Solution If the two prisoners could talk to each other, then it would be to their *mutual* advantage for *neither* to confess: $\alpha_2\beta_2 = (C^c, C^c)$. However, the prisoners are being held in separate isolation cells and *cannot* confer.

Writing the payoffs to each prisoner as a separate matrix, we have:

$$
\text{I} \quad \begin{array}{c} \\ \alpha_1 = C \\ \alpha_2 = C^c \end{array} \overset{\overset{\displaystyle \text{II}}{\beta_1 = C \quad \beta_2 = C^c}}{\begin{bmatrix} -10 & 0 \\ -20 & -1 \end{bmatrix}}
\qquad
\begin{array}{c} \\ \alpha_1 = C \\ \alpha_2 = C^c \end{array} \overset{\overset{\displaystyle \text{II}}{\beta_1 = C \quad \beta_2 = C^c}}{\begin{bmatrix} -10 & -20 \\ 0 & -1 \end{bmatrix}}
$$

$$\text{Payoff to I} \qquad\qquad\qquad \text{Payoff to II}$$

Looking at the matrix of the payoff to Prisoner I, we see that strategy α_1 dominates strategy α_2. Similarly, looking at the matrix of the payoff to Prisoner II, we see that strategy β_1 dominates strategy β_2. Thus, the solution is given by the strategy pair $\alpha_1\beta_1 = (C, C)$; that is, *both* confess.

If the prisoners could confer, it would be to their *mutual* advantage for them both *not* to confess. On the other hand, if they cannot confer, then it is to their *individual* advantage for them both to confess. The dilemma of this problem arises from the difference between these two solutions and psychological arguments about the logical basis for making decisions. While game theory says that both should confess, would this be psychologically satisfying to you as a prisoner knowing you would go to jail for 10 years?

DO EXERCISE 3.

Not all games can be reduced to a pair of pure strategies using the concept of dominance.

Example 5 Solve the following game.

$$
\text{I} \quad \begin{array}{c} \\ \alpha_1 \\ \alpha_2 \\ \alpha_3 \end{array} \overset{\overset{\displaystyle \text{II}}{\beta_1 \quad \beta_2 \quad \beta_3 \quad \beta_4}}{\begin{bmatrix} 7 & 2 & 1 & 4 \\ 4 & 3 & 4 & 5 \\ 1 & 2 & 8 & 3 \end{bmatrix}}
$$

Solution Here we can use dominance to eliminate pure strategy β_4, but that is as far as we can go toward the solution without using other techniques.

Now if player I chooses α_1, then the *minimum* he can win is 1 if

3. Compare the results of Margin Exercise 1 with the solution obtained in Example 4.

player II chooses β_3. We write this number to the right of the payoff matrix in the row corresponding to α_1. Thus,

			II			Row minimum	Maximum minimum
		β_1	β_2	β_3	β_4		
	α_1	7	2	1	4	1	
I	α_2	4	3	4	5	3	3
	α_3	1	2	8	3	1	
Column maximum		7	3	8	5		
Minimum maximum			3				

Continuing, if player I chooses α_2, then the *minimum* he can win is 3 if player II chooses β_2. If player I chooses α_3, then the *minimum* he can win is 1 if player II chooses β_1.

If player I now chooses the *maximum* of these *minima*, called the *maximin*, which is equal to 3, then player I can guarantee that, regardless of what player II does, he can win *at least* this maximum value of 3. This is player I's *security level*.

Now consider player II. Since the entries in the payoff matrix represent amounts that player II pays player I, player II wants to minimize his loss. Thus, if player II chooses β_1, then the *maximum* he can lose is 7 if player I chooses α_1. We write this number below the payoff matrix in the column corresponding to β_1.

Continuing, if player II chooses β_2, then the *maximum* he can lose is 3 if player I chooses α_2. If player II chooses β_3, then the *maximum* he can lose is 8 if player I chooses α_3. Also, if player II chooses β_4, then the *maximum* he can lose is 5 if player I chooses α_2. Here we have included β_4 even though it is a dominated strategy.

If player II now chooses the *minimum* of these *maxima*, called the *minimax*, which is equal to 3, then player II can *guarantee* that regardless of what player I does, he can lose *no more* than this minimax value of 3. This is player II's *security level*.

If the values of the maximin and minimax are *equal*, as they are here, then this is the value of the game and the corresponding pure strategies are optimum. Such strategies are said to be in *equilibrium* since neither player can gain by a *unilateral* change from them—that is, by one player changing his strategy while the other players do not. Thus, the solution to this game is

a) The optimum pure strategies, $\alpha_2\beta_2$;
b) The value of the game, $v = 3$.

A game with *zero* value v is called a fair game.

It is quite possible that there may be more than one equilibrium point for a given matrix game, although each will necessarily have the same game value. This causes no particular problem.

If the values of the maximin and minimax are *not* equal, then we must seek the solution in terms of mixed strategies, as in the next section.

Whether or not we had used dominance to eliminate β_4, we would have found the *same* equilibrium point. While dominance is not needed to find equilibrium points, it is useful in reducing the number of strategies that need be considered in the solution of a matrix game and will be useful in the next section.

DO EXERCISE 4.

Example 6 (Optional)   Solve the Prisoner's Dilemma Compounded Problem, Example 2, which in matrix form is:

$$
\begin{array}{cc}
 & \text{II} \\
\text{I} & \begin{array}{c|cccc}
 & \beta_1 = (C, H) & \beta_2 = (C, H^c) & \beta_3 = (C^c, H) & \beta_4 = (C^c, H^c) \\
\hline
\alpha_1 = (C, H) & (-10, -10) & (-10, -10) & (0, -20) & (-10, -10) \\
\alpha_2 = (C, H^c) & (-10, -10) & (-10, -10) & (-1, -1) & (-20, 0) \\
\alpha_3 = (C^c, H) & (-20, 0) & (-1, -1) & (-1, -1) & (-20, 0) \\
\alpha_4 = (C^c, H^c) & (-10, -10) & (0, -20) & (0, -20) & (-10, -10)
\end{array}
\end{array}
$$

Solution   We can obtain a solution using the concept of the equilibrium point. Let us start with the solution $\alpha_3\beta_3$ with payoff $(-1, -1)$ and see what happens if either prisoner changes his strategy unilaterally. If Prisoner I changes his strategy from α_3 to α_1 while Prisoner II holds with β_3, then we have $\alpha_1\beta_3$ with payoff $(0, -20)$, so that Prisoner I's sentence is *reduced* from 1 year to 0 years.

$\alpha_3\beta_3$ is *not* in equilibrium, since Prisoner I *increased* the payoff to himself by *reducing* his sentence from 1 to 0 years by unilaterally changing his strategy from α_3 to α_1.

Consider the strategy pair $\alpha_1\beta_3$ with payoff $(0, -20)$. If Prisoner II unilaterally changes his strategy from β_3 to β_1 (while Prisoner I holds with α_1), then we have $\alpha_1\beta_1$ with payoff $(-10, -10)$, so that Prisoner II has reduced his sentence from 20 to 10 years. Thus, $\alpha_1\beta_3$ is not in equilibrium.

Consider the strategy pair $\alpha_1\beta_1$, corresponding to *both* prisoners

4. Solve the following game by seeking the equilibrium solution.

$$
\begin{array}{cc}
 & \text{II} \\
\text{I} & \begin{array}{c|ccc}
 & \beta_1 & \beta_2 & \beta_3 \\
\hline
\alpha_1 & 7 & 2 & 1 \\
\alpha_2 & 3 & 4 & 4 \\
\alpha_3 & 3 & 1 & 8 \\
\alpha_4 & 6 & 5 & 7
\end{array}
\end{array}
$$

confessing and with payoff $(-10, -10)$. If *either* prisoner unilaterally changes his decision from "confess" to "not confess," then his sentence is *increased* from 10 to 20 years. Since an increase in sentence is a reduced payoff, this *is* an equilibrium-point solution.

Thus, we find that any strategy pair with payoff $(-10, -10)$ is an equilibrium-point solution:

$$\alpha_1\beta_1, \quad \alpha_1\beta_2, \quad \alpha_1\beta_4, \quad \alpha_2\beta_1, \quad \alpha_2\beta_2, \quad \alpha_4\beta_1, \quad \text{and} \quad \alpha_4\beta_4.$$

Each of these strategy pairs corresponds to a *final* decision of both prisoners to confess.

The assistant district attorney (being familiar with game theory) presents the arguments that we did in Example 4. Thus, it seems reasonable that he will have some measure of success in his attempt to get both prisoners to confess.

EXERCISE SET 11.1

Solve the following zero-sum matrix games for the optimum pure strategies and the value of the game. Player I chooses the row strategy α_i and player II chooses the column strategy β_j.

1. $\begin{bmatrix} 3 & 4 & 7 \\ 2 & 1 & 3 \\ 2 & 5 & 3 \end{bmatrix}$
2. $\begin{bmatrix} 2 & 4 & -1 & 5 \\ 3 & 7 & 2 & 4 \\ 5 & 6 & 8 & 9 \end{bmatrix}$
3. $\begin{bmatrix} 1 & -3 & 4 \\ 2 & 7 & 8 \\ -3 & 2 & 5 \\ 1 & 7 & 1 \end{bmatrix}$
4. $\begin{bmatrix} 7 & -8 & 1 & 2 \\ 2 & 5 & 5 & 4 \\ 9 & 7 & 8 & 5 \end{bmatrix}$

5. $\begin{bmatrix} 7 & 2 & 3 \\ 5 & 4 & 6 \\ 2 & 4 & 5 \\ 6 & 1 & 4 \end{bmatrix}$
6. $\begin{bmatrix} 5 & 3 & 3 & 5 \\ 3 & 2 & 7 & 1 \\ 2 & 3 & 8 & 0 \\ 6 & 1 & 0 & 5 \end{bmatrix}$
7. $\begin{bmatrix} 11 & 8 & 2 & 3 \\ 10 & 7 & 1 & 2 \\ 3 & 4 & 2 & 1 \\ 9 & 8 & 7 & 3 \end{bmatrix}$
8. $\begin{bmatrix} 3 & 4 & 11 & 9 & 7 \\ 6 & 4 & 8 & 3 & 2 \\ 8 & 5 & 7 & 10 & 6 \\ 9 & 2 & 7 & 3 & 4 \end{bmatrix}$

OBJECTIVE

You should be able to find the mixed strategies for a game that can be reduced to 2×2.

11.2 MIXED STRATEGIES: $m \times n$ AND 2×2 GAMES

Not all matrix games can be solved for an optimum strategy in terms of a pair of *pure* strategies.

Example 1 Solve the following matrix game.

$$\begin{array}{cc} & \text{II} \\ & \begin{array}{cc} \beta_1 & \beta_2 \end{array} \\ \text{I} \begin{array}{c} \alpha_1 \\ \alpha_2 \end{array} & \begin{bmatrix} 1 & 5 \\ 3 & 2 \end{bmatrix} \end{array}$$

Solution   It can be seen here that the number of strategies cannot be reduced using dominance. Furthermore, the maximin value is 2 while the minimax value is 3. Since these two values are *not equal*, we must seek a solution in terms of *mixed strategies*. A *mixed strategy* for a player is a *probability distribution* over his *pure* strategies.

Let

$$X = [x_1 \; x_2]$$

be a probability distribution over the pure strategies of player I:

$$S_{\mathrm{I}} = \{\alpha_1, \alpha_2\}.$$

Here x_i is the probability of choosing pure strategy α_i.

In general, $X = [x_1 \, x_2 \cdots x_i \cdots x_m]$ is the probability distribution over the pure strategies

$$S_{\mathrm{I}} = \{\alpha_1, \alpha_2, \ldots, \alpha_i, \ldots, \alpha_m\}.$$

Similarly, let

$$Y = [y_1 \; y_2]$$

be a probability distribution over the pure strategies of player II:

$$S_{\mathrm{II}} = \{\beta_1, \beta_2\}.$$

Here y_j is the probability of choosing pure strategy β_j. In general, $Y = [y_1 \, y_2 \cdots \dot{y}_j \cdots y_n]$ is the probability distribution over the pure strategies

$$S_{\mathrm{II}} = \{\beta_1, \beta_2, \ldots, \beta_j, \ldots, \beta_n\}.$$

For simplicity, we replace α_i and β_j by x_i and y_j and write:

$$
\begin{array}{c}
\mathrm{II} \\
\begin{array}{cc} y_1 & y_2 \end{array} \\
\mathrm{I} \begin{array}{c} x_1 \\ x_2 \end{array}
\begin{bmatrix} 1 & 5 \\ 3 & 2 \end{bmatrix}
\end{array}
$$

Player I seeks to determine X such that, if player II chooses β_1, that is, $Y = [1 \quad 0]$, then he (player I) gains a payoff of at least v_{I}; that is,

$$1x_1 + 3x_2 \geqslant v_{\mathrm{I}}.$$

Also, if player II chooses β_2, that is $Y = [0 \quad 1]$, then he (player I) again seeks to gain a payoff of at least v_{I}; that is,

$$5x_1 + 2x_2 \geqslant v_{\mathrm{I}}.$$

For any *convex combination** of player II's pure strategies β_1 and β_2, that is, $Y = [y_1 \ y_2]$,

$$(x_1 \cdot 1 + x_2 \cdot 3)y_1 + (x_1 \cdot 5 + x_2 \cdot 2)y_2 \geq v_I y_1 + v_I y_2 = v_I(y_1 + y_2) = v_I.$$

Thus, player I's minimum gain is still v_I. Note that since X and Y are probability distributions, we must have

$$x_1 + x_2 = y_1 + y_2 = 1.$$

The quantity v_I is the *minimum* amount that player I can win and he wants to *maximize* this value, so it is called the *maximin*.

Similarly, player II seeks to determine Y such that if player I chooses α_1, that is, $X = [1 \ \ 0]$, then he (player II) loses no more than v_{II}; that is,

$$1y_1 + 5y_2 \leq v_{II}.$$

Also, if player I chooses α_2, that is, $X = [0 \ \ 1]$, then he (player II) also loses no more than v_{II}; that is,

$$3y_1 + 2y_2 \leq v_{II}.$$

As before, for any convex combination of player I's pure strategies α_1 and α_2, that is, $X = [x_1 \ x_2]$,

$$x_1(1y_1 + 5y_2) + x_2(3y_1 + 2y_2) \leq v_{II}x_1 + v_{II}x_2 = v_{II}(x_1 + x_2) = v_{II}.$$

Thus, player II's maximum loss is v_{II}.

The quantity v_{II} is the *maximum* amount that player II can lose and he wants to *minimize* this value, so it is called the *minimax*.

The *von Neumann Minimax Theorem* states that† the inequality signs can be replaced by equal signs, so that $v_I = v_{II}$.

From a reasoning point of view, if the *strict* inequality held for some pure strategy of his opponent, then it would not be in the interest of his opponent to choose that pure strategy. Thus, the inequality signs can be replaced by equal signs. In either case, we obtain:

$$x_1 + 3x_2 = v, \tag{1}$$
$$5x_1 + 2x_2 = v, \tag{2}$$

and

$$y_1 + 5y_2 = v, \tag{3}$$
$$3y_1 + 2y_2 = v. \tag{4}$$

The value v of the game can be eliminated from each pair of

---

* Recall p. 197 for the definition of convex combination.

† There are many alternate formulations of the Minimax Theorem including an equivalence with the Duality Theorem of linear programming.

equations by subtraction, so that we obtain the following systems of equations:

$$x_1 + x_2 = 1, \qquad \text{(The sum of probability components must be 1.)}$$
$$4x_1 - x_2 = 0, \qquad \text{(Subtracting Eq. (1) from Eq. (2).)}$$

and

$$y_1 + y_2 = 1, \qquad \text{(The sum of probability components must be 1.)}$$
$$2y_1 - 3y_2 = 0. \qquad \text{(Subtracting Eq. (3) from Eq. (4).)}$$

Solving each of these systems, we obtain the optimum strategies:

$$X = [x_1 \quad x_2] = [\tfrac{1}{5} \quad \tfrac{4}{5}]$$

and

$$Y = [y_1 \quad y_2] = [\tfrac{3}{5} \quad \tfrac{2}{5}].$$

To obtain v, we substitute these values of x_1 and x_2 into Eq. (1) or into Eq. (2) or these values of y_1 and y_2 into Eq. (3) or into Eq. (4) and obtain $v = \tfrac{13}{5}$.

Thus, if player I adopts the strategy X, then he will win v independent of player II's strategy. If player II adopts the strategy Y, then he will lose v independent of player I's strategy. Only if *both* deviate from the optimum strategies can player I win more and player II lose more.

The optimum *mixed* strategy for player I, $X = [\tfrac{1}{5} \quad \tfrac{4}{5}]$, means that $\tfrac{1}{5}$ of the time he will choose *pure* strategy α_1 and $\tfrac{4}{5}$ of the time he will choose *pure* strategy α_2. However, if player II knows when player I is going to choose a particular pure strategy (regardless of the probability distribution), then player II may hold player I's winnings to less than v. Thus, in order for player I to win at least v, he must keep his opponent from knowing his choice. One way of doing this is for him to randomize his choices by using a spinning arrow pinned to a disk as illustrated below:

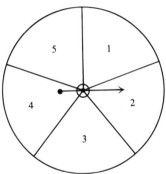

5. Solve the following matrix game for the optimum strategy for each player and the value of the game.

$$
\begin{array}{cc}
 & \text{II} \\
 & \begin{array}{cc} \beta_1 & \beta_2 \end{array} \\
\text{I} \begin{array}{c} \alpha_1 \\ \alpha_2 \end{array} & \begin{bmatrix} 7 & 2 \\ 1 & 4 \end{bmatrix}
\end{array}
$$

Here the circumference has been divided into 5 equal arcs. One of these is selected to represent x_1 (and α_1) and the other 4 to represent x_2 (and α_2). The arrow is spun by flipping with the finger. Where the pointer land determines which strategy is to be used. Why is this valid?

Similarly, a spinner can be used by player II, in this case the same spinner. Player II need only designate any 3 regions to represent y_1 (and β_1) and the remaining 2 to represent y_2 (and β_2).

How can a sweep second hand on a watch be used to randomize the pure strategies?

DO EXERCISE 5.

As we have solved 2×2 matrix games, so can we solve larger square-matrix games except that the amount of work involved is increased. Note that if the size of a matrix game can be reduced by eliminating dominated strategies, this should be done *before* setting up the equations for the optimum mixed strategies.

EXERCISE SET 11.2

Solve the following matrix games for the optimum strategy for each player and the value of the game. Player I chooses the row strategy α_i and player II chooses the column strategy β_j. (*Hint.* It may be necessary to use dominance to reduce these games to 2×2.)

1. $\begin{bmatrix} 1 & 3 \\ 4 & 2 \end{bmatrix}$ **2.** $\begin{bmatrix} 5 & 2 \\ 3 & 4 \end{bmatrix}$ **3.** $\begin{bmatrix} 2 & 4 & -1 & 5 \\ 3 & 7 & 2 & 4 \\ 2 & 3 & 8 & 9 \end{bmatrix}$ **4.** $\begin{bmatrix} 5 & 1 & 3 & 5 \\ 3 & 2 & 7 & -1 \\ 4 & 4 & 8 & 0 \\ 6 & 1 & 0 & 5 \end{bmatrix}$ **5.** $\begin{bmatrix} 1 & 7 & 2 & 9 \\ 1 & 8 & 4 & 7 \\ 2 & 4 & 7 & 3 \\ 6 & 6 & 1 & 7 \end{bmatrix}$ **6.** $\begin{bmatrix} 7 & -7 & 4 & -3 \\ 4 & 3 & -1 & 8 \\ 9 & -7 & 5 & -3 \\ 5 & 4 & -1 & 6 \end{bmatrix}$

7. *Business.* A new industrial plant is trying to minimize its costs by dumping its refuse rather than recycling it. Thus, they either truck the refuse to an isolated place in the country or dump it into the local stream. The government inspector can inspect only one site at a time and must choose between them. If the inspector catches them dumping far out in the country, there is a small fine; however, if he catches them dumping in the local stream, they will be put out of business. Representing this as a zero-sum game and assigning numerical values to various outcomes, we obtain:

$$
\begin{array}{cc}
 & \text{Inspector's visit} \\
 & \begin{array}{cc} \text{Far out} & \text{Local stream} \end{array}
\end{array}
$$

$$
\begin{array}{c}
\underline{\text{Plant's}} \\
\underline{\text{disposal}}
\end{array}
\begin{array}{c}
\text{Far out} \\
\text{Local stream}
\end{array}
\begin{bmatrix} -1 & 2 \\ 2 & -25 \end{bmatrix}
$$

What should the strategy of each be?

8. Two players individually and simultaneously show one or two fingers. If the numbers of fingers match, then player I wins from player II an amount equal to the *sum* of the number of fingers shown by each. Otherwise, player II wins from player I this sum. What should their strategies be? Is the game fair?

9. As in Exercise 7, what should the payoff be for *not* getting caught in order for the gain to be just worth the risk, that is, for the game to be fair ($v = 0$)?

10. As in Exercise 8, if *each* player chooses a $[\frac{1}{2} \quad \frac{1}{2}]$ strategy, what is the value of the game?

11. There are two poker players. At a certain point in the game the first player has a nothing hand and a choice of bluffing (that he holds good cards) or not bluffing. The other player can either call or not call. The payoffs can be represented by

$$
\begin{array}{cc}
 & \text{II} \\
 & \begin{array}{cc} C & C^c \end{array} \\
\text{I} \begin{array}{c} B \\ B^c \end{array} & \begin{bmatrix} -5 & 5 \\ 1 & 0 \end{bmatrix}
\end{array}
$$

What should each player do and what is the value of the game?

12. As in Exercise 8, if one player chooses the optimum strategy and the other player chooses a $[\frac{1}{2} \quad \frac{1}{2}]$ strategy, what is the value of the game?

11.3 MIXED STRATEGIES: $m \times 2$ and $2 \times n$ GAMES

OBJECTIVE

You should be able to solve $m \times 2$ and $2 \times n$ matrix games.

Matrix games with dimensions* $m \times 2$ and $2 \times n$ can be solved by considering all the possible 2×2 submatrix games.

Example 1 Solve the following matrix game.

$$
\begin{array}{cc}
 & \text{II} \\
 & \begin{array}{ccc} \beta_1 & \beta_2 & \beta_3 \end{array} \\
\text{I} \begin{array}{c} \alpha_1 \\ \alpha_2 \end{array} & \begin{bmatrix} 1 & 4 & 2 \\ 5 & 2 & 3 \end{bmatrix}
\end{array}
$$

Solution No strategy can be eliminated by dominance nor can a minimax solution (equilibrium point) be found using pure strategies. Thus, we proceed by writing:

$$
\begin{array}{cc}
 & \text{II} \\
 & \begin{array}{ccc} y_1 & y_2 & y_3 \end{array} \\
\text{I} \begin{array}{c} x_1 \\ x_2 \end{array} & \begin{bmatrix} 1 & 4 & 2 \\ 5 & 2 & 3 \end{bmatrix}
\end{array}
$$

The mixed strategy for player I must satisfy the *constraints*:

$$
\begin{aligned}
& x_1 + \quad x_2 = 1, \\
\beta_1: \quad & 1 \cdot x_1 + 5 \cdot x_2 \geqslant v, \\
\beta_2: \quad & 4 \cdot x_1 + 2 \cdot x_2 \geqslant v, \\
\beta_3: \quad & 2 \cdot x_1 + 3 \cdot x_2 \geqslant v;
\end{aligned}
$$

---
* Or those that can be reduced by dominance to $m \times 2$ or $2 \times n$.

and that for player II must satisfy

$$y_1 + \quad y_2 + \quad y_3 = 1,$$

$$\alpha_1: \quad 1 \cdot y_1 + 4 \cdot y_2 + 2 \cdot y_3 \leqslant v,$$

$$\alpha_2: \quad 5 \cdot y_1 + 2 \cdot y_2 + 3 \cdot y_3 \leqslant v.$$

Let us consider representing player I's constraints and player II's strategies graphically. For example, player II's pure strategy β_1 implies that player I's winnings, v, must satisfy the constraint:

$$\beta_1: \quad 1 \cdot x_1 + 5 \cdot x_2 \quad \geqslant v, \quad x_1 + x_2 = 1,$$

or

$$1 \cdot x_1 + 5 \cdot (1 - x_1) \geqslant v.$$

Graphing this constraint, we obtain:

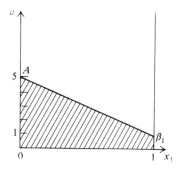

Thus, by choosing β_1, player II can hold player I's winnings to no more than the shaded region and player I can win a maximum of 5 at point A by choosing α_2 ($X = [0 \quad 1]$).

However, it is much more instructive to graph the constraints *two* at a time (for $m \times 2$ and $2 \times n$ games).

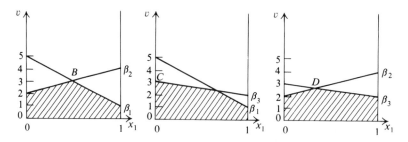

Let us solve each of these 2×2 games as in Section 10.2. Thus, taking β_1 and β_2 (that is, omitting β_3 or setting $y_3 = 0$), we have:

$$\begin{array}{c} \text{II} \\ \begin{array}{cc} y_1 & y_2 \end{array} \\ \text{I} \quad \begin{array}{c} x_1 \\ x_2 \end{array} \begin{bmatrix} 1 & 4 \\ 5 & 2 \end{bmatrix} \end{array}$$

with solution

$$X = [\tfrac{1}{2} \quad \tfrac{1}{2}],$$
$$Y = [\tfrac{1}{3} \quad \tfrac{2}{3} \quad 0],$$
$$v = 3,$$

indicated by point B in the *first* of the three preceding graphs. This represents the most that player I can win against β_1 and β_2. Player II can hold player I's winnings down to the shaded region and to this value.

Similarly, taking β_1 and β_3 (that is, omitting β_2 or setting $y_2 = 0$), we have:

$$\begin{array}{c} \text{II} \\ \begin{array}{cc} y_1 & y_3 \end{array} \\ \text{I} \quad \begin{array}{c} x_1 \\ x_2 \end{array} \begin{bmatrix} 1 & 2 \\ 5 & 3 \end{bmatrix} \end{array}$$

with solution (by dominance)

$$X = [0 \quad 1],$$
$$Y = [0 \quad 0 \quad 1],$$
$$v = 3,$$

indicated by point C in the *second* of the three preceding graphs. This represents the most that player I can win against β_1 and β_3. Player II can hold player I's winnings down to the shaded region and to this value.

Also, taking β_2 and β_3 (that is, omitting β_1 or setting $y_1 = 0$), we have:

$$\begin{array}{c} \text{II} \\ \begin{array}{cc} y_2 & y_3 \end{array} \\ \text{I} \quad \begin{array}{c} x_1 \\ x_2 \end{array} \begin{bmatrix} 4 & 2 \\ 2 & 3 \end{bmatrix} \end{array}$$

with solution

$$X = [\tfrac{1}{3} \quad \tfrac{2}{3}],$$
$$Y = [0 \quad \tfrac{1}{3} \quad \tfrac{2}{3}],$$
$$v = \tfrac{8}{3},$$

6. Solve the following matrix game both graphically and using 2×2 submatrix games.

$$
\begin{array}{c}
\quad\quad\quad \text{II} \\
\quad\quad \begin{array}{ccc} \beta_1 & \beta_2 & \beta_3 \end{array} \\
\text{I} \begin{array}{c} \alpha_1 \\ \alpha_2 \end{array}
\left[\begin{array}{ccc} 2 & 7 & 4 \\ 6 & 1 & 3 \end{array} \right]
\end{array}
$$

indicated by point D in the third of the three preceding graphs. This represents the most that player I can win against β_2 and β_3. Player II can hold player I's winnings down to the shaded region and to this value.

Now let us put all (three) constraints on one graph.

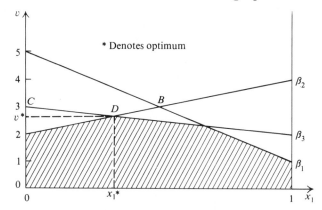

Thus, player II can hold player I's winnings *down* to the convex shaded region while player I can maximize his winnings by seeking the *highest* point in this region, that is, point D.

Alternately, we can examine the value of each possible 2×2 submatrix game. Since player II has the choice of which 2 pure strategies to use, he will choose those which *minimize* the value of all possible 2×2 submatrix games. As before, using β_2 and β_3 yields the minimum game value (corresponding to point D). Thus, the solution of the 2×3 game is

$$
\begin{aligned}
X &= \begin{bmatrix} \tfrac{1}{3} & \tfrac{2}{3} \end{bmatrix}, \\
Y &= \begin{bmatrix} 0 & \tfrac{1}{3} & \tfrac{2}{3} \end{bmatrix}, \\
v &= \tfrac{8}{3}.
\end{aligned}
$$

Note that the solution for any mixed strategy using all three of player II's pure strategies can be expressed as a convex combination of the 2×2 submatrix solutions and thus could not yield any further gain for player II.

In general, the solution to any $m \times 2$ or $2 \times n$ matrix game can be obtained by determining the optimum 2×2 submatrix game.

DO EXERCISE 6.

Now consider a game where player I has the choice of which strategies to omit.

Example 2 Solve the following matrix game both graphically and using 2×2 submatrix games.

$$
I \quad
\begin{array}{c}
 \\
 \\
\alpha_1 \\
\alpha_2 \\
\alpha_3 \\
\alpha_4
\end{array}
\begin{array}{c}
II \\
\begin{array}{cc}
\beta_1 & \beta_2
\end{array} \\
\begin{bmatrix}
2 & 5 \\
4 & 1 \\
3 & 4 \\
6 & 0
\end{bmatrix}
\end{array}
$$

Solution Alternately, we can write

$$
I \quad
\begin{array}{c}
 \\
 \\
x_1 \\
x_2 \\
x_3 \\
x_4
\end{array}
\begin{array}{c}
II \\
\begin{array}{cc}
y_1 & y_2
\end{array} \\
\begin{bmatrix}
2 & 5 \\
4 & 1 \\
3 & 4 \\
6 & 0
\end{bmatrix}
\end{array}
$$

Player II's mixed strategy must satisfy

$$
\begin{aligned}
\alpha_1: & \quad 2y_1 + 5y_2 \leq v, \\
\alpha_2: & \quad 4y_1 + y_2 \leq v, \\
\alpha_3: & \quad 3y_1 + 4y_2 \leq v, \\
\alpha_4: & \quad 6y_1 \leq v.
\end{aligned}
$$

These can all be graphed together, yielding

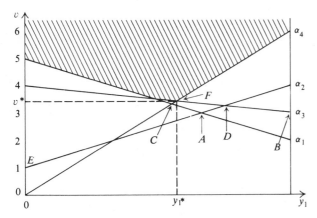

* Denotes optimum

Thus, player I can force player II's losses (player I's winnings) *up* to the convex shaded region while player II can *minimize* his losses by seeking the *lowest* point in this region, that is, point *F*.

Alternately, we can examine the value of each possible 2×2 submatrix game. Since player I has the choice of which 2 pure strategies to use, he will choose those which *maximize* the value of all possible 2×2 submatrix games.

Consider now the 2×2 submatrix games.

i)

$$
\begin{array}{c}
 \quad \text{II} \\
 \quad y_1 \quad y_2 \\
\text{I} \begin{array}{c} x_1 \\ x_2 \end{array} \begin{bmatrix} 2 & 5 \\ 4 & 1 \end{bmatrix}
\end{array}
\qquad
\begin{array}{l}
X = [\tfrac{1}{2} \quad \tfrac{1}{2} \quad 0 \quad 0], \\
v = 3. \quad \text{(See point } A.)
\end{array}
$$

ii)

$$
\begin{array}{c}
 \quad \text{II} \\
 \quad y_1 \quad y_2 \\
\text{I} \begin{array}{c} x_1 \\ x_3 \end{array} \begin{bmatrix} 2 & 5 \\ 3 & 4 \end{bmatrix}
\end{array}
\qquad
\begin{array}{l}
\text{By dominance} \\
X = [0 \quad 0 \quad 1 \quad 0], \\
Y = [1 \quad 0], \\
v = 3. \quad \text{(See Point } B.)
\end{array}
$$

iii)

$$
\begin{array}{c}
 \quad \text{II} \\
 \quad y_1 \quad y_2 \\
\text{I} \begin{array}{c} x_1 \\ x_4 \end{array} \begin{bmatrix} 2 & 5 \\ 6 & 0 \end{bmatrix}
\end{array}
\qquad
\begin{array}{l}
X = [\tfrac{2}{3} \quad 0 \quad 0 \quad \tfrac{1}{3}], \\
v = \tfrac{10}{3} = 3\tfrac{1}{3}. \quad \text{(See Point } C.)
\end{array}
$$

iv)

$$
\begin{array}{c}
 \quad \text{II} \\
 \quad y_1 \quad y_2 \\
\text{I} \begin{array}{c} x_2 \\ x_3 \end{array} \begin{bmatrix} 4 & 1 \\ 3 & 4 \end{bmatrix}
\end{array}
\qquad
\begin{array}{l}
X = [0 \quad \tfrac{1}{4} \quad \tfrac{3}{4} \quad 0], \\
v = \tfrac{13}{4} = 3\tfrac{1}{4}. \quad \text{(See point } D.)
\end{array}
$$

v)

$$
\begin{array}{c}
 \quad \text{II} \\
 \quad y_1 \quad y_2 \\
\text{I} \begin{array}{c} x_2 \\ x_4 \end{array} \begin{bmatrix} 4 & 1 \\ 6 & 0 \end{bmatrix}
\end{array}
\qquad
\begin{array}{l}
\text{By dominance} \\
X = [0 \quad 1 \quad 0 \quad 0], \\
Y = [0 \quad 1], \\
v = 1. \quad \text{(See point } E.)
\end{array}
$$

vi)

$$
\begin{array}{c}
 \quad \text{II} \\
 \quad y_1 \quad y_2 \\
\text{I} \begin{array}{c} x_3 \\ x_4 \end{array} \begin{bmatrix} 3 & 4 \\ 6 & 0 \end{bmatrix}
\end{array}
\qquad
\begin{array}{l}
X = [0 \quad 0 \quad \tfrac{6}{7} \quad \tfrac{1}{7}], \\
v = \tfrac{24}{7} = 3\tfrac{3}{7}. \quad \text{(See point } F.)
\end{array}
$$

Thus, to maximize his winnings, player I will choose to omit α_1 and α_2 as in subgame (vi). Solving for Y, we have

$$
X = [0 \quad 0 \quad \tfrac{6}{7} \quad \tfrac{1}{7}],
$$
$$
Y = [\tfrac{4}{7} \quad \tfrac{3}{7}],
$$
$$
v = \tfrac{24}{7}.
$$

Note that the solution for any mixed strategy using all four of player I's pure strategies can be expressed as a convex combination of the 2×2 submatrix solutions and thus could not yield a further gain for player I.

DO EXERCISE 7.

7. Solve the following matrix game both graphically and using 2×2 submatrix games.

$$\text{I}\quad \begin{array}{c} \\ \alpha_1 \\ \alpha_2 \\ \alpha_3 \end{array} \overset{\text{II}}{\begin{array}{cc} \beta_1 & \beta_2 \\ \hline \end{array}} \hspace{-3em}\begin{array}{c} \\ \begin{bmatrix} 7 & 3 \\ 1 & 4 \\ 0 & 5 \end{bmatrix} \end{array}$$

EXERCISE SET 11.3

Solve the following matrix games and illustrate the solution graphically. (*Hint.* Use dominance to reduce to $m \times 2$ or $2 \times n$ form, if necessary.)

1. $\begin{bmatrix} 1 & 4 & 7 & 5 \\ 2 & 1 & 8 & -1 \end{bmatrix}$

2. $\begin{bmatrix} 2 & -4 & 3 & -5 \\ -3 & 7 & -4 & 8 \end{bmatrix}$

3. $\begin{bmatrix} 3 & -2 \\ -1 & 5 \\ 6 & -7 \\ 2 & 3 \end{bmatrix}$

4. $\begin{bmatrix} 2 & 6 \\ 1 & 8 \\ 9 & 0 \\ 7 & 3 \end{bmatrix}$

5. $\begin{bmatrix} 2 & -8 & 1 & 7 \\ 4 & 5 & 3 & 2 \\ 3 & 7 & 8 & 9 \end{bmatrix}$

6. $\begin{bmatrix} 11 & 8 & 2 & 6 \\ 10 & 7 & 1 & 5 \\ 3 & 4 & 2 & 1 \\ 1 & 8 & 7 & 3 \end{bmatrix}$

7. $\begin{bmatrix} 3 & 7 & 2 & 9 \\ 1 & 8 & 4 & 7 \\ 2 & 4 & 7 & 3 \\ 4 & 5 & 1 & 6 \end{bmatrix}$

8. $\begin{bmatrix} 2 & -7 & 4 & -3 \\ 4 & -3 & 1 & 8 \\ 9 & -7 & 4 & -3 \\ 5 & 3 & -1 & 3 \end{bmatrix}$

9. *Business, Investing.* An investor is seeking to maximize the return on his money and has three possible types of investments: peace-industrial stocks, war-industrial stocks, and public utilities. The investor's opponent is the nebulous "force of circumstances" which may bring peace, war, or constant tension. The percent return to the investor is given by:

Investment	Peace	War	Tension
Peace-industrial stocks	7	2	4
War-industrial stocks	−2	18	8
Public utilities	6	2	2

Assuming that the investor has no control over the future, what would *his* strategy be?

10. *Business.* The buyer of a store is trying to decide how much of each of three lines of stock to buy: inexpensive, moderate, and expensive. He is trying to appeal to three groups of customers (low-income, moderate, and rich) without knowing how many of each he will have. If the number of sales is represented as the payoff of a matrix game, how much of each should the buyer stock to maximize his sales?

Lines	Customer's income levels: Low	Mod.	High
Inexp.	9	1	0
Mod.	11	9	2
Exp.	0	1	9

11.4 VOTING COALITIONS AND CHARACTERISTIC FORM

For games with more than 2 players, especially those for which we are more concerned with the formation of coalitions than with strategies, *characteristic* form is more convenient.

OBJECTIVE

You should be able to find the Shapley and Banzhaf values for the strength of a coalition.

Let $N = \{1, 2, \ldots, n\}$ be the set of n players and $S = \{1, 2, \ldots, s\} \subset N$ be any coalition of s of these players. Then the best that S can guarantee itself is obtained by considering its worst opposition, namely, the coalition $N - S$ of the remaining players. We now have a 2-person game of the coalition S versus the coalition $(N - S)$ with $v(S)$ the security level of the coalition S and $v(N - S)$ the security level of the coalition $(N - S)$. If the game is zero-sum, then $v(S) + v(N - S) = 0$, but this need not be the case.

How the quantity $v(S)$ is to be divided among the members of the coalition S is for *them* to decide and is not our concern here.

Since a coalition of *no* players cannot win anything, we can write

$$v(\emptyset) = 0.$$

Furthermore, we shall assume that the game is *proper;* that is, two disjoint cooperating coalitions can win at least as much together as separately. Then we may write:

$$v(S \cup T) \geqslant v(S) + v(T),$$

where $S \subset N$, $T \subset N$, and $S \cap T = \emptyset$.

If $v(S)$ is obtainable for each possible coalition $S \subset N$, then these $v(S)$ comprise the *characteristic function* form of a game.

The general solution of games in characteristic form is beyond the scope of this text. Instead, we consider two solutions for an "expected" value: the *Shapley* value and the *Banzhaf* value.

To derive the Shapley value, consider all possible $n!$ *arrangements* (permutations) of the n players. If each arrangement is equally probable, then each can be assigned a probability $1/n!$.

Suppose we let the players enter into coalitions randomly and focus our attention on those for which a particular player i is the *last* to form coalition S; that is, N is partitioned as follows:

$$N = \{\underbrace{j_1, j_2, \ldots, j_{s-1}}_{\substack{(s-1)\ \text{players} \\ \text{in} \\ \text{Coalition} \\ (S - \{i\})}},\ i,\ \underbrace{j_{s+1}, \ldots, j_n}_{(n-s)\ \text{players}}\}$$

$$\underbrace{\hspace{3cm}}_{\text{Coalition } S}\ \underbrace{\hspace{2.5cm}}_{\text{Coalition } (N - S)}$$

The first $(s - 1)$ players can enter the coalition S in $(s - 1)!$ ways.

Similarly, the last $(n - s)$ players can enter the coalition $(N - S)$ in $(n - s)!$ ways. Since the total number of possible orderings is $n!$, the probability that player i is the *last* member to enter the coalition S is:

$$p_n(s) = \frac{(s - 1)!(n - s)!}{n!}.$$

The contribution of player i to the coalition S is

$$v(S) - v(S - \{i\}).$$

Thus, the *expected* value of player i's contribution to coalition S as the last player to enter this coalition is:

$$\phi_i = \sum p_n(s)[v(S) - v(S - \{i\})] \qquad \text{for all } S \subset N \quad \text{and} \quad \text{for all } i \in S.$$

While this is a *formal* expression for ϕ_i, we will soon present a simple way to obtain ϕ_i. This is the *Shapley value* for player i. The Shapley value for all players is written

$$\phi = [\phi_1, \phi_2, \ldots, \phi_n],$$

where the components satisfy

$$\sum_{i=1}^{n} \phi_i = v(N).$$

Example 1 *Business, Political Science.* Four stockholders have, respectively, 5, 5, 3, and 2 shares of stock. Each share of stock carries with it one vote. Each stockholder votes his shares as a bloc. A proposition requires a simple majority (8 votes) to pass. Find the Shapley value for each stockholder.

Solution Games of this sort are called *voting* or *quota* games and are represented by

$$[q; w_1, w_2, \ldots, w_n].$$

Here there are n voters with a voting weight w_i (nonnegative) for each voter i. A coalition W is winning if the sum of the voting weights w_i of all voters i in the coalition W *is* $\geq q$. Here the quota q must be such that

$$q > \tfrac{1}{2}(\text{Sum of } all \text{ the voting weights } w_i) = \tfrac{1}{2}\sum_i w_i.$$

If a coalition S is winning,

$$v(S) = 1;$$

otherwise, it is losing and

$$v(S) = 0.$$

Under these conditions, player i contributes to a coalition only when

$$v(S) - v(S - \{i\}) = 1,$$

so that his Shapley value ϕ_i is the sum of $p_n(s)$ for all coalitions in which voter i is pivotal; that is, for each coalition in which voter i changes coalition S from winning to losing.

This example can be written as the quota game

$$[8; 5, 5, 3, 2].$$

If we let the various players enter into coalitions in all 24 possible arrangements and indicate with an asterisk when a player is *pivotal* in changing a winning coalition to a losing coalition, we have:

12*34	21*34	31*24	412*3
12*43	21*43	31*42	413*2
13*24	23*14	32*14	421*3
13*42	23*41	32*41	423*1
142*3	241*3	341*2	431*2
143*2	243*1	342*1	432*1

Thus, of the 24 possible arrangements, player 1 is pivotal in establishing the quota in 8, player 2 in 8, player 3 in 8, and player 4 in none. That is,

$$\phi_1 = \phi_2 = \phi_3 = \tfrac{8}{24} = \tfrac{1}{3} \quad \text{and} \quad \phi_4 = 0,$$

so that

$$\phi = [\tfrac{1}{3}, \tfrac{1}{3}, \tfrac{1}{3}, 0].$$

Note that player 3 with 3 votes is pivotal in the same number of coalitions as players 1 and 2 and thus has the same Shapley value.

Player 4 with zero Shapley value contributes nothing to *any* coalition. Such a player is called a *dummy*.

Alternately, we can obtain the Shapley value from the formula:

$$\phi_i = \sum p_n(s) \quad \text{for all coalitions in which voter } i \text{ is pivotal.}$$

For $s = 1$ or 4, there are *no* winning coalitions with player 1 pivotal. For $s = 2$, there are 2 winning coalitions ($S = \{2, 1\}$ and $S = \{3, 1\}$) with player 1 pivotal and for $s = 3$, there are also 2 winning coalitions ($S = \{2, 4, 1\}$ and $S = \{3, 4, 1\}$) with player 1 pivotal. Thus,

$$\phi_1 = 0\left(\frac{0!3!}{4!}\right) + 2\left(\frac{1!2!}{4!}\right) + 2\left(\frac{2!1!}{4!}\right) + 0\left(\frac{3!0!}{4!}\right)$$

$$= 0 + 2 \cdot \tfrac{1}{12} + 2 \cdot \tfrac{1}{12} + 0$$

$$= \tfrac{1}{3}, \quad \text{as before.}$$

The other ϕ_i can be obtained similarly.

Note that if the quota is raised from 8 to 9, then the quota game

$$[9; 5, 5, 3, 2]$$

has no dummies. (See Margin Exercise 8 following.)

DO EXERCISE 8.

The Shapley value has been derived various ways by different mathematicians. We have used the foregoing derivation to facilitate comparison between the Shapley value and the Banzhaf value. Both value concepts measure voting power although they sometimes yield somewhat different *numerical* values. This difference does not seem to be *qualitatively* significant. However, Banzhaf is a lawyer and his value has been used in various legal arguments.[*]

While Shapley considered the *ordered arrangement* of players entering a coalition, Banzhaf considered *unordered arrangements*.

Each voter can vote *yea* or *nay* on a particular issue, so that n voters can vote 2^n possible ways, each assumed equally probable and assuming no abstentions. Since we are interested in *how* a voter votes regardless of the *order* in which he casts his vote, we are dealing with combinations rather than permutations. A voter is considered *critical* if in a given combination, *his vote can reverse the outcome* of the total vote. The Banzhaf value for voter i, β_i, is the fraction of votes for which voter i is critical. For all voters, we write

$$\beta = [\beta_1, \beta_2, \ldots, \beta_n],$$

where

$$\sum_{i=1}^{n} \beta_i = 1.$$

8. Find the Shapley value for the quota game:

$$[9; 5, 5, 3, 2].$$

---

[*] See Lucas, W. F., "Measuring Power in Weighted Voting Systems," Technical Report No. 227, Department of Operations Research, Cornell University, Ithaca, N.Y., September 1974.

Example 2 *Business, Political Science.* Determine the Banzhaf value for the quota game

$$[8; 7, 4, 2, 1],$$

and compare the results with the Shapley value.

Solution To solve for the Banzhaf value, we construct a table which lists the n players across the top and, below, each possible voting combination for which the issue would *pass* (since for each *passing* combination a reversal of *all* votes would yield a *losing* combination). Then we place an asterisk on each critical vote. Note that when there is more than one asterisk in a row, each is to be interpreted independently. Thus, we obtain:

Player	1	2	3	4	Total
Number of votes	7	4	2	1	Quota = 8
Vote: Y or N	Y*	Y	Y	Y	1
	Y*	Y	Y	N	1
	Y*	Y	N	Y	1
	Y*	N	Y	Y	1
	Y*	Y*	N	N	2
	Y*	N	Y*	N	2
	Y*	N	N	Y*	2
Number critical	7	1	1	1	10

Thus,

$$\beta = [7, 1, 1, 1]/10 \quad \text{or} \quad [\tfrac{7}{10}, \tfrac{1}{10}, \tfrac{1}{10}, \tfrac{1}{10}].$$

Solving for the Shapley value, using the table as before, we have:

$$
\begin{array}{llll}
12^*34 & 21^*34 & 31^*24 & 41^*23 \\
12^*43 & 21^*43 & 31^*42 & 41^*32 \\
13^*24 & 231^*4 & 321^*4 & 421^*3 \\
13^*42 & 2341^* & 3241^* & 4231^* \\
14^*23 & 241^*3 & 341^*2 & 431^*2 \\
14^*32 & 2431^* & 3421^* & 4321^*
\end{array}
$$

so that

$$\phi = [9, 1, 1, 1]/12 \quad \text{or} \quad [\tfrac{9}{12}, \tfrac{1}{12}, \tfrac{1}{12}, \tfrac{1}{12}].$$

Alternately, we can use the formula and obtain:

$$\phi_1 = 0 + 3 \cdot \tfrac{2}{24} + 3 \cdot \tfrac{2}{24} + 1 \cdot \tfrac{6}{24} = \tfrac{18}{24},$$

$$\phi_2 = 0 + 1 \cdot \tfrac{2}{24} + 0 + 0 \qquad\quad = \tfrac{2}{24},$$

$$\phi_3 = 0 + 1 \cdot \tfrac{2}{24} + 0 + 0 \qquad\quad = \tfrac{2}{24},$$

$$\phi_4 = 0 + 1 \cdot \tfrac{2}{24} + 0 + 0 \qquad\quad = \tfrac{2}{24},$$

so that

$$\phi = [9, 1, 1, 1]/12, \quad \text{as before.}$$

DO EXERCISE 9.

9. Show that for the quota game of Example 1, the Banzhaf value is the *same* as the Shapley value.

EXERCISE SET 11.4

1. Find the Shapley and Banzhaf values for the quota game $[7; 6, 4, 2, 1]$. Show that they are identical in this case.

2. Find the Shapley and Banzhaf values for the quota game $[6; 4, 3, 2, 1]$. Show that they are identical in this case.

3. Find the Shapley and Banzhaf values for the voting game $[7; 7, 3, 2, 1]$. Player 1 is called a *dictator*. Why?

4. Find the Shapley and Banzhaf values for the voting game $[3; 1, 1, 1, 1]$ in which there are no blocs of more than one vote per voter.

5. *Political Science.* Assume the Democrats have 49 votes, the Republicans have 48, the Independents have 3, and each party votes as a bloc. If a simple majority (51) is required for passage of a bill, what are the voting strengths (Shapley and Banzhaf values) for each party? Write as a quota game. Note the strength of the small third party.

6. *Political Science.* Find the Shapley and Banzhaf values for the quota game $[7; 6, 3, 2, 1]$. Note that no coalition can win without player 1. Such a player is said to have *veto* power. Compare with Exercise 3 where player 1 was a *dictator*.

7. *Political Science.* For the quota game of Example 1, player 4 is a dummy. What is the minimum quota required to eliminate any dummies from the game? Find the voting strength (Shapley and Banzhaf values) for each player under the revised conditions.

8. *Political Science.* A committee consists of 4 members, each with one vote. A simple majority (3) is required for a decision; however, the chairman can break ties. What is his voting strength? (Both values.)

9. *Political Science.* If the chairman of the committee has veto power instead of just the power to break ties, what is his voting strength?

10. *Political Science.* As in Exercise 9, if two of the members agree to pool their votes, how do they change the voting strengths? three of the members?

CHAPTER 11 TEST

Given the matrix game $\begin{bmatrix} 1 & 7 & 5 & 0 \\ -7 & 4 & 1 & -3 \\ 2 & 8 & 2 & 7 \\ 3 & 3 & 2 & 8 \end{bmatrix}$.

1. Find the pure-strategy solution if it exists. Show your work.

2. Indicate all dominating strategies.

3. Illustrate the solution graphically.

4. Solve the game for the strategies for each player, and the value of the game.

Given the quota game [10; 7, 5, 3, 1].

5. Find the Shapley value.

6. Find the Banzhaf value.

Final Examination

Final Examination

Chapter 1

Not covered since it consists of review material.

Chapter 2

Consider this situation: *Item:* Office machine

Cost = $3400,
Expected life = 5 years,
Salvage value = $200.

1. Using the straight-line method, find a formula for the book values V_n.

2. Using the double-declining balance method, find a formula for the book values V_n.

3. Find the effective annual yield 16%, compounded quarterly.

4. Find the amount of an annuity where $4000 is being invested semiannually at 13.6%, compounded semi-annually, for 10 years.

5. A family buys a house for $150,000. A down payment of $40,000 is made, and $110,000 is borrowed at 15%, compounded monthly. The loan is to be paid off by 300 equal payments over the next 25 years. How much is each payment?

Chapter 3

6. Solve using the echelon method.

$$4x_1 + x_2 + x_3 = 0$$
$$x_1 - x_2 - 3x_3 = 9$$
$$3x_1 + 2x_2 - 4x_3 = 7$$

7. Complete the solution using the echelon method.

x_1	x_2	x_3	x_4	1
1	0	4	0	10
0	1	-2	0	5
0	0	4	-4	8

Given $A = \begin{bmatrix} 2 & -3 \\ 1 & -5 \end{bmatrix}$ and $B = \begin{bmatrix} 1 & 0 \\ -1 & -1 \end{bmatrix}$, find

8. $A - B$ **9.** AB **10.** $A + B$ **11.** $-2A$

Chapter 4

12. Write the following linear program in matrix form.

$$3x_1 + 2x_2 \leqslant 7,$$
$$2x_1 - 3x_2 \geqslant -2,$$
$$x_1 + 7x_2 \leqslant 11,$$
$$\max f: f = x_1 + 2x_2; \quad x_1, x_2 \geqslant 0.$$

Use the following linear program to do Questions 13 through 21:

$$7x_1 + 8x_2 \leqslant 56,$$
$$2x_1 + x_2 \leqslant 8,$$
$$3x_1 + 2x_2 \leqslant 18,$$
$$\max x_0: x_0 = x_1 + x_2; \quad x_1, x_2 \geqslant 0.$$

13. Graph the constraints and shade the feasible solution.

14. Find the optimum feasible solution, showing all work.

Chapter 5

15. Set up the initial simplex tableau and indicate the first pivot element.

16. Solve using the simplex algorithm.

17. Read off the optimum primal solution.

18. Write the dual linear program.

19. Read off the optimum dual solution.

20. Write the equations you should use to check your solution.

21. Solve the program of Problem 18 using the two-phase method.

Interpret the following tableaux:

22.

x_1	x_2	x_3	y_1	y_2	y_3	1
4	0	5	1	-7	0	1
3	1	-1	0	-3	0	4
-1	0	3	0	-2	1	3
4	0	-2	0	-5	0	18

23.

x_1	x_2	x_3	y_1	y_2	y_3	1
0	0	0	-1	1	1	4
1	1	0	3	0	-1	0
0	7	1	3	0	4	1
0	4	0	1	0	1	5

24. What is the minimum-cost shipping schedule for the following transportation problem?

		j			
		1	2	3	s_i
i	1	11	10	13	80
	2	17	14	9	100
	r_j	90	60	30	Min. cost

25. Solve the following assignment problem given "goodness of fit."

		j				
		1	2	3	4	5
i	1	9	11	13	3	17
	2	4	8	7	9	13
	3	12	13	9	7	11
	4	9	11	7	5	12
	5	13	9	10	6	16

Chapter 6

26. Given the following network, find the minimum spanning tree.

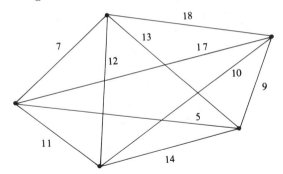

27. Find the minimum path from node A to node G through the following network.

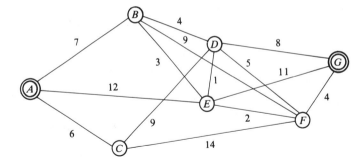

28. Find the maximum flow from node A to node G through the following network.

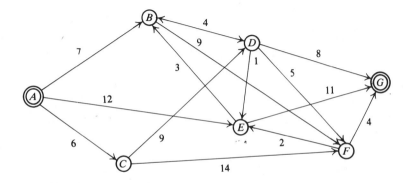

Chapter 7

29. Evaluate $P(8, 3)$.

30. How many different words can be formed from all the letters of the word CALCULUS?

31. A union consists of 58 women and 42 men. How many bargaining units can be formed consisting of 12 women and 18 men? Do not simplify!

32. How many ways can 8 students be seated at a circular table?

33. Expand: $(m + n)^7$.

Chapter 8

34. If $p(E) = \frac{10}{19}$, what are the odds *against* the event E?

35. Find $p(E_1 \cup E_2)$ where E_1 and E_2 are mutually exclusive and $p(E_1) = 0.26$ and $p(E_2) = 0.69$.

36. From a deck of 52 cards you draw 3 at random without replacement. What is the probability that all 3 are queens?

37. Three manufacturers supply all the chairs to a particular furniture store. A supplies 10 with 2% defective, B supplies 30 with 1% defective, and C supplies 40 with 5% defective. If a chair is purchased at random, what is the probability that if it is defective it came from B?

Chapter 9

38. A cage contains 4 white mice and 2 black mice. If 3 unidentified mice escape, what is the probability function for the number of white mice that escaped?

39. In Question 38, what is the expected value of the number of white mice that escaped?

40. In Question 38, what are the variance and standard deviation of the number of white mice that escaped?

41. An automobile safety patrol is checking seatbelt wearers. If 2 people in 5 wear seatbelts, what is the probability that of 6 cars stopped at random at least half the drivers are wearing seatbelts?

42. An automobile safety patrol is checking seatbelt wearers. If 2 people in 5 wear seatbelts, what is the probability that the third driver checked is the first to wear a seatbelt?

43. In Question 42, what is the probability that the fifth driver checked is the second to wear a seatbelt?

Chapter 10

44. Given the transition matrix

$$T = \begin{bmatrix} 0 & 1 & 0 \\ \frac{1}{2} & 0 & \frac{1}{2} \\ \frac{1}{3} & \frac{1}{3} & \frac{1}{3} \end{bmatrix},$$

draw the transition diagram.

45. Given the transition matrix of Question 44 and $P_0 = [0 \quad \frac{1}{2} \quad \frac{1}{2}]$, find P_1.

46. Demonstrate the regularity of T in Question 44.

47. Determine the fixed point of T in Question 44.

Chapter 11

48. Solve the following matrix game:

$$\begin{bmatrix} 3 & 1 & 4 & 1 \\ 5 & 9 & 7 & 3 \\ 4 & 3 & 4 & 6 \end{bmatrix}$$

49. Find the Shapley and Banzhaf values for the quota game:

$$[11; \quad 7, \quad 6, \quad 4, \quad 3]$$

APPENDIX A

Logic

APPENDIX A
Logic

A.1 INTRODUCTION

Logic is a study of language and reasoning. Just as everyone uses sentences to convey ideas, so do sentences convey ideas in mathematics. For example,

$$2x_1 - x_2 + x_3 = -1,$$
$$4 + 7 = 11,$$
$$L \perp M,$$

and so on.

STATEMENTS.   A declarative sentence which is either true or false, but not both at the same time, is called a *statement*.

The following are statements:

$$2 - 5 + 19 = 16$$

All math professors are mortal.

The American Stock Exchange lost 2 points on September 14, 1978.

$$4 + 2 = 0$$

The following are *not* statements:

Why are you a business major? (A question is neither true nor false)

He is a psychologist.   ("He" is undefined)

$x + 3 = 5$ (Cannot be judged true or false until we know what "x" is)

$2x_1 - x_2 + x_3 = -1$ (We must know what x_1, x_2, and x_3 are.)

OBJECTIVES

You should be able to

a) Determine whether a sentence is a statement.
b) Determine the truth value of a statement.
c) Translate a sentence to symbolic form.
d) State the truth tables for $\wedge$, $\vee$, $\rightarrow$, $\leftrightarrow$, and $\sim$.

Consider the following sentences for Margin Exercises 1 through 3:

a) $x < 2$

b) $x + y = y + x$

c) $2x_1 - x_2 + x_3 = -1$

d) $1 < 2$

e) $2 \cdot 5 - 3 + 6 = -1$

f) This sentence is false.

g) Right on, baby!

1. Which of the above are statements?

2. Identify the variables in each sentence.

3. Which will become statements when the variables are replaced by numbers?

Variables

The sentence

He is a psychologist

cannot be judged true or false because we do not know who *He* is. If the word "He" is replaced by "Sigmund Freud," forming the sentence

Sigmund Freud is a psychologist

the sentence becomes a (true) statement. Similarly, if "x" in the sentence

$$x + 1 = 5$$

is replaced by "3," forming the sentence

$$3 + 1 = 5,$$

the sentence becomes a (false) statement, but if x is replaced by "4," forming the sentence

$$4 + 1 = 5,$$

the sentence becomes a (true) statement.

The letter "x" is a *variable* in the sentence $x + 1 = 5$. A letter (or other symbol) that can represent various elements of a set under consideration is called a *variable*. Thus, "He" is a variable in the sentence

He is a psychologist.

DO EXERCISES 1 THROUGH 3.

Sentence Connectives

CONJUNCTION. If p and q are sentences, then the sentence "p and q" is called the *conjunction* of p and q, symbolized:

$$p \wedge q.$$

A system of equations like

$$2x_1 + 6x_2 = 26,$$
$$8x_1 - 3x_2 = -31,$$

is a conjunction

$$(2x_1 + 6x_2 = 26) \wedge (8x_1 - 3x_2 = -31).$$

For any statement there are just two possible truth values, true (T) *or false* (F). If p and q are both true, then $p \wedge q$ is true. If one or both of

p and q are false, then $p \wedge q$ is false. The truth table below defines the truth values of $p \wedge q$ for all possible values of p and q.

p	q	$p \wedge q$
T	T	T
T	F	F
F	T	F
F	F	F

[Memorize the table.]

Example 1 Determine the truth value of each sentence.

a) $(2 + 2 = 4) \wedge (3 + 2 = 7)$
b) $(2 + 2 = 4) \wedge (3 + 2 = 5)$

Solution

a) False, because $3 + 2 = 7$ is false.
b) True, because both statements are true.

DO EXERCISE 4.

We know the ordered pair $(-2, 5)$ is a solution of the system, or conjunction, of equations

$$(2x_1 + 6x_2 = 26) \wedge (8x_1 - 3x_2 = -31)$$

because when we replace x_1 in each equation by -2, and x_2 in each equation by 5, we get a true statement

$$[2(-2) + 6 \cdot 5 = 26] \wedge [8(-2) - 3 \cdot 5 = -31].$$

Similarly, the ordered pair $(13, 0)$ is not a solution of the conjunction because when we replace x_1 in each equation by 13, and x_2 in each equation by 0, we get a false statement

$$(2 \cdot 13 + 6 \cdot 0 = 26) \wedge (8 \cdot 13 - 3 \cdot 0 = -31).$$

DO EXERCISE 5.

DISJUNCTION. If p and q are sentences, then the sentence "p or q" is called the *disjunction* of p and q, symbolized:

$$p \vee q.$$

One way in which disjunctions are unlike conjunctions is that there are at least two uses of "or" in English. One use is *exclusive*, meaning

4. Determine the truth value of each sentence.

a) $(5 < 2) \wedge (4 < 6)$

b) $(2 - 3 = -4) \wedge (7 = 8)$

c) $(2 - 3 = -1) \wedge (\frac{16}{2} = 8)$

5. Determine whether $(-2, 1)$ is a solution of each conjunction.

a) $\quad x_1 + x_2 = -1$
$\quad -3x_1 - x_2 = 5$

b) $2x_1 - x_2 = -5$
$\quad 3x_1 + 2x_2 = 3$

6. Determine the truth value of each sentence.

a) $(5 < 2) \lor (4 < 6)$

b) $(2 - 3 = -4) \lor (7 = 8)$

c) $(2 - 3 = -1) \lor (\frac{15}{3} = 5)$

d) $(2 \text{ is odd}) \lor (\frac{1}{4} = \frac{1}{2})$

"one or the other but not both." For example, if the statement

Sharon is awake or asleep

is true, then this would mean that Sharon is awake or she is asleep, but she cannot be both. Another use of "or" is *inclusive*, meaning "and/or." For example, if the statement

Sharon is wearing a sweater or blouse

is true, then this would mean that she is wearing a sweater, or a blouse, or both.

Here the mathematical use of the word "or" is inclusive unless the context indicates otherwise. That is, $p \lor q$ is true when p is true, when q is true, or when *both* are true, and $p \lor q$ is false only when both are false. The truth table below defines the truth values of $p \lor q$ for all possible truth combinations of p and q.

p	q	$p \lor q$
T	T	T
T	F	T
F	T	T
F	F	F

[Memorize the table.]

Example 2 Determine the truth value of each sentence.

a) $(2 + 2 = 4) \lor (3 + 2 = 7)$
b) $(2 + 2 = 4) \lor (3 + 2 = 5)$
c) $(2 + 2 = 0) \lor (3 - 2 = 6)$

Solution

a) True, because one part is true.
b) True, because both parts are true.
c) False, because both parts are false.

DO EXERCISE 6.

Negation

A *negation*, or denial, of a sentence is formed in many ways. For example, if p is

Smoking causes lung cancer,

the negation of p is represented by each of the following

$\sim p$ (read "not p")

It is false that smoking causes lung cancer.

It is not true that smoking causes lung cancer.

Smoking does not cause lung cancer.

If p is a statement, then $\sim p$ is true when p is false and $\sim p$ is false when p is true. The truth table for negation is defined as follows:

p	$\sim p$
T	F
F	T

[Memorize the table.]

Note that

$$\sim(a = b) \quad \text{is written} \quad a \neq b,$$
$$\sim(a < b) \quad \text{is written} \quad a \geq b,$$
$$\sim(a \leq b) \quad \text{is written} \quad a > b.$$

DO EXERCISES 7 AND 8.

Conditional

If p and q are sentences, the sentence

If p, then q

is symbolized

$$p \rightarrow q$$

There is a truth-table definition for $p \rightarrow q$ agreed upon by logicians and mathematicians, but the definition is not as obvious as it was in the other definitions.

Suppose a fellow student says:

"If I get an A in mathematics, then I will take the next course."

When is he telling the truth and when is he lying? Examine the following four cases, where:

p: I get an A in mathematics.

and

q: I will take the next course.

7. Write four different representations of the negation of each.

a) p: $2 = 3$

b) p: Profit is wholesome.

8. Determine the truth value of each.

a) $2 \neq 3$

b) $\sim(7 \text{ is an odd number})$

1. p(true)   He gets an A in mathematics.
   q(true)   He takes the next course.

2. p(true)   He gets and A in mathematics.
   q(false)   He does not take the next course.

3. p(false)   He does not get the A.
   q(true)   He takes the next course.

4. p(false)   He does not get the A.
   q(false)   He does not take the next course.

In (1) it is reasonable to agree that the student was telling the truth; his claim is true. In (2) it is easy to agree that he lied, and his claim was false. In (3) you could not call him a liar since he takes the next course even though he did not get the A. In (4) you likewise could not call him a liar since he did not get the A and did not take the next course.

The truth-table definition for $p \rightarrow q$ conforms to the preceding example.

	p	q	$p \rightarrow q$
1)	T	T	T
2)	T	F	F
3)	F	T	T
4)	F	F	T

(The numbers refer to the preceding example.)

[Memorize the table.]

While other definitions seem possible, this one is made for reasons of consistency in mathematics which we cannot go into here.

The sentence $p \rightarrow q$ is called a *conditional*, with

p the *antecedent*   and q the *consequent*.

To summarize, a conditional is true when the antecedent is false or the consequent is true. A conditional is false only when the antecedent is true and the consequent is false.

Example 3   Determine the truth value of each of the following.

a) $(2 + 2 = 4) \rightarrow (3 + 2 = 5)$.
b) (3 is an odd number) $\rightarrow (4 < 0)$.
c) $(5 = 8) \rightarrow (6 = 6)$.
d) (6 is an odd number) $\rightarrow$ (8 is an odd number).

Solution

a) True, because the statement is in the form $T \rightarrow T$.

b) False, because the statement is in the form T → F.

c) True, because the statement is in the form F → T.

d) True, because the statement is in the form F → F.

DO EXERCISE 9.

In mathematics, $p \to q$ is encountered in many forms. You should be familiar with each. The following have the same meaning.[*]

$$p \to q$$
If p, then q
p implies q
q, provided p
q if p
q, given that p

Example 4 Translate this sentence to the form $p \to q$:

Your love life will improve provided you take Vitamin E.

Solution

(You take Vitamin E) → (Your love life will improve)

DO EXERCISE 10.

Biconditional

Suppose statements p and q have identical truth values. Then we say "p and q are *equivalent*" and this is symbolized

$$p \leftrightarrow q$$

We call $p \leftrightarrow q$ a *biconditional*. The truth table for $p \leftrightarrow q$ is as follows:

p	q	$p \leftrightarrow q$
T	T	T
T	F	F
F	T	F
F	F	T

[Memorize the table.]

Thus, $p \leftrightarrow q$ is true when p and q are both true or both false, and $p \leftrightarrow q$ is false when p and q have differing truth values.

---

[*] Other forms are "p is a sufficient condition for q" and "q is a necessary condition for p," but these will not be used in this text.

9. Determine the truth value of each.

a) $(7 < 6) \to (9 \text{ is odd})$

b) $(\sqrt{4} = 2) \to (8 \text{ is odd})$

c) $(\sqrt{4} = 2) \to (7 \text{ is odd})$

d) $(\sqrt{4} = 3) \to (16 \text{ is even})$

10. Translate each sentence to the form $p \to q$.

a) You will stop wetness given that you spray with DriasKanby.

b) If $a = b$, then $a^n = b^n$.

c) The card is a queen provided it is a face card.

d) Democrat A wins implies Democrat B wins.

11. Determine the truth value of each.

a) $(7 < 6) \leftrightarrow (9 \text{ is odd})$

b) $(\sqrt{9} = 3) \leftrightarrow (8 \text{ is odd})$

c) $(\sqrt{9} = 3) \leftrightarrow (7 \text{ is odd})$

d) $(\sqrt{9} = 8) \leftrightarrow (1 \text{ is even})$

12. Translate each sentence to the form $p \leftrightarrow q$.

a) x is in set A if and only if x is in set B.

b) $x + c = y + c$ is equivalent to $x = y$.

c) Sharon gets A's if and only if Sharon studies.

Example 5 Determine the truth value of each.

a) $(2 < 1) \leftrightarrow (2 < 3)$

b) $(2 \text{ is odd}) \leftrightarrow (3 \text{ is even})$

c) $(2 \text{ is even}) \leftrightarrow (3 \text{ is odd})$

Solution

a) False because the truth values do not agree.

b) True because the truth values (both false) agree.

c) True because the truth values (both true) agree.

DO EXERCISE 11.

In mathematics $p \leftrightarrow q$ is encountered in many forms. The following have the same meaning.*

$$p \leftrightarrow q$$

p is equivalent to q

p if and only if q

DO EXERCISE 12

EXERCISE SET A.1

Which are statements?

1. All numbers are even. **2.** All grass is green. **3.** $3x = 50$ **4.** $4 + 4y = 7$

5. The Dow Jones average was above 1000 on March 4, 1977. **6.** No psychologist has ever gone to a psychiatrist.

Determine the truth value of each statement.

7. $\sim(50 = 2 \cdot 25)$ **8.** $\sim(4 > 18)$ **9.** $(2 = 3) \rightarrow (2 + 5 = 3 + 5)$

10. $(2 = 3) \leftrightarrow (2 + 5 = 3 + 5)$ **11.** $(2 = 3) \vee (2 + 5 = 3 + 5)$ **12.** $(2 = 3) \wedge (2 + 5 = 3 + 5)$

Suppose p is true and q is false. Find the truth value of each statement.

13. $\sim p$ **14.** $\sim q$ **15.** $p \wedge \sim q$ **16.** $\sim p \vee q$

17. $p \rightarrow \sim q$ **18.** $\sim p \rightarrow q$ **19.** $p \leftrightarrow \sim q$ **20.** $p \rightarrow q$

21. $\sim p \rightarrow \sim q$ **22.** $\sim p \leftrightarrow \sim q$ **23.** $(p \wedge q) \rightarrow q$ **24.** $(p \vee q) \rightarrow q$

---

* The meaning "p is a necessary and sufficient condition for q" is also used, but not in this text.

Translate each sentence to symbolic form, as in the following example:

Example Translate "If a is perpendicular to c and b is perpendicular to c, then a is parallel to c" given

> p: a is perpendicular to c,
> q: b is perpendicular to c,
> r: a is parallel to c.

Solution The translation is $(p \wedge q) \to r$.

25. If $3x = 15$, then $x = 5$.

> p: $3x = 15$
> q: $x = 5$

26. $2x - 6y = 8$ if and only if $x - 3y = 4$

> p: $2x - 6y = 8$
> q: $x - 3y = 4$

27. The card is a jack and a face card.

> p: The card is a jack.
> q: The card is a face card.

28. The card is a jack or a face card.

> p: The card is a jack.
> q: The card is a face card.

29. $a < b$ or $a > b$ or $a = b$

> p: $a < b$
> q: $a > b$
> r: $a = b$

30. $x < 0$ or $x = 0$ or $x > 0$

> p: $x < 0$
> q: $x > 0$
> r: $x = 0$

31. The second ball is white, given that the first ball is red.

> p: The first ball is red.
> q: The second ball is white.

32. The second transistor is defective, given that the first transistor is defective.

> p: The second transistor is defective.
> q: The first transistor is defective.

33. If Sharon does not study, then she does not pass.

> p: Sharon studies.
> q: Sharon passes.

34. If Richard does not share, then Richard is not liked.

> p: Richard shares.
> q: Richard is liked.

35. He will pass both courses, neither course, or one of the courses.

> p: He passes both courses.
> q: He will pass neither course.
> r: He will pass one of the courses.

36. Either both will be alive, or neither will be alive, or one will be alive.

> p: Both will be alive.
> q: Neither will be alive.
> r: One will be alive.

A.2 TRUTH TABLES AND TAUTOLOGIES

We have defined truth tables for $p \wedge q$, $p \vee q$, $p \to q$, $p \leftrightarrow q$, and $\sim p$. We now construct truth tables for more complicated sentences. We do this by first listing *all possible* truth-value combinations. Then we determine successive truth values.

Example 1 Construct a truth table for $p \to \sim q$.

OBJECTIVES

You should be able to construct a truth table for a given statement and decide whether it is a tautology.

13. Construct the truth table
for $\sim p \rightarrow q$.

p	q	$\sim p$	$\sim p \rightarrow q$
T	T		
T	F		
F	T		
F	F		

Solution

①	①	②	③

p	q	$\sim q$	$p \rightarrow \sim q$
T	T	F	F
T	F	T	T
F	T	F	T
F	F	T	T

For example, $T \rightarrow T$ is T

① Write down all possible arrangements of the truth values of p and q.

② We use the column headed q (the second column headed ①) to find truth values for $\sim q$.

③ We use the column headed p (the first column headed ①) and the column headed ② to compute truth values for $p \rightarrow \sim q$.

DO EXERCISE 13.

Example 2 Construct a truth table for $(p \rightarrow q) \leftrightarrow (\sim q \rightarrow \sim p)$.

Solution

①	①	②	②	③	③	④

p	q	$\sim p$	$\sim q$	$p \rightarrow q$	$\sim q \rightarrow \sim p$	$(p \rightarrow q) \leftrightarrow (\sim q \rightarrow \sim p)$
T	T	F	F	T	T	T
T	F	F	T	F	F	T
F	T	T	F	T	T	T
F	F	T	T	T	T	T

① Write down all possible arrangements of the truth values of p and q.

② Use the values in the columns headed ① to find the truth values for $\sim p$ and $\sim q$.

③ Use the columns headed ① to find the truth values for $p \rightarrow q$, and the columns headed ② to find the truth values for $\sim q \rightarrow \sim p$.

④ Use the columns headed ③ to find the truth values for the original sentence $(p \rightarrow q) \leftrightarrow (\sim q \rightarrow \sim p)$.

Here is a condensed version.

p q	① ③ ① ④ ② ① ③ ② ① (p → q) ↔ (~q → ~p)
T T	T T T T F T T F T
T F	T F F T T F F F T
F T	F T T T F T T T F
F F	F T F T T F T T F

① Write down all possible arrangements of the truth values of p and q, and repeat them under each occurrence of p and q.

② Write down the truth values of $\sim q$ and $\sim p$ under the $\sim$ signs.

③ Use the columns headed ① to find the truth values for $p \to q$, and the columns headed ② to find the truth values for $\sim q \to \sim p$.

④ Use the columns headed ③ to find the truth values for the original sentence.

Note that the statement in Example 2 is true regardless of the truth values of p and q. Such statements are important in logic and are called *tautologies*. Another example of a tautology is

It is snowing or it is not snowing

symbolized by

$$p \lor \sim p$$

Its truth table is given by:

p	$\sim p$	$p \lor \sim p$
T	F	T
F	T	T

DO EXERCISES 14 AND 15.

Now let us consider a truth table with 3 basic statements. In such cases there are 2^3, or 8 combinations at the outset.

Example 3

a) Construct the truth table for

$$[(p \to q) \land (q \to r)] \to (p \to r)$$

b) Is the statement a tautology?

14. a) Construct the truth table for $[(p \to q) \land p] \to q$

b) Is the statement a tautology?

p	q	$p \to q$	$(p \to q) \land p$	$[(p \to q) \land p] \to q$
T	T			
T	F			
F	T			
F	F			

15. a) Construct the truth table for $p \land \sim p$.

b) Is the statement a tautology?

p	$\sim p$	$p \land \sim p$
T		
F		

| **Solution** a) |

p	q	r	$p \to q$	$q \to r$	$p \to r$	$(p \to q) \wedge (q \to r)$	$[(p \to q) \wedge (q \to r)] \to (p \to r)$
T	T	T	T	T	T	T	T
T	T	F	T	F	F	F	T
T	F	T	F	T	T	F	T
T	F	F	F	T	F	F	T
F	T	T	T	T	T	T	T
F	T	F	T	F	T	F	T
F	F	T	T	T	T	T	T
F	F	F	T	T	T	T	T

16. a) Construct the truth table for

$$p \to (q \vee r).$$

b) Is the statement a tautology?

Here is a condensed version.

$p\,q\,r$	① ② ① $[(p \to q)$	③ $\wedge$	① ② ① $(p \to r)]$	④ $\to$	① ② ① $(p \to r)$
T T T	T T T	T	T T T	T	T T T
T T F	T T T	F	T F F	T	T F F
T F T	T F F	F	F T T	T	T T T
T F F	T F F	F	F T F	T	T F F
F T T	F T T	T	T T T	T	F T T
F T F	F T T	F	T F F	T	F T F
F F T	F T F	T	F T T	T	F T T
F F F	F T F	T	F T F	T	F T F

b) Yes, it is a tautology because the statement has all T's as truth values.

DO EXERCISE 16.

EXERCISE SET A.2

For each statement, a) Construct a truth table. b) Determine whether the statement is a tautology.

1. $p \to p$ **2.** $p \to {\sim}p$ **3.** ${\sim}p \wedge q$ **4.** $p \vee {\sim}q$

5. $p \to (p \vee q)$ **6.** $p \to (p \wedge q)$ **7.** $p \leftrightarrow {\sim}{\sim}p$ **8.** $(p \wedge p) \leftrightarrow p$

9. $(p \wedge q) \leftrightarrow (q \wedge p)$ **10.** $(p \vee q) \leftrightarrow (q \vee p)$

11. $[{\sim}(p \wedge q)] \leftrightarrow [({\sim}p) \vee ({\sim}q)]$ **12.** $[{\sim}(p \vee q)] \leftrightarrow [({\sim}p) \wedge ({\sim}q)]$

13. $(p \to q) \to (q \to p)$ **14.** $(p \to q) \leftrightarrow ({\sim}q \to {\sim}p)$

15. $p \to (q \to r)$ **16.** $(p \to q) \to r$

17. $[p \wedge (q \vee r)] \leftrightarrow [(p \wedge q) \vee (p \wedge r)]$ **18.** $[p \vee (q \wedge r)] \leftrightarrow [(p \vee q) \wedge (p \vee r)]$

A.3 VALID ARGUMENTS

An *argument* is an assertion that from a certain set of sentences, called *hypotheses*, one can deduce another sentence q called a *conclusion*. For example,

$$p_1, p_2, p_3 \qquad \therefore \quad q \qquad (\therefore \text{ is read "therefore")}$$

denotes that from the sentences p_1, p_2, and p_3 we can deduce q. An argument is said to be *valid* when q is true whenever p_1, p_2, and p_3 are true. It can be shown (but not here) that an argument such as that above is valid when the statement

$$(p_1 \wedge p_2 \wedge p_3) \to q$$

is a tautology. If an argument is not valid, it is said to be *invalid*. Thus an argument is invalid any time we can produce a situation where the hypotheses are true, but the conclusion is false.

As an example of a valid argument, consider:

> If you work hard, then you will succeed.
> You work hard.
> _____
>
> $\therefore$ You succeed.

This takes the symbolic form:

$$
\begin{array}{l}
p \to q \\
p \\
\hline
\therefore q
\end{array}
$$

There are at least two ways to verify that this is a valid argument. One is to show that the statement

$$[(p \to q) \wedge p] \to q$$

is a tautology. You, in fact, did this in Margin Exercise 14. For more complicated arguments, a slightly different form and the use of a truth table can ease the work.

Remember, an argument is valid as long as we *cannot* produce a situation where the hypotheses are true and the conclusion is false. Consider again

$$
\begin{array}{l}
p \to q \\
p \\
\hline
\therefore q
\end{array}
$$

OBJECTIVE

You should be able to determine the validity of an argument.

We write a truth table which contains the basic statements plus the hypotheses and conclusion, plus any other statements necessary to compute truth values of these statements. We put an "H" above any hypothesis and a "C" above the conclusion. (Compare this with the length of the truth table in Margin Exercise 14.)

H	C	H

p	q	$p \to q$
T	T	T
T	F	F
F	T	T
F	F	T

Next we draw a line through any row where a hypothesis is false. The rows that remain are all the situations where the hypotheses are true. We check to see whether the conclusion is false in any of these rows. Since the conclusion is not false in any row, we conclude that the argument is valid.

As an example of an invalid argument, consider:

> If you work hard, then you will succeed.
> You succeed.
> ———————————————
> ∴ You work hard

This takes the symbolic form.

$$p \to q$$
$$q$$
$$\overline{}$$
$$\therefore q$$

We write a truth table which contains the basic statements plus the hypotheses and conclusion, plus any statements necessary to compute the truth values of these statements. We put an "H" above any hypotheses and a "C" above the conclusion.

C	H	H

p	q	$p \to q$
T	T	T
T	F	F
F	T	T
F	F	T

Next we draw a line through any row where a hypothesis is false. The rows that remain are all the situations where the hypotheses are true. We check to see whether the conclusion is false in any of these rows, and since it *is* in the third one, we conclude that the argument is *invalid*.

DO EXERCISES 17 AND 18.

Example Determine the validity of the argument

$$\sim p \to q, \quad q \to \sim r, \quad r, \qquad \therefore p$$

Solution

	C	H			H	H
p	q	r	~p	~r	~p → q	q → ~r
T	T	T	F	F	T	F
T	T	F	F	T	T	T
T	F	T	F	F	T	T
T	F	F	F	T	T	T
F	T	T	T	F	T	F
F	T	F	T	T	T	T
F	F	T	T	F	F	T
F	F	F	T	T	F	T

The argument is valid.

DO EXERCISE 19.

17. Determine the validity of the argument

$$p, \quad p \leftrightarrow q, \quad \therefore q$$

18. Determine the validity of the argument

$$p \vee q, \qquad \therefore q$$

19. Determine the validity of the argument

$$p \to r, \quad q \to p, \quad \sim r, \qquad \therefore \sim q$$

EXERCISE SET A.3

Determine the validity of each argument.

1. $p \wedge q, \qquad \therefore q$

2. $p \wedge q, \qquad \therefore p$

3. $p \to q, \quad \sim q, \qquad \therefore \sim p$

4. $p, \qquad \therefore p \vee q$

5. $p, \qquad \therefore p \wedge q$

6. $p \vee q, \quad \sim p, \qquad \therefore q$

7. $p \vee q, \quad \sim q, \qquad \therefore p$

8. $p \to q, \quad q \to r, \qquad \therefore p \to r$

9. $p \to q, \quad q \to p, \qquad \therefore p \leftrightarrow q$

10. $p \vee q, \quad p \vee \sim q, \qquad \therefore q$

11. $p \to q, \quad p \to r, \qquad \therefore p$

12. $p \vee q, \quad r \to q, \quad \sim p, \qquad \therefore r$

13. $p \to \sim q, \quad p \to r, \quad q, \qquad \therefore r$

14. $p \to q, \quad p \to r, \qquad \therefore q \wedge r$

Translate each argument to symbolic form and determine its validity.

15. If you smoke, then you will get lung cancer. You have lung cancer. Therefore, you smoke.

16. If you brush with Supergu, then your lovelife will improve. You brush with Supergu. Therefore, your lovelife improves.

17. If I pass mathematics, then I will take the next course. If I take the next course, then I will graduate. I do not graduate. Therefore, I do not pass mathematics.

19. Prudence gets promoted to president of the company or Prudence goes out on a date. Prudence is not promoted. Therefore, Prudence goes out on a date.

21. If mathematics is fun, then the moon is made of cheese. If the moon is not made out of rubber, then it is not made out of cheese. The moon is not made out of cheese. Therefore, mathematics is fun.

23. If you like Bach, then honk. You honk. Therefore, you like Bach.

18. Elton passes mathematics if and only if Elton studies. Elton studies or goes out with his girl friend. Elton does not go out with his girl friend. Therefore, Elton does not pass mathematics.

20. If you smoke, then you will get lung cancer. If you get lung cancer, then you will die. You do not get lung cancer. Therefore, you do not die.

22. If it rains, then the clothes will get wet. If the clothes get wet, then I will have nothing to wear. Therefore, if I have nothing to wear, then it does not rain.

OBJECTIVES

You should be able to write logical symbolism for a network and simplify it, if possible.

A.4 APPLICATION TO SWITCHING CIRCUITS

Logic can be applied to switching circuits. Note in the following illustration that the current cannot flow unless the switch is closed.

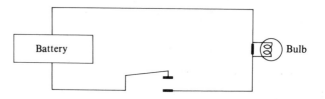

We associate a "closed switch" with a "true" statement, and an "open switch" with a "false" statement.

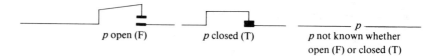

p open (F)	p closed (T)	p not known whether open (F) or closed (T)

To continue the analogy between logic and networks, note that two switches connected in *series* are like two statements forming a *conjunction*.

$$p \qquad q \qquad\qquad p \wedge q$$
Series

The only way current will flow through the network is for both switches to be closed (true).

Two switches connected in *parallel* are like two statements forming a *disjunction*.

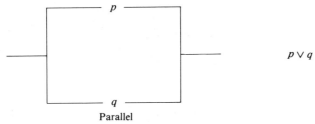

$p \lor q$

Parallel

Suppose negation were involved in a network as in the following

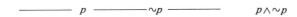

$p \land \sim p$

When this happens, we assume that when p is open (**F**), then $\sim p$ is closed (**T**), and when p is closed (**T**), then $\sim p$ is open (**F**). Note that when a switch and its "negation" are connected in series, current will never flow. This is because the statement $p \land \sim p$ is always false. Suppose a switch and its negation are connected in parallel as in the following:

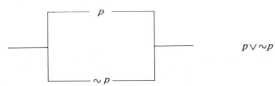

$p \lor \sim p$

Note that the current will always flow. This is because the statement $p \lor \sim p$ is a tautology.

Let us now consider a more complicated statement.

Example 1 Write logical symbolism for the following network.

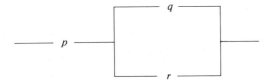

Solution The switches q and r are in parallel. This can be written as $q \lor r$. Then p and $q \lor r$ are in series. This can be written as

$$p \land (q \lor r)$$

DO EXERCISES 20 AND 21.

Write logical symbolism for each network.

20.

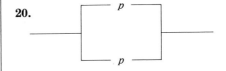

21.

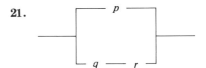

Consider the following network.

———————— p ———————— p ———————— $p \wedge p$

If you worked for an electronics firm and this were part of your network, your employer might be upset with you for wasting switches, because the network

———————— p ———————— p

would have the same effect as the preceding upon the flow of current.

Note that

$$(p \wedge p) \leftrightarrow p$$

is a tautology. Such tautologies can allow us to find "equivalent" networks with fewer switches.

Example 2   Write logical symbolism for the following network and find a simpler equivalent network.

Solution   At the top of the network we have p and q in series. This can be written as $p \wedge q$. At the bottom of the network we have p and r in series. This can be written as $p \wedge r$. Then $p \wedge q$ and $p \wedge r$ are in parallel. This can be written as

$$(p \wedge q) \vee (p \wedge r).$$

It turns out that

$$(p \wedge q) \vee (p \wedge r) \leftrightarrow p \wedge (q \vee r)$$

is a tautology. Thus the network of Example 1 is simpler and equivalent to this one.

The following is a list of tautologies that can be helpful in finding simpler equivalent networks.

1. $(p \wedge p) \leftrightarrow p$ ⎫
2. $(p \vee p) \leftrightarrow p$ ⎬   **Idempotent Laws**

3. $(p \wedge q) \vee (p \wedge r) \leftrightarrow p \wedge (q \vee r)$ ⎫
4. $(p \vee q) \wedge (p \vee r) \leftrightarrow p \vee (q \wedge r)$ ⎬   **Distributive Laws**

5. $(p \wedge q) \leftrightarrow (q \wedge p)$ ⎫
6. $(p \vee q) \leftrightarrow (q \vee p)$ ⎬   **Commutative Laws**

7. $\sim \sim p \leftrightarrow p$

8. $\sim(p \wedge q) \leftrightarrow (\sim p) \vee (\sim q)$ ⎫
9. $\sim(p \vee q) \leftrightarrow (\sim p) \wedge (\sim q)$ ⎭   **DeMorgan Laws**

10. $p \wedge (q \wedge r) \leftrightarrow (p \wedge q) \wedge r$ ⎫
11. $p \vee (q \vee r) \leftrightarrow (p \vee q) \vee r$ ⎭   **Associative Laws**

12. $(T \wedge p) \leftrightarrow p$ The conjunction of a tautology and any sentence p has the same truth value as p.

13. $(F \vee p) \leftrightarrow p$ The disjunction of a sentence that is always false and any sentence p has the same truth value as p.

14. $p \vee \sim p$

DO EXERCISES 22 AND 23.

Example 3   Write logical symbolism and find a simpler network.

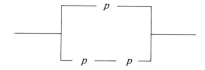

Solution   The logical symbolism is

$$p \vee (p \wedge p)$$

Then

$$[p \vee (p \wedge p)] \leftrightarrow (p \vee p) \qquad \text{by Idempotent Law (1)}$$
$$\leftrightarrow p \qquad \text{by Idempotent Law (2)}$$

DO EXERCISE 24.

Example 4   Write logical symbolism and find a simpler network.

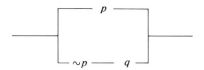

Solution   The logical symbolism is

$$p \vee (\sim p \wedge q)$$

Then

$$p \vee (\sim p \wedge q) \leftrightarrow (p \vee \sim p) \wedge (p \vee q) \qquad \text{by the Distributive Law (4)}$$
$$\leftrightarrow T \wedge (p \vee q) \qquad \text{by (14)}$$
$$\leftrightarrow p \vee q. \qquad \text{by (12)}$$

DO EXERCISE 25.

Write logical symbolism and find a simpler network.

22.

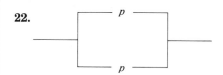

23.

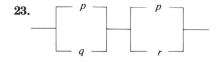

24. Write logical symbolism and simplify.

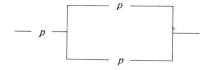

25. Write logical symbolism and simplify.

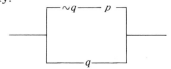

EXERCISE SET A.4

Write logical symbolism and simplify.

1. ——— p ——— p ——— q ———

2.

3.

4.

5.

6.

7.

8.

9.

10.

11.

12. Draw a network for $p \rightarrow q$.

13. A house has a staircase which rises to a long hall. There is a light switch at the bottom of the staircase, one at the top of the staircase at the beginning of the hall, and a third at the end of the hall. Design a circuit so that the light at the top of the stairs can be turned *on* or *off* from any of the three switches.

14. A committee of three is supposed to vote on a new corporate sales policy. A majority vote is needed to begin the new policy. Design a circuit which corresponds to current flowing with a "yes" vote.

APPENDIX B

Programming

APPENDIX B
Programming

B.1 FLOW CHARTING

Computations on high-speed computing machines are part of the current business and industrial scene. We have described computations mathematically using formulas and algorithms. The next step before programming a calculation for a computer is to *flow-chart* it; that is, to represent the calculation *graphically*.

A flow chart serves two purposes. First, it is an *aid* since once a calculation has been *adequately* flow-charted, the actual programming is relatively easy. Second, the flow chart provides *documentation* of the program for business and industrial applications.

Flow charts are essentially *independent* of programming language or computing machine. However, sometimes a programming language or computing machine has peculiarities which must be indicated on the flow chart. We shall point out such situations as they arise.

Example 1   Draw a flow chart for a square-root calculation. Given a number N, we want to input this number to the computer, have the computer find its square root $R = \sqrt{N}$, and print out N and R.

Solution   On a flow chart, we represent this as follows:

We have written simply $R := \sqrt{N}$ for the purposes of this example. Consider handheld computers. Some have a square-root button, so

OBJECTIVE

You should be able to flow-chart a computation, indicating all relevant decisions.

469

that this operation can be performed simply. Others require further instructions which we shall not pursue.

The essential features of this flow chart are

1. We *start* with the word "START" and a *name* inside an *oval*. Here the algorithm is named "ROOT".

2. An *arrow* points to the *next* instruction.

3. The *data* that must be *input* to the computer is put in a *parallelogram*-shaped box following the word "INPUT".

4. The *computation* is put in a *rectangular* box using *ordinary algebra*. The symbol ":=" is read "is assigned the value". The *assign* sign ":=" is used rather than the *equal* sign "=" since the assign sign can be used not only in place of an equal sign but also where an equal sign might be confusing. Thus, we write $n := n + 1$ rather than $n = n + 1$ to mean that n is assigned the value $n + 1$ or simply n is increased by 1.

5. The *information* to be *printed out* from the computer is put in a *parallelogram*-shaped box following the word "PRINT". Here we have *chosen* to print out just the numerical values of N and R.

6. We end with the word "END" and the *start name* inside an *oval*.

For each starting oval there is one and only one ending oval.

We use *ovals* to indicate the *start* and *end* of a program, *parallelograms* to indicate *input* and *output*, and *rectangles* to *indicate computations*. However, the *shape* of a particular box is *redundant* information used only to *emphasize* a particular feature of the computation, since we could use rectangular boxes throughout (but don't).

A computation such as ROOT is frequently part of a larger computation which outputs the number N used as input to ROOT. Thus, a *negative* value of N may be input to ROOT. In the real-number system the square root of a negative number does not exist. The computer might ignore the negative sign or might print out an error message depending on how it is *internally* programmed. We often prefer to make such decisions ourselves and show this on the flow chart.

Assume that if N is negative, we want to know this and end the calculation. Such a situation is shown on the following flow chart:

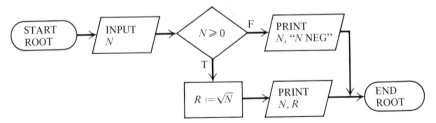

A *diamond*-shaped box is used for *decisions*. We can have a *logical* test such as on the statement "$N \geq 0$" illustrated above with each of the *two* possibilities, true (T) or false (F), having a different exit. We could also have a *numerical* test such as on "$N \gtreqless 0$" with each of the *three* possibilities ($<, =, >$) having a different exit:

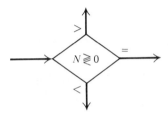

Continuing with the flow chart, for $N < 0$ we have chosen to print out the numerical value of N and the *statement* "N NEG".

Flow charts are useful as documentation in describing decisions such as what to do with negative values of N.

Note that starting and ending ovals have only one arrow leading in *or* one arrow leading out. Input, output, and computation boxes have only *one* arrow leading *in*. Only decision diamonds have more than one arrow leading out. This is done to aid in identification with the machine program.

DO EXERCISES 1 AND 2.

Example 2   Draw a flow chart to evaluate the permutation $P(n, k)$.

Solution   From Section 6.4 we obtain the expression

$$P(n, k) = n(n - 1) \cdots [n - (k - 1)].$$

Thus, $P(n, k)$ is a product of k factors. The first is n and each succeeding factor is one less than the preceding.

1. Modify the flow chart for ROOT to allow for the restriction $0 \leq N \leq 1$ (that is, $N \geq 0$ and $N \leq 1$) and provide for a printout if either of these constraints is violated. Use *logical* tests only.

2. Modify the flow chart for Margin Exercise 1 to use *numerical* tests with at least one of these having three different destinations.

Consider how you would evaluate $P(4, 3)$ on a handheld computer using the preceding formula:

$$P(4, 3) = 4(4 - 1)(4 - 2) = 4 \cdot 3 \cdot 2.$$

First, we start with a quantity (named P, for convenience) equal to 4:

$$P := 4$$

Next, we multiply this quantity by the second factor, 3, (and still call the result P):

$$P := P \cdot 3$$
$$= 4 \cdot 3 = 12$$

Then, we multiply this result by the third factor, 2, and obtain the result:

$$P(4, 3) := P := P \cdot 2 \ = 12 \cdot 2 = 24$$

We use this same procedure to obtain $P(n, k)$ and show how to flow-chart this calculation now that we can visualize the procedure in our minds. We construct flow charts by starting at a convenient place and working in both directions.

In the present example we start by naming the program PERM and inputting n and k:

The next step requires us to note that the calculation of $P(4, 3)$ was done in three steps:

1. $\qquad\qquad\qquad P := 4$
2. $\qquad\qquad\qquad P := P \cdot 3$
3. $\qquad\qquad\qquad P := P \cdot 2$

In general the calculation of $P(n, k)$ will require k steps. Let i count the number of steps, so that *initially*

$$i := 1$$

At each step we can write

$$P := Pm$$

where m is the ith factor of $P(n, k)$ and P is the product of the first i factors, so that *initially*

$$m := n$$

and

$$P := n$$

Adding these initial assignments to the flow chart, we obtain

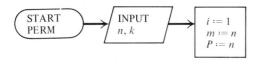

Within a calculation box the calculation starts at the top and proceeds stepwise to the bottom regardless of where the exit arrow leaves the box. The values of i, m, and P on the righthand side of the ":=" represent "old" values and the corresponding values on the lefthand side represent "new" values.

Since the calculation must be ended when $k = i$ and initially $i = 1$, we must next test k. If $k = i$, then we want to print n, k, and P ($= P(n, k)$). If $k > 1$, we continue with the calculation until $k = i$ but if $k < 1$, we *must* decide what to do. Since when $k < 1$, $P(n, k)$ makes no sense, let us simply print out k and the statement "$k < 1$" and end the calculation. At this point the flow chart is:

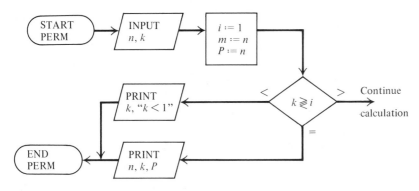

Now to continue the flow chart we must increase the step counter i by one

$$i := i + 1$$

decrease the factor m by one

$$m := m - 1$$

and calculate the new value of P

$$P := Pm$$

3. Modify the flow chart of Example 2 to evaluate $C(n, k)$ and follow step-by-step the evaluation of $C(4, 2)$.

Having done this, we must again test k to see if the calculation is to be ended. Thus, we obtain the completed flow chart:

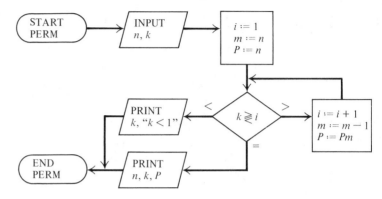

Let us follow step-by-step what happens when we evaluate $P(4, 2)$.

Input: $n = 4$, $k = 2$
$i = 1$, $m = 4$, $P = 4$
$2 > 1$, test
$i = 1 + 1 = 2$, $m = 4 - 1 = 3$, $P = 4 \cdot 3 = 12$
$2 = 2$, test
Output: $n = 4$, $k = 2$, $P = 12$.

DO EXERCISE 3.

Frequently, tests such as to determine if $k < 1$ are done immediately following the input, so that we write:

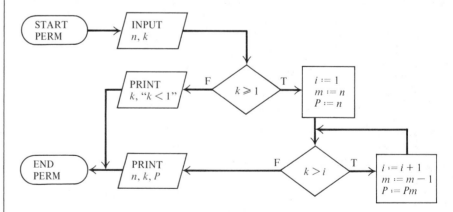

Note that for present purposes it was convenient to replace the three-exit numerical test by a two-exit logical test.

We may, for lack of room or to avoid "cluttering up" the flow chart, use *connectors* and write:

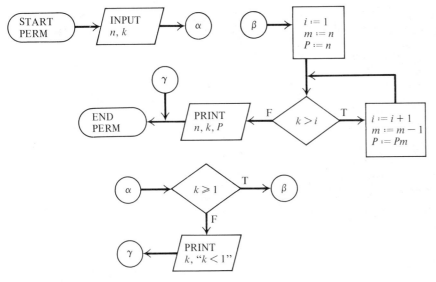

Here the connector $\rightarrow\text{\textcircled{α}}$
instructs one to "go to α" to continue the computation
and the connector $\text{\textcircled{$\alpha$}}\rightarrow$
instructs one to resume at this place the computation interrupted at α. A *connector* is indicated by a *circle* enclosing a name. Here the name is simply the Greek letter α (alpha).

If now we want to insure that both $k \geqslant 1$ and $n \geqslant k$ when the computation resumes at β, then we need only modify the "test" part of the flow chart:

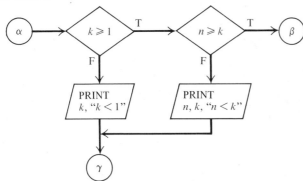

The computation between β and the output box contains a *loop*; that is, a part of the calculation that is computed *repeatedly* until $i = k$.

Such loops appear frequently and we have a special notation to indicate them:

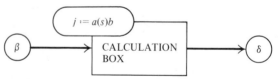

This *loop* means that j assumes all values from a in steps of s until b, where the step s may be negative. The flow chart for a *loop* can be written:

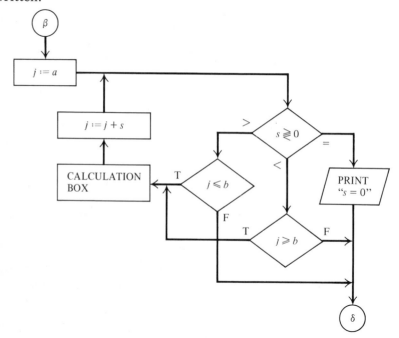

Note that to prevent confusion when lines cross, arrowheads are put at junctions only. Using a calculation loop, we can rewrite the flow chart for the present example:

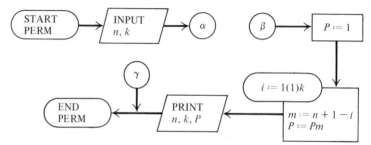

Note how the steps of the calculation have been modified to fit the requirements of the loop notation. Also note that here to initialize P, we simply write $P := 1$.

Let us follow again the computation of $P(4, 2)$.

> Input: $n = 4$, $k = 2$
> $P = 1$
> $a = 1$, $s = 1$, $b = k = 2$, $j = i = 1$
> $1 > 0$, test
> $1 < 2$, test
> $m = 4$, $P = 4$
> $i = 2$
> $1 > 0$, test
> $2 = 2$, test
> $m = 3$, $P = 12$
> $i = 3$
> $1 > 0$, test
> $3 > 2$, test
> Print: $n = 4$, $k = 2$, $P = 12$.

DO EXERCISE 4.

4. Modify the flow chart of Exercise 3 to use connectors to separate the input test from the remainder of the computation, modify the input test to insure that $k \geqslant 1$ and $k \leqslant 10$, and use loop notation. Verify step by step that $C(4, 2)$ is computed correctly.

Example 3 Write a flow chart to input n ordered pairs of data: $\{x_i, p_i \mid 1 \leqslant i \leqslant n\}$. Check the input to insure that all the p_i values are between 0 and 1, and compute the expected value. Note that we have written the ordered pair (x_i, p_i) as simply x_i, p_i *without* parentheses since in programming *all* quantities are assumed to be ordered unless specified otherwise.

Solution First, we input the data and write this much of the flow chart:

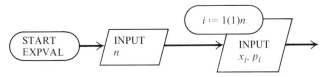

Here we first input n and then set up a loop to input the n data pairs. Note that we have written the loop with a parallelogram rather than a rectangle. Inputting the data thus permits us to input *varying* amounts of data in a simple manner.

Next we must test the input p_i to insure that $p_i \geqslant 0$ and $p_i \leqslant 1$. This

5. The "TEST" routine of Example 3 ends if *one* value of p_i is either <0 or >1 and prints out this *one* value. Modify this flow chart to print out *all* values of p_i that are either <0 or >1.

is done in a box marked "TEST" for which we can write

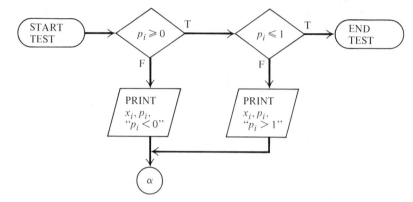

and continue the flow chart to obtain:

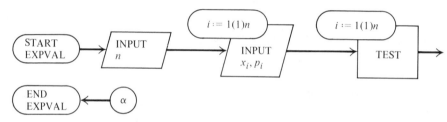

We have written the subflow-chart "TEST" to avoid cluttering up the main flow chart "EXPVAL" and to avoid having to recopy it.

Now from Section 9.2 the expected value is given by

$$E(X) = \sum_{i=1}^{n} x_i p_i.$$

Using this, we conclude the flow chart:

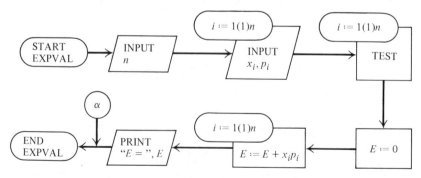

Note that here we must initialize E.

DO EXERCISE 5.

EXERCISE SET B.1

Write flow charts to do what is asked for.

1. $S = \sum_{i=0}^{5} i^2 = 0^2 + 1^2 + 2^2 + 3^2 + 4^2 + 5^2.$

2. $S = \sum_{i=0}^{4} (i + 1)^3.$

3. $S = \sum_{i=m}^{n} i^2$, where m and n are input. Check to insure that $n \geq m$; otherwise end.

4. $S = \sum_{i=0}^{n} (i + a)^3$, where n and a are input. Check to insure that $n \geq 0$; otherwise end.

5. Modify Exercise 1 to sum only odd values of i.

6. Modify Exercise 2 to sum only even values of i.

7. Use the EXPVAL program of Example 3 and compute the standard deviation of the given data.

8. Use the EXPVAL program of Example 3 and compute the quantity

$$Q = \frac{1}{\bar{x}^3} \cdot \sum_{i=1}^{n} (x_i - \bar{x})^3 p_i.$$

9. The integral $I = \int_{x_0}^{x_n} y(x) \, dx$ can be approximated by using the trapezoidal rule:

$$I_T = \frac{h}{2}\left[y_0 + 2 \sum_{i=1}^{n-1} y_i + y_n \right],$$

where

$$h = \frac{x_n - x_0}{n} \quad \text{and} \quad y_i = y(x_i).$$

Draw a flow chart for the trapezoidal rule.

10. The integral of Exercise 9 can also be approximated by Simpson's rule:

$$I_S = \frac{h}{3}[(y_0 + 4y_1 + y_2)$$
$$+ (y_2 + 4y_3 + y_4)$$
$$+ \cdots + (y_{n-2} + 4y_{n-1} + y_n)],$$

where n must be even and h and y_i are defined the same way as in Exercise 9.

11. Input a set of n different numbers x_i and print out a list of these numbers in algebraically increasing order. Verify the flow chart step by step with $n = 5$ and $x_1 = 5$, $x_2 = 1$, $x_3 = 2$, $x_4 = 9$, $x_5 = 4$. *Hint.* Aside from an input loop, flow-chart the calculation without loops. Answers may vary.

12. Write a flow chart to determine whether a given inputted number is an integer or not. Assume that the magnitude of the number is no larger than 10^{10}.

B.2 PROGRAMMING IN "BASIC"*

As a person can communicate with other people in various languages such as English, French, or Spanish, a person can communicate with high-speed computing machines in various languages such as BASIC, FORTRAN, COBOL, or Pascal. The particular language understood by the machine depends upon the "software" or internal programming of the machine.

OBJECTIVE

You should be able to program a simple algorithm in the BASIC programming language.

---

*BASIC is an acronym for Beginner's All-purpose Symbolic Instruction Code. For further details see "A Guide to BASIC Programming," 2nd Edition, by Donald D. Spencer (Addison-Wesley Publishing Company, Reading, Massachusetts, 1975).

When we communicate with another person, we are primarily interested in how to express our thoughts clearly and only secondarily in the details of how the other person hears and understands us. Similarly, our interest here is primarily in how we can express an algorithm so that the computing machine can understand it. By "understand" we mean that the machine is able to execute an inputted program as intended by the programmer.

BASIC, one of the simplest languages, is usually run on a typewriter-like *terminal* connected to the computer by wires or phone. Many terminals may share the same computer. This is called *time-sharing* and permits one to *interact* with the computer in a "conversational" mode.

Each institution has its own way of accessing the computer via the terminal. This the student is encouraged to learn. Our concern here is with the BASIC programming language common to all institutions.

Example 1 Write a program in BASIC to calculate

$$y = \frac{x^3 - 3}{A^2 + abx}$$

and print x and y where A, a, and b are given constants and x is input. Let $A = 2$, $a = -1{,}121.3$, and $b = 2 \cdot 10^{-3}$.

Solution First, we flow-chart this calculation even though it is simple:

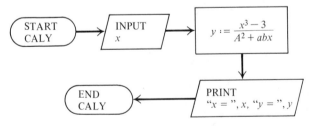

First, let us *start* the program with an *identification:*

100 REM CALY

Each instruction must start with a statement number (any integer between 1 and 99999), in this case 100. We use increments of 100 (rather than 1) for succeeding statement numbers to permit insertion of new statements between current ones later on if desired.

Next the instruction "REMARK" or "REM," for short, although ignored by the machine, is included to permit the programmer to make appropriate remarks. In this case we simply remark that this program is *named* "CALY".

A machine programmed for BASIC does not recognize spaces, so that we could just as well write the above instruction

100REMCALY

However, we use spaces in programming as we do in personal communications. While in personal communications spaces aid the reader, in BASIC spaces aid the programmer.

BASIC (along with most programming languages) uses only *capital* letters. Hence, in the equation of this example we can let

$$X \text{ represent } x$$

and

$$B \text{ represent } b.$$

But what do we do with a and A? In BASIC a variable or parameter can be represented by a one- or two-character name. A *one-character* name is simply a *letter* of the alphabet. A *two-character* name is a letter of the alphabet followed by a one-digit number.

Thus, we can let

$$A \text{ represent } a$$

and

$$A1 \text{ represent } A.$$

Other variable names could have been used; for example

$$A1 \text{ to represent } a$$

and

$$A2 \text{ to represent } A,$$

or whatever name one chooses.

In any case we might want to make note of this representation:

200 REM A1 = CAPITAL A
300 REM A = SMALL A
400 REM B = SMALL B

Such remarks are frequently used to avoid confusion in naming constants and variables.

Since the capital letter O and the number 0 look so much alike, some people and/or institutions adopt the convention of putting a / through

6. Given the function

$$z = \frac{(y - a)^2 + b}{B + y^4}$$

with $a = 1.1$, $b = -4$, $B = 3 \cdot 10^2$, y is inputted, and y and z are to be printed out.

a) Draw a flow chart for the calculation of z.

b) Identify the program with a REM instruction.

c) Indicate the BASIC naming of the constants with REM instructions.

Answers may vary.

7. Continuing with Margin Exercise 6, code the constants using LET instructions.

8. Continuing with Margin Exercise 6, assign to y a value, say $y = 5$.

the letter O, e.g., Ø, and not through the number 0. Others put a / through the number 0, e.g., Ø, and not through the letter O. Since in print the letter O is *fatter* than the number 0, *we* use these two characters as is without a /, and leave it to each individual to use whatever convention is desired for this and other look-alikes.

BEFORE CONTINUING WITH EXAMPLE 1, DO EXERCISE 6.

Continuing with Example 1:

Since A, a, and b are given constants, we know their values:

$$A = 2$$
$$a = -1,121.3$$
$$b = 2 \cdot 10^{-3}.$$

We code this

500 LET A1 = 2
600 LET A = −1121.3
700 LET B = 2E−3

where commas in numbers are omitted.

Note that BASIC requires the instruction "LET" before the assignment of numerical values.

Numbers are written as in ordinary algebra except for "scientific" or "exponential" notation as in $b = 2 \cdot 10^{-3}$ which in BASIC is written B = 2E−3. Integers can be written with or without decimal points. Negative signs for numbers or exponents must be used when needed but positive signs may be omitted.

DO EXERCISE 7.

Continuing with Example 1:

"Inputting" x can be done in a variety of ways. One way is simply to use the LET instruction to assign to x a value, say $x = 1$:

800 LET X = 1

We shall consider other means of input later on.

DO EXERCISE 8.

Continuing with Example 1:

We are now ready to code the calculation of y. There are three factors which must be kept in mind: first, the operational symbols

$(+, -, *, /, \uparrow)$; second, the priority of operation; and third, the appropriate use of parentheses.

The fundamental BASIC operations are

1. Addition: +
   The *sum A plus B* is written $A + B$ as in algebra.

2. Subtraction: −
   The *difference A minus B* is written $A - B$ as in algebra.

3. Multiplication: *
   The *product A times B* is written $A * B$. *Note.* We cannot imply the product by writing AB (omitting the *). Some one operational symbol must always be used between two BASIC quantities (constants and/or variables). Thus, the product of A and $-B$ can be written $-B * A$ but not $A * -B$.

4. Division: /
   The *quotient A divided by B* is written A/B.

5. Exponentiation: $\uparrow$
   The quantity *A raised to the power B*, A^B, is written $A \uparrow B$. Some machines use $* *$ in place of $\uparrow$. This is the only exception to the rule that two operational symbols cannot be used adjacently.

All operations are carried out from *left to right* and in algebraic order; that is, in the order

1. Exponentiation ($\uparrow$),
2. Multiplication (*) and division (/), and
3. Addition (+) and subtraction (−).

Thus, $A * B \uparrow C + D$ corresponds to $AB^C + D$ since this calculation is done as follows:

1. $B \uparrow C$: raise B to the power C,
2. $A * B \uparrow C$: multiply the result of (1) by A, and
3. $A * B \uparrow C + D$: add D to the result of (2).

We use *parentheses* as in ordinary algebra to group together quantities to be treated as a single entity, so that the priority of operations is changed. For example, consider $A * B + C$ and $A * (B + C)$:

1. $A * B + C$: In this case we multiply A and B, then add C.
2. $A * (B + C)$: In this case we add B and C, then multiply by A.

Parentheses *may* also be used for emphasis or clarity. For example, $A \uparrow B \uparrow C$ can be written $(A \uparrow B) \uparrow C$ to avoid confusion with $A \uparrow (B \uparrow C)$.

9. Continuing with Margin Exercise 6, code the variable z:

a) Code the numerator and denominator separately.

b) Code in one instruction.

Before coding y using the expression

$$y = \frac{x^3 - 3}{A^2 + abx},$$

recall that if one were doing this calculation by hand, one might calculate the numerator (N) and the denominator (D) separately and then divide N by D to obtain y:

$$N = x^3 - 3$$
$$D = A^2 + abx$$
$$y = \frac{N}{D}$$

Each of these expressions can be coded:

900 LET N = X↑3 − 3
1000 LET D = A1↑2 + A ∗ B ∗ X
1100 LET Y = N/D

Or these three instructions can be combined into a single instruction using parentheses:

900 LET Y = (X↑3 − 3)/(A1↑2 + A ∗ B ∗ X)

Note that in the preceding instructions the quantity between LET and the = sign is the *assigned* value of a single BASIC variable. This corresponds to our use of := (assign) in the flow charts. BASIC does not use an := sign (actually a combination of the *two* signs : and =); consequently, it uses the = sign to mean "assign."

The "assign" nature of the operation can be illustrated by noting that while we *can* write the *equation*

$$x^3 = y,$$

we *cannot* write a corresponding BASIC *instruction*

LET X↑3 = Y

since X↑3 is not a single BASIC variable.

DO EXERCISE 9.

Continuing with Example 1:

We have so far computed y and its numerical value is stored in the memory of the machine. Next we must instruct the machine to *output* the information we want.

Since BASIC is usually used on a typewriter-like terminal in *interactive* or *conversational* mode, we can write

1000 PRINT X, Y

Where the quantities to be printed out are separated by *commas*. In order to identify the numbers printed out, we label the output by replacing the preceding PRINT instruction by

1000 PRINT "X=", X, "Y=", Y

The characters in quotes will be printed just as they appear (including blank spaces). Thus, this instruction will print "X=" followed by the numerical value of X, then "Y=" followed by the numerical value of Y.

The *last* instruction of any BASIC program must always be an END instruction which tells the machine that the physical end of the program has been reached. Thus, we write

1100 END

At this point we can type into the terminal (with no statement number)

LIST

and the machine will print out our list of instructions:

```
100 REM CALY
200 REM A1 = CAPITAL A
300 REM A = SMALL A
400 REM B = SMALL B
500 LET A1 = 2
600 LET A = −1121.3
700 LET B = 2E−3
800 LET X = 1
900 LET Y = (X↑3 − 3)/(A1↑2 + A * B * X)
1000 PRINT "X=", X, "Y=", Y
1100 END
```

Now we type into the terminal (with no statement number)

RUN

and the machine will do our computation and print out

X= 1   Y= −1.13804

where BASIC rounds numbers to six significant figures and omits trailing zeroes.

10. Continuing with Margin Exercise 6,

a) Code the print instruction, identifying the output, and terminate the program.

b) LIST the program

c) If you have a terminal available, run the program and obtain a printout.

11. Modify the program of Margin Exercises 6 through 9 to input y in the interactive or conversational mode with the machine. Indicate the machine response and your response to the machine.

DO EXERCISE 10.

Example 2 Modify the program of Example 1 to *input x* in *interactive* or *conversational* mode with the machine.

Solution The LET instruction of Example 1

800 LET X = 1

is replaced by the INPUT instruction

800 INPUT X

Then, when we type

RUN

the machine prints out

?

This is how the machine asks for the numerical value of the input.

We respond by typing the input following the question mark:

? 1

If the input consists of more than one number, we use commas to separate them. Now the machine will proceed as before and give the same printout.

If we want another calculation, we type RUN again and supply the required input.

DO EXERCISE 11.

Example 3 Modify the program of Example 1 to input a *set* of values for *x* and print the output in a *table*.

Solution Replace the LET instruction of Example 1

800 LET X = 1

by the READ instruction

800 READ X

and the DATA instruction

1050 DATA 2.2, 3.3, 4.4, 5.5, 6.6, 7.7

where the data could be any set of desired *x*-values.

The DATA instruction may be placed *anywhere* in the program (before or after the READ instruction) but is usually inserted just before the END instruction.

The READ instruction causes the quantities following READ to be assigned sequentially the numerical values given in the DATA instruction. Thus, the *given* DATA instruction can be used to assign one variable six values or, if the READ instruction were followed by two variable names, these would be assigned three sets of values (for a total of $2 \times 3 = 6$ values).

The present PRINT instruction will yield results. However, if we want the output in tabular form, let us insert

750 PRINT "X", "Y"
760 PRINT

This causes the table headings X and Y to be printed *before* the calculation starts, followed by a blank line. Then we replace instruction 1000 with

1000 PRINT X, Y

This causes the numerical values of X and Y to be printed under the printed headings for each value of X in the data.

Listing the program, we now have

$$\left.\begin{array}{l} 100 \\ \quad \cdot \\ \quad \cdot \\ \quad \cdot \\ 700 \end{array}\right\} \text{no change}$$

750 PRINT "X", "Y"
760 PRINT
800 READ X
900 LET Y = (X↑3 − 3)/(A1↑2 + A * B * X)
1000 PRINT X, Y
1050 DATA 2.2, 3.3, 4.4, 5.5, 6.6, 7.7
1100 END

DO EXERCISE 12.

Example 4   Modify the program of Example 3 to handle the data internally allowing for the constant-increment nature of the input.

Solution   First, let us revise the flow chart of Example 1 to allow

12. Modify the program of Margin Exercises 6 through 9 to input the set of values for y = 2.4, 4.8, 6.0, 8.4 and print the output in a table. LIST your program.

for the constant-increment nature of the input.

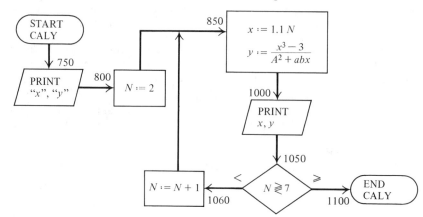

We have introduced a variable N to use the *test* since it is advisable to *test on an integer* if possible (to avoid round-off error problems).

Let us start with a list of the program of Example 3:

```
100 REM CALY
200 REM A1 = CAPITAL A
300 REM A = SMALL A
400 REM B = SMALL B
500 LET A1 = 2
600 LET A = −1121.3
700 LET B = 2E−3
750 PRINT "X", "Y"
760 PRINT
800 READ X
900 LET Y = (X↑3 − 3)/(A1↑2 + A * B * X)
1000 PRINT X, Y
1050 DATA 2.2, 3.3, 4.4, 5.5, 6.6, 7.7
1100 END
```

Now replace 800 by

```
800 LET N = 2
```

and add

```
850 LET X = 1.1 * N
```

Delete 1050 data and add

```
1050 IF N >= 7 THEN 1100
1060 LET N = N + 1
1070 GO TO 850
```

For convenience in following the instructions, let us list the modified program before explaining it.

$$
\left. \begin{array}{l} 100 \\ \quad . \\ \quad . \\ \quad . \\ 700 \end{array} \right\} \text{ as before}
$$

```
750 PRINT "X", "Y"
760 PRINT
800 LET N = 2
850 LET X = 1.1 * N
900 LET Y = (X↑3 − 3)/(A1↑2 + A * B * X)
1000 PRINT X, Y
1050 IF N >= 7 THEN 1100
1060 LET N = N + 1
1070 GO TO 850
1100 END
```

The IF–THEN instruction (1050) is interpreted "If $N \geq 7$, then go to instruction 1100, otherwise continue with the next (1060) instruction".

The BASIC relational symbols used in the IF–THEN instruction are similar to the usual mathematical symbols:

BASIC symbol	Mathematical symbol
=	=
>	>
<	<
>=	$\geq$
<=	$\leq$
< >	$\neq$

The instruction

1060 LET N = N + 1

emphasizes the assign (:=) nature of the BASIC = sign.

The instruction

1070 GO TO 850

means just what it says: "850 is the next instruction to be *executed*".

The statement numbers from the BASIC program have been added to the flow chart to aid identification.

The printout from this will be the same as for Example 3.

13. Modify the program of Margin Exercise 12 to handle the data internally, allowing for the constant-increment nature of the input. Draw a flow chart and identify the essential features with statement numbers.

DO EXERCISE 13

Example 5   Modify the program of Example 4 to use a *loop* since the flow chart of Example 4 can also be written:

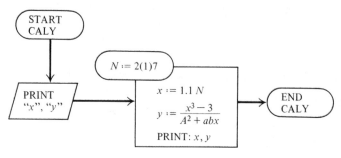

Solution   This program is written

$$\left.\begin{array}{l} 100 \\ \cdot \\ \cdot \\ \cdot \\ 700 \end{array}\right\} \text{as before}$$

```
750 PRINT "X", "Y"
760 PRINT
800 FOR N = 2 TO 7
850 LET X = 1.1 * N
900 LET Y = (X ↑ 3 − 3)/(A1 ↑ 2 + A * B * X)
1000 PRINT X, Y
1050 NEXT N
1100 END
```

The loop indicated on the flow chart by

N := F(S)L

is written in BASIC

FOR N = F TO L STEP S
  .
  .
  .
NEXT N

with appropriate statement numbers.

The step size S can have any nonzero value. The convention has been adopted to omit the STEP S part of the FOR instruction when $S = 1$, as in the present case. The instruction NEXT N signals the end of the calculation in the loop.

DO EXERCISE 14.

Some problems are conveniently expressed in terms of *subscripted variables*.

Example 6 Modify the program of Example 5 to use subscripted variables as indicated in the following flow chart:

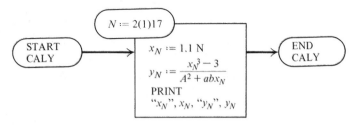

Note that the upper limit of the loop has been increased from 7 to 17.

Solution The subscripted variable x_N is written in BASIC as $X(N)$ which can take on the 11 values $X(0)$, $X(1)$, ..., $X(10)$. Whenever we need more than 11 values for a subscripted variable, we use a DIMENSION statement:

150 DIM X(17), Y(17)

This allows for 18 values for each of $X(N)$ and $Y(N)$ where $N = 0, 1, \ldots, 17$. The DIM statement (or instruction) can be placed anywhere in the program but is usually placed at the beginning.

Continuing with the program, we write

$$\left.\begin{array}{l} 100 \\ \cdot \\ \cdot \\ \cdot \\ 700 \end{array}\right\} \text{ as before}$$

800 FOR N = 2 TO 17
850 LET X(N) = 1.1 * N
900 LET Y = (X(N) ↑ 3 − 3)/(A1 ↑ 2 + A * B * X(N))
1000 PRINT "X(N)", X(N), "Y(N)", Y(N)
1050 NEXT N
1100 END

DO EXERCISE 15.

14. Modify the program of Margin Exercise 13 to use a loop.

15. Modify the program of Margin Exercise 14 to use subscripted variables and $N := 1(1)\ 14$ in the loop.

EXERCISE SET B.2

In Exercises 1 through 8 assign BASIC variable names to the indicated quantities. Answers may vary.

1. a, a_1, b, B **2.** c, C, c_1 **3.** e, e_1, e' **4.** h, h', H

5. u, u_{init}, u_{fin} **6.** v, v_{lo}, v_{hi} **7.** $s, s_{01}, s_{02}, s_{12}$ **8.** $t, t_{01}, t_{11}, t_{12}$

In Exercises 9 through 24 write the given algebraic assignments in BASIC. Answers may vary.

9. $x := 5bB + .71 \times 10^{-2}$

10. $y := 17.3a + .014b$

11. $u := \dfrac{3.7}{b'} + 4.27 \times 10^3$

12. $v := \dfrac{4a + 1.8 \times 10^3}{c}$

13. $s := a^2 b + AB^2$

14. $t := aa_1^2 A^3$

15. $a := \dfrac{x_1^2}{y_1} + \dfrac{y_2}{x_2^2}$

16. $c := \dfrac{x_1^2 + y_1}{x_2 + y_2^2}$

17. $y := \dfrac{x^2}{ab}$

18. $u := \dfrac{ab}{6xy}$

19. $z := \sqrt{x^2 + \dfrac{1}{y^2}}$

20. $y := (c^2 + d^3)^{1/3}$

21. $u := [(5 + .3x)^2 + (.02 + 1.9y)^2]^{1/2}$

22. $v := \dfrac{(.321x + .07y^2)u}{.9x^2 + 47y}$

23. $X := \left(\dfrac{3x^4 - 7ay^2}{1.42xy + 1.3b}\right)^{1/3}$

24. $Y := \sqrt{\dfrac{4a^2b - 1.207 \times 10^3\, c^3}{a^4 + 1.7c}}$

In Exercises 25 through 32 write a BASIC program to do what is asked for. Use REM instructions to identify the program and to distinguish BASIC variable names when appropriate. Print out (with identification) all input and the calculated quantity. Assume numerical values for parameters where appropriate.

25. Calculate $A := A_0(1 + 1/n)^p$ where

a) n and p are coded in (LET instructions) and A_0 is read from DATA.
b) n, p, and A_0 are read from DATA.
c) n and p are read from DATA and A_0 is input interactively.

26. Calculate $P := \dfrac{P_0}{1 + ax^2}$ where

a) P_0 and a are coded in and x is read from DATA.
b) P_0, a, and x are read from DATA.
c) P_0 and a are read from DATA and x is input interactively.

27. Calculate $x' := \dfrac{x_1^2}{a + x_2^3}$ where

a) a is coded in and x_1 and x_2 are read from DATA.
b) a, x_1, and x_2 are read in from DATA.
c) a is read from DATA and x_1 and x_2 are input interactively.
d) a and three sets of x_1, x_2 values are read from one DATA statement.

28. Calculate $z := c(x^2 + y)^n$ where

a) c and n are coded in and x and y are read from DATA.
b) c, n, x, and y are read from DATA.
c) c and n are read from DATA and x and y are input interactively.
d) c is coded in, n is input interactively, and three sets of x, y values are read from DATA.

29. Continuing with Exercise 25, revise the program to print the output in a table where

a) n and p are coded in and $A_0 :=$ 150, 300, 450, 600, 750, 900 are read from DATA.

b) The A_0 values from (a) are determined using an IF–THEN instruction (see Example 4).

c) The A_0 values from (a) are determined using a FOR–TO–STEP loop (see Example 5).

31. Continuing with Example 27, revise the program to print the output in a table where

a) a is coded in and $(x_1, x_2) :=$ (1, 2), (2, 4), (3, 6), (4, 8) are read from DATA.

b) The x_1, x_2 values from (a) are determined using an IF–THEN instruction.

c) The x_1, x_2 values from (a) are determined using a FOR–TO–STEP instruction.

d) We want to calculate x' such that for *each* value of $x_1 :=$ 1, 2, 3, 4, we set $x_2 :=$ 2, 4, 6, 8.

30. Continuing with Exercise 26, revise the program to print the output in a table where

a) P_0 and a are coded in and $x :=$ 7, 14, 21, 28, 35 are read from DATA.

b) The x values from (a) are determined using an IF–THEN instruction (see Example 4).

c) The x values from (a) are determined using a FOR–TO–STEP instruction (see Example 5).

32. Continuing with Exercise 28, revise the program to print the output in a table where

a) c and n are coded in and $(x, y) :=$ (1, 5), (2, 1), (3, 1.5), (4, 2) are read from DATA.

b) The x, y-values of (a) are determined using an IF–THEN instruction.

c) The x, y-values of (a) are determined using a FOR–TO–STEP instruction.

d) We want to calculate P such that for *each* value of $x :=$ 1, 2, 3, 4, we set $y :=$ .5, 1, 1.5, 2.

Using flow charts from Exercise set B.1, program the following using appropriate printouts:

33. (Exercise B.1:1) $S = \sum_{i=0}^{5} i^2$.

34. (Exercise B.1:2) $S = \sum_{i=0}^{4} (i + 1)^3$.

35. (Exercise B.1:3) $S = \sum_{i=m}^{n} i^2$; terminate if $m > n$.

36. (Exercise B.1:4) $S = \sum_{i=0}^{n} (i + a)^3$. Terminate if $n < 0$.

37. (Exercise B.1:5) $S = \sum_{i=0}^{5} i^2$, for odd i.

38. (Exercise B.1:6) $S = \sum_{i=0}^{4} (i + 1)^3$, for even i.

39. Write a program to evaluate $P(n, k)$. See Example 2, Section B.1.

40. Write a program to evaluate $C(n, k)$. See Marginal Exercise 3, Section B.1.

41. Write a program for the Trapezoidal Rule, allowing for $n \leq 100$. (Exercise B.1:9).

42. Write a program for Simpson's Rule (Exercise B.1:10). Check to insure that n is even and allow for $n \leq 100$. *Hint.* The BASIC instruction INT(M) yields the integer part of M.

43. Write a program to order a set of numbers where $n \leq 100$. See Exercise B.1:11.

44. Write a program to determine whether a given number is integer without using the BASIC INT instruction. See Exercise B.1:12.

Tables

Table 1
Compound Interest

Period	1¼%	1½%	1¾%	2%	2½%	3%
1	1.012500	1.015000	1.017500	1.020000	1.025000	1.030000
2	1.025156	1.030225	1.035306	1.040400	1.050625	1.060900
3	1.037970	1.045678	1.053424	1.061208	1.076891	1.092727
4	1.050945	1.061363	1.071859	1.082432	1.103813	1.125509
5	1.064082	1.077283	1.090617	1.104081	1.131408	1.159274
6	1.077383	1.093442	1.109703	1.126163	1.159693	1.194052
7	1.090850	1.109844	1.129123	1.148686	1.188685	1.229874
8	1.104486	1.126492	1.148883	1.171660	1.218402	1.266770
9	1.118292	1.143389	1.168988	1.195093	1.248862	1.304773
10	1.132271	1.160540	1.189445	1.218995	1.280084	1.343916
11	1.146424	1.177948	1.210260	1.243375	1.312086	1.384233
12	1.160754	1.195617	1.231440	1.268243	1.344888	1.425760
13	1.175263	1.213551	1.252990	1.293608	1.378510	1.468533
14	1.189954	1.231754	1.274917	1.319480	1.412973	1.512589
15	1.204828	1.250230	1.297228	1.345870	1.448297	1.557967
16	1.219888	1.268983	1.319929	1.372787	1.484504	1.604706
17	1.235137	1.288018	1.343028	1.400243	1.521617	1.652847
18	1.250576	1.307338	1.366531	1.428248	1.559657	1.702432
19	1.266208	1.326948	1.390445	1.456813	1.598648	1.753505
20	1.282036	1.346852	1.414778	1.485949	1.638614	1.806110
21	1.298061	1.367055	1.439537	1.515668	1.679579	1.860293
22	1.314287	1.387561	1.464729	1.545981	1.721568	1.916102
23	1.330716	1.408374	1.490362	1.576901	1.764607	1.973585
24	1.347350	1.429500	1.516443	1.608439	1.808722	2.032793
25	1.364192	1.450943	1.542981	1.640608	1.853940	2.093777
26	1.381244	1.472707	1.569983	1.673420	1.900289	2.156590
27	1.398510	1.494798	1.597458	1.706888	1.947796	2.221288
28	1.415991	1.517220	1.625414	1.741026	1.996491	2.287927
29	1.433691	1.539978	1.653859	1.775847	2.046403	2.356565
30	1.451612	1.563078	1.682802	1.811364	2.097563	2.427262
31	1.469757	1.586524	1.712251	1.847591	2.150002	2.500080
32	1.488129	1.610322	1.742215	1.884543	2.203752	2.575082
33	1.506731	1.634477	1.772704	1.922234	2.258846	2.652334
34	1.525565	1.658994	1.803726	1.960679	2.315317	2.731904
35	1.544635	1.683879	1.835291	1.999893	2.373200	2.813861
36	1.563943	1.709137	1.867409	2.039891	2.432530	2.898277
37	1.583492	1.734774	1.900089	2.080689	2.493343	2.985225
38	1.603286	1.760796	1.933341	2.122303	2.555677	3.074782
39	1.623327	1.787208	1.967174	2.164749	2.619569	3.167025
40	1.643619	1.814016	2.001600	2.208044	2.685058	3.262036
41	1.664164	1.841226	2.036628	2.252205	2.752184	3.359897
42	1.684966	1.868844	2.072269	2.297249	2.820989	3.460694
43	1.706028	1.896877	2.108534	2.343194	2.891514	3.564515
44	1.727353	1.925330	2.145433	2.390058	2.963802	3.671450
45	1.748945	1.954210	2.182978	2.437859	3.037897	3.781594
46	1.770807	1.983523	2.221180	2.486616	3.113844	3.895042
47	1.792942	2.013276	2.260051	2.536348	3.191690	4.011893
48	1.815354	2.043475	2.299602	2.587075	3.271482	4.132250
49	1.838046	2.074127	2.339845	2.638817	3.353269	4.256218
50	1.861022	2.105239	2.380792	2.691593	3.437101	4.383905

TABLE 1 **499**

Table 1 (*Continued*)

Period	1¼%	1½%	1¾%	2%	2½%	3%
51	1.884285	2.136818	2.422456	2.745425	3.523029	4.515422
52	1.907839	2.168870	2.464849	2.800334	3.611105	4.650885
53	1.931687	2.201403	2.507984	2.856341	3.701383	4.790412
54	1.955833	2.234424	2.551874	2.913468	3.793918	4.934124
55	1.980281	2.267940	2.596532	2.971737	3.888766	5.082148
56	2.005035	2.301959	2.641971	3.031172	3.985985	5.234612
57	2.030098	2.336488	2.688205	3.091795	4.085635	5.391650
58	2.055474	2.371535	2.735249	3.153631	4.187776	5.553400
59	2.081167	2.407108	2.783116	3.216704	4.292470	5.720002
60	2.107182	2.443215	2.831821	3.281038	4.399782	5.891602
61	2.133522	2.479863	2.881378	3.346659	4.509777	6.068350
62	2.160191	2.517061	2.931802	3.413592	4.622521	6.250401
63	2.187193	2.554817	2.983109	3.481864	4.738084	6.437913
64	2.214533	2.593139	3.035313	3.551501	4.856536	6.631050
65	2.242215	2.632036	3.088431	3.622531	4.977949	6.829982
66	2.270243	2.671517	3.142479	3.694982	5.102398	7.034881
67	2.298621	2.711590	3.197472	3.768882	5.229958	7.245927
68	2.327354	2.752264	3.253428	3.844260	5.360707	7.463305
69	2.356446	2.793548	3.310363	3.921145	5.494725	7.687204
70	2.385902	2.835451	3.368294	3.999568	5.632093	7.917820
71	2.415726	2.877983	3.427239	4.079559	5.772895	8.155355
72	2.445923	2.921153	3.487216	4.161150	5.917217	8.400016
73	2.476497	2.964970	3.548242	4.244373	6.065147	8.652016
74	2.507453	3.009445	3.610336	4.329260	6.216776	8.911576
75	2.538796	3.054587	3.673517	4.415845	6.372195	9.178923
76	2.570531	3.100406	3.737804	4.504162	6.531500	9.454291
77	2.602663	3.146912	3.803216	4.594245	6.694788	9.737920
78	2.635196	3.194116	3.869772	4.686130	6.862158	10.030058
79	2.668136	3.242028	3.937493	4.779853	7.033712	10.330960
80	2.701488	3.290658	4.006399	4.875450	7.209555	10.640889
81	2.735257	3.340018	4.076511	4.972959	7.389794	10.960116
82	2.769448	3.390118	4.147850	5.072418	7.574539	11.288919
83	2.804066	3.440970	4.220437	5.173866	7.763902	11.627587
84	2.839117	3.492585	4.294295	5.277343	7.958000	11.976415
85	2.874606	3.544974	4.369445	5.382890	8.156950	12.335707
86	2.910539	3.598149	4.445910	5.490548	8.360874	12.705778
87	2.946921	3.652121	4.523713	5.600359	8.569896	13.086951
88	2.983758	3.706903	4.602878	5.712366	8.784143	13.479560
89	3.021055	3.762507	4.683428	5.826613	9.003747	13.883947
90	3.058818	3.818945	4.765388	5.943145	9.228841	14.300465
91	3.097053	3.876229	4.848782	6.062008	9.459562	14.729479
92	3.135766	3.934372	4.933636	6.183248	9.696051	15.171363
93	3.174963	3.993388	5.019975	6.306913	9.938452	15.626504
94	3.214650	4.053289	5.107825	6.433051	10.186913	16.095299
95	3.254833	4.114088	5.197212	6.561712	10.441586	16.578158
96	3.295518	4.175799	5.288163	6.692946	10.702626	17.075503
97	3.336712	4.238436	5.380706	6.826805	10.970192	17.587768
98	3.378421	4.302013	5.474868	6.963341	11.244447	18.115401
99	3.420651	4.366543	5.570678	7.102608	11.525558	18.658863
100	3.463409	4.432041	5.668165	7.244660	11.813697	19.218629

(*Continued*)

Table 1 (*Continued*)

Period	3½%	4%	5%	6%	7%	8%
1	1.035000	1.040000	1.050000	1.060000	1.070000	1.080000
2	1.071225	1.081600	1.102500	1.123600	1.144900	1.166400
3	1.108718	1.124864	1.157625	1.191016	1.225043	1.259712
4	1.147523	1.169859	1.215506	1.262477	1.310796	1.360489
5	1.187686	1.216653	1.276281	1.338226	1.402552	1.469328
6	1.229255	1.265319	1.340095	1.418520	1.500731	1.586874
7	1.272279	1.315932	1.407100	1.503631	1.605782	1.713824
8	1.316809	1.368569	1.477455	1.593849	1.718187	1.850930
9	1.362897	1.423312	1.551328	1.689480	1.838460	1.999004
10	1.410598	1.480244	1.628894	1.790849	1.967152	2.158924
11	1.459969	1.539454	1.710339	1.898300	2.104853	2.331638
12	1.511068	1.601032	1.795856	2.012198	2.252193	2.518169
13	1.563955	1.665073	1.885649	2.132930	2.409847	2.719623
14	1.618693	1.731676	1.979931	2.260906	2.578536	2,937193
15	1.675347	1.800943	2.078928	2.396560	2.759034	3.172168
16	1.733984	1.872981	2.182874	2.540354	2.952166	3.425941
17	1.794673	1.947900	2.292018	2.692775	3.158818	3.700016
18	1.857487	2.025816	2.406619	2.854342	3.379935	3.996017
19	1.922499	2.106849	2.526950	3.025603	3.616530	4.315698
20	1.989786	2.191123	2.653298	3.207139	3.869687	4.660954
21	2.059429	2.278768	2.785963	3.399567	4.140565	5.033830
22	2.131509	2.369919	2.925261	3.603541	4.430405	5.436536
23	2.206112	2.464716	3.071524	3.819753	4.740533	5.871459
24	2.283326	2.563305	3.225100	4.048938	5.072370	6.341176
25	2.363242	2.665837	3.386355	4.291874	5.427436	6.848470
26	2.445955	2.772470	3.555673	4.549386	5.807357	7.396348
27	2.531563	2.883369	3.733457	4.822349	6.213872	7.988056
28	2.620168	2.998704	3.920130	5.111690	6.648843	8.627100
29	2.711874	3.118652	4.116137	5.418391	7.114262	9.317268
30	2.806790	3.243395	4.321944	5.743494	7.612260	10.062649
31	2.905028	3.373134	4.538041	6.088104	8.145118	10.867661
32	3.006704	3.508059	4.764943	6.453390	8.715276	11.737074
33	3.111939	3.648381	5.003190	6.840593	9.325345	12.676040
34	3.220857	3.794316	5.253350	7.251029	9.978119	13.690123
35	3.333587	3.946089	5.516018	7.686091	10.676587	14.785333
36	3.450263	4.103933	5.791819	8.147256	11.423948	15.968160
37	3.571022	4.268090	6.081410	8.636091	12.223624	17.245613
38	3.696008	4.438814	6.385481	9.154256	13.079278	18.625262
39	3.825368	4.616367	6.704755	9.703511	13.994827	20.115283
40	3.959256	4.801022	7.039993	10.285722	14.974465	21.724506
41	4.097830	4.993063	7.391993	10.902865	16.022678	23.462466
42	4.241254	5.192786	7.761593	11.557037	17.144265	25.339463
43	4.389698	5.400497	8.149673	12.250459	18.344364	27.366620
44	4.543337	5.616517	8.557157	12.985487	19.628469	29.555950
45	4.702354	5.841178	8.985015	13.764616	21.002462	31.920426
46	4.866936	6.074825	9.434266	14.590493	22.472634	34.474060
47	5.037279	6.317818	9.905979	15.465923	24.045718	37.231985
48	5.213584	6.570531	10.401278	16.393878	25.728918	40.210544
49	5.396059	6.833352	10.921342	17.377511	27.529942	43.427388
50	5.584921	7.106686	11.467409	18.420162	29.457038	46.901579

TABLE 1 **501**

Table 1 (*Continued*)

Period	3¹/₂%	4%	5%	6%	7%	8%
51	5.780393	7.390953	12.040779	19.525372	31.519031	50.653705
52	5.982707	7.686591	12.642818	20.696894	33.725363	54.706001
53	6.192102	7.994055	13.274959	21.938708	36.086138	59.082481
54	6.408826	8.313817	13.938707	23.255030	38.612168	63.809079
55	6.633135	8.646370	14.635642	24.650332	41.315020	68.913805
56	6.865295	8.992225	15.367424	26.129352	44.207071	74.426909
57	7.105580	9.351914	16.135795	27.697113	47.301566	80.381062
58	7.354275	9.725991	16.942585	29.358940	50.612676	86.811547
59	7.611675	10.115031	17.789714	31.120476	54.155563	93.756471
60	7.878084	10.519632	18.679200	32.987705	57.946452	101.256989
61	8.153817	10.940417	19.613160	34.966967	62.002704	109.357548
62	8.439201	11.378034	20.593818	37.064985	66.342893	118.106152
63	8.734573	11.833155	21.623509	39.288884	70.986896	127.554644
64	9.040283	12.306481	22.704684	41.646217	75.955979	137.759016
65	9.356693	12.798740	23.839918	44.144990	81.272898	148.779737
66	9.684177	13.310690	25.031914	46.793689	86.962001	160.682116
67	10.023123	13.843118	26.283510	49.601310	93.049341	173.536685
68	10.373932	14.396843	27.597686	52.577389	99.562795	187.419620
69	10.737020	14.972717	28.977570	55.732032	106.532191	202.413190
70	11.112816	15.571626	30.426449	59.075954	113.989444	218.606245
71	11.501765	16.194491	31.947771	62.620511	121.968705	236.094745
72	11.904327	16.842271	33.545160	66.377742	130.506514	254.982325
73	12.320978	17.515962	35.222418	70.360407	139.641970	275.380911
74	12.752212	18.216600	36.983539	74.582031	149.416908	297.411384
75	13.198539	18.945264	38.832716	79.056953	159.876092	321.204295
76	13.660488	19.703075	40.774352	83.800370	171.067418	346.900639
77	14.138605	20.491198	42.813096	88.828392	183.042137	374.652690
78	14.633456	21.310846	44.953724	94.158096	195.855087	404.624905
79	15.145627	22.163280	47.201410	99.807582	209.564943	436.994897
80	15.675724	23.049811	49.561481	105.796037	224.234489	471.954489
81	16.224374	23.971803	52.039555	112.143799	239.930903	509.710848
82	16.792227	24.930675	54.641533	118.872427	256.726066	550.487716
83	17.379955	25.927902	57.373610	126.004773	274.696891	594.526733
84	17.988253	26.965018	60.242291	133.565059	293.925673	642.088872
85	18.617842	28.043619	63.254406	141.578963	314.500470	693.455982
86	19.269466	29.165364	66.417126	150.073701	336.515503	748.932461
87	19.943897	30.331979	69.737982	159.078123	360.071588	808.847058
88	20.641933	31.545258	73.224881	168.622810	385.276599	873.554823
89	21.364401	32.807068	76.886125	178.740179	412.245961	943.439209
90	22.112155	34.119351	80.730431	189.464590	441.103178	1018.914346
91	22.886080	35.484125	84.766953	200.832465	471.980400	1100.427494
92	23.687093	36.903490	89.005301	212.882413	505.019028	1188.461694
93	24.516141	38.379630	93.455566	225.655358	540.370360	1283.538630
94	25.374206	39.914815	98.128344	239.194679	578.196285	1386.221720
95	26.262303	41.511408	103.034761	253.546360	618.670025	1497.119458
96	27.181484	43.171864	108.186499	268.759142	661.976927	1616.889015
97	28.132836	44.898739	113.595824	284.884691	708.315312	1746.240136
98	29.117485	46.694689	119.275615	301.977772	757.897384	1885.939347
99	30.136597	48.562477	125.239396	320.096438	810.950201	2036.814495
100	31.191378	50.504976	131.501366	339.302224	867.716715	2199.759655

Table 2
Present Value

Period	1¼%	1½%	1¾%	2%	2½%	3%
1	.987654	.985221	.982801	.980392	.975609	.970873
2	.975461	.970661	.965898	.961168	.951814	.942595
3	.963419	.956316	.949285	.942322	.928599	.915141
4	.951525	.942184	.932959	.923845	.905950	.888487
5	.939777	.928260	.916912	.905730	.883854	.862608
6	.928175	.914542	.901142	.887971	.862296	.837484
7	.916716	.901026	.885643	.870560	.841265	.813091
8	.905399	.887711	.870411	.853490	.820746	.789409
9	.894221	.874592	.855441	.836755	.800728	.766416
10	.883181	.861667	.840728	.820348	.781198	.744093
11	.872278	.848933	.826269	.804263	.762144	.722421
12	.861509	.836387	.812057	.788493	.743555	.701379
13	.850873	.824027	.798091	.773032	.725420	.680951
14	.840369	.811849	.784365	.757875	.707727	.661117
15	.829994	.799851	.770875	.743014	.690465	.641861
16	.819747	.788031	.757617	.728445	.673624	.623166
17	.809627	.776385	.744586	.714162	.657195	.605016
18	.799632	.764911	.731780	.700159	.641165	.587394
19	.789760	.753607	.719194	.686430	.625527	.570286
20	.780009	.742470	.706825	.672971	.610270	.553675
21	.770380	.731497	.694668	.659775	.595386	.537549
22	.760869	.720687	.682720	.646839	.580864	.521892
23	.751475	.710037	.670978	.634155	.566697	.506691
24	.742198	.699543	.659438	.621721	.552875	.491933
25	.733035	.689205	.648096	.609530	.539390	.477605
26	.723985	.679020	.636950	.597579	.526234	.463694
27	.715047	.668985	.625995	.585862	.513399	.450189
28	.706219	.659099	.615228	.574374	.500877	.437076
29	.697500	.649358	.604646	.563112	.488661	.424346
30	.688889	.639762	.594247	.552070	.476742	.411986
31	.680385	.630307	.584027	.541245	.465114	.399987
32	.671985	.620992	.573982	.530633	.453770	.388337
33	.663688	.611815	.564110	.520228	.442702	.377026
34	.655495	.602774	.554408	.510028	.431905	.366044
35	.647402	.593866	.544873	.500027	.421371	.355383
36	.639409	.585089	.535501	.490223	.411093	.345032
37	.631516	.576443	.526291	.480610	.401067	.334982
38	.623719	.567924	.517239	.471187	.391284	.325226
39	.616019	.559531	.508343	.461948	.381741	.315753
40	.608414	.551262	.499600	.452890	.372430	.306556
41	.600902	.543115	.491008	.444010	.363346	.297628
42	.593484	.535089	.482563	.435304	.354484	.288959
43	.586157	.527181	.474263	.426768	.345838	.280542
44	.578920	.519390	.466106	.418400	.337403	.272371
45	.571773	.511714	.458090	.410196	.329174	.264438
46	.564714	.504152	.450211	.402153	.321145	.256736
47	.557743	.496702	.442468	.394268	.313312	.249258
48	.550857	.489361	.434858	.386627	.305671	.241998
49	.544056	.482129	.427379	.378958	.298215	.234950
50	.537339	.475004	.420028	.371527	.290942	.228107

TABLE 2 **503**

Table 2 (*Continued*)

Period	1¹/₄%	1¹/₂%	1³/₄%	2%	2¹/₂%	3%
51	.530705	.467984	.412804	.364243	.283846	.221463
52	.524153	.461068	.405704	.357101	.276923	.215012
53	.517682	.454255	.398727	.350099	.270168	.208750
54	.511291	.447541	.391869	.343234	.263579	.202670
55	.504979	.440928	.385129	.336504	.257150	.196767
56	.498744	.434411	.378505	.329906	.250878	.191036
57	.492587	.427991	.371995	.323437	.244759	.185471
58	.486506	.421666	.365597	.317095	.238789	.180069
59	.480500	.415435	.359309	.310877	.232965	.174825
60	.474567	.409295	.353130	.304782	.227283	.169733
61	.468709	.403247	.347056	.298806	.221740	.164789
62	.462922	.397287	.341087	.292947	.216331	.159989
63	.457207	.391416	.335221	.287203	.211055	.155329
64	.451562	.385632	.329455	.281571	.205907	.150805
65	.445988	.379933	.323789	.276050	.200885	.146413
66	.440481	.374318	.318220	.270637	.195985	.142148
67	.435043	.368786	.312747	.265331	.191205	.138008
68	.429672	.363336	.307369	.260128	.186542	.133988
69	.424368	.357967	.302082	.255028	.181992	.130086
70	.419129	.352676	.296886	.250027	.177553	.126297
71	.413954	.347464	.291780	.245125	.173223	.122618
72	.408844	.342330	.286762	.240318	.168998	.119047
73	.403796	.337270	.281830	.235606	.164876	.115579
74	.398811	.332286	.276983	.230986	.160854	.112213
75	.393887	.327376	.272219	.226457	.156931	.108945
76	.389025	.322537	.267537	.222017	.153103	.105772
77	.384222	.317771	.262935	.217664	.149369	.102691
78	.379478	.313075	.258413	.213396	.145726	.099700
79	.374793	.308448	.253969	.209211	.142172	.096796
80	.370166	.303890	.249601	.205109	.138704	.093977
81	.365596	.299399	.245308	.201087	.135321	.091239
82	.361083	.294974	.241089	.197145	.132021	.088582
83	.356625	.290615	.236942	.193279	.128800	.086002
84	.352222	.286320	.232867	.189489	.125659	.083497
85	.347874	.282089	.228862	.185774	.122594	.081065
86	.343579	.277920	.224926	.182131	.119604	.078704
87	.339337	.273813	.221057	.178560	.116687	.076411
88	.335148	.269766	.217255	.175059	.113841	.074186
89	.331010	.265779	.213519	.171626	.111064	.072025
90	.326924	.261852	.209847	.168261	.108355	.069927
91	.322888	.257982	.206237	.164962	.105712	.067891
92	.318901	.254169	.202690	.161727	.103134	.065913
93	.314964	.250413	.199204	.158556	.100619	.063993
94	.311076	.246713	.195778	.155447	.098165	.062129
95	.307235	.243067	.192411	.152399	.095770	.060320
96	.303442	.239474	.189102	.149411	.093434	.058563
97	.299696	.235935	.185849	.146481	.091155	.056857
98	.295996	.232449	.182653	.143609	.088932	.055201
99	.292342	.229013	.179511	.140793	.086763	.053593
100	.288733	.225629	.176424	.138032	.084647	.052032

(*Continued*)

Table 2 (*Continued*)

Period	3½%	4%	5%	6%	7%	8%
1	.966183	.961538	.952380	.943396	.934579	.925925
2	.933510	.924556	.907029	.889996	.873438	.857338
3	.901942	.888996	.863837	.839619	.816297	.793832
4	.871442	.854804	.822702	.792093	.762895	.735029
5	.841973	.821927	.783526	.747258	.712986	.680583
6	.813500	.790314	.746215	.704960	.666342	.630169
7	.785990	.759917	.710681	.665057	.622749	.583490
8	.759411	.730690	.676839	.627412	.582009	.540268
9	.733731	.702586	.644608	.591898	.543933	.500248
10	.708918	.675564	.613913	.558394	.508349	.463193
11	.684945	.649580	.584679	.526787	.475092	.428882
12	.661783	.624597	.556837	.496969	.444011	.397113
13	.639404	.600574	.530321	.468839	.414964	.367697
14	.617781	.577475	.505067	.442300	.387817	.340461
15	.596890	.555264	.481017	.417265	.362446	.315241
16	.576705	.533908	.458111	.393646	.338734	.291890
17	.557203	.513373	.436296	.371364	.316574	.270268
18	.538361	.493628	.415520	.350343	.295863	.250249
19	.520155	.474642	.395733	.330512	.276508	.231712
20	.502565	.456386	.376889	.311804	.258418	.214548
21	.485570	.438833	.358942	.294155	.241513	.198655
22	.469150	.421955	.341849	.277505	.225713	.183940
23	.453285	.405726	.325571	.261797	.210946	.170315
24	.437957	.390121	.310067	.246978	.197146	.157699
25	.423147	.375116	.295302	.232998	.184249	.146017
26	.408837	.360689	.281240	.219810	.172195	.135201
27	.395012	.346816	.267848	.207367	.160930	.125186
28	.381654	.333477	.255093	.195630	.150402	.115913
29	.368748	.320651	.242946	.184556	.140562	.107327
30	.356278	.308318	.231377	.174110	.131367	.099377
31	.344230	.296460	.220359	.164254	.122773	.092016
32	.332589	.285057	.209866	.154957	.114741	.085200
33	.321342	.274094	.199872	.146186	.107234	.078888
34	.310476	.263552	.190354	.137911	.100219	.073045
35	.299976	.253415	.181290	.130105	.093662	.067634
36	.289832	.243668	.172657	.122740	.087535	.062624
37	.280031	.234296	.164435	.115793	.081808	.057985
38	.270561	.225285	.156605	.109238	.076456	.053690
39	.261412	.216620	.149147	.103055	.071455	.049713
40	.252572	.208289	.142045	.097222	.066780	.046030
41	.244031	.200277	.135281	.091719	.062411	.042621
42	.235779	.192574	.128839	.086527	.058328	.039464
43	.227805	.185168	.122704	.081629	.054512	.036540
44	.220102	.178046	.116861	.077009	.050946	.033834
45	.212659	.171198	.111296	.072650	.047613	.031327
46	.205467	.164613	.105996	.068537	.044498	.029007
47	.198519	.158282	.100949	.064658	.041587	.026858
48	.191806	.152194	.096142	.060998	.038866	.024869
49	.185320	.146341	.091563	.057545	.036324	.023026
50	.179053	.140712	.087203	.054288	.033947	.021321

TABLE 2 **505**

Table 2 (*Continued*)

Period	3½%	4%	5%	6%	7%	8%
51	.172998	.135300	.083051	.051215	.031726	.019741
52	.167148	.130096	.079096	.048316	.029651	.018279
53	.161495	.125093	.075329	.045581	.027711	.016925
54	.156034	.120281	.071742	.043001	.025898	.015671
55	.150758	.115655	.068326	.040567	.024204	.014510
56	.145660	.111207	.065072	.038271	.022620	.013435
57	.140734	.106930	.061974	.036104	.021140	.012440
58	.135975	.102817	.059022	.034061	.019757	.011519
59	.131377	.098862	.056212	.032133	.018465	.010665
60	.126934	.095060	.053535	.030314	.017257	.009875
61	.122641	.091404	.050986	.028598	.016128	.009144
62	.118494	.087888	.048558	.026979	.015073	.008466
63	.114487	.084508	.046245	.025452	.014087	.007839
64	.110615	.081258	.044043	.024011	.013165	.007259
65	.106875	.078132	.041946	.022652	.012304	.006721
66	.103261	.075127	.039949	.021370	.011499	.006223
67	.099769	.072238	.038046	.020160	.010746	.005762
68	.096395	.069459	.036234	.019019	.010043	.005335
69	.093135	.066788	.034509	.017943	.009386	.004940
70	.089986	.064219	.032866	.016927	.008772	.004574
71	.086943	.061749	.031301	.015969	.008198	.004235
72	.084003	.059374	.029810	.015065	.007662	.003921
73	.081162	.057090	.028391	.014212	.007161	.003631
74	.078417	.054895	.027039	.013408	.006692	.003362
75	.075765	.052783	.025751	.012649	.006254	.003113
76	.073203	.050753	.024525	.011933	.005845	.002882
77	.070728	.048801	.023357	.011257	.005463	.002669
78	.068336	.046924	.022245	.010620	.005105	.002471
79	.066025	.045119	.021185	.010019	.004771	.002288
80	.063792	.043384	.020176	.009452	.004459	.002118
81	.061635	.041715	.019216	.008917	.004167	.001961
82	.059551	.040111	.018301	.008412	.003895	.001816
83	.057537	.038568	.017429	.007936	.003640	.001682
84	.055591	.037085	.016599	.007486	.003402	.001557
85	.053711	.035658	.015809	.007063	.003179	.001442
86	.051895	.034287	.015056	.006663	.002971	.001335
87	.050140	.032968	.014339	.006286	.002777	.001236
88	.048445	.031700	.013656	.005930	.002595	.001144
89	.046806	.030481	.013006	.005594	.002425	.001059
90	.045223	.029308	.012386	.005278	.002267	.000981
91	.043694	.028181	.011797	.004979	.002118	.000908
92	.042217	.027097	.011235	.004697	.001980	.000841
93	.040789	.026055	.010700	.004431	.001850	.000779
94	.039410	.025053	.010190	.004180	.001729	.000721
95	.038077	.024089	.009705	.003944	.001616	.000667
96	.036789	.023163	.009243	.003720	.001510	.000618
97	.035545	.022272	.008803	.003510	.001411	.000572
98	.034343	.021415	.008383	.003311	.001319	.000530
99	.033182	.020592	.007984	.003124	.001233	.000490
100	.032060	.019800	.007604	.002947	.001152	.000454

Answers

Answers

1.

2. $\frac{56}{9}$ **3.** $x > \frac{15}{4}$ **4.** Consistent, dependent **5.** No **6.** $(3, 2)$ **7.** $(2, 7)$

8. $(-4, -\frac{1}{3})$ **9.** No solution

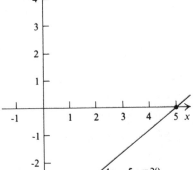

CHAPTER 1

MARGIN EXERCISES

1. 6 **2.** 660 lb **3.** $x > \frac{14}{5}$ **4.** $x \leq \frac{19}{17}$ **5.** $x > 3497.5$, or 3498 suits, or more

6.

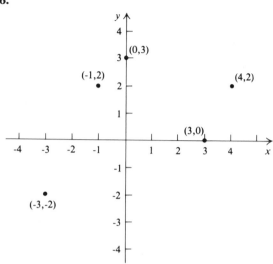

7. A point in the 2nd quadrant has first coordinate negative, second coordinate positive. A point in the 3rd quadrant has both coordinates negative.

8. a) Yes   b) No   **9.** Line through $(0, 1)$ and $(1, 3)$

10. a) y-intercept: $(0, -2)$; x-intercept: $(3, 0)$ b) y-intercept and x-intercept both $(0, 0)$ c) y-intercept: $(0, -5)$; x-intercept: $(-3, 0)$ **11.** Line through $(0, -2)$ and $(3, 0)$ **12.** Line through $(0, 0)$ and $(1, 2)$ **13.** Line through $(0, -2)$ and $(3, 0)$ **14.** a) Line through $(0, 0)$ and $(1, 2)$ b) Line through $(0, -4)$ and $(2, 0)$ **15.** Horizontal line through $(0, -2)$ **16.** Vertical line through $(3, 0)$

17.

18. Yes   **19.** No   **20.** Consistent, dependent   **21.** Inconsistent, independent   **22.** Consistent, independent   **23.** a) $(1, 2)$ b) $(-2, 2)$ **24.** $(\frac{4}{3}, \frac{14}{3})$ **25.** $(4, -2)$ **26.** $(\frac{1}{2}, \frac{1}{3})$ **27.** $(-2, 5)$ **28.** $(-7, 10)$ **29.** No solution   **30.** An infinite number of solutions   **31.** $-8, 0, t^3, x_1^3, k$ **32.** $f(-2) = -8$, $f(5) = 125$, $f(\frac{1}{2}) = \frac{1}{8}$, $f(t) = t^3$, $f(\sqrt[3]{t}) = t$ **33.** $f(2) = 12$, $f(-5) = 5$, $f(x_1) = 4x_1 + x_1^2$ **34.** a) 160; the total profit from the sale of 25 units of the first product and 10 units of the second product is \$160   b) 108; the total profit from the sale of 0 units of the first product and 18 units of the second product is \$108   **35.** $f(4, -5) = -13$, $f(1, 2) = -4$ **36.** 19¢, 37¢; 18¢ more   **37.** a) $P(x) = 25x - 100{,}000$ b) The company will break even by producing and selling 4000 calculators   c) $x > 4000$ d) $x < 4000$ **38.** $(\$12.50, 250)$

EXERCISE SET 1.1, p. 9

1. $\frac{79}{12}$ **3.** $\frac{4}{5}$ **5.** 140 **7.** 200 **9.** 340 lb **11.** \$800 **13.** $x \leq 3$ **15.** $x < -\frac{5}{13}$ **17.** $x \geq -\frac{6}{5}$

19. $x > 4380$ **21** $x \geq 50\%$

EXERCISE SET 1.2, p. 17

1. Line through $(0, 4)$ and (5.0) **3.** Line through $(0, -4)$ and $(2, 0)$ **5.** Line through $(0, 4)$ and (3.0)

7. Line through $(0, 0)$ and $(1, 1)$

9.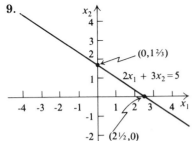

11. Horizontal line through $(0, 5)$

13. Vertical line through $(-2, 0)$

15. The horizontal x_1-axis

17. Horizontal line through $(0, -3.5)$

19.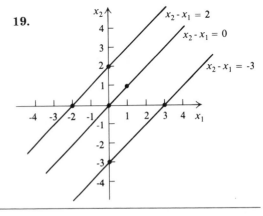

EXERCISE SET 1.3, p. 21

1. No **3.** Consistent, independent **5.** Consistent, dependent **7.** Inconsistent, independent **9.** $(2, -1)$

11. No solution **13.** $(-5, 4)$

EXERCISE SET 1.4, pp. 25–26

1. $(2, -3)$ **3.** $(2, -2)$ **5.** $(6, 2)$ **7.** $\left(\frac{9}{19}, \frac{51}{38}\right)$ **9.** $\left(\frac{3}{2}, \frac{5}{2}\right)$ **11.** $\left(-\frac{1}{3}, -4\right)$ **13.** An infinite number of solutions

15. No solution **17.** 13 and 16 **19.** 5 pairs of cloth gloves and 15 pairs of pigskin gloves **21.** 42 L of 2%,

18 L of 6% **23.** \$2000 at 12%, \$2800 at 13% **25.** 20 km/h, 3 km/h **27.** $\left(\dfrac{\pi + 3\sqrt{2}}{\pi^2 + 2}, \dfrac{3\pi - \sqrt{2}}{\pi^2 + 2}\right)$

29. $(0.924, -0.833)$

EXERCISE SET 1.5, pp. 31–32

1. a) 29, 19.5, 19.05, 19.005, 19.0005 b) 34, -6, -21, $5k - 1$ **3.** 13, 4, 5, 85, $u^2 + 4$ **5.** 49, 4, 16, $20\frac{1}{4}$

7. $-3, -3, 12$ **9.** 112 **11.** \$70, \$250, \$20,050 **13.** \$216, \$1080 **15.** $-26, 12, 54$ **17.** 0, 22, -50

19. $-5, 25, -25$ **21** 9, 196, 64

EXERCISE SET 1.6, pp. 36–37

1. a) $P(x) = 45x - 360,000$ b) 8000 c) $x > 8000$ d) $x < 8000$ **3.** a) $P(x) = 55x - 49,500$ b) 900 c) $x > 900$ d) $x < 900$ **5.** a) $C(x) = 20x + 10,000$ b) $R(x) = 100x$ c) $P(x) = 80x - 10,000$ d) Profit of \$150,000 e) 125 f) $x > 125$ g) $x < 125$ **7.** (\$50, 500) **9.** (\$10, 370) **11.** (\$40, 7600)

CHAPTER 1 TEST, pp. 38–39

1.

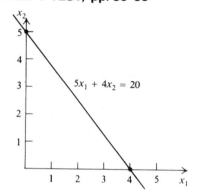

$5x_1 + 4x_2 = 20$

2. -10 **3.** $x \leqslant -\frac{4}{7}$ **4.** \$550

5. Inconsistent, independent **6.** Yes

7. $(3, -1)$ **8.** $(4, 1)$ **9.** $(\frac{3}{2}, \frac{5}{2})$

10. An infinite number of solutions

11. 4, 4 **12.** -110 **13.** (\$70, 7350)

CHAPTER 2

MARGIN EXERCISES

1. 4, 7, 12, 19; 228 **2.** 2; 3 **3.** 19; -5 **4.** \$6300; $-\$346.75$ **5.** 2; $\frac{1}{2}$ **6.** 70 **7.** \$2832.50 **8.** 20,100 **9.** 256 **10.** \$22,100 **11.** 5 **12.** -3 **13.** 0.85 **14.** 1.09 **15.** $\frac{1}{2}$ **16.** 256 **17.** $\frac{1}{81}$ **18.** 510 **19.** \$21,470,000 approx. **20.** No **21.** No **22.** 1 **23.** $\frac{3125}{3}$, or $1041\frac{2}{3}$ **24.** \$20,000 **25.** \$7100, \$1420, 20% **26.**

Year	Rate of depreciation	Annual depreciation	Book value	Total depreciation
0			\$8700	
1	$\frac{1}{5}$ or 20%	\$1420	7280	\$1420
2	20%	1420	5860	2840
3	20%	1420	4440	4260
4	20%	1420	3020	5680
5	20%	1420	1600	7100

27. a) $V_n = \$8700 - (\$1420)n$ b) $-\$1420$

28.

Year	Rate of depreciation	Annual depreciation	Book value	Total depreciation
0			$8700	
1	$\frac{2}{5}$ or 40%	$3480	5220	$3480
2	40%	2088	3132	5568
3	40%	1252.80	1879.20	6820.80
4		279.20	1600	7100
5		0	1600	7100

29. $V_n = \$8700(0.60)^n$ **30.** a) $\frac{5}{15}, \frac{4}{15}, \frac{3}{15}, \frac{2}{15}, \frac{1}{15}$ b) $2366.67, $6333.33; $1893.33, $4440.00; $1420.00, $3020.00

31.

Year	Rate of depreciation	Annual depreciation	Book value	Total depreciation
0			$8700	
1	$\frac{5}{15}$	$2366.67	6333.33	$2366.67
2	$\frac{4}{15}$	1893.33	4440.00	4260.00
3	$\frac{3}{15}$	1420.00	3020.00	5680.00
4	$\frac{2}{15}$	946.67	2073.33	6626.67
5	$\frac{1}{15}$	473.33	1600.00	7100.00

32. $1015   **33.** $1725.50   **34.** $2524.95   **35.** a) $1120 b) $1123.60 c) $1125.51 d) $1126.49 e) $1127.49
36. $2969.88   **37.** 9.203%   **38.** 15.2%   **39.** $5134.51   **40.** $21,241.93   **41.** $723.78   **42.** $19,636.29
43. $13,677.74   **44.** $404.08

EXERCISE SET 2.1, pp. 48–49

1. $\frac{1}{2}, \frac{2}{3}, \frac{3}{4}, \frac{4}{5}, \frac{15}{16}$ **3.** $0, \frac{1}{3}, \frac{2}{7}, \frac{3}{13}; \frac{14}{211}$ **5.** 2; 5   **7.** $1.06; $0.06   **9.** 5; $-\frac{2}{3}$ **11.** 47   **13.** $-\$1628.16   **15.** 45,150

17. 690   **19.** $\dfrac{n(n+1)}{2}$ **21.** $4.96   **23.** $18,450

EXERCISE SET 2.2, pp. 52–53

1. 2 **3.** $-\frac{1}{3}$ **5.** 0.95 **7.** 2187 **9.** $\frac{1}{5^6}$, or $\frac{1}{15,625}$ **11.** \$2331.64 **13.** 1016 **15.** \$5866.60 **17.** \$2,684,000 approx. **19.** No **21.** $12\frac{1}{2}$ **23.** 486 **25.** \$12,500 **27.** \$4,000,000,000 **29.** $\approx$ \$3,333,333; $66\frac{2}{3}\%$

EXERCISE SET 2.3, p. 61

1. a)

Year	Rate of depreciation	Annual depreciation	Book value	Total depreciation
0			\$8000	
1	$\frac{1}{4}$ or 25%	\$1500	6500	\$1500
2	25%	1500	5000	3000
3	25%	1500	3500	4500
4	25%	1500	2000	6000

b) $V_n = \$8000 - (\$1500)n$ c) $-\$1500$

3. a)

Year	Rate of depreciation	Annual depreciation	Book value	Total depreciation
0			\$450	
1	$\frac{1}{8}$ or 12.5%	\$56.25	393.75	\$56.25
2	12.5%	56.25	337.50	112.50
3	12.5%	56.25	281.25	168.75
4	12.5%	56.25	225.00	225.00
5	12.5%	56.25	168.75	281.25
6	12.5%	56.25	112.50	337.50
7	12.5%	56.25	56.25	393.75
8	12.5%	56.25	0	450.00

b) $V_n = \$450 - (\$56.25)n$ c) $-\$56.25$

5. a)

Year	Rate of depreciation	Annual depreciation	Book value	Total depreciation
0			$8000	
1	$\frac{2}{4}$ or 50%	$4000	4000	$4000
2	50%	2000	2000	6000
3		0	2000	6000
4		0	2000	6000

b) $V_n = \$800(0.50)^n$

7. a)

Year	Rate of depreciation	Annual depreciation	Book value	Total depreciation
0			$450	
1	$\frac{2}{8}$ or 25%	$112.50	337.50	$112.50
2	25%	84.38	252.12	196.88
3	25%	63.28	189.84	260.16
4	25%	47.46	142.38	307.62
5	25%	35.60	106.78	343.22
6	25%	26.70	80.08	369.92
7	25%	20.02	60.06	389.94
8	25%	15.02	45.04	404.96

b) $V_n = \$450(0.75)^n$

9. a) $\frac{4}{10}, \frac{3}{10}, \frac{2}{10}, \frac{1}{10}$ b)

Year	Rate of depreciation	Annual depreciation	Book value	Total depreciation
0			$8000	
1	$\frac{4}{10}$	$2400	5600	$2400
2	$\frac{3}{10}$	1800	3800	4200
3	$\frac{2}{10}$	1200	2600	5400
4	$\frac{1}{10}$	600	2000	6000

11. a) $\frac{8}{36}, \frac{7}{36}, \frac{6}{36}, \frac{5}{36}, \frac{4}{36}, \frac{3}{36}, \frac{2}{36}, \frac{1}{36}$ b)

Year	Rate of depreciation	Annual depreciation	Book value	Total depreciation
0			$450	
1	$\frac{8}{36}$	$100	350	$100.00
2	$\frac{7}{36}$	87.50	262.50	187.50
3	$\frac{6}{36}$	75.00	187.50	262.50
4	$\frac{5}{36}$	62.50	125.00	325.00
5	$\frac{4}{36}$	50.00	75.00	375.00
6	$\frac{3}{36}$	37.50	37.50	412.50
7	$\frac{2}{36}$	25.00	12.50	437.50
8	$\frac{1}{36}$	12.50	0	450.00

EXERCISE SET 2.4, p. 67

1. $2060   **3.** $2560   **5.** a) $2280 b) $2289.80 c) $2295.05 d) $2297.76 e) $2300.52   **7.** $793.83
9. $788.49   **11.** $3450.32   **13.** $1144.90   **15.** $2.41   **17.** 6.25%

EXERCISE SET 2.5, p. 70

1. 8.16%   **3.** 9.308%   **5.** 8.271%   **7.** 8.328%   **9.** 8.329%   **11.** 13.7%   **13.** 16.8%

EXERCISE SET 2.6, pp. 73–74

1. $4439.94 **3.** $13,816.45 **5.** $48,594.74 **7.** $1228.29 **9.** $103,399.40 **11.** $3673.84 **13.** $992.37

15. $100.33 **17.** $P = \dfrac{Vi}{[(1 + i)^N - 1]}$

EXERCISE SET 2.7, pp. 77–78

1. $3387.21 **3.** $7023.58 **5.** $32,702.87 **7.** $760.95 **9.** $139.50 **11.** $548.84 $P = \dfrac{Si}{[1 - (1 + i)^{-N}]}$

15 $12,500; you want $1000 of interest each year, so just solve $I = Pit$ for P when $I = \$1000$, $i = 0.08$, and $t = 1$.

CHAPTER 2 TEST, p. 79

1. 5; 3   **2.** 62   **3.** $31.40   **4.** 1.05   **5.** $155.13   **6.** $1257.79   **7.** Yes, $1086.96

8. a)

Year	Rate of depreciation	Annual depreciation	Book value	Total depreciation
0			$8500	
1	$\frac{1}{4}$ or 25%	$1487.50	7012.50	$1487.50
2	25%	1487.50	5525	2975
3	25%	1487.50	4037.50	4462.50
4	25%	1487.50	2550	5950

b) $V_n = \$8500 - (\$1487.50)n$ c) $-\$1487.50$

9. a)

Year	Rate of depreciation	Annual depreciation	Book value	Total depreciation
0			$8500	
1	$\frac{2}{4}$ or 50%	$4250	4250	$4250
2		1700	2550	5950
3		0	2550	5950
4		0	2550	5950

b) $V_n = \$8500(0.50)^n$

10. a) $\frac{4}{10}, \frac{3}{10}, \frac{2}{10}, \frac{1}{10}$ b)

Year	Rate of depreciation	Annual depreciation	Book value	Total depreciation
0			$8500	
1	$\frac{4}{10}$	$2380	6120	$2380
2	$\frac{3}{10}$	1785	4335	4165
3	$\frac{2}{10}$	1190	3145	5355
4	$\frac{1}{10}$	595	2550	5950

11. a) $1180   **b)** $1191.02   **c)** $1194.05   **d)** $1195.62   **12.** 9.308%   **13.** 18.3%   **14.** $37,089.80   **15.** $1090.52
16. $103,796.58   **17.** $723.77

CHAPTER 3

MARGIN EXERCISES

1. $2x_1 - 6x_2 = \frac{1}{4}$
   $-3x_1 + x_2 = 8$

2.
x_1	x_2	1
2	5	-24
5	-2	-2

3. $(-2, -4)$ **4.** $(-\frac{1}{3}, \frac{1}{2})$

5. $x_1 = 2$, $x_2 = \frac{1}{2}$, $x_3 = -2$, or $(2, \frac{1}{2}, -2)$ **6.** $1800 @ 7%, $1900 @ 9%   **7.** No solution   **8.** $x_1 = \frac{3}{2} + \frac{1}{2}x_2$, $x_2 = $ any
number; $x_2 = 0$, $x_1 = \frac{3}{2}$; $x_2 = 1$, $x_1 = 2$; $x_2 = -4$, $x_1 = -\frac{1}{2}$

9. a)
x_1	x_2	x_3	x_4	1
1	4	0	0	-2
0	0	1	0	8
0	0	0	1	-3

b) $x_1 = -2 - 4x_2$, $x_2 = $ any number, $x_3 = 8$, $x_4 = -3$ **c)** $x_1 = -2$, $x_2 = 0$, $x_3 = 8$, $x_4 = -3$; $x_1 = -6$, $x_2 = 1$, $x_3 = 8$, $x_4 = -3$;
$x_1 = 6$, $x_2 = -2$, $x_3 = 8$, $x_4 = -3$

10. No solution   **11.** $x_1 = -10$, $x_2 = \frac{11}{3}$, $x_3 = \frac{1}{6}$, or $(-10, \frac{11}{3}, \frac{1}{6})$ **12.** 2×3 **13.** 1×3 **14.** 3×1 **15.** 1×1
16. 2×2 **17. a)** $a_{12} = 0$ **b)** $a_{22} = -6$ **c)** $a_{21} = 4$ **d)** $a_{32} = -2$ **18. a)** No   **b)** No   **19.** $a = -6$, $b = \frac{1}{2}$

20. Row vectors are B and C,   Column vectors are A and D **21.** $A^T = \begin{bmatrix} -8 & -4 & 6 \\ 1 & 0 & 7 \\ -2 & -1 & 8 \end{bmatrix}$, $B^T = \begin{bmatrix} -7 \\ 9 \\ 10 \\ \frac{1}{4} \end{bmatrix}$,

$C^T = [-20 \quad 41]$, $D^T = \begin{bmatrix} -4 & 1 & 0 \\ 5 & 0 & 1 \end{bmatrix}$ **22. a)** $\begin{bmatrix} -3 & -6 \\ 9 & 0 \end{bmatrix}$ **b)** Same as (a)   **c)** Yes   **23.** $\begin{bmatrix} 0 & 15 \\ 1 & 12 \\ 1 & \frac{1}{2} \end{bmatrix}$

24. a) A b) A c) Yes **25.** a) $\begin{bmatrix} 36 & 6 & 0 \\ 12 & 6 & -30 \\ -18 & 54 & 3 \end{bmatrix}$ b) $\begin{bmatrix} -6 & -1 & 0 \\ -2 & -1 & 5 \\ 3 & -9 & -\frac{1}{2} \end{bmatrix}$ c) O d) O e) $\begin{bmatrix} -\frac{1}{5} & -\frac{1}{30} & 0 \\ -\frac{1}{15} & -\frac{1}{30} & \frac{1}{6} \\ \frac{1}{10} & -\frac{3}{10} & -\frac{1}{60} \end{bmatrix}$

f) $\begin{bmatrix} 6t & t & 0 \\ 2t & t & -5t \\ -3t & 9t & \frac{1}{2}t \end{bmatrix}$ **26.** a) $\begin{bmatrix} 13 & 2 \\ 5 & -8 \end{bmatrix}$ b) $\begin{bmatrix} -13 & -2 \\ -5 & 8 \end{bmatrix}$ c) No **27.** $\sum\limits_{i=1}^{6} i^2$ **28.** $\sum\limits_{i=1}^{4} t^i$ **29.** $\sum\limits_{i=1}^{38} p_i$ **30.** $2 + 2^2 + 2^3$

31. $p_1q_1 + p_2q_2 + \cdots + p_{20}q_{20}$ **32.** $t + 2t^2 + 3t^3 + 4t^4 + 5t^5$ **33.** $[4a + 5b]$, or $4a + 5b$ **34.** $[7]$, or 7

35. $[-36]$, or -36 **36.** $\begin{bmatrix} 11 & 2 \\ 7 & -13 \end{bmatrix}\begin{bmatrix} x_1 \\ x_2 \end{bmatrix} = \begin{bmatrix} -1 \\ 8 \end{bmatrix}$ **37.** $\begin{aligned} 3y_1 - 7y_2 &= 4 \\ -2y_1 + y_2 &= 5 \end{aligned}$

38. a) $[5c + 6d \quad 5g + 6h]$ b) $[-3 \quad -24]$ **39.** a) $\begin{bmatrix} 5c + 6d & 5g + 6h \\ 3c - d & 3g - h \end{bmatrix}$ b) $\begin{bmatrix} -3 & -24 \\ -11 & 4 \end{bmatrix}$

40. a) $AB = \begin{bmatrix} -23 & 16 & 39 \\ 6 & 4 & 14 \end{bmatrix}$, BA not possible b) $AB = [31]$, or 31; $BA = \begin{bmatrix} 12 & 18 & -6 & 6 \\ 16 & 24 & -8 & 8 \\ 0 & 0 & 0 & 0 \\ -10 & -15 & 5 & -5 \end{bmatrix}$

41. a) $AB = \begin{bmatrix} -32 & -3 \\ -20 & -1 \end{bmatrix}$ b) $BA = \begin{bmatrix} -1 & 5 \\ 12 & -32 \end{bmatrix}$ c) No d) No, we have a counterexample with AB and BA

42. a) $(AB)C = A(BC) = \begin{bmatrix} -43 & -19 \\ 99 & 39 \end{bmatrix}$ b) $A(B + C) = AB + AC = \begin{bmatrix} 5 & 1 \\ 30 & 6 \end{bmatrix}$

43. a) $AI = A$ b) $IA = A$ c) Same $= A$ d) $IX = X$ e) XI cannot be found f) Same **44.** $A^{-1}A = AA^{-1} = I$; that is, A and A^{-1} commute and their product is the identity.

45. a) $\begin{bmatrix} 3 & 5 \\ 1 & 2 \end{bmatrix}\begin{bmatrix} x_1 \\ x_2 \end{bmatrix} = \begin{bmatrix} -1 \\ 4 \end{bmatrix}$ b) $A = \begin{bmatrix} 3 & 5 \\ 1 & 2 \end{bmatrix}$ c) $x_1 = -22, x_2 = 13$

46. a) $[X \quad Y] = \begin{bmatrix} x_1 & y_1 \\ x_2 & y_2 \\ x_3 & y_3 \end{bmatrix}$ b) $\begin{bmatrix} X \\ Y \end{bmatrix} = \begin{bmatrix} x_1 \\ x_2 \\ x_3 \\ y_1 \\ y_2 \\ y_3 \end{bmatrix}$ **47.** $[A \quad I] = \begin{bmatrix} -2 & 3 & 1 & 0 \\ 4 & 5 & 0 & 1 \end{bmatrix}$, $\begin{bmatrix} A \\ I \end{bmatrix} = \begin{bmatrix} -2 & 3 \\ 4 & 5 \\ 1 & 0 \\ 0 & 1 \end{bmatrix}$

48. a) $[A \quad I]^T = \begin{bmatrix} 5 & -2 \\ 7 & 0 \\ \hline 1 & 0 \\ 0 & 1 \end{bmatrix}$ b) $\begin{bmatrix} A \\ I \end{bmatrix}^T = \begin{bmatrix} 5 & -2 & \vdots & 1 & 0 \\ 7 & 0 & \vdots & 0 & 1 \end{bmatrix}$ **49.** $A^{-1} = \begin{bmatrix} \frac{5}{22} & \frac{3}{22} \\ -\frac{2}{11} & \frac{1}{11} \end{bmatrix}$

EXERCISE SET 3.1, pp. 90–91

1. $(-3, 2)$ **3.** $(7, 3)$ **5.** $(\frac{5}{2}, -1)$ **7.** $(-3, 5, 7)$ **9.** $(0, 2, 1)$ **11.** $(2, \frac{1}{2}, -2)$ **13.** $(4, 5, 6, 7)$ **15.** $4100 @ 7\%,

$4700 @ 8% **17.** $30,000 @ 5%, $40,000 @ 6% **19.** 8 white, 22 yellow **21.** 150 lb soybean meal; 200 lb corn meal
23. $400 @ 7%, $500 @ 8%, $1600 @ 9% **25.** $y = 2x^2 + 3x - 1$ **27.** a) $y = 2500x^2 - 6500x + 5000$ b) $19,000

EXERCISE SET 3.2, p. 97

1. $x_1 = 2 + 3x_2$, $x_2 = $ any number **3.** No solution **5.** $x_1 = 1 - x_3$, $x_2 = -x_3$, $x_3 = $ any number **7.** $x_1 = \frac{1}{5}x_3$,
$x_2 = -\frac{7}{5}x_3$, $x_3 = $ any number **9.** $x_1 = -\frac{7}{2}x_3$, $x_2 = -\frac{19}{2}x_3$, $x_3 = $ any number **11.** $x_1 = 1 + 3x_3$, $x_2 = 4 - 5x_3$,
$x_3 = $ any number, $x_4 = -2$ **13.** No solution **15.** $x_1 = 19 - 2x_3 - 16x_4$, $x_2 = -6 + 3x_3 + 4x_4$, $x_3 = $ any number,
$x_4 = $ any number **17.** $x_1 = -1 + 4x_3$, $x_2 = 4 + 2x_3$, $x_3 = $ any number **19.** $x_1 = 6 - 8x_3 + 3x_4$, $x_2 = 4 - 4x_3 - 2x_4$,
$x_3 = $ any number, $x_4 = $ any number, $x_5 = -5$

EXERCISE SET 3.3, pp. 102–103

1. 2×2 **3.** 2×3 **5.** $\begin{bmatrix} -1 & 3 \\ 2 & 1 \end{bmatrix}$ **7.** $\begin{bmatrix} -1 & 6 & -7 \\ 4 & 2 & 0 \end{bmatrix}$ **9.** $\begin{bmatrix} 3 & 9 \\ 12 & 6 \end{bmatrix}$ **11.** $\begin{bmatrix} 5 & 10 & 15 \\ -15 & -10 & -5 \end{bmatrix}$ **13.** $\begin{bmatrix} 3 & 3 \\ 6 & 3 \end{bmatrix}$

15. $\begin{bmatrix} -1 & -10 & 1 \\ 2 & 2 & 2 \end{bmatrix}$ **17.** Not possible **19.** $\begin{bmatrix} -k & -2k & -3k \\ 3k & 2k & k \end{bmatrix}$ **21.** A **23.** $\begin{bmatrix} 1 & 4 \\ 3 & 2 \end{bmatrix}$ **25.** $\begin{bmatrix} -1 & 3 \\ -2 & 2 \\ -3 & 1 \end{bmatrix}$ **27.** $\begin{bmatrix} -1 & 2 \\ 3 & 1 \end{bmatrix}$

29. $a_{11} = -4$, $a_{12} = 5$, $a_{31} = 1$, $a_{22} = 9$, $a_{32} = 3$, $a_{21} = 0$ **31.** $X^T = [x_1 \ x_2 \ x_3 \ x_4]$

EXERCISE SET 3.4, pp. 114–115

1. $AB = [-10]$, or -10; $BA = \begin{bmatrix} -6 & 3 \\ 8 & -4 \end{bmatrix}$ **3.** $AB = [17]$, or 17; $BA = \begin{bmatrix} 18 & 0 & -36 \\ -10 & 0 & 20 \\ \frac{1}{2} & 0 & -1 \end{bmatrix}$

5. $AB = \begin{bmatrix} 7 & -6 & 1 \\ -15 & 12 & 3 \\ -2 & -1 & 8 \end{bmatrix}$, $BA = \begin{bmatrix} 9 & 11 & -10 \\ 3 & 4 & -4 \\ -3 & -1 & 14 \end{bmatrix}$ **7.** $AB = [35 \ -35]$, BA not possible **9.** $AB = 0$

11. $\begin{bmatrix} 11 & 3 \\ 7 & 2 \end{bmatrix}\begin{bmatrix} x_1 \\ x_2 \end{bmatrix} = \begin{bmatrix} -4 \\ 5 \end{bmatrix}$ **13.** $\begin{bmatrix} 3 & 1 & 0 \\ 1 & -1 & 2 \\ 1 & 1 & 1 \end{bmatrix}\begin{bmatrix} x_1 \\ x_2 \\ x_3 \end{bmatrix} = \begin{bmatrix} 2 \\ -4 \\ 5 \end{bmatrix}$ **15.** $\begin{array}{r} x_1 + 2x_2 = -1 \\ 4x_1 - 3x_2 = 2 \end{array}$

17. $x_1 = -23$, $x_2 = 83$ **19.** $x_1 = \frac{1}{11}$, $x_2 = -\frac{6}{11}$ **21.** $x_1 = -1$, $x_2 = 5$, $x_3 = 1$

23. $(A + B)(A + B) = \begin{bmatrix} 1 & 0 \\ 4 & 4 \end{bmatrix}$, $A^2 + 2AB + B^2 = \begin{bmatrix} -3 & -3 \\ 9 & 8 \end{bmatrix}$ **25.** $\begin{bmatrix} X \\ Y \end{bmatrix}^T = [a \ b \ c \ e \ f \ g]$

27. $[A \ B] = \begin{bmatrix} 2 & -1 & 3 & \vdots & 0 & 1 & -2 \\ 4 & 1 & 0 & \vdots & 1 & -3 & 7 \end{bmatrix}$, $[A \ B]^T = \begin{bmatrix} 2 & 4 \\ -1 & 1 \\ 3 & 0 \\ \hline 0 & 1 \\ 1 & -3 \\ -2 & 7 \end{bmatrix}$ **29.** Yes

EXERCISE SET 3.5, p. 118

1. $A^{-1} = \begin{bmatrix} -3 & -2 \\ 5 & -3 \end{bmatrix}$ **3.** $A^{-1} = \begin{bmatrix} 2 & -3 \\ -7 & 11 \end{bmatrix}$ **5.** $A^{-1} = \begin{bmatrix} \frac{2}{11} & \frac{3}{11} \\ -\frac{1}{11} & \frac{4}{11} \end{bmatrix}$ **7.** $A^{-1} = \begin{bmatrix} \frac{3}{8} & \frac{1}{8} & -\frac{1}{4} \\ -\frac{1}{8} & -\frac{3}{8} & \frac{3}{4} \\ -\frac{1}{4} & \frac{1}{4} & \frac{1}{2} \end{bmatrix}$ **9.** $A^{-1} = \begin{bmatrix} \frac{1}{3} & 0 & \frac{1}{3} \\ -\frac{2}{5} & \frac{2}{5} & \frac{1}{5} \\ \frac{2}{15} & \frac{1}{5} & -\frac{1}{15} \end{bmatrix}$

CHAPTER 3 TEST, p. 119

1. $\begin{bmatrix} 7 & 4 \\ 3 & 1 \end{bmatrix}\begin{bmatrix} x_1 \\ x_2 \end{bmatrix} = \begin{bmatrix} -21 \\ -9 \end{bmatrix}$, $x_1 = -3$, $x_2 = 0$ **2.** $\begin{bmatrix} 3 & -2 & 3 \\ 1 & 1 & -1 \\ 2 & 3 & -5 \end{bmatrix}\begin{bmatrix} x_1 \\ x_2 \\ x_3 \end{bmatrix} = \begin{bmatrix} 24 \\ -7 \\ -32 \end{bmatrix}$, $x_1 = 1$, $x_2 = -3$, $x_3 = 5$

3. $\begin{array}{l} 4x_1 - 8x_2 = -20 \\ 3x_1 - 6x_2 = -15 \end{array}$, $x_1 = -5 + 2x_2$, $x_1 =$ any number   **4.** $\begin{array}{l} 8x_1 - 4x_2 = 20 \\ 6x_1 - 3x_2 = 16 \end{array}$, no solution

5. $x_1 = 5 - 6x_3$, $x_2 = 3 + 2x_3$, $x_3 =$ any number, $x_4 = 2$ **6.** No solution   **7.** $\begin{bmatrix} -3 & 1 \\ -4 & 1 \end{bmatrix}$ **8.** $\begin{bmatrix} 2 & 3 \\ 1 & 5 \end{bmatrix}$

9. $\begin{bmatrix} -3 & 3 \\ -6 & 1 \end{bmatrix}$ **10.** $\begin{bmatrix} 12 & -8 \\ 20 & -4 \end{bmatrix}$ **11.** $C^T = [2 \quad -3 \quad 4]$ **12.** AB not possible, $BA = \begin{bmatrix} 5 \\ 7 \end{bmatrix}$ **13.** $x_1 = \frac{1}{2}$, $x_2 = \frac{2}{3}$

14. $x_1 = -2$, $x_2 = 3$, $x_3 = 9$ **15.** $\begin{bmatrix} X \\ Y \end{bmatrix}^T = [p \quad q \quad t \quad u]$ **16.** \$4000 @ 8%, \$5300 @ 10%   **17.** $A^{-1} = \begin{bmatrix} 2 & 0 & 1 \\ 3 & 1 & 2 \\ 1 & 0 & 1 \end{bmatrix}$

CHAPTER 4

MARGIN EXERCISES

1.

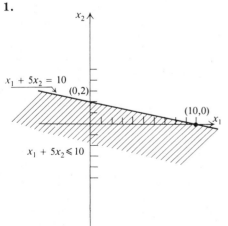

2.

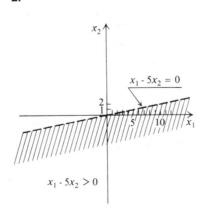

3.

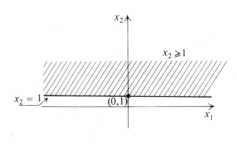

4, 5, 6, 7, 8.

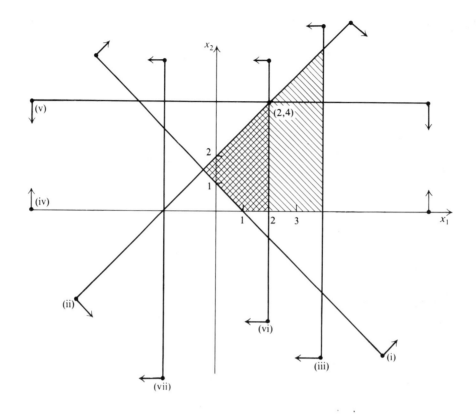

7. a) $(2, 4)$ degenerate; intersection of ii, v, and vi   b) Constraints (iii) and (v) redundant   **8.** Solution set empty

9. a, b, e **10.** $(\frac{20}{7}, \frac{12}{7})$ **11.** min: $f(-\frac{4}{3}, -\frac{2}{3}) = -\frac{14}{3}$,   max: $f(2, 2) = 10$ **12.** min: $f(-\frac{4}{3}, -\frac{2}{3}) = f(2, 1) = 0$ [any point on the line segment between $(-\frac{4}{3}, -\frac{2}{3})$ and $(2, 1)$],   max: $f(0, 2) = 4$.

13.

	Composition		Supply available
	Chairs	Sofas	
Number of units	x_1	x_2	
Wood (feet)	20	100	1900
Foam (lbs)	1	50	500
Material (yds)	2	20	240
Unit price ($)	20	300	Maximize income

$$20x_1 + 100x_2 \leq 1900,$$
$$x_1 + 50x_2 \leq 500,$$
$$2x_1 + 20x_2 \leq 240,$$
$$\max f\colon f = 20x_1 + 300x_2; \quad x_1, x_2 \geq 0.$$

14. $\begin{bmatrix} 20 & 100 \\ 1 & 50 \\ 2 & 50 \end{bmatrix} \begin{bmatrix} x_1 \\ x_2 \end{bmatrix} \le \begin{bmatrix} 1900 \\ 500 \\ 240 \end{bmatrix}$ $\max f: f = \begin{bmatrix} 20 & 300 \end{bmatrix} \begin{bmatrix} x_1 \\ x_2 \end{bmatrix}; \quad \begin{bmatrix} x_1 \\ x_2 \end{bmatrix} \ge \begin{bmatrix} 0 \\ 0 \end{bmatrix}.$

15. $x_1 = 25$, $x_2 = \frac{19}{2}$, $f = 3350$

16. a)

	Composition (tons)		Amount required (tons)
	Ore A	Ore B	
Number of units (tons)	y_1	y_2	
Iron	0.10	0	200
Aluminum	0	0.20	500
Copper	0.02	0.01	100
Cost ($) per ton	10	15	Minimize cost

$$0.1y_1 + 0y_2 \ge 200,$$
$$0y_1 + 0.2y_2 \ge 500,$$
$$0.02y_1 + 0.01y_2 \ge 100,$$
$$\min f: f = 10y_1 + 15y_2; \quad y_1, y_2 \ge 0.$$

b) $\begin{bmatrix} 0.1 & 0 \\ 0 & 0.2 \\ 0.02 & 0.01 \end{bmatrix} \begin{bmatrix} y_1 \\ y_2 \end{bmatrix} \ge \begin{bmatrix} 200 \\ 500 \\ 100 \end{bmatrix}$

$\min f: f = \begin{bmatrix} 10 & 15 \end{bmatrix} \begin{bmatrix} y_1 \\ y_2 \end{bmatrix}; \quad \begin{bmatrix} y_1 \\ y_2 \end{bmatrix} \ge \begin{bmatrix} 0 \\ 0 \end{bmatrix}.$

17. $f = 75{,}000$ at $(y, y_2) = (3750, 2500)$

18. $20, point a

EXERCISE SET 4.1, p. 133

1.

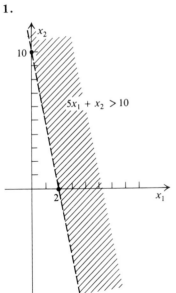

$5x_1 + x_2 > 10$

3.

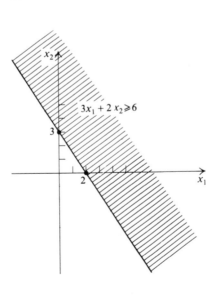

$3x_1 + 2x_2 \ge 6$

5.

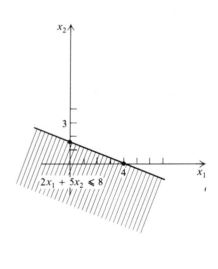

$2x_1 + 5x_2 \le 8$

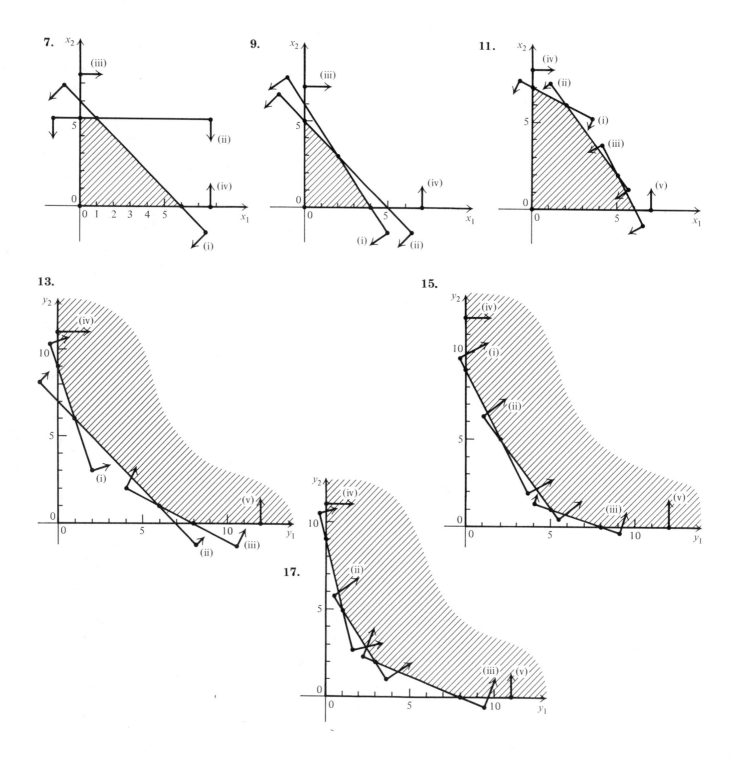

19. a) i) $x_1 + 2x_2 \leqslant 6$
  ii) $0 \leqslant x_1 \leqslant 5$
  iii) $x_2 \leqslant -2$
 b) Nonempty
 c) Bounded
 d) No redundancies
 e) No degeneracies

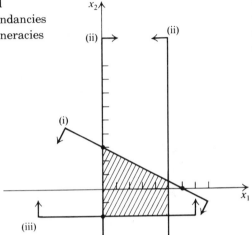

21. a) i) $x_1 \geqslant -3$
  ii) $x_1 - 2x_2 \leqslant 4$
  iii) $x_2 - 3x_1 \leqslant 9$
  iv) $3x_1 + x_2 \leqslant 10$
 b) Nonempty
 c) Bounded
 d) No redundancies
 e) No degeneracies

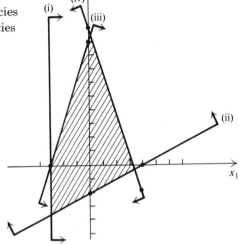

23. a) i) $-3x_1 + 2x_2 \geqslant 6$
  ii) $2x_1 + x_2 \leqslant -2$
  iii) $x_1 + x_2 \geqslant 4$
  iv) $2x_1 + 7x_2 \leqslant 21$
 b) Empty
 c) Bounded
 d) (i) or (ii)
 e) No degeneracies

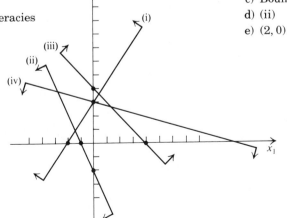

25. a) i) $x_1 \geqslant 0$
  ii) $x_2 \geqslant 0$
  iii) $x_1 + x_2 \geqslant 2$
  iv) $x_1 - x_2 \leqslant 2$
  v) $x_2 \leqslant 6$
 b) Nonempty
 c) Bounded
 d) (ii)
 e) $(2, 0)$

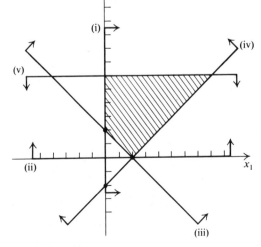

27. a) i) $x_1 + x_2 \leqslant 0$
ii) $2x_1 - 3x_2 \leqslant 15$
iii) $x_2 \leqslant 5$
iv) $x_1 \geqslant 0$
v) $2x_1 + x_2 \geqslant 3$
b) Nonempty ⎫ one point,
c) Bounded ⎭ $(3, -3)$
d) (iii), (iv)
e) $(3, -3)$

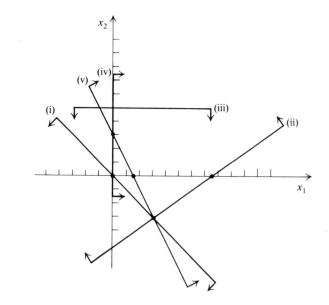

29. a) i) $x_1 \geqslant 0$
ii) $x_2 \geqslant 0$
iii) $5x_2 - 3x_1 \leqslant 15$
iv) $x_1 \leqslant 4x_2$
v) $2x_1 - 5x_2 \leqslant 10$
b) Nonempty
c) Unbounded
d) (ii)
e) $(0, 0)$

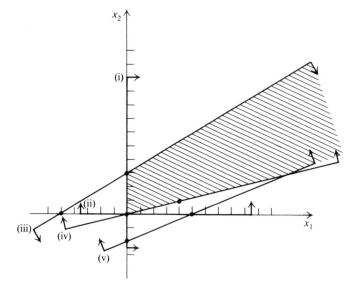

EXERCISE SET 4.2, pp. 138–140

1. $x_1 = 1, x_2 = 5; f = 11$ **3.** $x_1 = 2, x_2 = 3; f = 22$ **5.** $x_1 = 2, x_2 = 6; f = 30$

7. $y_1 = 6, y_2 = 1; f = 22$ **9.** $y_1 = 5, y_2 = 1; f = 15$ **11.** $y_1 = 3, y_2 = 2; f = 19$

13. max: $f = 7$ at $(5, -2)$; min: $f = -3$ at $(0, 3)$ **15.** max: $f = \frac{26}{7}$ at $\left(\frac{24}{7}, \frac{-2}{7}\right)$; min: $f = -\frac{28}{3}$ at $\left(\frac{1}{6}, \frac{19}{2}\right)$ **17.** max: $f = \frac{64}{3}$

at $(8, \frac{8}{3})$; min: $f = 0$ for any point on the line segment between $\left(\frac{24}{11}, \frac{72}{11}\right)$ and $\left(\frac{5}{4}, \frac{15}{4}\right)$ **19.** No feasible solution, ∴ no optimum feasible solution, max or min.   **21.** max: unbounded; min: $f = 0$ at $(0, 0)$ **23.** max and min: both have $f = 3$ at $(3, -3)$

EXERCISE SET 4.3, pp. 145–146

1. Let x_1 = number of suits to be made,
  x_2 = number of dresses to be made,
  f = income.
Then
$$x_1 + 2x_2 \leq 60,$$
$$4x_1 + 3x_2 \leq 120,$$
$$\max f: f = 120x_1 + 75x_2; \quad x_1, x_2 \geq 0.$$

b) $\begin{bmatrix} 1 & 2 \\ 4 & 3 \end{bmatrix} \begin{bmatrix} x_1 \\ x_2 \end{bmatrix} \leq \begin{bmatrix} 60 \\ 120 \end{bmatrix}$

$\max f: f = \begin{bmatrix} 120 & 75 \end{bmatrix} \begin{bmatrix} x_1 \\ x_2 \end{bmatrix}; \quad \begin{bmatrix} x_1 \\ x_2 \end{bmatrix} \geq \begin{bmatrix} 0 \\ 0 \end{bmatrix}$

c) max income = \$3600 for $x_1 = 30$ suits and $x_2 = 0$ dresses.

5. a) Let x_1 = number of animals of species A1,
  x_2 = number of animals of species A2,
  f = total number of animals;

  F1: $x_1 + 1.2x_2 \leq 600,$
  F2: $2x_1 + 1.8x_2 \leq 960,$
  F3: $2x_1 + 0.6x_2 \leq 720.$
  $\max f: f = x_1 + x_2; \quad x_1, x_2 \geq 0.$

b) $\begin{bmatrix} 1 & 1.2 \\ 2 & 1.8 \\ 2 & 0.6 \end{bmatrix} \begin{bmatrix} x_1 \\ x_2 \end{bmatrix} \leq \begin{bmatrix} 600 \\ 960 \\ 720 \end{bmatrix}$

$\max f: f = \begin{bmatrix} 1 & 1 \end{bmatrix} \begin{bmatrix} x_1 \\ x_2 \end{bmatrix}; \quad \begin{bmatrix} x_1 \\ x_2 \end{bmatrix} \geq \begin{bmatrix} 0 \\ 0 \end{bmatrix}$

c) max number = 520 for $x_1 = 120$ and $x_2 = 400$.

3. a) Let x_1 = number of lbs of Mixture I,
  x_2 = number of lbs of Mixture II,
  f = income.
Then
$$0.6x_1 + 0.2x_2 \leq 1800,$$
$$0.3x_1 + 0.5x_2 \leq 1500,$$
$$0.1x_1 + 0.3x_2 \leq 750.$$
$$\max f: f = 0.75x_1 + 2x_2; \quad x_1, x_2 \geq 0.$$

b) $\begin{bmatrix} 0.6 & 0.2 \\ 0.3 & 0.5 \\ 0.1 & 0.3 \end{bmatrix} \begin{bmatrix} x_1 \\ x_2 \end{bmatrix} \leq \begin{bmatrix} 1800 \\ 1500 \\ 750 \end{bmatrix}$

$\max f: f = \begin{bmatrix} 0.75 & 2 \end{bmatrix} \begin{bmatrix} x_1 \\ x_2 \end{bmatrix}; \quad \begin{bmatrix} x_1 \\ x_2 \end{bmatrix} \geq \begin{bmatrix} 0 \\ 0 \end{bmatrix}$

c) max income = \$5,156.25 for $x_1 = 1875$ lbs and $x_2 = 1875$ lbs.

7. a)   F1: $x_1 + 1.2x_2 \leq 720,$
  F2: $2x_1 + 1.8x_2 \leq 960,$
  F3: $2x_1 + 0.6x_2 \leq 600.$
  $\max f: f = x_1 + x_2; \quad x_1 x_2 \geq 0.$

c) max number = $\frac{1600}{3}$ (ignore fractions) for $x_1 = 0$ and $x_2 = \frac{1600}{3}$. Species A1 would become extinct (in that area).

EXERCISE SET 4.4, pp. 151–152

1. a) Let y_1 = number of sacks of soybean meal,
  y_2 = number of sacks of oats,
  f = cost.
Then $50y_1 + 15y_2 \geq 120,$
    $8y_1 + 5y_2 \geq 24,$
    $5y_1 + y_2 \geq 10.$
$\min f: f = 15y_1 + 5y_2; \quad y_1, y_2 \geq 0.$

b) $\begin{bmatrix} 50 & 15 \\ 8 & 5 \\ 5 & 1 \end{bmatrix} \begin{bmatrix} y_1 \\ y_2 \end{bmatrix} \geq \begin{bmatrix} 120 \\ 24 \\ 10 \end{bmatrix}$

$\min f: f = \begin{bmatrix} 15 & 5 \end{bmatrix} \begin{bmatrix} y_1 \\ y_2 \end{bmatrix}; \quad \begin{bmatrix} y_1 \\ y_2 \end{bmatrix} \geq \begin{bmatrix} 0 \\ 0 \end{bmatrix}$

c) min cost = \$$\frac{480}{13}$ for $y_1 = \frac{24}{13}$ and $y_2 = \frac{24}{13}$.

3. a) Third constraint $5y_1 + 8y_3 \geqslant 10$, becomes $5y_1 + 8y_3 \geqslant 20$.   c) min cost = $\$\frac{276}{7}$ for $y_1 = \frac{12}{7}$ and $y_3 = \frac{12}{7}$

5. a) Let y_1 = number of P1 airplanes, y_2 = number of P2 airplanes, f = cost in $ thousands. Then

$$40y_1 + 80y_2 \geqslant 2000,$$
$$40y_1 + 30y_2 \geqslant 1500,$$
$$120y_1 + 40y_2 \geqslant 2400,$$
$$\min f\colon f = 12y_1 + 10y_2; \quad y_1, y_2 \geqslant 0.$$

b) $\begin{bmatrix} 40 & 80 \\ 40 & 30 \\ 120 & 40 \end{bmatrix} \begin{bmatrix} y_1 \\ y_2 \end{bmatrix} \geqslant \begin{bmatrix} 2000 \\ 1500 \\ 240 \end{bmatrix}$,

$$\min f\colon f = [12 \quad 10]\begin{bmatrix} y_1 \\ y_2 \end{bmatrix}; \quad \begin{bmatrix} y_1 \\ y_2 \end{bmatrix} \geqslant \begin{bmatrix} 0 \\ 0 \end{bmatrix}.$$

c) min cost = **$460 thousand for y_1 = 30 and y_2 = 10.**

7. a)
$$40y_1 + 40y_3 \geqslant 2000,$$
$$40y_1 + 80y_3 \geqslant 1500,$$
$$120y_1 + 80y_3 \geqslant 2400,$$
$$\min f\colon f = 12y_1 + 15y_3; \quad y_1, y_3 \geqslant 0.$$

b) $\begin{bmatrix} 40 & 40 \\ 40 & 80 \\ 120 & 80 \end{bmatrix} \begin{bmatrix} y_1 \\ y_3 \end{bmatrix} \geqslant \begin{bmatrix} 2000 \\ 1500 \\ 2400 \end{bmatrix}$,

$$\min f\colon f = [12 \quad 15]\begin{bmatrix} y_1 \\ y_2 \end{bmatrix}; \quad \begin{bmatrix} y_1 \\ y_3 \end{bmatrix} \geqslant \begin{bmatrix} 0 \\ 0 \end{bmatrix}.$$

c) min cost = **$600 thousand for y_1 = 50 and y_3 = 0.**

9. Replace $\begin{bmatrix} 2000 \\ 1500 \\ 2400 \end{bmatrix}$ by $\begin{bmatrix} 1600 \\ 2100 \\ 2400 \end{bmatrix}$.   a) Set $y_3 = 0$.   c) min cost = **$630 thousand for y_1 = 52.5 and y_2 = 0.**

11. a) Set $y_2 = 0$.   c) min cost = **$517.5 thousand for $y_i = \frac{55}{2}$ and $y_3 = \frac{25}{2}$.**

13. $90.   **15.** 75 lbs   **17.** $16\frac{2}{3}$ **19.** $9 per mile.

CHAPTER 4 TEST, p. 153

1. Let x_1 = number of lbs of mixture I,

    x_2 = number of lbs of mixture II,

    f = income.

Then

$$0.8x_1 + 0.6x_2 \leqslant 100,$$
$$0.01x_1 + 0.03x_2 \leqslant 10,$$
$$0x_1 + 0.04x_2 \leqslant 5,$$
$$0.12x_1 + 0.24x_2 \leqslant 25,$$
$$0.07x_1 + 0.09x_2 \leqslant 15,$$
$$\max f\colon f = 0.95x_1 + 1.35x_2; \quad x_1, x_2 \geqslant 0.$$

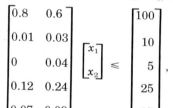

$$\begin{bmatrix} 0.8 & 0.6 \\ 0.01 & 0.03 \\ 0 & 0.04 \\ 0.12 & 0.24 \\ 0.07 & 0.09 \end{bmatrix} \begin{bmatrix} x_1 \\ x_2 \end{bmatrix} \leqslant \begin{bmatrix} 100 \\ 10 \\ 5 \\ 25 \\ 15 \end{bmatrix},$$

$$\max f\colon f = [0.95 \quad 1.35]\begin{bmatrix} x_1 \\ x_2 \end{bmatrix}; \quad \begin{bmatrix} x_1 \\ x_2 \end{bmatrix} \geqslant \begin{bmatrix} 0 \\ 0 \end{bmatrix}.$$

2.

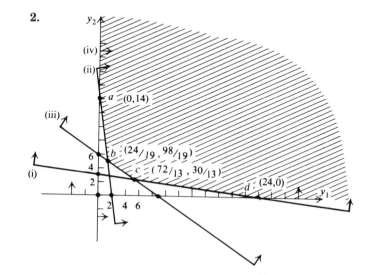

3. Minimum is $6\frac{8}{19}$ when $y_1 = \frac{24}{19}$ and $y_2 = \frac{98}{19}$.   **4.** 6

CHAPTER 5

MARGIN EXERCISES

1.
$$\begin{aligned}
20x_1 + 100x_2 + \; y_1 \qquad\qquad\quad &= 1900 \\
x_1 + \; 50x_2 \qquad + \; y_2 \qquad\quad &= 500, \\
2x_1 + \; 20x_2 \qquad\qquad + \; y_3 &= 240,
\end{aligned}$$

max x_0: $x_0 = 20x_1 + 300x_2 + 0y_1 + 0 \cdot y_2 + 0 \cdot y_3$; $x_i \geq 0$ for all i, $y_i \geq 0$ for all i.

$$\begin{bmatrix} 20 & 100 & 1 & 0 & 0 \\ 1 & 50 & 0 & 1 & 0 \\ 2 & 20 & 0 & 0 & 1 \end{bmatrix} \begin{bmatrix} x_1 \\ x_2 \\ y_1 \\ y_2 \\ y_3 \end{bmatrix} = \begin{bmatrix} 1900 \\ 500 \\ 240 \end{bmatrix},$$

max x_0: $x_0 = [20 \;\; 300 \;\; 0 \;\; 0 \;\; 0][x_1 \;\; x_2 \;\; y_1 \;\; y_2 \;\; y_3]^T$, $[x_1 \;\; x_2 \;\; y_1 \;\; y_2 \;\; y_3]^T = [0 \;\; 0 \;\; 0 \;\; 0 \;\; 0 \;\; 0]^T$.

2.

x_1	x_2	y_1	y_2	y_3	1
20	100	1	0	0	1900
1	50	0	1	0	500
2	20	0	0	1	240
−20	−300	0	0	0	0

3. $x_1 = x_2 = 0$; $y_1 = 1900$, $y_2 = 500$, $y_3 = 240$; $x_0 = 0$.

4. Second column, second row

5. $x_1 = 25$, $x_2 = \frac{19}{2}$, $y_1 = 450$, $y_2 = y_3 = 0$; $x_0 = 3350$

6.
$$\begin{aligned}
0.1y_1 + \quad 0y_2 - x_1 &= 200, \\
0y_1 + \; 0.2y_2 - x_2 &= 500, \\
0.02y_1 + 0.01y_2 - x_3 &= 100, \\
y_1, y_2, x_1, x_2, x_3 &\geq \quad 0,
\end{aligned}$$

max $-y_0$: $-y_0 = -10y_1 - \; 15y_2$.

7. a)
$$\begin{aligned}
0.1y_1 + \quad 0y_2 - x_1 + z_1 &= 200, \\
0y_1 + \; 0.2y_2 - x_2 + z_2 &= 500, \\
0.02y_1 + 0.01y_2 - x_3 + z_3 &= 100, \\
y_1, y_2, x_1, x_2, x_3, z_1, z_2, z_3 &\geq \quad 0,
\end{aligned}$$

max $-y_0$: $-y_0 = -10y_1 - \; 15y_2$,

b) max z_0: $z_0 = -z_1 - z_2 - z_3$.

8.

	y_1	y_2	x_1	x_2	x_3	z_1	z_2	z_3	1
	0.1	0	−1	0	0	1	0	0	200
	0	0.2	0	−1	0	0	1	0	500
	0.02	0.01	0	0	−1	0	0	1	100
$-y_0$	−10	−15	0	0	0	0	0	0	0
z_0	0	0	0	0	0	1	1	1	0

9. Same tableau as ME 8 except for z_0:

$$z_0: [-0.12 \quad -0.21 \quad 1 \quad 1 \quad 1 \quad 0 \quad 0 \quad 0 \quad -800]$$

Initial feasible solution: $y_1 = y_2 = x_1 = x_2 = x_3 = 0$, $z_1 = 200$, $z_2 = 500$, $z_3 = 100$

10. $y_1 = 3750$, $y_2 = 2500$, $x_1 = 175$, $x_2 = x_3 = 0$; $x_0 = 75{,}000$

11.
$$\begin{aligned}
20y_1 + \quad y_2 + \quad 2y_3 &\geq 20, \\
100y_1 + 50y_2 + 20y_3 &\geq 300, \\
y_1, \quad y_2, \quad y_3 &\geq 0,
\end{aligned}$$
$$\min y_0: y_0 = 1900y_1 + 500y_2 + 240y_3$$

12.
$$\begin{aligned}
0.1x_1 + 0 \cdot x_2 + 0.02x_3 &\leq 10, \\
0 \cdot x_1 + 0.2x_2 + 0.01x_3 &\leq 15, \\
x_1, x_2, x_3 &\geq 0,
\end{aligned}$$
$$\max x_0: x_0 = 200x_1 + 500x_2 + 100x_3$$

13.

x_1	x_2	y_1	y_2	y_3	1
20	100	1	0	0	1900
1	50	0	1	0	500
2	20	0	0	1	240
−20	−300	0	0	0	0

Initial primal sol.: $x_0 = x_1 = x_2 = 0$, $y_1 = 1900$, $y_2 = 500$, $y_3 = 240$

Initial dual sol.: $y_0 = y_1 = y_2 = y_3 = 0$, $x_1 = -20$, $x_2 = -300$

14.
$$\begin{aligned}
0.1x_1 + 0 \cdot x_2 + 0.02x_3 + \quad y_1 \quad\quad &= 10, \\
0 \cdot x_1 + 0.2x_2 + 0.01x_3 \quad\quad + \quad y_2 &= 15, \\
x_1, x_2, x_3, y_1, y_2 &\geq 0,
\end{aligned}$$
$$\max x_0: x_0 = 200x_1 + 500x_2 + 100x_3 + 0 \cdot y_1 + 0 \cdot y_2$$

x_1	x_2	x_3	y_1	y_2	1
0.1	2	0.02	1	0	10
0	0.2	0.01	0	1	15
−200	−500	−100	0	0	0

Initial primal sol.: $x_0 = x_1 = x_2 = x_3 = 0$, $y_1 = 10$, $y_2 = 15$

Initial dual sol.: $y_0 = y_1 = y_2 = 0$, $x_1 = -200$, $x_2 = -500$, $x_3 = -100$

15. Dual (min) solution: $y_1 = 3750$, $y_2 = 2500$, $x_1 = 175$, $x_2 = x_3 = 0$; $y_0 = 75{,}000$

16. Primal (max) solution: $x_1 = 0$, $x_2 = 75$, $x_3 = 500$, $y_1 = y_2 = 0$; $x_0 = 75{,}000$

17.

x_1	x_2	x_3	y_1	y_2	y_3	1
$-\frac{11}{7}$	0	1	$\frac{1}{7}$	$\frac{2}{7}$	0	$\frac{4}{7}$
$-\frac{16}{7}$	1	0	$-\frac{3}{7}$	$\frac{1}{7}$	0	$\frac{2}{7}$
$\frac{128}{7}$	0	0	$\frac{3}{7}$	$-\frac{15}{7}$	1	$\frac{19}{7}$
$\frac{68}{7}$	0	0	$\frac{11}{7}$	$\frac{1}{7}$	0	$\frac{2}{7}$

19.

x_1	x_2	x_3	y_1	y_2	y_3	1
4	1	−2	1	0	0	2
7	0	−2	1	1	0	13
14	0	−1	2	0	1	12
6	0	−2	4	0	0	8

Note all negative entries in x_3 column

18. $x_0 = 10$; $x_1 = 1$, $x_2 = 0$, $x_3 = 1$

$x_0 = 10$; $x_1 = 0$, $x_2 = \frac{5}{2}$, $x_3 = \frac{1}{4}$

20. $x_{11} = 40$, $x_{13} = 80$, $x_{21} = 20$, $x_{22} = 50$, $x_{12} = x_{23} = 0$; $x_0 = 1410$

21. $x_{11} = 60$, $x_{13} = 60$, $x_{21} = 50$, $x_{22} = 20$, $x_{12} = x_{23} = 0$; $x_0 = 1370$

22. Same as Example 1. **23.** $x_{13} = x_{22} = x_{35} = x_{41} = x_{54} = 1$

EXERCISE SET 5.1, pp. 171–172

1. $x_1 = 1$, $x_2 = 5$, $y_1 = y_2 = 0$; $x_0 = 11$ **3.** $x_1 = 2$, $x_2 = 3$, $y_1 = y_2 = 0$; $x_0 = 22$ **5.** $x_1 = 2$, $x_2 = 6$, $y_1 = y_2 = 0$, $y_3 = 2$; $x_0 = 30$ **7.** $x_1 = 30$, $x_2 = y_1 = 0$, $y_2 = 20$, $x_0 = 3600$ **9.** $x_1 = 0$, $x_2 = 4$, $y_1 = 1$, $y_2 = y_3 = 0$; $x_0 = 90$ **11.** $x_1 = 8$, $x_2 = 3$, $y_1 = 2$, $y_2 = y_3 = 0$; $x_0 = 19$ **13.** $x_1 = 9$, $x_2 = 4$, $y_1 = 5$, $y_2 = y_3 = 0$; $x_0 = 79$ **15.** $x_1 = \frac{6}{5}$, $x_2 = 0$, $x_3 = \frac{22}{5}$, $y_1 = y_2 = 0$, $y_3 = \frac{144}{5}$; $x_0 = 22$ **17.** $x_1 = 0$, $x_2 = \frac{9}{4}$, $x_3 = \frac{19}{4}$, $y_1 = \frac{11}{4}$, $y_2 = y_3 = 0$; $x_0 = \frac{75}{4}$ **19.** No bookcases, 12 desks, 65 tables; max sales = \$6400.

EXERCISE SET 5.2, pp. 179–181

1. $y_1 = 6$, $y_2 = 1$, $x_1 = 10$, $x_2 = x_3 = 0$; $y_0 = 22$ **3.** $y_1 = 5$, $y_2 = 1$, $x_1 = 2$, $x_2 = x_3 = 0$; $y_0 = 15$ **5.** $y_1 = 3$, $y_2 = 2$, $x_1 = 5$, $x_2 = x_3 = 0$; $y_0 = 19$ **7.** $y_1 = y_2 = \frac{24}{13}$, $x_1 = x_2 = 0$, $x_3 = \frac{14}{13}$; $y_0 = \frac{480}{13}$ **9.** $y_1 = y_3 = \frac{12}{7}$, $x_1 = x_2 = 0$, $x_3 = 4$; $y_0 = \frac{276}{7}$ **11.** $y_1 = 30$, $y_2 = 10$, $x_1 = x_2 = 0$, $x_3 = 1600$; $y_0 = 460$ **13.** $y_1 = 50$, $y_3 = x_1 = 0$, $x_2 = 500$, $x_3 = 3600$; $y_0 = 600$ **15.** $y_1 = 52.5$, $y_2 = 0$, $x_1 = 500$, $x_2 = 0$, $x_3 = 3900$; $y_0 = 630$ **17.** $y_1 = 27.5$, $y_3 = 12.5$, $x_1 = x_2 = 0$, $x_3 = 1900$; $y_0 = 517.5$. **19.** $x_1 = 2$, $x_2 = 6$, $y_1 = y_2 = 0$, $y_3 = 2$, $y_4 = 18$; $x_0 = 30$ **21.** $x_1 = 8$, $x_2 = 3$, $y_1 = 2$, $y_2 = y_3 = 0$, $y_4 = 36$, $y_5 = 24$; $x_0 = 19$ **23.** $x_1 = \frac{21}{2}$, $x_2 = 1$, $y_1 = \frac{25}{2}$, $y_2 = \frac{3}{2}$, $y_3 = 0$, $y_4 = \frac{51}{2}$; $x_0 = \frac{155}{2}$ **25.** $x_1 = x_2 = 0$, $x_3 = 5$, $y_1 = 3$, $y_2 = 9$, $y_3 = 36$, $y_4 = 4$; $x_0 = 25$

EXERCISE SET 5.3, pp. 191–194

For solutions to Exercise 1 through 17, see solutions to Exercises 1 through 17 of Set 5.2.

19. $y_1 = 4$, $y_2 = 6$, $x_1 = 4$, $x_2 = x_3 = 0$; $y_0 = 76$ (P: $x_1 = 0$, $x_2 = \frac{4}{7}$, $x_3 = \frac{11}{7}$, $y_1 = y_2 = 0$)

21. $y_1 = 5$, $y_2 = 7$, $x_1 = x_2 = 0$, $x_3 = 1$; $y_0 = 53$ (P: $x_1 = \frac{9}{7}$, $x_2 = \frac{2}{7}$, $x_3 = y_1 = y_2 = 0$)

23. $x_1 = 0$, $x_2 = \frac{7}{4}$, $x_3 = \frac{5}{4}$, $y_1 = \frac{59}{4}$, $y_2 = y_3 = 0$; $x_0 = \frac{19}{4}$ (P: $y_1 = 0$, $y_2 = \frac{1}{8}$, $y_3 = \frac{5}{4}$, $x_1 = \frac{27}{8}$, $x_2 = x_3 = 0$)

25. $y_1 = \frac{8}{11}$, $y_2 = \frac{2}{11}$, $y_3 = x_1 = x_2 = 0$, $x_3 = \frac{13}{11}$; $y_0 = \frac{82}{11}$ (P: $x_1 = \frac{29}{11}$, $x_2 = \frac{6}{11}$, $x_3 = y_1 = y_2 = 0$, $y_3 = 4$)

27. Cost \$0.29 with all soybeans = 288 gms **29.** Cost \$0.91 with 130 gms cheese and 138 gms beef.

EXERCISE SET 5.4, p. 204

1. Primal degenerate: $x_0 = 12$, $x_1 = 0$, $x_2 = 0$, $x_3 = 4$ Dual nonunique: (1) $y_0 = 12$, $y_1 = 0$, $y_2 = 3$, $y_3 = 0$, (2) $y_0 = 12$, $y_1 = 0$, $y_2 = \frac{9}{4}$, $y_3 = \frac{1}{4}$ **3.** Primal unbounded, dual infeasible. **5.** Primal degenerate: $x_0 = 12$, $x_1 = 0$, $x_2 = 0$, $x_3 = 3$ Dual nonunique: (1) $y_0 = 12$, $y_1 = 0$, $y_2 = 1$, $y_3 = 1$ (2) $y_0 = 12$, $y_1 = 0$, $y_2 = 0$, $y_3 = 2$.

EXERCISE SET 5.5, pp. 211–213

Values of basic variables are given; values of nonbasic variables are zero.

1. $x_{11} = 70$, $x_{12} = 80$, $x_{22} = 10$, $x_{23} = 70$; $x_0 = 2230$ **3.** $x_{11} = 70$, $x_{22} = 80$, $x_{31} = x_{32} = 50$; $x_0 = 2850$ **5.** $x_{11} = 15$, $x_{12} = 60$, $x_{14} = 75$, $x_{21} = 30$, $x_{23} = 70$; $x_0 = 2495$ **7.** $x_{11} = 70$, $x_{12} = 10$, $x_{23} = 90$, $x_{31} = 0$, $x_{32} = 110$; $x_0 = 3510$ **9.** $x_{13} = 20$, $x_{14} = 80$, $x_{21} = 25$, $x_{22} = 100$, $x_{31} = 50$, $x_{33} = 100$; $x_0 = 3905$

EXERCISE SET 5.6, pp. 219–220

Values of basic variables are given; values of nonbasic variables are zero.

1. $x_{12} = x_{25} = x_{31} = x_{44} = x_{53} = 1$; $a_0 = 432$ **3.** $x_{15} = x_{22} = x_{31} = x_{43} = x_{54} = 1$; $a_0 = 439$ **5.** $x_{11} = x_{25} = x_{34} = x_{43} = x_{52} = 1$ or $x_{13} = x_{25} = x_{34} = x_{41} = x_{52} = 1$; $a_0 = 440$ **7.** $x_{12} = x_{25} = x_{33} = x_{41} = x_{56} = x_{64} = 1$; $a_0 = 496$.

CHAPTER 5 TEST, p. 221

1. $x_1 = 11$, $x_2 = 5$, $y_1 = 5$, $y_2 = y_3 = 0$; $x_0 = 59$. **2.** $y_1 = \frac{62}{7}$, $y_2 = \frac{60}{7}$, $x_1 = 0$, $x_2 = \frac{10}{7}$, $x_3 = 0$; $y_0 = \frac{428}{7}$

3. a)
$$y_1 + y_2 + 5y_3 \geq 4,$$
$$2y_1 + y_2 + 3y_3 \geq 3,$$
$$\text{min } y_0: y_0 = 26y_1 + 16y_2 + 70y_3;$$
$$y_1, y_2, y_3 \geq 0.$$

b) $y_1 = 0$, $y_2 = \frac{3}{2}$, $y_3 = \frac{1}{2}$, $x_1 = x_2 = 0$; $y_0 = 59$

c)
$$x_1 + 2x_2 + y_1 = 26,$$
$$x_1 + x_2 + y_2 = 16,$$
$$5x_1 + 3x_2 + y_3 = 70,$$
$$x_0 = 4x_1 + 3x_2;$$
$$x_1, x_2; y_1, y_2, y_3 \geq 0.$$

d)
$$y_1 + y_2 + 5y_3 - x_1 = 4,$$
$$2y_1 + y_2 + 3y_3 - x_2 = 3,$$
$$y_0 = 26y_1 + 16y_2 + 70y_3 = x_0;$$
$$y_1, y_2, y_3; x_1, x_2 \geq 0.$$

4.

x_1	x_2	x_3	y_1	y_2	y_3	1
16*	0	−11	1	−5	0	2
−3	1	2	0	1	0	0
5	0	−4	0	−3	1	5
−1	0	3	0	1	0	0

5. a) Primal degenerate: $x_1 = x_2 = x_3 = 0$, $x_4 = 2$, $x_5 = 0$, $x_6 = 5$; $x_0 = 10$.
Dual nonunique: $x_1 = 2$, $x_2 = 1$, $x_3 = 1$, $x_4 = x_5 = x_6 = 0$; $x_0 = 10$.
b) Primal unbounded; dual infeasible.

6. $x_{12} = 35$, $x_{13} = 10$, $x_{23} = 50$, $x_{31} = 30$, $x_{32} = 35$; $x_0 = 2625$

7. $x_{11} = x_{23} = x_{32} = x_{44} = x_{55} = 1$; $a_0 = 35$

CHAPTER 6

MARGIN EXERCISES

1. a) No b) Yes   **2.** a) *ABCA, ACDFA, ABCDFA, GHJG* b) *DEGD, DFGD, DFGED*   **3.** Drawing only the *part* of the network showing *deletion* of arcs, we obtain 8 spanning trees:

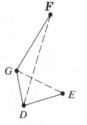

4.

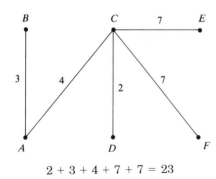

$$2 + 3 + 4 + 7 + 7 = 23$$

5.

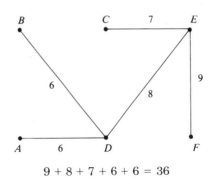

$$9 + 8 + 7 + 6 + 6 = 36$$

6. *ACDFGJ* and *ACEFGJ*, each with minimum path length 22.   **7.** *ABEDG* and *ABEG*, each with minimum path length 20.   **8.** Uppermost: *ABEG* with 8 units saturating *EG*, *ABEFG* with 2 units saturating *BE*, *ABDEFG* with 2 units saturating *AB*, *ACBDEFG* with 2 units saturating *EF*, *ACBDFG* with 1 unit saturating *CB*, *ACDFG* with 2 units saturating *CD*, *ACFG* with 1 unit saturating *FG*; Total = 18 units, saturating cut {*EG*, *FG*}

Bottommost; *ACFG* with 5 units saturating *CF*, *ACDFG* with 2 units saturating *AC* and *CD*, *ABDFG* with 3 units saturating *FG*, *ABDFEG*, with 4 units saturating *DF*, *ABDEG* with 1 unit saturating *BD*, *ABEG* with 3 units saturating *EG*: Total = 18 units; saturating cut {*EG*, *FG*}

EXERCISE SET 6.1, pp. 230–231

1.

min- - -

max——

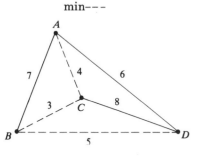

min 12
max 21

3.

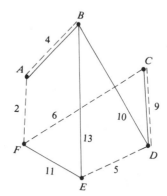

min 26
max 47

5.

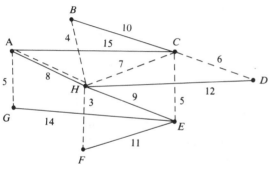

min 38
max 79

EXERCISE SET 6.2, pp. 239–240

1. *ACG*: length 12   **3.** *GCA*: length 12   **5.** *ACBEFDHJ*: length 29   **7.** *ADCGFHLMN*: length 31

EXERCISE SET 6.3, pp. 246–247

1. Uppermost: *ABFH* with 7 units saturating *BF*, *ABDFH* with 1 unit saturating *AB*, *ACEDFH* with 1 unit saturating *FH*, *ACEGH* with 4 units saturating *CE* and *EG*, *ACGH* with 6 units saturating *AC*; Total = 19 units; saturating cut {*AB*, *AC*}

Bottommost: *ACGH* with 8 units saturating *CG*, *ACEGH* with 3 units saturating *AC*, *ABDEGH* with 1 unit saturating *EG* and *GH*, *ABDFH* with 2 units saturating *BD*, *ABFH* with 5 units saturating *AB*; Total = 19 units; saturating cut {*AB*, *AC*}

3. Uppermost:

			Bottommost:		
ABFJ:	11	*FJ*,	*ACGJ*:	12	*CG*,
ABFHJ:	2	*AB*,	*ACEGJ*:	5	*AC*,
AEDFHJ:	4	*ED*,	*AEGJ*:	1	*EG*,
AEFHJ:	1	*FH*,	*AEDGJ*:	1	*GJ*,
AEGHJ:	1	*HJ*,	*AEDGHJ*:	3	*ED*,
AEGJ:	3	*AE*,	*AEFHJ*:	4	*AE*,
ACDGJ:	8	*CD*, *DG*,	*ABDFHJ*:	1	*HJ*,
ACEGJ:	2	*EG*,	*ABDFJ*:	6	*BD*,
ACGJ:	6	*GJ*;	*ABFJ*:	5	*FJ*;
Total:	38	{*FJ*, *HJ*, *GJ*}	Total:	38	{*FJ*, *HJ*, *GJ*}

CHAPTER 6 TEST, p. 248

1. a) 33 b) 72   **2.** *ABFHJ*: 18   **3.** *ABFJ*, 4, *BF*; *ABCDFJ*, 3, *BC*, *FJ*; *ACDFHJ*, 1, *CD*, *HJ*; *ADGJ*, 2, *GJ*; Total, 10, {*FJ*, *HJ*, *GJ*}.

CHAPTER 7

MARGIN EXERCISES

1. a) Yes, no b) No, yes   **2.** a) $D = \{x \mid x = 2n + 1, n \text{ is an integer}, n \geq 0\}$ b) $F = \{1, 4, 7, 10, 13, \ldots\}$
3. {a}, {b}, {c}, {d}, {a, b}, {a, c}, {a, d}, {b, c}, {b, d}, {c, d}, {a, b, c}, {a, b, d}, {b, c, d}, {a, c, d}, {a, b, c, d}, ∅
4. a) Yes b) No, $0 \notin A$ **5.** a) 4 b) 0   **6.** *Before* he buys, the universal set is the set of articles which can be bought for $10, or less. *After* he buys, it is that set of purchases.   **7.** $A^c = \{2, 4, 6, 8\}$ **8.** $A \times B =$
{(a, 1), (a, 2), (a, 3), (a, 4), (b, 1), (b, 2), (b, 3), (b, 4), (c, 1), (c, 2), (c, 3), (c, 4)}, $\mathcal{N}(A \times B) = 12$

9. $A \cup B = \{a, b, c, e, g, f, s\}$, $A \cap B = \{b, c, e\}$.

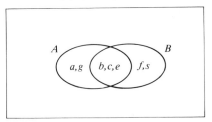

10. a) $\{\{1, 2, 3, 4\}, \{5\}\}$ | 1, 2, 3, 4 | 5 |

     b) $\{\{1, 2, 3\}, \{4, 5\}\}$ | 1, 2, 3 | 4, 5 |

     c) $\{\{1\}, \{2\}, \{3\}, \{4, 5\}\}$ | 1 | 2 | 3 | 4, 5 |

There are other possibilities.

11. $A - B = \{a, e, g\}$, $B - A = \{f, h\}$, $A^c = \{f, h, i, j, k, \ldots, z\}$, $B^c = \{a, g, i, j, k, l, \ldots z\}$, $B^c - A^c = \{a, g\}$, shaded as in the accompanying figure.

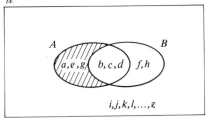

12. 60   **13.** 53   **14.** $S = \{(0, 3), (1, 2), (2, 1), (3, 0)\}$ where each ordered pair = (no. G, no. S)

15.

```
                                              G
                                              • (3, 0)
                                    (2, 0)  G
                                           •
                                              • S
                                              (2, 1)
                         (1, 0)  G
                                •
                                              • G
                                              (2, 1)
                                    (1, 1)  S
                                           •
                                              • S
                                              (1, 2)
              (0, 0)  •
                                              • G
                                              (2, 1)
                                    (1, 1)  G
                                           •
                                              • S
                                              (1, 2)
                         (0, 1)  S
                                •
                                              • G
                                              (1, 2)
                                    (0, 2)  S
                                           •
                                              • S
                                              (0, 3)
```

First draw   Second draw   Third draw

16. $3 \cdot 2 \cdot 1$, or 6   **17.** $5 \cdot 4 \cdot 3 \cdot 2 \cdot 1$, or 120   **18.** $4 \cdot 3 \cdot 2 \cdot 1$, or 24   **19.** 2^{10}, or 1024   **20.** $5 \cdot 40 \cdot 8$, or 1600

21. $5 \cdot 4 \cdot 3 \cdot 2 \cdot 1$, or 120   **22.** 120   **23.** 720, 120   **24.** 40,320   **25.** 362,880   **26.** 6!   **27.** a) $8! = 8 \cdot 7!$

b) $38! = 38 \cdot 37!$ **28.** 360   **29.** a) 5040   b) 362,880   **30.** a) 5040   b) 2520   c) 840   d) 210

31. $(6-1)!$, or 120   **32.** a) $52 \cdot 51$, or 2652   b) $52 \cdot 52$, or 2704   **33.** a) 720   b) 240   c) 144   d) 576

34. a) 5   b) 1   c) 10   d) 5   e) 1   **35.** a) $P(4,3) = 4 \cdot 3 \cdot 2 = 24$ b) *ABC, ACB, ABD, ADB, ACD, ADC,*

BAC, BCA, BAD, BDA, BCD, BDC, CAB, CBA, CAD, CDA, CBD, CDB, DAB, DBA, DAC, DCA, DBC, DCB

c) $\binom{4}{3} = \dfrac{4 \cdot 3 \cdot 2}{3 \cdot 2 \cdot 1} = 4$ d) $\{A, B, C\}, \{A, C, D\}, \{B, C, D\}, \{A, B, D\}$ **36.** a) 120   b) 120   c) 126   d) 126   **37.** a) 56

b) $\binom{8}{3}$ **38.** $\binom{10}{8} = 45$ **39.** $\binom{7}{4} \cdot \binom{5}{3} = 350$ **40.** $\binom{3}{1}\binom{4}{1} + \binom{4}{2}$, or 18   **41.** $\binom{7}{1}\binom{6}{2}\binom{4}{2}\binom{2}{2} = \dfrac{7!}{1!2!2!2!} = 630$

42. a) 9   b) 1   c) 4   d) 2   e) 2   f) 3780   g) 3780   **43.** $-1512x^5$ **44.** $5670x^4$ **45.** $8064y^5$ **46.** $243x^5$

47. $x^{10} - 5x^8 + 10x^6 - 10x^4 + 5x^2 - 1$ **48.** $16x^4 + 32x^3/y + 24x^2/y^2 + 8x/y^3 + 1/y^4$

49. $x^6 - 6\sqrt{2}x^5 + 30x^4 - 40\sqrt{2}x^3 + 60x^2 - 24\sqrt{2}x + 8$ **50.** $\binom{n}{0}2^n + \binom{n}{1}2^{n-1} + \cdots + \binom{n}{n}2^0$

51. 1   6   15   20   15   6   1

EXERCISE SET 7.1, pp. 257–258

1. $A = \{0, 1, 2, 3, 4, 5, 6, 7, 8, 9, 10\}$, $B = \{1, 2, 3, 4, 5, 6, 7, 8, 9\}$, $C = \{0, 2, 4, 6, 8, 10, 12\}$ **3.** $A^c = \{11, 12\}$,

$B^c = \{0, 10, 11, 12\}$, $C^c = \{1, 3, 5, 7, 9, 11\}$ **5.** True   **7.** False   **9.** True   **11.** True   **13.** False   **15.** False

17. $A^c = \{m, n, r, t, c\}$, $B^c = \{a, e, i, o, u\}$, $C^c = \{i, o, u, m, n, r, t\}$ **19.** False   **21.** True   **23.** False   **25.** False

27. $E \times F = \{(e, r), (e, t), (i, r), (i, t)\}$ **29.** Yes   **31.** $C \times E = \{(a, e), (a, i), (c, e), (c, i), (e, e), (e, i)\}$ **33.** 6

35. a) 0   b) $\emptyset$ c) 1   **37.** a) 2   b) $\emptyset, \{a\}, \{b\}, \{a, b\}$ c) 4   **39.** $1, 2, 4, 8, 2^4, 2^n$

EXERCISE SET 7.2, pp. 269–270

1. a) A b) B c) $\{0, 10\}$ d) $\emptyset$ e) A f) $\{2, 4, 8, 10\}$ g) $\{1, 5, 7, 11\}$ h) $\{0, 2, 4, 8, 10, 12\}$ i) $\{10\}$ j) $\{1, 5, 7\}$

3.

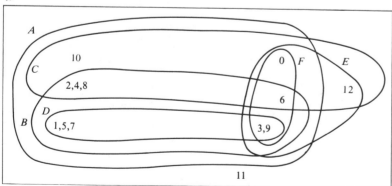

5. a) $\{e, i, r, t\}$ b) $\emptyset$ c) $\{c, r\}$ d) $\{i\}$ e) $\emptyset$ f) $\{m, n, r, t\}$ g) $\emptyset$ h) $\{a, e\}$ **7.** 5   **9.** 4   **11.** 85   **13.** 47%, 36%, 5%

EXERCISE SET 7.3, pp. 273–274

1. $S = \{I, II, III, \text{none}\}$

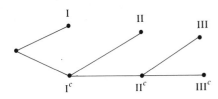

5. $S = \{(M, S, <18), (M, S, \geq 18), (M, M, <18), (M, M, \geq 18),$
$(F, S, <18), (F, S, \geq 18), (F, M, <18), (F, M, \geq 18)\}$

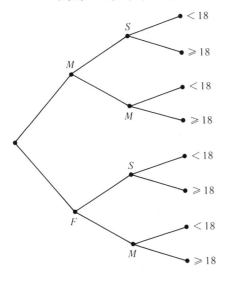

9. $n - 1$

3. $S = \{(AN, CLR), (AN, CLD), (AN, PR), (N, CLR), (N, CLD), (N, PR),$
$(BN, CLR), (BN, CLD), (BN, PR)\}$

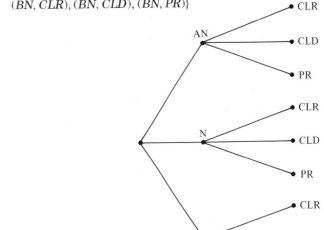

7.

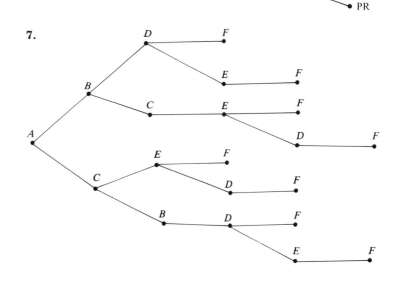

EXERCISE SET 7.4, pp. 282–283

1. 120   **3.** 1   **5.** 720   **7.** 380   **9.** 2520   **11.** $n(n - 1)(n - 2)$ **13.** $3 \cdot 4 \cdot 2$, or 24   **15.** $P(4, 4) = 4! = 24$

17. $P(4, 2) = 4 \cdot 3 = 12$ **19.** $P(4, 4) = 24$ **21.** $(4 - 1)! = 3! = 6$; no   **23.** $5 \cdot 5 \cdot 5 \cdot 5 = 625$, $5 \cdot 4 \cdot 3 \cdot 2 = 120$

25. a) $5 \cdot 5 \cdot 5 = 125$ b) $5 \cdot 4 \cdot 3 = 60$ **27.** $7^4 = 2401$, $7 \cdot 6 \cdot 5 \cdot 4 = 840$ **29.** $7^3 = 343$ **31.** 5040, 144, 1440

33. $6! = 720$, $2 \cdot 2 \cdot 2 \cdot 4 \cdot 3! = 192$

EXERCISE SET 7.5, pp. 288–292

1. 78   **3.** 78   **5.** 7   **7.** $\dfrac{n(n-1)}{2}$ **9.** $\dbinom{6}{4} = 15$ **11.** $\dbinom{10}{7} \cdot \dbinom{8}{5} = 6720$ **13.** a) $\dbinom{14}{5} = 2002$

b) $\dbinom{6}{2}\dbinom{8}{3} + \dbinom{6}{1}\dbinom{8}{4} + \dbinom{8}{5} = 1316$ c) $\dbinom{6}{1}\dbinom{8}{4} + \dbinom{8}{5} = 476$ d) $\dbinom{6}{4}\dbinom{8}{1} + \dbinom{6}{3}\dbinom{8}{2} + \dbinom{6}{2}\dbinom{8}{3} + \dbinom{6}{1}\dbinom{8}{4} + \dbinom{8}{5} = 1996$

15. $\dfrac{11!}{3!\,4!\,2!\,2!} = 69{,}300$ **17.** $\dfrac{10!}{2!\,3!\,3!\,1!\,1!} = 50{,}400$ **19.** $\dfrac{20!}{2!\,5!\,8!\,3!\,2!} = 20{,}951{,}330{,}400$ **21.** $\dfrac{9!}{3!\,2!\,4!} = 1260$

23. a) $\dbinom{20}{3} = 1140$ b) $\dbinom{20}{3}\dbinom{3}{1} = 3420$ **25.** $5 \cdot 4 \cdot 3 = 60, \dfrac{5 \cdot 4 \cdot 3}{3 \cdot 2 \cdot 1} = 10$ **27.** a) $\dbinom{30}{6} = 593{,}775$

b) $\dbinom{10}{4}\dbinom{20}{2} = 39{,}900$ **29.** $\dbinom{12}{4}\dbinom{8}{4}\dbinom{4}{4} = 34{,}650$ **31.** $\dfrac{34{,}650}{3!} = 5775$ **33.** 2520, 2520, 1320

35. 240   **37.** a) $\dbinom{8}{0} + \dbinom{8}{1} + \cdots + \dbinom{8}{8} = \displaystyle\sum_{i=0}^{8}\dbinom{8}{i}$ b) $\displaystyle\sum_{i=0}^{8}\dbinom{8}{i} = 2^8 = 256$ c) $\displaystyle\sum_{i=0}^{9}\dbinom{9}{i} = 2^9 = 512$

39. 4   **41.** $13 \cdot 48 = 624$ **43.** $13 \cdot \dbinom{4}{2}\dbinom{12}{3}\dbinom{4}{1}\dbinom{4}{1}\dbinom{4}{1} = 1{,}098{,}240$ **45.** $\dbinom{13}{5}4 - 36 - 4 = 5108$

47. $10 \cdot 4^5 - 36 - 4 = 10{,}200$ **49.** $P(10, 6) = 10 \cdot 9 \cdot 8 \cdot 7 \cdot 6 \cdot 5 = 151{,}200$ **51.** $C(8, 2) = 28, 2C(8, 2) = 56$

53. MATH: $4! = 24$; BUSINESS: $\dfrac{8!}{3!} = 6720$; PHILOSOPHICAL: $\dfrac{13!}{2!2!2!2!2!} = 194{,}594{,}400$ **55.** $2^7 - 1 = 127$

57. $80 \cdot 26 \cdot 9999 = 20{,}797{,}920$ **59.** $C(n, 2) - n\text{\textbackslash} = \dfrac{n^2 - 3n}{2}$ **61.** a) $P(6, 5) = 6 \cdot 5 \cdot 4 \cdot 3 \cdot 2 = 720$ b) $6^5 = 7776$

c) $P(5, 4) = 5 \cdot 4 \cdot 3 \cdot 2 = 120$ d) $P(3, 2) = 3 \cdot 2 = 6$ **63.** $\dbinom{n}{4} = \dfrac{n(n-1)(n-2)(n-3)}{24}$ **65.** 5   **67.** 8   **69.** 6

71. 11

EXERCISE SET 7.6, pp. 296–297

1. $15a^4b^2$ **3.** $-745{,}472a^3$ **5.** $1120x^{12}y^2$ **7.** $-1{,}959{,}552u^5v^{10}$ **9.** $m^5 + 5m^4n + 10m^3n^2 + 10m^2n^3 + 5mn^4 + n^5$

11. $x^{10} - 15x^8y + 90x^6y^2 - 270x^4y^3 + 405x^2y^4 - 243y^5$ **13.** $x^{-8} + 4x^{-4} + 6 + 4x^4 + x^8$ **15.** $\dbinom{n}{0} - \dbinom{n}{1} + \dbinom{n}{2} - \dbinom{n}{3} +$

$\cdots + (-1)^n\dbinom{n}{n}$ **17.** $140\sqrt{2}$ **19.** $9 - 12\sqrt{3}t + 18t^2 - 4\sqrt{3}t^3 + t^4$ **21.** -3 **23.** $-5 + \sqrt{8}$

CHAPTER 7, TEST p. 298

1. B **2.** A **3.** {(a, d), (a, e), (b, d), (b, e), (c, d), (c, e)}   **4.** C **5.** Yes   **6.** No   **7.** 840   **8.** 35   **9.** 720

10. 1   **11.** $6 \cdot 6 \cdot 6 = 216$ **12.** $\dbinom{52}{2} = 1326$ **13.** $5! = 120$ **14.** $4 \cdot 4 \cdot 4 = 64, 4 \cdot 3 \cdot 2 = 24$ **15.** $2 \cdot 4! \cdot 4! = 1152$

16. $\dfrac{8!}{1!2!2!1!1!1!} = 10{,}080$ **17.** $\dbinom{12}{3} + \dbinom{9}{1}\dbinom{12}{2} = 814$ **18.** 35   **19.** $x^8 + 12x^6y + 54x^4y^2 + 108x^2y^3 + 81y^4$

CHAPTER 8

MARGIN EXERCISES

1. a) $\frac{1}{6}$ b) Answers may vary c) Answers may vary **2.** a) $\frac{4}{52}$, or $\frac{1}{13}$ b) $\frac{2}{52}$, or $\frac{1}{26}$ c) $\frac{26}{52}$, or $\frac{1}{2}$ d) $\frac{12}{52}$, or $\frac{3}{13}$ **3.** $\frac{5}{8}$

4. 0 **5.** 1 **6.** $\frac{1}{17}$ **7.** $\frac{4}{13}$ **8.** $\frac{6}{13}$ **9.** a) $\frac{5}{36}$ b) $\frac{6}{36}$ c) $\frac{2}{36}$ d) 0 **10.** a) $\frac{23}{34}$ b) 0.37 **11.** a) $\frac{1}{26}$ b) $\frac{25}{26}$

c) $\frac{1}{4}$ d) $\frac{3}{4}$ **12.** a) 1:25 b) 25:1 **13.** 0.63 **14.** 0.76 **15.** $\frac{45}{91}$ **16.** a) $\frac{1}{6}$ b) $\frac{1}{2}$ c) $\frac{1}{12}$ d) Yes

17. a) $\frac{1}{2}$ b) $\frac{1}{2}$ c) $\frac{1}{4}$ **18.** a) $\frac{1}{6}$ b) 0.132 **19.** $\frac{1}{2} \cdot \frac{1}{2} \cdot \frac{1}{6}$, or $\frac{1}{24}$ **20.** $\frac{1}{6} \cdot \frac{1}{6} \cdot \frac{1}{6}$, or $\frac{1}{216}$ **21.** $(\frac{1}{2})^4$, $\frac{1}{16}$

22. a) $\frac{4}{5} \cdot \frac{2}{3}$, or $\frac{8}{15}$ b) $(1 - \frac{4}{5})(1 - \frac{2}{3})$, or $\frac{1}{15}$ c) $\frac{6}{15}$, or $\frac{2}{5}$ **23.** Same answers **24.** $\frac{5}{32}$ **25.** Answers may vary **26.** 0.9

27. a) 0.37 b) 0.63 **28.** 0.06 **29.** a) $p_4 = 0.625$, $p_5 = 0.63333$, $p_6 = 0.63194$, $p_7 = 0.63213$ **30.** 0.2

31. $\frac{2}{20} \div \frac{8}{20} = \frac{1}{4}$ **32.** a) 0.5 b) 0.47 c) 0.56 d) No; $0.47 \neq 0.56$ e) 0.38 **33.** No; $\frac{4}{5} \cdot \frac{2}{3} \neq \frac{3}{5}$ **34.** $\frac{45}{91}$ **35.** $\frac{26}{91}$, or $\frac{2}{7}$

36. a) $\frac{200}{450}(0.03) + \frac{150}{450}(0.04) + \frac{100}{450}(0.05)$, or $\frac{17}{450}$ b) $\frac{6}{17}, \frac{6}{17}, \frac{5}{17}$

EXERCISE 8.1, pp. 306–307

1. $\frac{4}{52}$, or $\frac{1}{13}$ **3.** $\frac{13}{52}$, or $\frac{1}{4}$ **5.** $\frac{4}{52}$, or $\frac{1}{13}$ **7.** $\frac{1}{2}$ **9.** $\frac{8}{52}$, or $\frac{2}{13}$ **11.** $\frac{6}{16}$, or $\frac{3}{8}$ **13.** 0 **15.** $\dfrac{\binom{7}{2} \cdot \binom{8}{2}}{\binom{15}{4}} = \dfrac{28}{65}$

17. $\frac{4}{36}$, or $\frac{1}{9}$ **19.** $\frac{1}{36}$ **21.** 0 **23.** $\dfrac{\binom{7}{3}\binom{8}{2}\binom{10}{2}}{\binom{25}{7}} = \dfrac{441}{4807}$ **25.** $\dfrac{\binom{4}{3} \cdot \binom{4}{2}}{\binom{52}{5}} = \dfrac{1}{108{,}290}$ **27.** $\dfrac{\binom{4}{4}\binom{4}{1}}{\binom{52}{5}} = \dfrac{1}{649{,}740}$

29. 0 **31.** $\dfrac{\binom{10}{2}\binom{10}{2}}{\binom{20}{4}} = \dfrac{135}{323}$ **33.** a) $\dfrac{\binom{5}{1}\binom{3}{1}}{\binom{8}{2}} = \dfrac{15}{28}$ b) $\dfrac{\binom{5}{2}}{\binom{8}{2}} = \dfrac{5}{14}$ c) $\dfrac{\binom{3}{2}}{\binom{8}{2}} = \dfrac{3}{28}$ d) $\frac{5}{14} + \frac{3}{28} = \frac{13}{28}$

35. a) $\frac{28}{45}$ b) 17:28 c) 28:17 **37.** a) $\frac{1}{201}$ b) $\frac{3}{250{,}003}$ c) $\frac{3}{500{,}003}$ d) $\frac{2}{201}$

EXERCISE SET 8.2, p. 313

1. a) 0.77 b) $\frac{9}{14}$ **3.** $\frac{1}{6}$ **5.** a) $\frac{68}{95}$ b) $\frac{27}{95}$ c) $\frac{51}{190}$ d) $\frac{3}{190}$ **7.** a) $\frac{5}{9}$ b) $\frac{4}{9}$ c) $\frac{1}{9}$ **9.** a) $\frac{1}{221}$ b) $\frac{1}{17}$ c) $\frac{4}{17}$ d) $\frac{12}{17}$

11. $\frac{1}{3}$ **13.** $\frac{1}{30}$

EXERCISE SET 8.3, pp. 327–329

1. $\frac{11}{18}$ **3.** 0.1584 **5.** a) $\frac{1}{4}$ b) $\frac{1}{13}$ c) $\frac{1}{52}$ **7.** a) $\frac{1}{4}$ b) $\frac{13}{51}$ c) $\frac{13}{204}$ **9.** $\frac{1}{3} \cdot \frac{2}{3} \cdot \frac{2}{3} \cdot \frac{1}{3} \cdot \frac{2}{3}$, or $\frac{8}{243}$ **11.** 0.2

13. $(0.90)^6$, or 0.531441; $(0.10)^6$ or 0.000001 **15.** 0.2 **17.** a) $\frac{15}{32}$ b) $\frac{3}{4} \cdot \frac{3}{8} + \frac{1}{4} \cdot \frac{5}{8}$, or $\frac{7}{16}$ c) $\frac{3}{32}$ **19.** $\frac{37}{64}$

21. Left to the student **23.** $\frac{4}{15}$, $\frac{23}{45}$ **25.** a) $\dfrac{\binom{13}{1}\binom{4}{2}\binom{12}{3}\binom{4}{1}\binom{4}{1}}{\binom{52}{5}} = \dfrac{1760}{4165}$ b) $\dfrac{\binom{13}{1}\binom{4}{3}\binom{12}{2}\binom{4}{1}\binom{4}{1}}{\binom{52}{5}} = \dfrac{88}{4165}$

(Continued on next page.)

25. (*Continued*)

c) $\dfrac{\binom{13}{2}\binom{4}{2}\binom{4}{2}\binom{44}{1}}{\binom{52}{5}} = \dfrac{198}{4165}$ d) $\dfrac{\binom{13}{1}\binom{4}{2}\binom{12}{1}\binom{4}{3}}{\binom{52}{5}} = \dfrac{6}{4165}$ e) $\dfrac{\binom{13}{1}\binom{4}{4}\binom{48}{1}}{\binom{52}{5}} = \dfrac{1}{4165}$ **27.** $\frac{29}{54}$ **29.** $\frac{20}{27}$

31. $p_n^T(2) = p_{n-1}(1)[1 + A_n^T(1)(1 + A_n^T(2)(1 + \cdots (1 + A_n^T(m_L))\cdots]$, where $A_n^T(m) = \dfrac{(n - 2m)(n - 2m - 1)}{2m(365 + m + 1 - n)}$ and

m_L = the largest integer $\leqslant \dfrac{(n - 2)}{2}$.

EXERCISE SET 8.4, p. 336

1. 0.71 **3.** 0.8 **5.** $\frac{11}{15}$ **7.** 0.70018 **9.** 51% (100% − 49%) **11.** (0.4)(0.4) = 0.16, (0.6)(0.4) + (0.4)(0.6) = **0.48**
13. 5%

EXERCISE SET 8.5, pp. 345–347

1. 0.5 **3.** $\frac{12}{13}$ **5.** $\frac{1}{13}$ **7.** a) $\frac{3}{7}$ b) $\frac{4}{7}$ **9.** $\frac{3}{4}$; no **11.** 51% should show up, ∴ not independent; O.K.
13. Yes **15.** $\frac{7}{30}$ **17.** 0.0215 **19.** 14%, 14.5% **21.** $\frac{3}{4}$ **23.** $\frac{2}{3}$ **25.** $\frac{281}{480}, \frac{7}{96}$ **27.** 0.875 **29.** 0.19

EXERCISE SET 8.6, pp. 351–353

1. $\frac{3}{43}$ **3.** $\frac{4}{7}, \frac{8}{29}$ **5.** $\frac{2}{3}$ **7.** $\frac{49}{89}$ **9.** $\frac{1}{32}$ **11.** 0.16, $\frac{5}{16}$ **13.** $\frac{1}{2}$ **15.** $\frac{17}{30}, \frac{5}{8}$ **17.** $\frac{1}{2}, \frac{297}{400}, \frac{13}{33}$ **19.** $\frac{5}{16}, \frac{1}{2}, \frac{1}{8}$

CHAPTER 8, TEST, p. 354

1. a) $p(E^c) = \frac{16}{39}$ b) 23:16 c) 16:23 **2.** 0.9 **3.** $\frac{2}{5}$ **4.** 0.4838 **5.** a) 0.0344 b) No **6.** $\frac{1}{3}$
7. a) $\frac{1}{775}$ b) $\frac{60}{775}$ c) $\frac{714}{775}$ **8.** $\frac{3}{5} \cdot \frac{2}{5} \cdot \frac{2}{5} \cdot \frac{2}{5} \cdot \frac{3}{5} = \frac{72}{3125}$ **9.** 0.31 **10.** 0.8; no **11.** a) $\frac{39}{1000}$ b) $\frac{10}{39}, \frac{21}{39}, \frac{8}{39}$

CHAPTER 9

MARGIN EXERCISES

1.

x	0	1	2	3	4	5
p	$\frac{6}{36}$	$\frac{10}{36}$	$\frac{8}{36}$	$\frac{6}{36}$	$\frac{4}{36}$	$\frac{2}{36}$

2.

x	0	1	2	3
p	$\frac{24}{91}$	$\frac{45}{91}$	$\frac{20}{91}$	$\frac{2}{91}$

3. 78.08 **4.** a) 2.277 b) Easy **5.** $\frac{70}{36}$ **6.** 1 **7.** $E(\bar{X})$ = \$1.76; to charity \$8.20 per ticket
8. Example 6. **9.** Betting *with* shooter $E(\bar{X}) = -\frac{7}{495} = -\frac{28}{1980}$ to win. Betting *against* shooter $E(\bar{X}) = -\frac{3}{220} = -\frac{27}{1980}$
to win. ∴ Better to bet *against* shooter. **10.** 64.91 **11.** $\frac{52}{91}$ **12.** 8.06 **13.** 0.75593

14. $\binom{6}{2} \cdot \frac{1}{2^6} = \frac{15}{64}$

TTHHHH	HHTTHH	HHHHTT	HTHHTH
HTTHHH	HHTHTH	**HHHTTH**	THHHHT
THTHHH	HTHHHT	HHHTHT	THHHTH
HTHTHH	THHTHH	HHTHHT	

15. $\binom{6}{4}(0.8)^4(0.2)^2 = 0.24576$

16. K is the number cured.

k	p_k
0	0.000064
1	0.001536
2	0.015360
3	0.081920
4	0.245760
5	0.393216
6	0.262144

17. 4.8   **18.** $\sigma^2 = 0.96$, $\sigma = 0.97980$ **19.** $p = \frac{20}{91}$ (= p_2 in Margin Exercise 2). Now $p_3 = \binom{4}{3}\left(\frac{20}{91}\right)^3\left(\frac{71}{91}\right)^1 =$

0.033132   **20.** $(0.2)^3(0.8)^1 = 0.0064$ **21.** 3   **22.** 1.25   **23.** $\sigma^2 = \frac{5}{16}$, $\sigma = 0.55902$

24. $p_{4,5} = \binom{4}{3}(0.2)^1(0.8)^4 = 0.32768$; $p_4 = \binom{5}{4}(0.2)^1(0.8)^4 = 0.40960$ **25.** 5   **26.** $\sigma^2 = 1.25$, $\sigma = 1.11803$

27. 0.86   **28.** $z = 1$, $x = 86.14$; $z = -1$, $x = 70.02$; $z = 2$, $x = 94.20$; $z = -2$, $x = 61.96$ **29.** 58%

30. 96%; actual class = 100%   **31.** 240

EXERCISE SET 9.1, pp. 359–360

1.

n	0	1	2	3	4
p	$\frac{14}{323}$	$\frac{80}{323}$	$\frac{135}{323}$	$\frac{80}{323}$	$\frac{14}{323}$

3.

n	0	1	2	3
p	$\frac{1}{14}$	$\frac{6}{14}$	$\frac{6}{14}$	$\frac{1}{14}$

5.

n	0	1	2	3	4	5
p	$\frac{1001}{7752}$	$\frac{3003}{7752}$	$\frac{2730}{7752}$	$\frac{910}{7752}$	$\frac{105}{7752}$	$\frac{3}{7752}$

7.

n	0	1	2	3
p	$\frac{1}{8}$	$\frac{3}{8}$	$\frac{3}{8}$	$\frac{1}{8}$

9.

n_{White}	0	1	2
n_{Red}	3	2	1
p	$\frac{5}{12}$	$\frac{6}{12}$	$\frac{1}{12}$

11.

n	0	1	2	3
p	$\frac{1}{20}$	$\frac{9}{20}$	$\frac{9}{20}$	$\frac{1}{20}$

13.

n	0	1	2	3
p	$\frac{8}{100}$	$\frac{54}{100}$	$\frac{36}{100}$	$\frac{2}{100}$

15.

n	4	5	6	7
p	$\frac{1}{8}$	$\frac{1}{4}$	$\frac{5}{16}$	$\frac{5}{16}$

17.

n	1	2	3	4	5	6	$\cdots$
p	0	$\frac{1}{2}$	$\frac{1}{4}$	$\frac{1}{8}$	$\frac{1}{16}$	$\frac{1}{32}$	$\cdots$

EXERCISE 9.2, pp. 368–369

1. 86.533 mph   **3.** $5.89   **5.** 2 women, 1 man   **7.** 2   **9.** $\frac{3}{5}$ **11.** $\frac{3}{2}$ **13.** $\frac{3}{4}$, 1   **15.** $\frac{3}{2}$ **17.** $\frac{132}{100}$ **19.** $\frac{93}{16}$

EXERCISE SET 9.3, pp. 371–372

1. 11.530, 3.396 mph   **3.** 1.53, $1.24   **5.** $\frac{114}{190} = 0.6$, 0.77460   **7.** 0.44, 0.66332

EXERCISE SET 9.4, pp. 378–379

1.

n	0	1	2	3	4	5
p	$\frac{1}{32}$	$\frac{5}{32}$	$\frac{10}{32}$	$\frac{10}{32}$	$\frac{5}{32}$	$\frac{1}{32}$

; $\frac{5}{2}, \frac{5}{4}$, 1.11803   **3.** $3, 3, \frac{11}{16}$ **5.** $\frac{3}{16}, \frac{5}{2}, \frac{5}{4}$, 1.11803   **7.** 0.885735, 0.6, 0.54, 0.73485

9.

n	0	1	2	3	4	5
p	0.00032	0.00640	0.05120	0.20480	0.40960	0.32768

;   4, 0.67232

11.

n	0	1	2	3	4	5
10%p	0.59049	0.32805	0.07290	0.00810	0.00045	0.00001
20%p	0.32768	0.40960	0.20480	0.05120	0.00640	0.00032
30%p	0.16807	0.36015	0.30870	0.13230	0.02835	0.00243

10%, 20%, 30%, 30%

13. $\frac{3}{16}, \frac{15}{16}$ **15.** 0.26272, 0.15053

EXERCISE SET 9.5, pp. 383–384

1. $\frac{3}{16}$ **3.** 0.081, 0.271   **5.** 7   **7.** 10, 9.48683   **9.** $\frac{46656}{823543} = \binom{6}{7}^6\left(\frac{1}{7}\right)$ **11.** $\frac{46656}{823543} = \binom{6}{1}\left(\frac{6}{7}\right)^5\left(\frac{1}{7}\right)^2$

13. $\frac{405}{4096}, \frac{1215}{4096}$ **15.**

n	2	3	4	5	6	$\cdots$
p	$\frac{1}{4}$	$\frac{1}{8}$	$\frac{1}{8}$	$\frac{3}{2}$	$\frac{5}{64}$	$\cdots$

17.

n	1	2	3	4	5	6	7	8	9	$\cdots$
p	$\frac{1}{2}$	0	$\frac{1}{8}$	0	$\frac{2}{32}$	0	$\frac{5}{128}$	0	$\frac{14}{512}$	$\cdots$

19. 1   **21.** $\frac{1}{2}$ **23.** $> \frac{1}{2}$

EXERCISE SET 9.6, pp. 387–388

1. a) +0.294, −0.294   b) 88.231 mph, 84.835 mph   c) 37.5%   d) 38%   **3.** 0.1%, 1   **5.** a) 5%   b) 8%
7. a) 27%   b) 34%

CHAPTER 9, TEST, p. 389

1.

n	0	1	2	3	4	5
p	$\binom{5}{0}\left(\frac{3}{4}\right)^5\left(\frac{1}{4}\right)^0$ $= \frac{243}{1024}$	$\binom{5}{1}\left(\frac{3}{4}\right)^4\left(\frac{1}{4}\right)^1$ $= \frac{405}{1024}$	$\binom{5}{2}\left(\frac{3}{4}\right)^3\left(\frac{1}{4}\right)^2$ $= \frac{270}{1024}$	$\binom{5}{3}\left(\frac{3}{4}\right)^2\left(\frac{1}{4}\right)^3$ $= \frac{90}{1024}$	$\binom{5}{4}\left(\frac{3}{4}\right)\left(\frac{1}{4}\right)^4$ $= \frac{15}{1024}$	$\binom{5}{5}\left(\frac{3}{4}\right)^0\left(\frac{1}{4}\right)^5$ $= \frac{1}{1024}$

2. $\binom{5}{4}\left(\frac{1}{4}\right)^1\left(\frac{3}{4}\right)^4 + \binom{5}{5}\left(\frac{1}{4}\right)^0\left(\frac{3}{4}\right)^5 = \frac{648}{1024} = \frac{81}{128}$ **3.** $5 \cdot \frac{3}{4} = \frac{15}{4}$ **4.** $\mu = 2$ **5.** $\sigma^2 = \frac{4}{3}$

6. $\sigma = \frac{2}{3}\sqrt{3} \approx 1.155$ **7.** $z = \dfrac{1-2}{\sqrt{\frac{4}{3}}} = -\sqrt{\frac{3}{4}}$ **8.** $\left(\frac{4}{5}\right)^2\left(\frac{1}{5}\right) = \frac{16}{125}$

CHAPTER 10

MARGIN EXERCISES

1. a)   b)

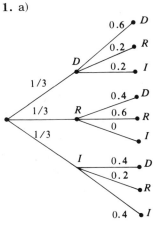

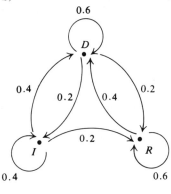

2.

	D	R	I
D	0.6	0.2	0.2
R	0.4	0.6	0
I	0.4	0.2	0.4

3. a) Qualifies

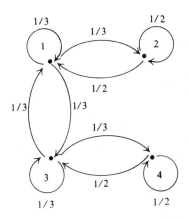

b) **Does not qualify.** Sum of elements in second row now is less than one.

4. $P_0 = \begin{bmatrix}\frac{1}{3} & \frac{1}{3} & \frac{1}{3}\end{bmatrix}$ **5.** Regular

$P_1 = \begin{bmatrix}\frac{7}{15} & \frac{5}{15} & \frac{3}{15}\end{bmatrix}$

$P_2 = \begin{bmatrix}\frac{37}{75} & \frac{25}{75} & \frac{13}{75}\end{bmatrix}$

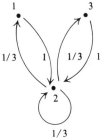

6. Not regular, second state absorbing

7. Regular. There is a nonzero element in the main diagonal.

8. Regular

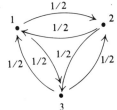

9. $P_8 = [0.235349 \quad 0.305860 \quad 0.458791]$ for $P_0 = [1 \quad 0 \quad 0]$;
$\bar{P} = [0.230769 \quad 0.307692 \quad 0.461538]$

10. $\bar{P} = \begin{bmatrix}\frac{1}{5} & \frac{3}{5} & \frac{1}{5}\end{bmatrix}$ **11.** Chain is cyclic: $T = T^3 = T^5 = \cdots$, ∴ not regular. Since chain is ergodic, we can obtain $\bar{P} = \begin{bmatrix}\frac{1}{2} & \frac{1}{3} & \frac{1}{6}\end{bmatrix}$. But this is *not* a "long-run" probability vector since chain is cyclic.

EXERCISE SET 10.1, pp. 399–400

1.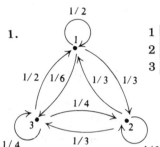

$$\begin{array}{cc} & \begin{array}{ccc} 1 & 2 & 3 \end{array} \\ \begin{array}{c} 1 \\ 2 \\ 3 \end{array} & \left[\begin{array}{ccc} \frac{1}{2} & \frac{1}{3} & \frac{1}{6} \\ \frac{1}{3} & \frac{1}{3} & \frac{1}{3} \\ \frac{1}{2} & \frac{1}{4} & \frac{1}{4} \end{array}\right] \end{array}$$

3. No; negative element in first row.

5.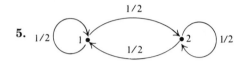

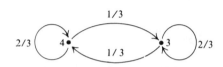

7.

9.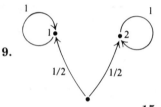

11. $\overset{1}{\bullet} \xrightarrow{\quad 1 \quad} \overset{2}{\bullet} \xrightarrow{\quad 1 \quad} \overset{3}{\bullet}\,1$

13.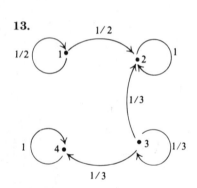

15. First zone: $P_0 = [0\ \ 1\ \ 0]$, $P_1 = [\frac{1}{3}\ \ \frac{1}{3}\ \ \frac{1}{3}]$, $P_2 = [\frac{16}{36}\ \ \frac{11}{36}\ \ \frac{9}{36}]$

17. $P_0 = [0\ \ 0\ \ 1]$ $P_1 = [\frac{1}{2}\ \ \frac{1}{4}\ \ \frac{1}{4}]$ $P_2 = [\frac{22}{48}\ \ \frac{15}{48}\ \ \frac{11}{48}]$

19. $P_2 = [\frac{9}{25}\ \ \frac{14}{25}\ \ \frac{2}{25}]$ **21.** $P_2 = [\frac{1}{4}\ \ \frac{1}{4}\ \ \frac{1}{4}\ \ \frac{1}{4}]$ **23.** $P_1 = [\frac{1}{2}\ \ \frac{1}{2}]$

25. $P_2 = [1\ \ 0]$ **27.** $P_2 = [\frac{1}{2}\ \ \frac{1}{2}\ \ 0]$ **29.** $P_3 = [0\ \ 0\ \ 1]$

31. $P_2 = [\frac{1}{16}\ \ \frac{3}{16}\ \ \frac{7}{16}\ \ \frac{5}{16}]$ **33.** $P_2 = [\frac{9}{144}\ \ \frac{79}{144}\ \ \frac{4}{144}\ \ \frac{52}{144}]$

35. $P_2 = [\frac{5}{36}\ \ \frac{13}{36}\ \ \frac{13}{36}\ \ \frac{5}{36}]$

EXERCISE SET 10.2, pp. 404–405

1. Not regular; second state absorbing **3.** Not regular; absorbing states **5.** Not regular; absorbing states

7. Not regular; absorbing state **9.** Not regular; absorbing states **11.** Regular **13.** Regular **15.** Not regular

17. Regular **19.** Not regular

EXERCISE SET 10.3, pp. 409–410

1. $[\frac{1}{3},\ \frac{2}{3}]$ **3.** $[\frac{2}{12}\ \ \frac{5}{12}\ \ \frac{5}{12}]$ **5.** $[\frac{2}{10}\ \ \frac{3}{10}\ \ \frac{5}{10}]$ **7.** $[\frac{5}{17}\ \ \frac{2}{17}\ \ \frac{5}{17}\ \ \frac{5}{17}]$ **9.** $[\frac{1}{10}\ \ \frac{2}{10}\ \ \frac{4}{10}\ \ \frac{3}{10}]$ **11.** $[\frac{30}{67}\ \ \frac{21}{67}\ \ \frac{16}{67}]$

13. $T = \begin{bmatrix} 1 & 0 & 0 & 0 & 0 \\ \frac{1}{2} & 0 & \frac{1}{2} & 0 & 0 \\ 0 & \frac{1}{2} & 0 & \frac{1}{2} & 0 \\ 0 & 0 & \frac{1}{2} & 0 & \frac{1}{2} \\ 0 & 0 & 0 & 0 & 1 \end{bmatrix}$

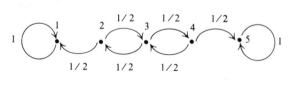

(See top of next page.)

The two fixed points are his home and the bar.
$P_0 = [0 \; 0 \; 1 \; 0 \; 0]$, $P_1 = [0 \; \frac{1}{2} \; 0 \; \frac{1}{2} \; 0]$, $P_2 = [\frac{1}{4} \; 0 \; \frac{1}{2} \; 0 \; \frac{1}{4}]$, $P_3 = [\frac{1}{4} \; \frac{1}{4} \; 0 \; \frac{1}{4} \; \frac{1}{4}]$, $P_4 = [\frac{3}{8} \; 0 \; \frac{2}{8} \; 0 \; \frac{3}{8}] \cdots$
For large n, P_n approaches $[\frac{1}{2} \; 0 \; 0 \; 0 \; \frac{1}{2}]$; p (home) $= \frac{1}{2}$.

15. $T = \begin{bmatrix} 0 & 1 & 0 & 0 & 0 \\ \frac{1}{2} & 0 & \frac{1}{2} & 0 & 0 \\ 0 & \frac{1}{2} & 0 & \frac{1}{2} & 0 \\ 0 & 0 & \frac{1}{2} & 0 & \frac{1}{2} \\ 0 & 0 & 0 & 0 & 1 \end{bmatrix}$

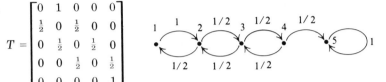

The fixed point is his home.
$P_0 = [0 \; 0 \; 1 \; 0 \; 0]$
$P_1 = [0 \; \frac{1}{2} \; 0 \; \frac{1}{2} \; 0]$
$P_2 = [\frac{1}{4} \; 0 \; \frac{1}{2} \; 0 \; \frac{1}{4}]$
$P_3 = [0 \; \frac{1}{2} \; 0 \; \frac{1}{4} \; \frac{1}{4}]$
$P_4 = [\frac{2}{8} \; 0 \; \frac{3}{8} \; 0 \; \frac{3}{8}]$

For large n, P_n approaches $[0 \; 0 \; 0 \; 0 \; 1]$; p (home) $= 1$.

17. $T = \begin{bmatrix} 0 & \frac{2}{3} & 0 & \frac{1}{3} \\ \frac{2}{3} & 0 & \frac{1}{3} & 0 \\ 0 & \frac{1}{2} & 0 & \frac{1}{2} \\ \frac{1}{2} & 0 & \frac{1}{2} & 0 \end{bmatrix}$ $\bar{P} = [\frac{3}{10} \; \frac{3}{10} \; \frac{2}{10} \; \frac{2}{10}]$. Thus, the mouse keeps moving.

CHAPTER 10 TEST, p. 411

1.

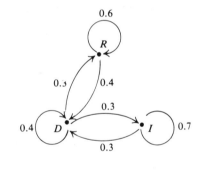

$\begin{array}{c} \\ D \\ R \\ I \end{array}\begin{array}{ccc} D & R & I \\ \begin{bmatrix} 0.4 & 0.3 & 0.3 \\ 0.4 & 0.6 & 0 \\ 0.3 & 0 & 0.7 \end{bmatrix} \end{array}$

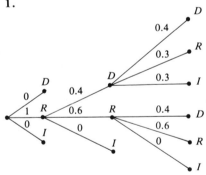

2. a) No, has absorbing states) b) No, cyclic c) No, not ergodic) d) Yes, taking powers
3. $P_2 = [\frac{3}{12} \; \frac{7}{12} \; \frac{2}{12}]$, $\bar{P} = [\frac{2}{9} \; \frac{3}{9} \; \frac{4}{9}]$

CHAPTER 11

MARGIN EXERCISES

1. Answers may vary   **2.** $\alpha_3\beta_2, v = 4$ **3.** Answers may vary   **4.** $\alpha_4\beta_2, v = 5$
5. $X = [\frac{3}{8} \; \frac{5}{8}]$, $Y = [\frac{1}{4} \; \frac{3}{4}]$, $v = \frac{13}{4}$ **6.** $X = [\frac{3}{5} \; \frac{2}{5}]$, $Y = [\frac{1}{5} \; 0 \; \frac{4}{5}]$, $v = \frac{18}{5}$
7. $X = [\frac{5}{9} \; 0 \; \frac{4}{9}]$, $Y = [\frac{2}{9} \; \frac{7}{9}]$, $v = \frac{35}{9}$ **8.** $\phi = [2, 2, 1, 1]/6$ **9.** $\phi = \beta = [1, 1, 1, 0]/3$

EXERCISE SET 11.1, p. 424

1. (α, β_1), 3   **3.** (α_2, β_1), 2   **5.** (α_2, β_2), 4   **7.** (α_4, β_4), 3

EXERCISE SET 11.2, pp. 428–429

1. $X = [\frac{1}{2} \quad \frac{1}{2}]$, $Y = [\frac{1}{4} \quad \frac{3}{4}]$, $v = \frac{5}{2}$ **3.** $X = [0 \quad \frac{6}{7} \quad \frac{1}{7}]$, $Y = [\frac{6}{7} \quad 0 \quad \frac{1}{7} \quad 0]$, $v = \frac{20}{7}$ **5.** $X = [0 \quad 0 \quad \frac{1}{2} \quad \frac{1}{2}]$, $Y = [\frac{3}{5} \quad 0 \quad \frac{2}{5} \quad 0]$, $v = 4$ **7.** Plant: $[\frac{9}{10} \quad \frac{1}{10}]$; inspector: $[\frac{9}{10} \quad \frac{1}{10}]$; $v = -\frac{7}{10}$, where first strategy is "far out" and second is "local stream."   **9.** 5   **11.** $X = [\frac{1}{11} \quad \frac{10}{11}]$, $Y = [\frac{5}{11} \quad \frac{6}{11}]$, $v = \frac{5}{11}$; $\therefore$ for *one* trial: take $X = [0 \quad 1]$, $Y = [0 \quad 1]$; then $v = 0$.

EXERCISE SET 11.3, p. 435

1. $X = [\frac{3}{7} \quad \frac{4}{7}]$, $Y = [\frac{6}{7} \quad 0 \quad 0 \quad \frac{1}{7}]$, $v = \frac{11}{7}$ **3.** $X = [0 \quad 0 \quad \frac{1}{14} \quad \frac{13}{14}]$, $Y = [\frac{5}{7} \quad \frac{2}{7}]$, $v = \frac{16}{7}$ **5.** $X = [0 \quad \frac{3}{4} \quad \frac{1}{4}]$, $Y = [\frac{7}{8} \quad 0 \quad 0 \quad \frac{1}{8}]$, $v = \frac{15}{4}$ **7.** $X = [0 \quad 0 \quad \frac{3}{8} \quad \frac{5}{8}]$, $Y = [\frac{3}{4} \quad 0 \quad \frac{1}{4} \quad 0]$, $v = \frac{13}{4}$ **9.** $X = [\frac{10}{13} \quad \frac{3}{13} \quad 0]$, $Y = [\frac{4}{13} \quad 0 \quad \frac{9}{13}]$, $v = \frac{64}{13}$

EXERCISE SET 11.4, p. 441

1. $\phi = \beta = [3, 1, 1, 1]/6$ **3.** $\phi = \beta = [1, 0, 0, 0]$; no one else can influence a decision   **5.** $[51; 49, 48, 3]$, $\phi = \beta = [1, 1, 1]/3$ **7.** $\phi = \beta = [2, 2, 1, 1]/6$ for $[9; 5, 5, 3, 2]$ so that quota must be raised from 8 to 9   **9.** $[4; 3, 1, 1, 1]$; $\phi = [9, 1, 1, 1]/12$, $\beta = [7, 1, 1, 1]/10$

CHAPTER 11 TEST, p. 442

1. Minimax $\neq$ maximin, $\therefore$ no pure-strategy solution

2. β_1 dominates β_2
α_1 dominates α_2
α_4 dominates α_3

$$\begin{array}{c} \\ \alpha_1 \\ \alpha_4 \end{array} \begin{array}{ccc} \beta_1 & \beta_3 & \beta_4 \\ \left[\begin{array}{ccc} 1 & 5 & 0 \\ 3 & 2 & 8 \end{array}\right. & & \left.\vphantom{\begin{array}{c}1\\3\end{array}}\right] \end{array}$$

3.

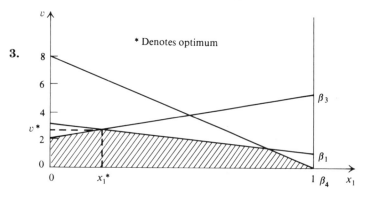

* Denotes optimum

4. $X = [\frac{1}{5} \quad 0 \quad 0 \quad \frac{4}{5}]$, $Y = [\frac{3}{5} \quad 0 \quad \frac{2}{5} \quad 0]$, $v = \frac{13}{5}$

5. $\phi = [4, 1, 1, 0]/6$ **6.** $\beta = [3, 1, 1, 0]/5$

FINAL EXAM, pp. 443–447

1. $V_n = \$3400 - (\$440)n$ **2.** $V_n = \$3400 \, (0.60)^n$ **3.** 16.986%   **4.** \$160,441.91   **5.** \$1408.91

6. $x_1 = 1$, $x_2 = -2$, $x_3 = -2$ **7.** $x_1 = 2 - 4x_4$, $x_2 = 9 + 2x_4$, $x_3 = 2 + x_4$, $x_4 =$ any number   **8.** $\begin{bmatrix} 1 & -3 \\ 2 & -4 \end{bmatrix}$

9. $\begin{bmatrix} 5 & 3 \\ 6 & 5 \end{bmatrix}$ **10.** $\begin{bmatrix} 3 & -3 \\ 0 & -6 \end{bmatrix}$ **11.** $\begin{bmatrix} -4 & 6 \\ -2 & 10 \end{bmatrix}$ **12.** $\begin{bmatrix} 3 & 2 \\ -2 & 3 \\ 1 & 7 \end{bmatrix} \begin{bmatrix} x_1 \\ x_2 \end{bmatrix} \leq \begin{bmatrix} 7 \\ 2 \\ 11 \end{bmatrix}$, $\max f: f = \begin{bmatrix} 1 & 2 \end{bmatrix} \begin{bmatrix} x_1 \\ x_2 \end{bmatrix}$; $\begin{bmatrix} x_1 \\ x_2 \end{bmatrix} \geq \begin{bmatrix} 0 \\ 0 \end{bmatrix}$

13 and 14.

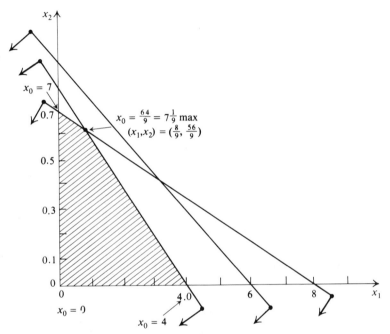

15.

x_1	x_2	y_1	y_2	y_3	1
7	8*	1	0	0	56
2*	1	0	1	0	8
3	2	0	0	1	18
-1	-1	0	0	0	0

Pivot element: *either* of starred (*) elements.

16. Final tableau:

x_1	x_2	y_1	y_2	y_3	1
0	1	$\frac{2}{9}$	$-\frac{7}{9}$	0	$\frac{56}{9}$
1	0	$-\frac{1}{9}$	$\frac{8}{9}$	0	$\frac{8}{9}$
0	0	$-\frac{1}{9}$	$-\frac{10}{9}$	1	$\frac{26}{9}$
0	0	$\frac{1}{9}$	$\frac{1}{9}$	0	$\frac{64}{9}$

$x_1 = \frac{8}{9}$, $x_2 = \frac{56}{9}$, $y_1 = 0$, $y_2 = 0$, $y_3 = \frac{26}{9}$; $x_0 = \frac{64}{9}$

17. $x_1 = \frac{8}{9}$, $x_2 = \frac{56}{9}$, $y_1 = 0$, $y_2 = 0$, $y_3 = \frac{26}{9}$; $x_0 = \frac{64}{9}$

18. $7y_1 + 2y_2 + 3y_3 \geq 1$, $8y_1 + y_2 + 2y_3 \geq 1$, $\min y_0$: $y_0 = 56y_1 + 8y_2 + 18y_3$; $y_1, y_2, y_3 \geq 0$.

19. $x_1 = 0$, $x_2 = 0$, $y_1 = \frac{1}{9}$, $y_2 = \frac{1}{9}$, $y_3 = 0$; $y_0 = \frac{64}{9}$

20.
$$7x_1 + 8x_2 + y_1 = 56,$$
$$2x_1 + x_2 + y_2 = 8,$$
$$3x_1 + 2x_2 + y_3 = 18,$$
$$x_0 = x_1 + x_2; \; x_1, x_2; \; y_1, y_2, y_3 \geq 0.$$

$$7y_1 + 2y_2 + 3y_3 - x_1 = 1,$$
$$8y_1 + y_2 + 2y_3 - x_2 = 1,$$
$$y_0 = 56y_1 + 8y_2 + 18y_3 = x_0,$$
$$y_1, y_2, y_3; \; x_1, x_2 \geq 0.$$

21. Same as 19.   **22.** Primal solution unbounded, dual solution infeasible.   **23.** Primal solution degenerate, dual solution nonunique.   **24.** $x_{11} = 80$, $x_{21} = 10$, $x_{22} = 60$, $x_{23} = 30$; $x_0 = 2160$ **25.** $x_{13} = x_{24} = x_{31} = x_{42} = x_{55} = 1$, $a_0 = 61$

26. 31

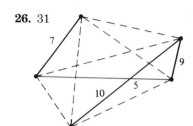

7   9   10   5

27. 16: $ABEFG$ **28.** 23   **29.** 336   **30.** 5040   **31.** $\binom{58}{12} \cdot \binom{42}{18}$ **32.** 5040

33. $m^7 + 7m^6n + 21m^5n^2 + 35m^4n^3 + 35m^3n^4 + 21m^2n^5 + 7mn^6 + n^7$

34. 9:10   **35.** 0.95   **36.** $\frac{1}{5525}$ **37.** $\frac{3}{25}$ **38.**

n	0	1	2	3
p	0	$\frac{1}{5}$	$\frac{3}{5}$	$\frac{1}{5}$

   39. 2

40. $\sigma^2 = \frac{2}{5}, \sigma = \sqrt{\frac{2}{5}}$ **41.** $\frac{1424}{3125}$ **42.** $\frac{18}{125}$ **43.** $\frac{432}{3125}$

44.

1   1   2   1/2   1/3   1/3   1/2   3   1/3

45. $\begin{bmatrix} \frac{5}{12} & \frac{2}{12} & \frac{5}{12} \end{bmatrix}$ **46.** T^4 has all positive elements $[5, 3, 3, 1]/12$

47. $\begin{bmatrix} \frac{3}{10} & \frac{4}{10} & \frac{3}{10} \end{bmatrix}$ **48.** $X = \begin{bmatrix} 0 & \frac{1}{2} & \frac{1}{2} \end{bmatrix}$, $Y = \begin{bmatrix} \frac{3}{4} & 0 & 0 & \frac{1}{4} \end{bmatrix}$, $v = \frac{9}{2}$

49. $\phi = \beta = [5, 3, 3, 1]/12$

APPENDIX A

MARGIN EXERCISES

1. (d), (e)   **2.** a) x b) x, y c) x_1, x_2, x_3 d) None   e) None   f) None   g) None   **3.** (a), (b), (c)   **4.** a) F   b) F   c) T
5. a) Yes   b) No   **6.** a) T   b) F   c) T   d) F   **7.** a) $\sim p$, $2 \neq 3$, $\sim(2 = 3)$; 2 does not equal 3   b) $\sim p$; Profit is not wholesome, It is false that profit is wholesome, It is not true that profit is wholesome   **8.** a) T   b) F   **9.** a) T
b) F   c) T   d) T   **10.** a) (You spray with DriasKanby) $\rightarrow$ (You will stop wetness)   b) $(a = b) \rightarrow (a^n = b^n)$ c) (The card is a face card) $\rightarrow$ (The card is a queen)   d) (Democrat A wins) $\rightarrow$ (Democrat B wins)   **11.** a) F   b) F   c) T
d) T   **12.** a) (x is in set A) $\leftrightarrow$ (x is in set B)   b) $(x + c = y + c) \leftrightarrow (x = y)$ c) (Sharon gets A's) $\leftrightarrow$ (Sharon studies)

13.

p	q	$\sim p$	$\sim p \rightarrow q$
T	T	F	T
T	F	F	T
F	T	T	T
F	F	T	F

14.

p	q	$p \rightarrow q$	$(p \rightarrow q) \wedge p$	$[(p \rightarrow q) \wedge p] \rightarrow q$
T	T	T	T	T
T	F	F	F	T
F	T	T	F	T
F	F	T	F	T

a)   b) Yes

15. a)

p	$\sim p$	$p \wedge \sim p$
T	F	F
F	T	F

b) Not a tautology

16. a)

p	q	r	$q \vee r$	$p \rightarrow (q \vee r)$
T	T	T	T	T
T	T	F	T	T
T	F	T	T	T
T	F	F	F	F
F	T	T	T	T
F	T	F	T	T
F	F	T	T	T
F	F	F	F	T

b) Not a tautology

17. Valid   **18.** Invalid   **19.** Valid   **20.** $p \vee p$ **21.** $p \vee (q \wedge r)$ **22.** $p \vee p, p$ **23.** $(p \vee q) \wedge (p \vee r)$,
$p \vee (q \wedge r)$ **24.** $p \wedge (p \vee p), p$ **25.** $(p \wedge \sim p) \vee q, p \vee q$

EXERCISE SET A.1, pp. 456–457

1. Statement   **3.** Not a statement   **5.** Statement   **7.** F   **9.** T   **11.** F   **13.** F   **15.** T   **17.** T
19. T   **21.** T   **23.** T   **25.** $p \rightarrow q$ **27.** $p \wedge q$ **29.** $p \vee q \vee r$ **31.** $p \rightarrow q$ **33.** $\sim p \rightarrow \sim q$ **35.** $p \vee q \vee r$

EXERCISE SET A.2, p. 460

1. a)

p	$p \rightarrow p$
T	T
F	T

b) Yes

3. a)

p	q	$\sim p$	$\sim p \wedge q$
T	T	F	F
T	F	F	F
F	T	T	T
F	F	T	F

b) No

5. a)

p	q	$p \vee q$	$p \rightarrow (p \vee q)$
T	T	T	T
T	F	T	T
F	T	T	T
F	F	F	T

b) Yes

7. a)

p	$\sim p$	$\sim \sim p$	$p \leftrightarrow \sim \sim p$
T	F	T	T
F	T	F	T

b) Yes

9. a)

p	q	$p \wedge q$	$q \wedge p$	$(p \wedge q) \leftrightarrow (q \wedge p)$
T	T	T	T	T
T	F	F	F	T
F	T	F	F	T
F	F	F	F	T

b) Yes

11. a)

p	q	$\sim p$	$\sim q$	$p \wedge q$	$\sim(p \wedge q)$	$(\sim p) \vee (\sim q)$	$[\sim(p \wedge q)] \leftrightarrow [(\sim p) \vee (\sim q)]$
T	T	F	F	T	F	F	T
T	F	F	T	F	T	T	T
F	T	T	F	F	T	T	T
F	F	T	T	F	T	T	T

b) Yes

13. a)

p	q	$p \rightarrow q$	$q \rightarrow p$	$(p \rightarrow q) \rightarrow (q \rightarrow p)$
T	T	T	T	T
T	F	F	T	T
F	T	T	F	F
F	F	T	T	T

b) No

15. a)

p	q	r	$q \rightarrow r$	$p \rightarrow (q \rightarrow r)$
T	T	T	T	T
T	T	F	F	F
T	F	T	T	T
T	F	F	T	T
F	T	T	T	T
F	T	F	F	T
F	F	T	T	T
F	F	F	T	T

b) No

17. a)

p	q	r	$p \wedge q$	$p \wedge r$	$q \vee r$	$A: p \wedge (q \vee r)$	$B: (p \wedge q) \vee (p \wedge r)$	$A \leftrightarrow B$
T	T	T	T	T	T	T	T	T
T	T	F	T	F	T	T	T	T
T	F	T	F	T	T	T	T	T
T	F	F	F	F	F	F	F	T
F	T	T	F	F	T	F	F	T
F	T	F	F	F	T	F	F	T
F	F	T	F	F	T	F	F	T
F	F	F	F	F	F	F	F	T

b) Yes

EXERCISE SET A.3, pp. 463–464

1. Valid   **3.** Valid   **5.** Invalid   **7.** Valid   **9.** Valid   **11.** Invalid   **13.** Invalid   **15.** $p \rightarrow q, q \therefore p$; invalid   **17.** $p \rightarrow q, q \rightarrow r, \sim r \therefore \sim p$; valid   **19.** $p \vee q, \sim p \therefore q$; valid   **21.** $p \rightarrow q, \sim r \rightarrow \sim q, \sim 1 \therefore p$; invalid **23.** $p \rightarrow q, q \therefore p$; invalid

---

EXERCISE SET A.4, p. 468

1. $(p \wedge p) \wedge q$, or $p \wedge (p \wedge q)$; $p \wedge q$ **3.** $p \vee (p \wedge q), p$ **5.** $(p \vee q) \vee (q \vee p), p \vee q$ **7.** $(p \vee \sim p) \vee q$; current always flows, so do not use any switches.   **9.** $(p \vee q) \vee (p \wedge \sim q), p \vee q$ **11.** $[(p \wedge \sim q) \vee (\sim q \wedge r)] \wedge p \wedge (q \vee r)$, $p \wedge (\sim q \wedge r)$

13.   **14.**

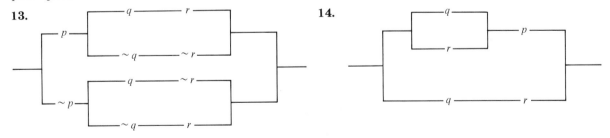

APPENDIX A TEST, p. 469

1. T   **2.** F   **3.** T   **4.** F   **5.** F   **6.** T   **7.** T, F, T, T   **8.** a) $p \wedge q$: T, T, F, F, F, F, F, F; $(p \wedge q) \rightarrow r$: T, F, T, T, T, T, T, T   b) No   **9.** $p \rightarrow q$ **10.** Valid   **11.** Invalid   **12.** $(\sim q \wedge p) \vee q, p \vee q$

APPENDIX B

MARGIN EXERCISES

1.

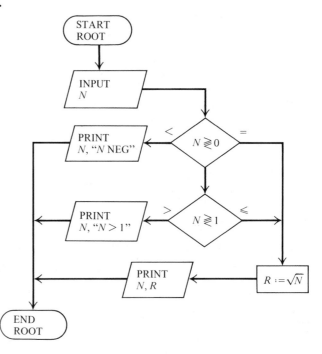

2.

3.

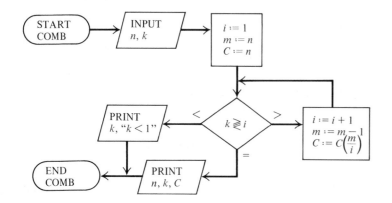

Input: $n = 4$, $k = 2$;
$i = 1$, $m = 4$, $C = 4$;
$2 > 1$, **test**;
$i = 2$, $m = 3$, $C = 6$;
$2 = 2$, **test**;
Output: $n = 4$, $k = 2$, $C = 6$.

4.

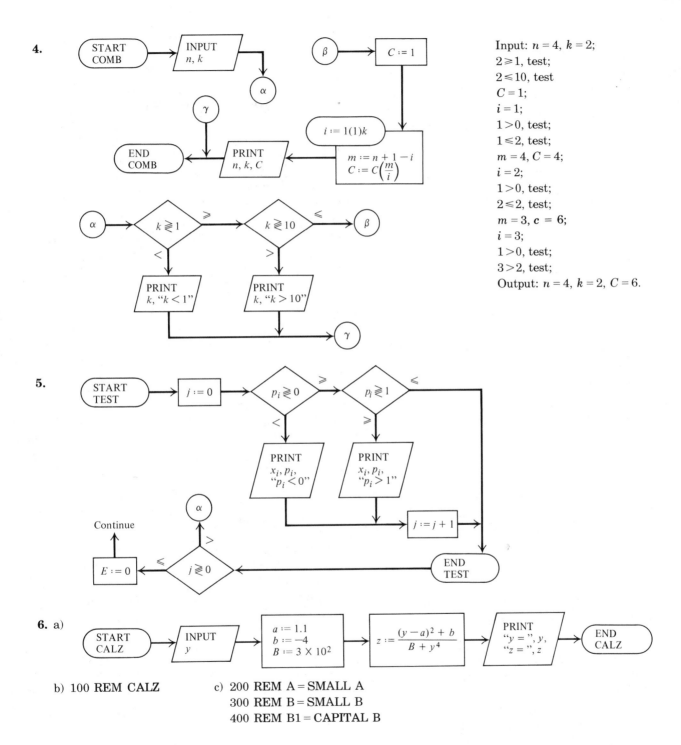

Input: $n = 4$, $k = 2$;
$2 \geqslant 1$, test;
$2 \leqslant 10$, test
$C = 1$;
$i = 1$;
$1 > 0$, test;
$1 \leqslant 2$, test;
$m = 4$, $C = 4$;
$i = 2$;
$1 > 0$, test;
$2 \leqslant 2$, test;
$m = 3$, $c = 6$;
$i = 3$;
$1 > 0$, test;
$3 > 2$, test;
Output: $n = 4$, $k = 2$, $C = 6$.

5.

6. a)

b) 100 **REM CALZ**

c) 200 **REM A = SMALL A**
   300 **REM B = SMALL B**
   400 **REM B1 = CAPITAL B**

7. 500 LET A = 1.1
600 LET B = −4
700 LET B1 = 3E2

8. 800 LET Y = 5

9. a) 900 LET N = (Y − A)↑2 + B
1000 LET D = B1 + Y↑4
1100 LET Z = N/D

b) 900 LET Z = ((Y − A)↑2 + B)/(B1 + Y↑4)

10. a) 1000 PRINT "Y =", Y, "Z =", Z
1100 END

b) 100 REM CALZ
200 REM A = SMALL A
300 REM B = SMALL B
400 REM B1 = CAPITAL B
500 LET A = 1.1
600 LET B = −4
700 LET B1 = 3E2
800 LET Y = 5
900 LET Z = ((Y − A)↑2 + B)/(B1 + Y↑4)
1000 PRINT "Y =", Y, "Z =", Z
1100 END

11. Replace 800 LET Y = 5
by 800 INPUT Y
Machine: ?
You continue: ?5

12. 100 to 700 same as Exercise 10, then
750 PRINT "Y", "Z"
760 PRINT
800 READ Y
900 same as Exercise 10.
1000 PRINT Y, Z
1050 DATA 2.4, 4.8, 6.0, 8.4
1100 END

13.

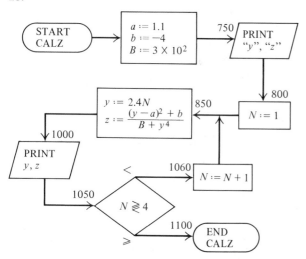

100 to 700 same as Exercise 10.
750 PRINT "Y", "Z"
760 PRINT
800 LET Y = 2.4∗N
900 same as Exercise 10.
1000 PRINT Y, Z
1050 IF N >= 4 THEN 1100
1060 LET N = N + 1
1070 GO TO 850
1100 END

14. 100 to 700 same as Exercise 10.
    750 PRINT "Y", "Z"
    760 PRINT
    800 FOR N = 1 TO 4
    850 LET Y = 2.4 ∗ N
    900 same as Exercise 10.
    1000 PRINT Y, Z
    1050 NEXT N
    1100 END

15. 100 to 700 same as Exercise 10.
    150 DIM Y(14), Z(14)
    800 FOR N = 1 TO 14
    850 LET Y(N) = 2.4 ∗ N
    900 LET Z(N) = ((Y(N) − A)↑2 + B)/(B1 + Y(N)↑4)
    1000 PRINT "Y(N)", Y(N), "Z(N)", Z(N)
    1050 NEXT N
    1100 END

EXERCISE SET B.1, p. 481

1.

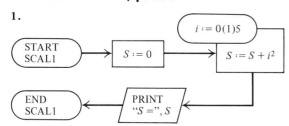

3.

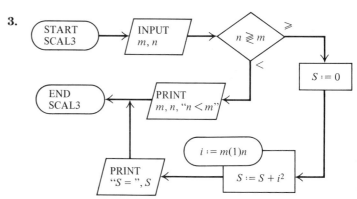

5. Replace $i := 0(1)5$ loop of Exercise 1 by $i := 0(2)5$.

7.

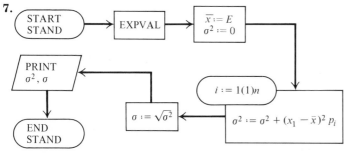

9.

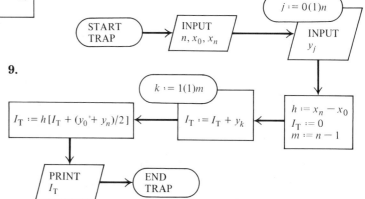

11.

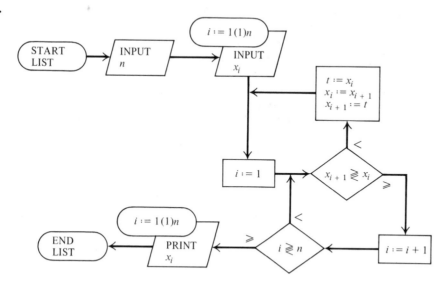

EXERCISE SET B.2, pp. 494–495

1. A, A1, B, B1   **3.** E, E1, E2   **5.** U, U1, U2   **7.** S, S1, S2, S3   **9.** $X = 5*B*B1 + .71E-2$

11. $U = 3.7/B1 + 4.27E3$ **13.** $S = A\uparrow2*B + A1*B1\uparrow2$ **15.** $A = X1\uparrow2/Y1 + Y2/X2\uparrow2$

17. $Y = X\uparrow2/A/B$ or $Y = X\uparrow2/(A*B)$ **19.** $Z = (X\uparrow2 + 1/Y\uparrow2)\uparrow.5$

21. $U = ((5 + .3*X)\uparrow2 + (.02 + 1.9*Y)\uparrow2)\uparrow.5$ **23.** $X1 = ((3 + X\uparrow4 - 7*A*Y\uparrow2)/(1.42*X*Y + 1.3*B))\uparrow(1/3)$

25. a) 100 REM EX 25 A
200 LET N = 4
300 LET P = .06
400 READ A0
500 LET A = A0*(1 + 1/N)↑P
600 PRINT "N = ", N, "P = ", P, "A0 = ", A0, "A = ", A
700 DATA 1000
800 END

b) 100 REM EX 25 B
200 READ N, P, A0
300 LET A = A0*(1 + 1/N)↑P
400 PRINT "N = ", N, "P = ", P, "A0 = ", A0, "A = ", A
500 DATA 4, .06, 1000
600 END

c) 100 REM EX 25 C
200 READ N, P
300 INPUT A0
400 LET A = A0*(1 + 1/N)↑P
500 PRINT "N = ", N, "P = ", P, "A0 = ", A0, "A = ", A
600 DATA 4, .06
700 END

27. a) 100 REM EX 27 A
200 REM X0 = X′
300 LET A = 1.3
400 READ X1, X2
500 LET X0 = X1↑2/(A + X2↑3)
600 PRINT "A = ", A, "X1 = ", X1, "X2 = ", X2, "X0 = ", X0
700 DATA .3, .04
800 END

b) 100 REM EX 27 B
200 REM X0 = X′
300 omit
400 READ A, X1, X2
500 same as (a)
600 same as (a)
700 DATA 1.3, .3, .04
800 END

c) 100 REM EX 27 C
200 REM X0 = X′
300 READ A
400 INPUT X1, X2
500 same as (a)
600 same as (a)
700 DATA 1.3
800 END

d) 100 REM EX 27 D
200 REM X0 = X′
300 READ A
400 READ X1, X2
500 same as (a)
600 same as (a)
650 GO TO 400
700 DATA 1.3, .3, .04, .4, .06, .5, .08
800 END

29. a) 100 REM EX 29 A
200 LET N = 4
300 LET P = .06
400 PRINT "N = ", N, "P = ", P
500 PRINT
600 PRINT "A0 = ", "A"
700 PRINT
800 READ A0
900 LET A = A0 * (1 + 1/N)→P
1000 PRINT A0, A
1100 GO TO 800
1200 DATA 150, 300, 450, 600, 750, 900
1300 END

b) 100 REM EX 29 B
200–700 same as (a)
800 LET A0 = 150
900 same as (a)
1000 same as (a)
1050 LET A0 = A0 + 150
1060 IF A0 > 900 THEN 1300
1100 GO TO 800
1300 END

c) 100 REM 29 C
200–700 same as (a)
800 FOR A0 = 150 TO 900 STEP 150
900 same as (a)
1000 same as (a)
1100 NEXT A0
1300 END

31. a) 100 REM 31 A
200 REM X0 = X′
300 LET A = 1.3
400 PRINT "A = ", A
500 PRINT
600 PRINT "X1", "X2", "X′"
700 PRINT
800 READ X1, X2
900 LET X0 = X1↑2/(A + X2↑3)
1000 PRINT X1, X2, X0
1100 GO TO 800
1200 DATA 1, 2, 2, 4, 3, 6, 4, 8
1300 END

b) 100 REM 31 B
200–700 same as (a)
800 LET X1 = 1
850 LET X2 = 2 * X1
900–1000 same as (a)
1050 LET X1 = X1 + 1
1060 IF X1 > 4 THEN 1300
1100 GO TO 850
1300 END

c) 100 REM 31 C
200–700 same as (a)
800 FOR X1 = 1 TO 4
850 LET X2 = 2 * X1
900–1000 same as (a)
1100 NEXT X1
1300 END

d) 100 REM EX 31 D
200–700 same as (a)
800 FOR X1 = 1 TO 4
850 FOR X2 = 2 TO 8 STEP 2
900–1000 same as (a)
1050 NEXT X2
1100 NEXT X1
1300 END

33. 100 REM SCAL1
200 LET S = 0
300 FOR I = 0 TO 5
400 LET S = S + I↑2
500 NEXT I
600 PRINT "S = ", S
700 END

35. 100 REM SCAL3
200 INPUT M, N
300 IF N < M THEN 1000
400 LET S = 0
500 FOR I = M TO N
600 LET S = S + I↑2
700 NEXT I
800 PRINT "S = ", S
900 GO TO 1100
1000 PRINT M, N, "N < M"
1100 END

37. Replace

300 FOR I = 0 TO 5

of Exercise 33 by

300 FOR I = 0 TO 5 STEP 2

39. 100 REM PERM
200 INPUT N, K
300 LET I = 1
400 LET M = N
500 LET P = N
600 IF K < I THEN 1200
700 IF K = I THEN 1400
800 LET I = I + 1
900 LET M = M − 1
1000 LET P = P * M
1100 GO TO 600
1200 PRINT K, "K < 1"
1300 GO TO 1500
1400 PRINT "K = ", K, "N = ", N, "P = ", P
1500 END

41. 100 REM TRAP
200 DIM Y(100)
300 REM X1 = X(N)
400 READ N
500 If N > 100 THEN 1900
600 READ X0, X1
700 FOR J = 0 TO N
800 READ Y(J)
900 NEXT J
100 LET H = (X1 − X0)/N
1100 LET I = 0
1200 LET M = N − 1
1300 FOR K = 1 TO M
1400 I = I + Y(K)
1500 NEXT K
1600 I = H * (I + (Y(0) + Y(N))/2)
1700 PRINT "I(TRAP) = ", I
1800 GO TO 2100
1900 PRINT N, "N > 100"
2000 DATA
2100 END

43. 100 REM LIST
200 DIM X(100)
300 READ N
400 IF N > 100 THEN 1900
500 FOR I = 1 TO N
600 READ X(I)
700 NEXT I
800 LET J = 1
900 IF X(J + 1) > = X(J) THEN 1400
1000 LET X(0) = X(J)
1100 LET X(J) = X(J + 1)
1200 LET X(J + 1) = X(0)
1300 GO TO 800
1400 LET J = J + 1
1500 IF J < N THEN 900
1600 FOR I = 1 TO N
1700 PRINT X(I)
1800 GO TO 2100
1900 PRINT N, "N > 100"
2000 DATA
2100 END

Index